RENEWALS 458-457

Preparative Layer Chromatography

CHROMATOGRAPHIC SCIENCE SERIES

A Series of Textbooks and Reference Books

Editor: JACK CAZES

1. Dynamics of Chromatography: Principles and Theory, *J. Calvin Giddings*
2. Gas Chromatographic Analysis of Drugs and Pesticides, *Benjamin J. Gudzinowicz*
3. Principles of Adsorption Chromatography: The Separation of Nonionic Organic Compounds, *Lloyd R. Snyder*
4. Multicomponent Chromatography: Theory of Interference, *Friedrich Helfferich and Gerhard Klein*
5. Quantitative Analysis by Gas Chromatography, Josef Novák
6. High-Speed Liquid Chromatography, *Peter M. Rajcsanyi and Elisabeth Rajcsanyi*
7. Fundamentals of Integrated GC-MS (in three parts), *Benjamin J. Gudzinowicz, Michael J. Gudzinowicz, and Horace F. Martin*
8. Liquid Chromatography of Polymers and Related Materials, *Jack Cazes*
9. GLC and HPLC Determination of Therapeutic Agents (in three parts), Part 1 *edited by Kiyoshi Tsuji and Walter Morozowich*, Parts 2 and 3 *edited by Kiyoshi Tsuji*
10. Biological/Biomedical Applications of Liquid Chromatography, *edited by Gerald L. Hawk*
11. Chromatography in Petroleum Analysis, *edited by Klaus H. Altgelt and T. H. Gouw*
12. Biological/Biomedical Applications of Liquid Chromatography II, *edited by Gerald L. Hawk*
13. Liquid Chromatography of Polymers and Related Materials II, *edited by Jack Cazes and Xavier Delamare*
14. Introduction to Analytical Gas Chromatography: History, Principles, and Practice, *John A. Perry*
15. Applications of Glass Capillary Gas Chromatography, *edited by Walter G. Jennings*
16. Steroid Analysis by HPLC: Recent Applications, *edited by Marie P. Kautsky*
17. Thin-Layer Chromatography: Techniques and Applications, *Bernard Fried and Joseph Sherma*
18. Biological/Biomedical Applications of Liquid Chromatography III, *edited by Gerald L. Hawk*
19. Liquid Chromatography of Polymers and Related Materials III, *edited by Jack Cazes*
20. Biological/Biomedical Applications of Liquid Chromatography, *edited by Gerald L. Hawk*

21. Chromatographic Separation and Extraction with Foamed Plastics and Rubbers, *G. J. Moody and J. D. R. Thomas*
22. Analytical Pyrolysis: A Comprehensive Guide, *William J. Irwin*
23. Liquid Chromatography Detectors, *edited by Thomas M. Vickrey*
24. High-Performance Liquid Chromatography in Forensic Chemistry, *edited by Ira S. Lurie and John D. Wittwer, Jr.*
25. Steric Exclusion Liquid Chromatography of Polymers, *edited by Josef Janca*
26. HPLC Analysis of Biological Compounds: A Laboratory Guide, *William S. Hancock and James T. Sparrow*
27. Affinity Chromatography: Template Chromatography of Nucleic Acids and Proteins, *Herbert Schott*
28. HPLC in Nucleic Acid Research: Methods and Applications, *edited by Phyllis R. Brown*
29. Pyrolysis and GC in Polymer Analysis, *edited by S. A. Liebman and E. J. Levy*
30. Modern Chromatographic Analysis of the Vitamins, *edited by André P. De Leenheer, Willy E. Lambert, and Marcel G. M. De Ruyter*
31. Ion-Pair Chromatography, *edited by Milton T. W. Hearn*
32. Therapeutic Drug Monitoring and Toxicology by Liquid Chromatography, *edited by Steven H. Y. Wong*
33. Affinity Chromatography: Practical and Theoretical Aspects, *Peter Mohr and Klaus Pommerening*
34. Reaction Detection in Liquid Chromatography, *edited by Ira S. Krull*
35. Thin-Layer Chromatography: Techniques and Applications, Second Edition, Revised and Expanded, *Bernard Fried and Joseph Sherma*
36. Quantitative Thin-Layer Chromatography and Its Industrial Applications, *edited by Laszlo R. Treiber*
37. Ion Chromatography, *edited by James G. Tarter*
38. Chromatographic Theory and Basic Principles, *edited by Jan Åke Jönsson*
39. Field-Flow Fractionation: Analysis of Macromolecules and Particles, *Josef Janca*
40. Chromatographic Chiral Separations, *edited by Morris Zief and Laura J. Crane*
41. Quantitative Analysis by Gas Chromatography, Second Edition, Revised and Expanded, *Josef Novák*
42. Flow Perturbation Gas Chromatography, *N. A. Katsanos*
43. Ion-Exchange Chromatography of Proteins, *Shuichi Yamamoto, Kazuhiro Naka-nishi, and Ryuichi Matsuno*
44. Countercurrent Chromatography: Theory and Practice, *edited by N. Bhushan Man-dava and Yoichiro Ito*
45. Microbore Column Chromatography: A Unified Approach to Chromatography, *edited by Frank J. Yang*
46. Preparative-Scale Chromatography, *edited by Eli Grushka*
47. Packings and Stationary Phases in Chromatographic Techniques, *edited by Klaus K. Unger*
48. Detection-Oriented Derivatization Techniques in Liquid Chromatography, *edited by Henk Lingeman and Willy J. M. Underberg*
49. Chromatographic Analysis of Pharmaceuticals, *edited by John A. Adamovics*
50. Multidimensional Chromatography: Techniques and Applications, *edited by Hernan Cortes*

51. HPLC of Biological Macromolecules: Methods and Applications, *edited by Karen M. Gooding and Fred E. Regnier*
52. Modern Thin-Layer Chromatography, *edited by Nelu Grinberg*
53. Chromatographic Analysis of Alkaloids, *Milan Popl, Jan Fähnrich, and Vlastimil Tatar*
54. HPLC in Clinical Chemistry, *I. N. Papadoyannis*
55. Handbook of Thin-Layer Chromatography, *edited by Joseph Sherma and Bernard Fried*
56. Gas–Liquid–Solid Chromatography, *V. G. Berezkin*
57. Complexation Chromatography, *edited by D. Cagniant*
58. Liquid Chromatography–Mass Spectrometry, *W. M. A. Niessen and Jan van der Greef*
59. Trace Analysis with Microcolumn Liquid Chromatography, *Milos Krejcl*
60. Modern Chromatographic Analysis of Vitamins: Second Edition, *edited by André P. De Leenheer, Willy E. Lambert, and Hans J. Nelis*
61. Preparative and Production Scale Chromatography, *edited by G. Ganetsos and P. E. Barker*
62. Diode Array Detection in HPLC, *edited by Ludwig Huber and Stephan A. George*
63. Handbook of Affinity Chromatography, *edited by Toni Kline*
64. Capillary Electrophoresis Technology, *edited by Norberto A. Guzman*
65. Lipid Chromatographic Analysis, *edited by Takayuki Shibamoto*
66. Thin-Layer Chromatography: Techniques and Applications: Third Edition, Revised and Expanded, *Bernard Fried and Joseph Sherma*
67. Liquid Chromatography for the Analyst, *Raymond P. W. Scott*
68. Centrifugal Partition Chromatography, *edited by Alain P. Foucault*
69. Handbook of Size Exclusion Chromatography, *edited by Chi-San Wu*
70. Techniques and Practice of Chromatography, *Raymond P. W. Scott*
71. Handbook of Thin-Layer Chromatography: Second Edition, Revised and Expanded, *edited by Joseph Sherma and Bernard Fried*
72. Liquid Chromatography of Oligomers, *Constantin V. Uglea*
73. Chromatographic Detectors: Design, Function, and Operation, *Raymond P. W. Scott*
74. Chromatographic Analysis of Pharmaceuticals: Second Edition, Revised and Expanded, *edited by John A. Adamovics*
75. Supercritical Fluid Chromatography with Packed Columns: Techniques and Applications, *edited by Klaus Anton and Claire Berger*
76. Introduction to Analytical Gas Chromatography: Second Edition, Revised and Expanded, *Raymond P. W. Scott*
77. Chromatographic Analysis of Environmental and Food Toxicants, *edited by Takayuki Shibamoto*
78. Handbook of HPLC, *edited by Elena Katz, Roy Eksteen, Peter Schoenmakers, and Neil Miller*
79. Liquid Chromatography–Mass Spectrometry: Second Edition, Revised and Expanded, *Wilfried Niessen*
80. Capillary Electrophoresis of Proteins, *Tim Wehr, Roberto Rodríguez-Díaz, and Mingde Zhu*
81. Thin-Layer Chromatography: Fourth Edition, Revised and Expanded, *Bernard Fried and Joseph Sherma*
82. Countercurrent Chromatography, *edited by Jean-Michel Menet and Didier Thiébaut*

83. Micellar Liquid Chromatography, *Alain Berthod and Celia García-Alvarez-Coque*
84. Modern Chromatographic Analysis of Vitamins: Third Edition, Revised and Expanded, edited by *André P. De Leenheer, Willy E. Lambert, and Jan F. Van Bocxlaer*
85. Quantitative Chromatographic Analysis, *Thomas E. Beesley, Benjamin Buglio, and Raymond P. W. Scott*
86. Current Practice of Gas Chromatography–Mass Spectrometry, *edited by W. M. A. Niessen*
87. HPLC of Biological Macromolecules: Second Edition, Revised and Expanded, *edited by Karen M. Gooding and Fred E. Regnier*
88. Scale-Up and Optimization in Preparative Chromatography: Principles and Bio-pharmaceutical Applications, *edited by Anurag S. Rathore and Ajoy Velayudhan*
89. Handbook of Thin-Layer Chromatography: Third Edition, Revised and Expanded, *edited by Joseph Sherma and Bernard Fried*
90. Chiral Separations by Liquid Chromatography and Related Technologies, *Hassan Y. Aboul-Enein and Imran Ali*
91. Handbook of Size Exclusion Chromatography and Related Techniques: Second Edition, *edited by Chi-San Wu*
92. Handbook of Affinity Chromatography: Second Edition, *edited by David S. Hage*
93. Chromatographic Analysis of the Environment: Third Edition, *edited by Leo M. L. Nollet*
94. Microfluidic Lab-on-a-Chip for Chemical and Biological Analysis and Discovery, *Paul C.H. Li*
95. Preparative Layer Chromatography, *edited by Teresa Kowalska and Joseph Sherma*

Preparative Layer Chromatography

edited by
Teresa Kowalska
University of Silesia
Katowice, Poland

Joseph Sherma
Lafayette College
Easton, Pennsylvania

A CRC title, part of the Taylor & Francis imprint, a member of the Taylor & Francis Group, the academic division of T&F Informa plc.

Published in 2006 by
CRC Press
Taylor & Francis Group
6000 Broken Sound Parkway NW, Suite 300
Boca Raton, FL 33487-2742

© 2006 by Taylor & Francis Group, LLC
CRC Press is an imprint of Taylor & Francis Group

No claim to original U.S. Government works
Printed in the United States of America on acid-free paper
10 9 8 7 6 5 4 3 2 1

International Standard Book Number-10: 0-8493-4039-X (Hardcover)
International Standard Book Number-13: 978-0-8493-4039-0 (Hardcover)
Library of Congress Card Number 2005052138

This book contains information obtained from authentic and highly regarded sources. Reprinted material is quoted with permission, and sources are indicated. A wide variety of references are listed. Reasonable efforts have been made to publish reliable data and information, but the author and the publisher cannot assume responsibility for the validity of all materials or for the consequences of their use.

No part of this book may be reprinted, reproduced, transmitted, or utilized in any form by any electronic, mechanical, or other means, now known or hereafter invented, including photocopying, microfilming, and recording, or in any information storage or retrieval system, without written permission from the publishers.

For permission to photocopy or use material electronically from this work, please access www.copyright.com (http://www.copyright.com/) or contact the Copyright Clearance Center, Inc. (CCC) 222 Rosewood Drive, Danvers, MA 01923, 978-750-8400. CCC is a not-for-profit organization that provides licenses and registration for a variety of users. For organizations that have been granted a photocopy license by the CCC, a separate system of payment has been arranged.

Trademark Notice: Product or corporate names may be trademarks or registered trademarks, and are used only for identification and explanation without intent to infringe.

Library of Congress Cataloging-in-Publication Data

Preparative layer chromatography / edited by Teresa Kowalska and Joseph Sherma.
p. cm. -- (Chromatographic science series ; 95)
Includes bibliographical references and index.
ISBN 0-8493-4039-X (alk. paper)
1. Preparative layer chromatography. I. Kowalska, Teresa. II. Sherma, Joseph. III. Chromatographic science ; v. 95.

QD79.C52P74 2006
543'.84--dc22

2005052138

Visit the Taylor & Francis Web site at
http://www.taylorandfrancis.com

and the CRC Press Web site at
http://www.crcpress.com

Preface

This book has been designed as a practical, comprehensive source of information on the field of classical preparative layer chromatography (PLC). It is organized in two parts, the first of which covers the theory and up-to-date procedures of PLC (Chapter 1 to Chapter 8), while the second (Chapter 9 to Chapter 16) includes applications to a selection of the most important compound classes and samples types. Overall, the topics covered in the 16 chapters are evidence for the versatility and wide use of PLC at the current time. We have designed this first book ever published on PLC to be valuable for scientists with a high degree of experience in the separation sciences, but because most chapters include considerable introductory and background information, it is also appropriate for the relatively inexperienced chromatographer.

The contributors to the book are experts on the topics about which they write and include many of the best known and most knowledgeable workers in the field of thin-layer chromatography and PLC throughout the world. Rather than attempting to adopt a uniform style, we have allowed chapter authors the freedom to present their topics in a way that they considered most effective. They have used figures and tables as needed to augment the text, and selective reference lists include the most important new literature, as well significant older references, to set the basis of their chapters.

We had great cooperation from the authors in submitting their chapters in a timely fashion, so that the book has been completed about six months sooner than anticipated. None of the chapters was unduly delayed, so all are equally up to date in their coverage. The authors represent laboratories in Germany, Poland, Romania, Norway, Canada, Japan, India, and the U.S. and, therefore, have provided a global perspective for the book.

We hope that this book will be valuable for practitioners and teachers in diverse scientific fields that make use of chromatographic methods and that it will promote better understanding of the field and lead to its even wider utilization.

Teresa Kowalska
Joseph Sherma

Preface

About the Editors

Teresa Kowalska is currently a professor in the Department of the Physicochemical Basis of Chromatography at the University of Silesia (Katowice, Poland). Her scientific interests include the physicochemical foundations of liquid chromatography and gas chromatography, with special attention focused on modeling of planar chromatography both in its analytical and preparative mode. Over the past 37 yr Dr. Kowalska has directed the programs of over 70 M.Sc. degree students who have carried out their research on the theory and practice of different chromatographic and hyphenated techniques. She has also supervised the research in the separation science of 8 Ph.D. students. Dr. Kowalska is the author of more than 200 scientific papers, more than 300 scientific conference papers, and a vast number of the book chapters and encyclopedia entries in the field of chromatography. It is perhaps noteworthy that she has authored (and then updated) the chapter on "Theory and Mechanism of Thin-Layer Chromatography" for all three editions of the *Handbook of Thin-Layer Chromatography*, edited by professors J. Sherma and B. Fried, and published by Marcel Dekker.

Dr. Kowalska has acted as editor of *Acta Chromatographica*, the annual periodical published by the University of Silesia (Katowice, Poland) and devoted to all chromatographic and hyphenated techniques, right from its establishment in 1992. *Acta Chromatographica* appears as a hard copy journal and also online in the digital format. Its contributors originate from an international academic community, and it is meant to promote the development in separation sciences. It apparently serves its purpose well, as can be judged from a wide readership, abundant citations throughout the professional literature, and also from the ISI ranking quota.

Last but not least, in the course of the past almost 30 yr Dr. Kowalska has been active as organizer (and in recent years as a cochairperson, also) of the annual all-Polish chromatographic symposia with international participation, uninterruptedly held each year (since 1977) in the small mountain resort of Szczyrk in South Poland. Integration of an international community of chromatographers through these meetings has been regarded by Dr. Kowalska as a specific yet important contribution to chromatography.

Joseph Sherma is John D. and Frances H. Larkin Professor Emeritus of Chemistry at Lafayette College, Easton, Pennsylvania. He is author or coauthor of over 500 scientific papers and editor or coeditor of over 50 books and manuals in the areas of analytical chemistry and chromatography. Dr. Sherma is coauthor, with Bernard Fried, Kreider Professor Emeritus of Biology at Lafayette College, of *Thin Layer Chromatography* (editions 1 to 4) and coeditor with Professor Fried of the *Handbook of Thin Layer Chromatography* (editions 1 to 3), both published by Marcel Dekker, Inc. He served for 23 yr as the editor for residues and trace elements of the *Journal* *of AOAC International* and serves currently on the editorial advisory boards of the *Journal of Liquid Chromatography and Related Technologies*, the *Journal of Environmental Science and Health (Part B)*, the *Journal of Planar Chromatography–Modern TLC*, *Acta Chromatographica*, and *Acta Universitatis Cibiniensis, Seria F. Chemia*. Dr. Sherma received his Ph.D. degree (1958) from Rutgers State University, New Brunswick, New Jersey.

Contributor List

Dr. Simla Basar
Institute of Organic Chemistry
University of Hamburg
Hamburg, Germany

Prof. Jan Błądek
Institute of Chemistry
Military University of Technology
Warsaw, Poland

Dr. Virginia Coman
Raluca Ripan Institute for Research in Chemistry
Cluj-Napoca, Romania

Dr. Tadeusz H. Dzido
Faculty of Pharmacy
Medical Academy of Lublin
Lublin, Poland

Dr. Monika Fabiańska
Faculty of Earth Sciences
The University of Silesia
Sosnowiec, Poland

Prof. George W. Francis
Department of Chemistry
University of Bergen
Bergen, Norway

Dr. Michał Ł. Hajnos
Faculty of Pharmacy
Medical University of Lublin
Lublin, Poland

Dr. Heinz E. Hauck
LSA/R&D
Merck KGaA
Darmstadt, Germany

Dr. Weerasinghe Indrasena
Ocean Nutrition Canada Ltd.
Halifax, Nova Scotia, Canada

Grzegorz Jóźwiak
Faculty of Pharmacy
Medical University of Lublin
Lublin, Poland

Prof. Krzysztof Kaczmarski
Faculty of Chemistry
Technical University of Rzeszów
Rzeszów, Poland

Dr. Angelika Koch
Frohme Apotheke
Hamburg, Germany

Prof. Teresa Kowalska
Institute of Chemistry
The University of Silesia
Katowice, Poland

Dr. Emi Miyamoto
Department of Health Science
Kochi Women's University
Kochi, Japan

Dr. Ali Mohammad
Department of Applied Chemistry,
Faculty of Engineering and Technology
Aligarh Muslim University
Aligarh, India

Dr. Gertrud E. Morlock
University of Hohenheim
Institute of Food Chemistry
Stuttgart, Germany

Dr. Beata Polak
Faculty of Pharmacy
Medical University of Lublin
Lublin, Poland

Dr. Wojciech Prus
University of Bielsko-Biala
Faculty of Textile Engineering and
Environmental Protection
Department of Chemistry
Bielsko-Biala, Poland

Rita Richter
Institute of Organic Chemistry
University of Hamburg,
Hamburg, Germany

Dr. Mieczysław Sajewicz
Institute of Chemistry
The University of Silesia
Katowice, Poland

Michael Schulz
LSA/R&D
Merck KGaA
Darmstadt, Germany

Dr. Joseph Sherma
Department of Chemistry
Lafayette College
Easton, Pennsylvania

Prof. Bernd Spangenberg
Environmental Techniques Section
University of Applied Sciences
Offenburg, Germany

Anna Szymańczyk
Institute of Chemistry
Military University of Technology
Warsaw, Poland

Prof. Monika Waksmundzka-Hajnos
Faculty of Pharmacy
Medical University of Lublin
Lublin, Poland

Prof. Fumio Watanabe
Department of Health Science
Kochi Women's University
Kochi, Japan

Prof. Teresa Wawrzynowicz
Faculty of Pharmacy
Medical University of Lublin
Lublin, Poland

Contents

SECTION I

SECTION II

Section I

1 Introduction

Teresa Kowalska and Joseph Sherma

CONTENTS

1.1 CHROMATOGRAPHY BACKGROUND

The invention of chromatography can be traced to the milestone paper published in 1906 by the Russian botanist and plant physiologist Mikhail Semyonovitch Tswett (1872–1919) [1,2]. In the experiments reported in this paper, Tswett separated chloroplast pigments from leaves in a column of precipitated chalk washed with carbon disulfide mobile phase. During the 20th century and in the new millennium, chromatography has become an indispensable separation tool that is very widely used in natural and life science laboratories throughout the world.

Tswett's initial column liquid chromatography method was developed, tested, and applied in two parallel modes, liquid–solid adsorption and liquid–liquid partition. Adsorption chromatography, based on a purely physical principle of adsorption, considerably outperformed its partition counterpart with mechanically coated stationary phases to become the most important liquid chromatographic method. This remains true today in thin-layer chromatography (TLC), for which silica gel is by far the major stationary phase. In column chromatography, however, reversed-phase liquid chromatography using chemically bonded stationary phases is the most popular method.

Preparative layer chromatography (PLC) was apparently first reported by F.J. Ritter and G.M. Meyer in 1962 [3]. They used layers of 1-mm thickness. Earlier preparative work, e.g., that reported by J.M. Miller and R.G. Kirchner (the inventors of TLC as it is performed today by development of the layer in a closed tank, analogous to ascending paper chromatography) in 1951 and 1952 [4,5], was termed TLC but was carried out on adsorbent bars used as columns or on analytical layers after column chromatography. In his classic TLC laboratory handbook, originally published in German in 1962 and translated to English in 1965 [6], Egon Stahl made only a few statements about the method he called "micropreparative TLC." Layers of 0.5- and 0.7-mm thickness prepared with a spreading device were recommended for this method by Stahl, as well as streak (band) application of larger quantities of mixture using a "specially designed instrument" (described in the Ritter and Meyer paper [3]) or a "microspray gun" (used in Stahl's laboratory).

1.2 BASICS OF PLC

PLC is used to separate and isolate amounts of material (e.g., 10 to 1000 mg) larger than those used in analytical TLC. The purpose of PLC is to obtain pure compounds for further chromatographic or spectrometric analysis or for determination of biological activity. "Classical" PLC (CPLC), involving mobile phase migration by capillary action, requires relatively simple and inexpensive equipment, but thorough comprehension of the relevant chromatographic principles and techniques is critical. The required information for performing successful PLC is provided in this book.

The sample dissolved in a weak (nonpolar for silica gel), volatile solvent is applied as a narrow band across the plate. Manual application can be achieved with a pipet or syringe guided by a ruler, or round spots can be placed close together, side by side, in a line. Sample application instruments are available commercially, e.g., a mechanical streaker from Analtech and an automated spray-on apparatus from CAMAG.

Plates with 0.5- to 2-mm layer thickness are normally used for increased loading capacity. Layers can be self-made in the laboratory, or commercially precoated preparative plates are available with silica gel, alumina, cellulose, C-2 or C-18 bonded silica gel, and other sorbents. Resolution is lower than on thinner analytical layers having a smaller average particle size and particle size range. Precoated plates with a preadsorbent or concentrating zone facilitate application of sample bands.

The mobile phase is usually selected by trial-and-error guided by prior experience or by performing preliminary analytical separations of the sample in a saturated chamber. PLC separations will be inferior to analytical TLC separations using the same mobile phase because of the thicker layer, larger particle size, and overloaded sample conditions used for PLC. A good general rule is that analytical TLC should achieve separations with least 0.1 R_f value difference if the PLC separations are to be adequate with the transferred mobile phase. Isocratic development is usually used, but gradient development has been applied in certain situations for increased resolution.

Rectangular glass tanks (N chambers) with inner dimensions of $21 \times 21 \times 9$ cm are used most frequently for the ascending, capillary-flow development of PLC plates, which usually measure 20×20 cm. The tank is lined with thick chromatography paper (e.g., Whatman 3 MM) soaked in the mobile phase and allowed to equilibrate with the mobile phase vapor for up to 2 h prior to development over a maximum distance of 18 cm. A saturated chamber provides faster capillary flow of the mobile phase, more uniform bulk and alpha solvent fronts, and higher separation efficiency. A plate angle of 75° from horizontal is recommended for the fastest development with minimum zone distortion. Special taper plates (Analtech) with layer thicknesses ranging from 300 μm (bottom) to 1700 μm (top) provide increased mobile phase velocity compared to plates with uniform layer thickness. Circular, multiple, and two-dimensional development, as well as development at temperatures other than the ambient, are also used for PLC in special applications.

Zones containing separated compounds can be detected nondestructively after plate development and evaporation of the mobile phase by their natural color in white light, natural fluorescence under 254-nm or 366-nm ultraviolet (UV) light, or absorption of UV light on layers containing a fluorescent indicator (phosphor). This method, termed *fluorescence quenching*, gives dark zones on a fluorescent background. Postchromatographic chromogenic or fluorogenic reagents can be used to detect compounds that are not naturally visible or fluorescent and do not quench fluorescence. One of the most widely used reagents is iodine vapor, which reversibly detects many types of compounds as brown zones. Destructive chromogenic or fluorogenic reagents must be applied only to the side edges of the layer (the rest of the layer is covered with a glass plate) to locate the areas from which to recover the separated compounds.

The zones containing the desired compounds are scraped from the plate backing, the compounds are eluted with a strong solvent, any remaining sorbent particles are separated, and the solution is concentrated.

All of these steps are described in greater detail in the chapters in Section I of this book.

1.3 PRINCIPLES AND CHARACTERISTICS OF PLC

Successful separations in adsorption chromatography (adsorption TLC included) are due to the difference in the energies of adsorption between the two separated species. The phenomenon of adsorption on a solid–liquid interface can be characterized best by the empirical adsorption isotherm, which is unique for a given compound with a particular stationary phase–mobile phase combination. There is one feature, however, that all mixture components share in common, namely, the general nature of their adsorption isotherms. Each empirical isotherm consists of a linear and nonlinear part. The linear part corresponds to the stepwise saturation of active sites on the adsorbent surface, prior to its complete saturation. For a compound deposited on the adsorbent surface within the linear range of the isotherm, a circular TLC zone shape is expected and the densitometrically measured mass distribution (i.e., the concentration profile of the zone) should be regular (Gaussian). In the case of mass overload, the system operates within the nonlinear sector of the isotherm, with an oval zone shape and tailing of the skewed (non-Gaussian) concentration profile.

There is no rigid demarcation line between adsorption TLC operating within the linear or nonlinear range for the following reasons: (1) each individual solute is characterized by its own adsorption isotherm, (2) for most analytes in most chromatographic systems, the respective adsorption isotherms remain unknown, and (3) there is no need for steady control of the isotherm sector that is utilized in an experiment. When separating a compound mixture, it often happens that some of the constituents are in the linear range of their adsorption isotherms, whereas others are in the nonlinear (i.e., mass-overload) range.

However, mostly because of the intuitive, trial-and-error approach of thin-layer chromatographers, long ago the technique split into two subtechniques, one benefiting from the linear range of the adsorption isotherm and the other utilizing the advantages of the nonlinear range. Much less attention has been given to nonlinear TLC, and no monograph covering this subject has ever appeared on the market until now. The remainder of this section will be focused on this very important, yet considerably underestimated, method.

Linear TLC is well suited to perform traditional analytical tasks. With its use, one can successfully separate mixtures consisting of a limited number of analytes and, using more or less sophisticated supplementary techniques, identify the constituents of such mixtures on the layer. With more complex samples, linear TLC can at least help to fractionate samples into the separate classes of compounds. Owing to the proportionality between the amount of solute contained in zones and their *in situ* densitometric scan areas, a standard calibration plot that is useful for quantification of an analyte in a sample can be established. In summary, linear TLC conforms very well with many separation-, identification-, and quantification-oriented analytical strategies that are focused on the individual chemical species. Owing to its flexibility, economy, and relative simplicity, linear TLC sometimes outperforms certain other more expensive and less user-friendly analytical techniques. Optimization of an analytical result of linear TLC can be attained easily with a number of simple semiempirical, or even purely empirical, retention models (e.g., the Martin-Synge or Snyder-Soczewinski approaches); in more difficult cases, a variety of more advanced chemometric approaches are also available. Knowledge of the operating principles, techniques, and applications of linear TLC can be learned from books published in many languages throughout the world, and also from a selection of international scientific journals devoted exclusively, or partially, to the chromatographic sciences.

Contrary to the linear mode, nonlinear adsorption TLC was conceived, and to a large extent remained, an "unofficial" and almost "underground" separation technique. Among chromatography practitioners, it is known as PLC (or PTLC [preparative thin-layer chromatography] or PPC [preparative planar chromatography]), and it is generally accepted that this otherwise very useful separation tool does not serve an analytical purpose, at least in the sense discussed in the previous paragraph. As described in the last section, in PLC much higher amounts of the mixtures are applied to be separated than in linear TLC. Therefore, the layer has to be considerably thicker, and the optimization rules borrowed from the linear mode are either obeyed to a much lesser extent or do not hold at all. Thus, a simple statement that PLC lacks a theoretical basis of its own is essentially true. There is no book published in English or any other language from which this technique can be learned. In most cases, it is individually "discovered" by those who need a simple, rapid, inexpensive, and low-scale (i.e., microgram or milligram) separation tool and are fortunate enough to be familiar with the more common analytical TLC. The basic disadvantage of all of these "amateur discoveries," no matter how inquisitive and inventive their authors might be, is that they are usually made in an unorganized trial-and-error manner (sometimes aided by advice from a slightly more experienced colleague), and they eventually prove far less beneficial than they could be if properly introduced in book form by an expert, or group of experts.

All researchers involved in natural and life sciences who wish to isolate or purify microgram or milligram quantities of a given compound but, for whatever reason, cannot make use of a column technique would benefit from the correct use of PLC. For many scientists worldwide, the fully automated column approach can prove to be too expensive, too complex, or both. Nonpressurized open preparative column chromatography can be chosen, but this is not very effective with low levels of compounds, particularly not with difficult separations (i.e., in the case of closely migrating or partially overlapping peak profiles).

PLC is well suited for micropreparative separations. For example, consider the simple scenario of someone involved in a multistage organic synthesis and needing rapid spectrometric (e.g., mass, infrared, nuclear magnetic resonance, or x-ray) confirmation of each step of the synthetic procedure (available today for trace amounts of the compounds). In such cases, optimization of compound isolation with an instrumental technique will almost certainly prove to be much more time consuming and expensive than the planar mode, e.g., using short, narrow strips of aluminum- or plastic-backed adsorbent layers or microscope-slide-sized glass-backed plates in a pilot procedure.

Manufacturers of TLC materials and accessories are well prepared to satisfy the needs for professionally performed PLC. High-quality precoated preparative plates are available from a number of commercial sources. Alternatively, less expensive or specialty preparative plates can be "homemade" in the laboratory, and loose sorbents and coating devices can be purchased for this purpose. More-or-less-automated devices can also be purchased for band application of higher quantities of sample solutions to preparative layers. At least for some users, sophisticated densitometric and other instrumental techniques are available as nondestructive tools for preliminary detection and identification of separated compounds in order to enhance the efficiency of their isolation. The only aid still missing, and maybe the most important of all, is a comprehensive monograph on PLC that might encourage and instruct many potential users on how to fully benefit from this very versatile, efficient, relatively inexpensive, and rather easy to use isolation and purification technique. This book was planned to fill that void.

The oldest, simplest, and most frequently employed method for feeding the mobile phase to the layer in PLC is with the aid of capillary forces in the ascending direction. Forced-flow development can also be used for analytical TLC and PLC. The most popular forced-flow modes are overpressured layer chromatography (OPLC; sometimes termed *optimum performance laminar chromatography*), in which the mobile phase flow is due to mechanical force (pressure), and rotation planar chromatography (RPC), for which the transport of the mobile phase occurs due to a centrifugal force. In this book, coverage is restricted to CPLC, which operates with capillary flow and is accessible in all laboratories without the purchase of quite expensive and rather complex instrumentation needed for preparative OPLC and RPC. CPLC is technically compatible with capillary flow analytical TLC, the mode that is most widely used. This means that for the two methods (analytical and preparative), virtually the same laboratory equipment and procedures can be used (e.g., sorbent type, coating device [if commercial precoated plates are not used], chamber, sample application device, mobile phase, development mode [ascending], reagent sprayer, etc.), and personnel can easily switch from one method to the other or perform both in a parallel manner at the same time. This technical compatibility of the two modes and the ease of switching between them represent significant cost-saving advantages that for most laboratories throughout the world are difficult to overestimate.

Another reason for choosing CPLC as the sole option discussed in this book is that the chromatographic behavior of compounds in this method resembles most closely that observed in capillary flow analytical TLC. This is particularly important in view of the fact that, so far, the theoretical background of PLC has been practically nonexistent, while the semiempirical rules valid for capillary flow analytical TLC can be approximated relatively well for use in CPLC. This means that an experienced chromatographer well acquainted with the separation potential of capillary flow analytical TLC has a good chance to (intuitively, at least) select working parameters for CPLC that will produce a close to optimum result. The same goal cannot, however, be so easily reached with preparative OPLC and RPC. For example, the forced flow that operates in OPLC elevates quite drastically the chromatographic activity of the adsorbent layer [7]. This results from the mobile phase being pushed through the narrow pores of the layer, which are impenetrable when capillary forces alone are at play. The elevation of adsorbent activity is equivalent to a corresponding lowering of the mobile phase strength, leading to a considerable change in retention of solutes in a given layer-mobile phase system and making an intuitive, experience-based system optimization virtually impossible.

1.4 ORGANIZATION OF THE BOOK

Section I of this book includes chapters on the principles and practice of PLC. After this introductory Chapter 1, Chapter 2 provides information on efforts undertaken to date in order to establish the theoretical foundations of PLC. With growing availability and popularity of modern computer-aided densitometers, separation results can be obtained in digital form as a series of concentration profiles that can be relatively easily assessed and processed. From these, relevant conclusions can be drawn in exactly the same manner as in automated column chromatographic techniques. Efforts undertaken to build a theoretical foundation of PLC largely consist of adaptation of known strategies (with their validity confirmed in preparative column liquid chromatography) to the working conditions of PLC systems.

Chapter 3 through Chapter 8 deal with the basic aspects of the practical uses of PLC. Chapter 3 describes sorbent materials and precoated layers for normal or straight phase (adsorption) chromatography (silica gel and aluminum oxide 60) and partition chromatography (silica gel, aluminum oxide 150, and cellulose), and precoated layers for reversed-phase chromatography (RP-18 or C-18). Properties of the bulk sorbents and precoated layers, a survey of commercial products, and examples of substance classes that can be separated are given.

Chapter 4 discusses the selection and optimization of mobile phases for successful separations in PLC. Chapter 5 details procedures for sample application and development of layers, and Chapter 6 complements Chapter 5 by dealing specifically with the use of horizontal chambers for the development of preparative layers, including linear, continuous, two-dimensional, gradient, circular, and anticircular modes.

In Chapter 7, approaches for visualization of zones in chromatograms are discussed, including use of nondestructive and destructive dyeing reagents, fluorescence quenching on layers with a fluorescent indicator, and densitometry. In Chapter 8, additional detection methods, such as those used for biologically active and radioactive zones, as well as the recovery of separated, detected zones by scraping and elution techniques are covered.

Section II of the book, encompassing Chapter 9 through Chapter 16, presents practical applications of PLC in the chemical, biochemical, and life science fields. The great variety of these applications illustrates well the versatility and excellent performance of PLC in solving a wide spectrum of micropreparative problems.

Chapter 9 shows the importance of PLC in the critical field of medical research, with representative examples of the applications to amino acids, carbohydrates, lipids, and pharmacokinetic studies.

Chapter 10 is devoted to the preparation and purification of hydrophilic vitamins (C, B_1, B_2, B_6, B_{12}, nicotinic acid and nicotinamide, pantothenic acid, biotin, and folic acid) in pharmaceutical preparations, food products, and biological samples.

PLC of plant extracts is presented in Chapter 11, with sections on the choice of systems, sampling, choice of the sample solvent, detection, and development modes. These applications in the field of pharmacognosy play a key role in the investigation and understanding of the healing potential of the constituents of medicinal plants.

PLC of lipids is discussed in Chapter 12. Lipids play a vital role in virtually all aspects of human and animal life. Many studies of food quality, human health, metabolic and ageing processes, pheromone activity in animals, etc., benefit greatly from the use of PLC for the separation and isolation of lipids.

Chapter 13 is devoted to the PLC of natural pigments, which encompass flavonoids, anthocyanins, carotenoids, chlorophylls and chlorophyll derivatives, porphyrins, quinones, and betalains. Chromatography of pigments is especially difficult because many are photo- and air-sensitive and can degrade rapidly unless precautions are taken.

In Chapter 14, one of the least-used applications of TLC and PLC is described, namely inorganics and organometallics. These separations in the analytical mode often require quite unusual stationary phases (e.g., inorganic ion exchangers and impregnated and mixed layers) combined with a variety of diverse mobile phases. This means that the use of the analogous systems in the preparative mode represents an unusually difficult challenge.

Chapter 15 allows a fairly broad insight into the areas of experimental geochemistry that can benefit from PLC. Owing to the considerable complexity of geochemical samples having an organic origin, PLC plays the role of a pilot separation technique, enabling primary group fractionation of the respective natural mixtures, followed by secondary fractionation of these groups with the aid of automated column techniques, and, finally, followed by identification of the individual separated species. Examples related to oils, bitumens, coal liquids, and pyrolysates are given.

The final chapter (Chapter 16) shows how PLC can be used to isolate and identify unknown terpenoic compounds from the frankincense resin (olibanum) and to find marker diterpenes. The novel development at low temperatures is included in the PLC methods described.

1.5 EPILOGUE

The editors wish to express their deepest conviction that this first book ever published on PLC will prove to be a catalyst that inspires much wider use of the method and is instructive for understanding the theory and correct operation of the various steps. We believe it will fill a serious void in the available information on planar chromatography and that it will become an appreciated reference book or tutorial for all those in need of the simple and cost-saving, yet very efficient, PLC microseparation technique.

REFERENCES

1. Tswett, M., *Ber. D. Botan. Ges.*, 24, 384–393, 1906. (Translated from German into English by Strain, H.H. and Sherma, J., *J. Chem. Educ.*, 44, 238–242, 1967.)
2. Strain, H.H. and Sherma, J., *J. Chem. Educ.*, 44, 236–237, 1967.
3. Ritter, F.J. and Meyer, G.M., *Nature*, 193, 941–942, 1962.
4. Miller, J.M. and Kirchner, J.G., *Anal. Chem.*, 23, 428–430, 1951.
5. Miller, J.M. and Kirchner, J.G., *Anal. Chem.*, 24, 1480–1482, 1952.
6. Stahl, E., Ed., *Thin Layer Chromatography: A Laboratory Handbook*, Springer-Verlag, Berlin, 1965, pp. 6, 13, and 39.
7. Pieniak, A., Sajewicz, M., Kowalska, T., Kaczmarski, K., and Tyrpien, K., *J. Liq. Chromatogr. Relat. Technol.*, 28, 2479–2488, 2005.

2 Adsorption Planar Chromatography in the Nonlinear Range: Selected Drawbacks and Selected Guidelines

Krzysztof Kaczmarski, Wojciech Prus, Mieczysław Sajewicz, and Teresa Kowalska

CONTENTS

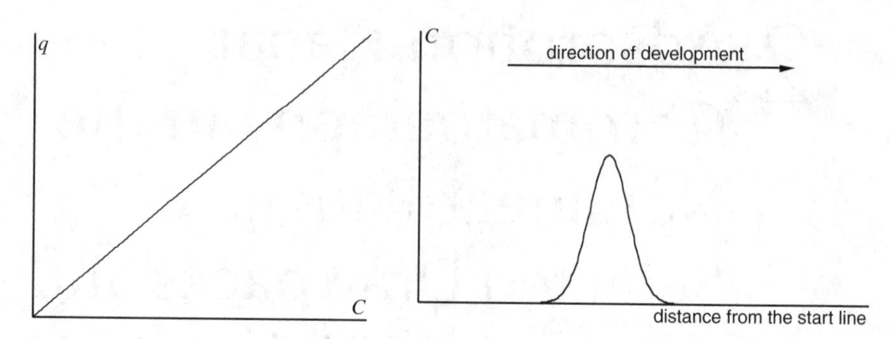

FIGURE 2.1 The linear isotherm of adsorption and the corresponding Gaussian distribution of the analyte's concentration in the chromatographic band.

2.1 ADSORPTION ON A SOLID–LIQUID INTERFACE AND THE EXPERIMENTAL ISOTHERMS OF ADSORPTION

Isotherm models reflect interactions between active sites on the adsorbent surface and the adsorbed species and, simultaneously, the interactions occurring exclusively among the adsorbed species. The dependence of the isotherm shapes on concentration profiles in thin-layer chromatography (TLC) is fully analogous to that observed in high-performance liquid chromatography (HPLC), and the relationship between column chromatographic peak profiles and the isotherm models has been discussed in depth by Guiochon et al. [1].

The simplest isotherm model is furnished by Henry's law

$$q = H \cdot C \tag{2.1}$$

where q is the concentration of the adsorbed species, H is Henry's constant, and C holds for the concentration of this species in the mobile phase. This isotherm is also called the *linear isotherm* and, in this case, the concentration profiles resemble that shown in Figure 2.1.

It should be stressed that in the case of linear isotherm, the peak broadening effect results from eddy diffusion and from resistance of the mass transfer only, and it does not depend on Henry's constant. In practice, such concentration profiles are observed for these analyte concentrations, which are low enough for the equilibrium isotherm to be regarded as linear.

One of the simplest nonlinear isotherm models is the Langmuir model. Its basic assumption is that adsorbate deposits on the adsorbent surface in the form of the monomolecular layer, owing to the delocalized interactions with the adsorbent surface. The Langmuir isotherm can be given by the following relationship:

$$q = \frac{q_s KC}{1 + KC} \tag{2.2}$$

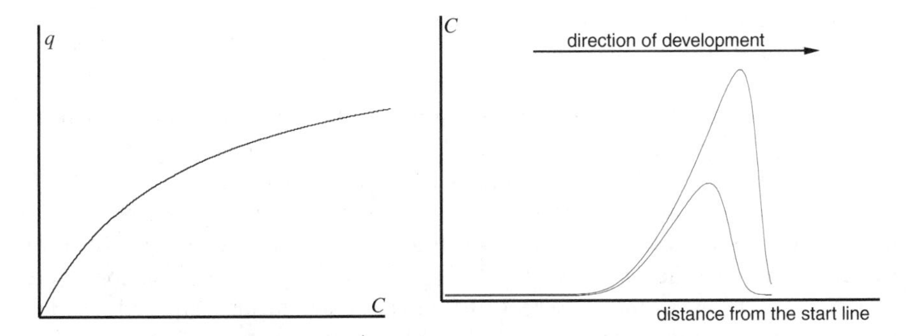

FIGURE 2.2 The Langmuir isotherm of adsorption and the corresponding nonsymmetrical distribution of the analyte's concentration in the chromatographic band.

where q_s is the saturation capacity and K the equilibrium constant. To make use of this particular model, ideality of the liquid mixture and of the adsorbed phase must be assumed. Concentration profiles of the chromatographic bands obtained when adsorption can be best modeled with aid of the Langmuir isotherm are similar to that presented in Figure 2.2. The larger the equilibrium constant, the more stretched is the concentration tail (and, automatically, the chromatographic band, also).

With the adsorbate concentration low enough, the Langmuir isotherm transforms into the linear equation and becomes the simplest isotherm of adsorption, as described by Henry's law.

The case mathematically reciprocal to the Langmuir isotherm is represented by the anti-Langmuir isotherm of adsorption, which, however, is not derived from any physical or physicochemical assumptions regarding the process of adsorption, as was the case with the Langmuir isotherm. The anti-Langmuir isotherm can be given by the following relationship (Figure 2.3):

$$q = \frac{q_s KC}{1 - KC} \tag{2.3}$$

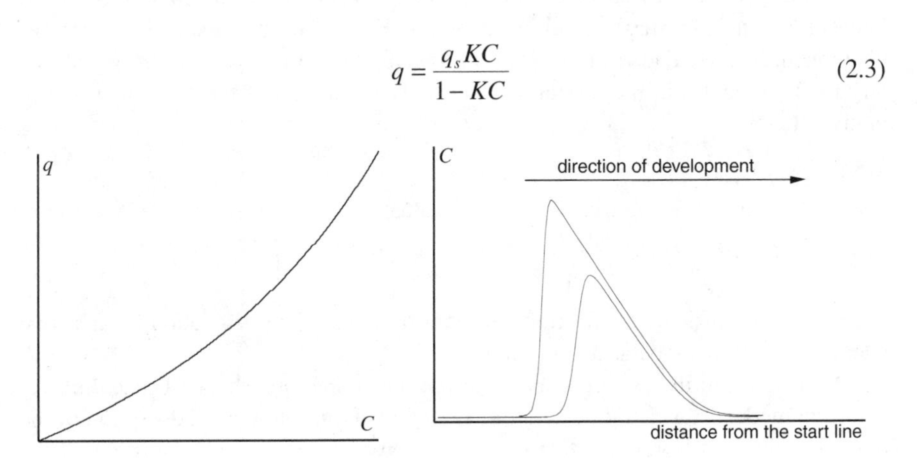

FIGURE 2.3 The anti-Langmuir isotherm of adsorption and the corresponding nonsymmetrical distribution of the analyte's concentration in the chromatographic band.

Also, in this case, with the adsorbate concentration low enough, the anti-Langmuir isotherm transforms into the linear equation and becomes the simplest isotherm of adsorption, as described by Henry's law.

There are several isotherm models for which the isotherm shapes and peak profiles are very similar to that for the anti-Langmuir case. One of these models was devised by Fowler and Guggenheim [2], and it assumes ideal adsorption on a set of localized active sites with weak interactions among the molecules adsorbed on the neighboring active sites. It also assumes that the energy of interactions between the two adsorbed molecules is so small that the principle of random distribution of the adsorbed molecules on the adsorbent surface is not significantly affected. For the liquid–solid equilibria, the Fowler-Guggenheim isotherm has been empirically extended, and it is written as:

$$KC = \frac{\theta}{1-\theta} e^{-\chi\theta} \tag{2.4}$$

where χ denotes the empirically found interaction energy between the two molecules adsorbed on the nearest-neighbor sites and $\theta = q/q_s$ is the degree of the surface coverage. For $\chi = 0$, the Fowler-Guggenheim isotherm simply becomes the Langmuir isotherm.

The Fowler-Guggenheim-Jovanovic model [3] assumes (as it was the earlier case also) the occurrence of intermolecular interactions among the molecules adsorbed as a monolayer but is based on the Jovanovic isotherm. The single-component isotherm is represented by the equation:

$$\theta = 1 - e^{-\left(aCe^{\chi\theta}\right)} \tag{2.5}$$

where a is a model constant.

Contrary to the last two isotherms, which take into the account interactions between the neighboring molecules only, the Kiselev model assumes the single-component localized adsorption, with the specific lateral interactions among all the adsorbed molecules in the monolayer [4–6]. The equation of the Kiselev isotherm is given below:

$$\frac{\theta}{(1-\theta)\cdot C} = \frac{K}{\left[1 - K \cdot K_a \cdot (1-\theta) \cdot C\right]^2} \tag{2.6}$$

where $\theta = q/q_s$, K is the equilibrium constant for the analyte adsorption on an active site, and K_a is the association constant.

Also, the multilayer isotherms have the anti-Langmuir shape. The multilayer isotherm models can easily be derived, assuming an infinitely fast adsorption of the adsorbate on the adsorbent active sites, followed by a subsequent adsorption of the molecules on the first, the second, and consecutive adsorbed layers [7,8].

Assuming that the equilibria between the adjacent layers are depicted by the same equilibrium constant K_p and that K is the equilibrium constant between the first layer and the active sites, the two-layer isotherm model can be expressed as:

$$q = q_s \frac{KC\left(1 + 2K_pC\right)}{1 + KC + KCK_pC} \tag{2.7a}$$

and the three-layer model by the following relationship:

$$q = q_s \frac{KC\left(1 + 2K_pC + 3\left(K_pC\right)^2\right)}{1 + KC + KCK_pC + KC\left(K_pC\right)^2} \tag{2.7b}$$

Expression for the higher numbers of layers originate directly from Equation 2.7a and Equation 2.7b.

The above discussed isotherm models are correct in the case of the single compounds chromatographed in a given chromatographic system. Considering, however, the practical applications of chromatography, multicomponent competitive isotherm models have to be taken into account. Only in an analytical case it can be assumed that the components are eluted independent from one another and that Henry's equation can be used to describe the retention process. In the overload conditions, such an assumption, though, would lead to serious mistakes.

In the simplest case to which the Langmuir isotherm is applicable, the two-component competitive model is given below [1]:

$$\begin{aligned} q_1 &= \frac{q_sK_1C_1}{1 + K_1C_1 + K_2C_2} \\ q_2 &= \frac{q_sK_2C_2}{1 + K_1C_1 + K_2C_2} \end{aligned} \tag{2.8}$$

The two-component Fowler-Guggenheim model can be given as follows [3]:

$$\begin{aligned} q_1 &= \frac{q_sK_1C_1\exp\left(\chi_1\Theta_1 + \chi_{12}\Theta_2\right)}{1 + K_1C_1\exp\left(\chi_1\Theta_1 + \chi_{12}\Theta_2\right) + K_2C_2\exp\left(\chi_{21}\Theta_1 + \chi_2\Theta_2\right)} \\ q_2 &= \frac{q_sK_2C_2\exp\left(\chi_{21}\Theta_1 + \chi_2\Theta_2\right)}{1 + K_1C_1\exp\left(\chi_1\Theta_1 + \chi_{12}\Theta_2\right) + K_2C_2\exp\left(\chi_{21}\Theta_1 + \chi_2\Theta_2\right)} \end{aligned} \tag{2.9}$$

where K_i is the isotherm parameter, Θ_i is the fractional coverage of the ith component, and $\Theta_i = q_i/q_s$. Terms χ_1 and χ_2 relate to the energy of lateral interactions between the molecules of the corresponding components. Terms χ_{12} and χ_{21} take into the account cross-interaction between the separated components.

Finally, for the two-component and the two-layer isotherm of adsorption, it is easy to obtain the following isotherm model (Equation 2.10) by the method described elsewhere [7]:

$$q_1 = q_s \frac{K_1C_1 \cdot (1+2K_{11}C_1 + K_{12}C_2) + C_1C_2K_2K_{21}}{D}$$
$$q_2 = q_s \frac{K_2C_2 \cdot (1+2K_{22}C_2 + K_{21}C_1) + K_1K_{12}C_1C_2}{D} \quad (2.10)$$

where $D = 1 + K_1C_1 + K_2C_2 + K_1K_{11}C_1^2 + K_2K_{22}C_2^2 + (K_2K_{21} + K_1K_{12})C_1C_2$, q_s is the maximum capacity of the adsorbent, C_i is the concentration of the component in mobile phase, q_i is the concentration of the adsorbed component, K_i is the equilibrium constant of adsorption of the ith component on the adsorbent surface, K_{ii} is the equilibrium constant of adsorption for the ith component on the same previously adsorbed ith component, and K_{ij} is the equilibrium constant of adsorption for the ith component on the jth component. It was also assumed that $K_{ij} = K_{ji}$.

2.2 TLC IN THE LINEAR AND THE NONLINEAR REGION OF THE ADSORPTION ISOTHERM

Chromatographic separations are often used for analytical purposes (in order to establish the qualitative and quantitative composition of a given mixture of compounds) and, occasionally, one refers to this technique as the analytical one, which is not correct, because chromatography has an indisputable importance as a versatile separation tool, also, enabling isolation of preparative amounts of substances. Such preparative chromatographic separations are most frequently carried out by means of liquid chromatography (LC) on the nonpressurized columns filled with an adsorbent and also with the aid of HPLC, TLC, and gas chromatography (GC).

The adsorption mechanism of solute retention is one of the two universal mechanisms in the chromatographic process (the other being the partition mechanism), and it operates on a purely physical principle. In fact, virtually all solutes can adsorb on a microporous solid surface (or be partitioned between the two immiscible liquids). If the analytes are applied to a chromatographic system in the low-enough aliquots, then the respective retention processes occur within the linear range of the empirical adsorption isotherms of the species involved and concentration profiles of the resulting chromatographic bands are Gaussian, as shown in Figure 2.1.

The adsorption TLC operating in the linear range of the adsorption isotherm (sometimes dubbed as the linear adsorption TLC or simply as the linear TLC) is utilized for purely analytical purposes (which include establishing of a qualitative composition of a given mixture of analytes, often followed by their quantification in the examined sample with aid of the calibration plot approach). In order to introduce certain amount of rationale to the linear adsorption TLC (and enable optimization of the separation result), several semiempirical models have been elaborated, and the most popular of them is going to be briefly summarized in the forthcoming paragraphs of this section.

It is the main aim of semiempirical chromatographic models to couple the empirical parameters of retention with the established thermodynamic quantities generally used in physical chemistry. The validity of a model for chromatographic practice can hardly be overestimated, because it often and successfully helps to overcome the old trial-and-error approach to running the analyses, especially when incorporated in the separation selectivity oriented optimization strategy.

Partition (liquid–liquid) TLC was first among the chromatographic techniques to gain thermodynamic foundations, owing to the pioneering work of Martin and Synge [9], the 1952 Nobel Prize winners in chemistry. The Martin and Synge model describes the idealized parameter R_F (i.e., the parameter R_F') in the following way:

$$R_F' = \underset{\text{(I)}}{\frac{t_m}{t_m + t_s}} = \underset{\text{(II)}}{\frac{n_m}{n_m + n_s}} = \underset{\text{(III)}}{\frac{m_m}{m_m + m_s}} \tag{2.11}$$

where t_m and t_s denote the time spent by a solute molecule in the mobile and stationary phases, respectively, n_m and n_s are the numbers of the solute molecules equilibrially contained in the mobile and stationary phases, and m_m and m_s are the respective mole numbers.

The most popular (and also most important) semiempirical model of the linear adsorption TLC, analogous to that of Martin and Synge, was established in the late 1960s by Snyder [10] and Soczewinski [11] independently, and it is often referred to as the *displacement model of solute retention*. The crucial assumption of this model is that the retention mechanism consists of a competition among the solute and the solvent molecules to the adsorbent active sites and, hence, in the virtually endless acts of the solvent molecules displacing those of the solute on the solid surface (and *vice versa*). Further, the authors assumed that certain part of the mobile phase rests adsorbed and stagnant on the adsorbent surface. This adsorbed mobile phase formally resembles the liquid stationary phase in partition chromatography. Thus, utilizing with imagination the main concept of the Martin and Synge model of partition chromatography, Snyder and Soczewinski managed to define the R_F parameter valid for the adsorption TLC as given below:

$$R_F' = \frac{t_m}{t_m + t_a} \equiv \frac{n_m}{n_m + n_a} = \frac{m_m}{m_m + m_a} = \frac{c_m\left(V_m - V_a W_a\right)}{c_m\left(V_m - V_a W_a\right) + c_a V_a W_a} \tag{2.12}$$

where t_m and t_a denote the time spent by a solute molecule in the mobile phase and on the adsorbent surface, respectively, n_m and n_a are the numbers of the solute molecules equilibrially contained in the mobile phase and on the adsorbent surface, m_m and m_a are the mole numbers of the solute molecules contained in the nonadsorbed and the adsorbed moieties of mobile phase, c_m and c_a are the molar concentrations of the solute in the nonadsorbed and the adsorbed moities of mobile phase, V_m is the total volume of the mobile phase, V_a is the volume of the adsorbed mobile phase per mass unit of an adsorbent, and W_a is the mass of the adsorbent considered.

Transformation of Equation 2.12 results in the following relationship:

$$R'_F = \frac{1}{1 + K_{th}\Phi} \quad (2.13)$$

where $K_{th} = \frac{c_a}{c_m}$ is the thermodynamic equilibrium constant of adsorption and $\Phi = V_a W_a / (V_m - V_a W_a)$.

From the general framework of the Snyder and Soczewinski model of the linear adsorption TLC, two very simple relationships were derived, which proved extremely useful for rapid prediction of solute retention in the thin-layer chromatographic systems employing binary mobile phases. One of them (known as the Soczewinski equation) proved successful in the case of the adsorption and the normal phase TLC modes. Another (known as the Snyder equation) proved similarly successful in the case of the reversed-phase TLC mode.

The following Soczewinski equation is a simple linear relationship with respect to $\log X_S$, linking the retention parameter (i.e., R_M) of a given solute with quantitative composition of the binary eluent applied:

$$R_M = C - n \log X_S \quad (2.14)$$

where C is in the first instance the equation constant (with a clear physicochemical explanation though), X_s is the molar fraction of the stronger solvent in the nonaqueous mobile phase, and n is the number of active sites on the surface of an adsorbent.

Apart from enabling rapid prediction of solute retention, the Soczewinski equation allows a molecular-level scrutiny of the solute — stationary phase interactions. The numerical value of the parameter n from Equation 2.14, which is at least approximately equal to unity ($n \approx 1$), gives evidence of the one-point attachment of the solute molecule to the stationary phase surface. The numerical values of n higher than unity prove that in a given chromatographic system, solute molecules interact with the stationary phase in more than one point (the so-called multipoint attachment).

The following Snyder equation is another simple linear relationship with respect to φ, which links the retention parameter (i.e., $\ln k$) of a given solute with the volume fraction of the organic modifier in the aqueous binary mobile phase (φ):

$$\ln k = \ln k_w - S\varphi \quad (2.15)$$

where k is the solute's retention coefficient ($k = (1 - R_F)/R_F$), k_w is the solute's retention coefficient extrapolated for pure water as mobile phase, and S is the constant characteristic of a given stationary phase.

Consequences of the Snyder and Soczewinski model are manifold, and their practical importance is very significant. The most spectacular conclusions of this model are (1) a possibility to quantify adsorbents' chromatographic activity and (2) a possibility to define and quantify "chromatographic polarity" of solvents (known as the solvents' elution strength). These two conclusions could only be drawn on the assumption as to the displacement mechanism of solute retention. An obvious necessity was to quantify the effect of displacement, which resulted in the following relationship for the thermodynamic equilibrium constant of adsorption, K_{th}, in the case of an active chromatographic adsorbent and of the monocomponent eluent:

$$\log K_{th} = \log V_a + \alpha(S^0 - A_S \in^0) \tag{2.16}$$

where α is the function of the adsorbent surface energy independent of the properties of the solute (known as the activity coefficient of the adsorbent; practical determination of its numerical values can be regarded as quantification of the adsorbent's chromatographic activity), S^0 is the adsorption energy of the solute chromatographed on an active adsorbent with aid of *n*-pentane as monocomponent mobile phase, A_S is the surface area of the adsorbent occupied by an adsorbed solute molecule, and $\in^0$ is the parameter usually referred to as the solvent elution strength, or simply the solvent strength (the adsorption energy of solvent per unit of the adsorbent surface area).

Assuming that the adsorbent surface is occupied by an adsorbed solute molecule (A_S) and a molecule of a stronger solvent (n_B) which are equal to one another, the elution strength of a binary eluent, $\in_{AB}$, shows the following dependence on its quantitative composition:

$$\in_{AB} = \in_A + \frac{\log\left[x_B 10^{\alpha \cdot n_B(\in_B - \in_A)} + 1 - x_B\right]}{\alpha \cdot n_B} \tag{2.17}$$

where $\in_A$ is the elution strength of the weaker component (A) of a given binary eluent, $\in_B$ is the elution strength of the stronger component (B) of the same eluent, and x_B is the molar volume of the component B.

Coupling Equation 2.16 and Equation 2.17, we can obtain the following relationship, which describes the dependence of the solute's retention parameter, R_M, on the quantitative composition of a given binary eluent:

$$\log k \equiv R_M = \log V_a + \alpha\left[S^0 - A_S \in_A\right] - \frac{A_S \log\left[x_B 10^{\alpha \cdot n_B(\in_B - \in_A)} + 1 - x_B\right]}{n_B} \tag{2.18}$$

Apart from the most widely utilized Snyder and Soczewinski semiempirical model of linear TLC, several other physicochemically grounded approaches to the same question exist as well [12]. Also, a choice of the empirical rules in mathematical

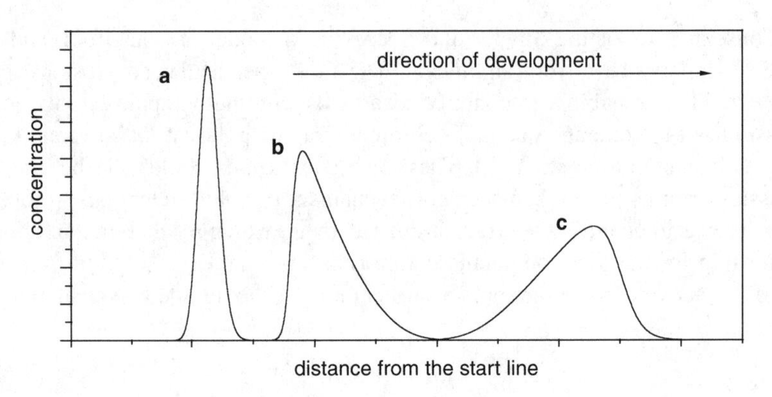

FIGURE 2.4 Three types of concentration profiles encountered among the thin-layer chromatographic bands: (a) symmetrical (Gaussian) without tailing, (b) skewed with front tailing, and (c) skewed with back tailing [14].

and nonmathematical form [13] exist, which prove very helpful in selecting proper experimental conditions for running the developments in the analytical TLC mode. All these approaches inevitably share one common trait, namely, an absolute negligence of intermolecular interactions of the analyte-analyte type (i.e., of the so-called lateral interactions).

The preparative (or nonlinear) adsorption TLC has never attracted enough attention from the side of theoreticians of the planar technique to result in a codified system of rules, helpful in an efficient carrying out of micropreparative isolation of individual compounds or compound groups. Normally, it is taken for granted that to the preparative (i.e., nonlinear) adsorption TLC, the same rules can be applied as to the analytical (i.e., linear) variant, although it is also known in advance that performance of these rules in the former case is considerably worse than in the latter one.

The primary reason that no valid rules for preparative TLC have so far been elaborated is due to a rather limited access to densitometric detection, which is still not commonplace in many thin-layer chromatographic laboratories throughout the world. And the demand for densitometric detection in preparative TLC has one very simple reason. Working in the nonlinear range of the experimental isotherms of adsorption of the analytes and because of the resulting mass overload of the respective sorbents, the chromatographic bands cannot be Gaussian in terms of their concentration profiles (or circular, if the spot application and the traditional visualization methods are considered). In the preparative TLC mode, the skewed concentration profiles are dealt with exclusively, which can either demonstrate the back tailing or the front tailing, thus negatively affecting resolution and separation, basically due to an increased chance of the bands overlapping.

Use of densitometric detection provides an insight into the concentration profiles of chromatographic bands, thus furnishing an indispensable prerequisite, needed for proper assessment of the retention mechanisms in the preparative adsorption TLC. Figure 2.4 shows three types of the band concentration profiles. The Gaussian peak (a) in this figure represents the linear isotherm of adsorption of a given species, peak

(b) is valid for the anti-Langmuir-type of the adsorption isotherm, and peak (c) results from the Langmuir-type of the adsorption isotherm.

It is generally assumed that the symmetrical concentration profiles without tailing (which can be approximated by the Gaussian function; see Figure 2.4a) give evidence of the fact that the adsorbent is not yet overloaded by the analyte (i.e., that the chromatographic band still appears within the linear range of the adsorption isotherm) and from this particular shape, the chemical structure of the analyte itself cannot be judged (i.e., the crucial question cannot be answered regarding whether or not the analyte is equipped with such functionalities that allow its participation in the lateral interactions).

From the asymmetrical concentration profile with front tailing (see Figure 2.4b), it can correctly be deduced that (1) the adsorbent layer is already overloaded by the analyte (i.e., the analysis is being run in the nonlinear range of the adsorption isotherm) and (2) the lateral interactions (i.e., those of the self-associative type) among the analyte molecules take place. The easiest way to approximate this type of concentration profile is by using the anti-Langmuir isotherm (which has no physicochemical explanation yet models the cases with lateral interactions in a fairly accurate manner).

From the asymmetrical concentration profile with back tailing (see Figure 2.4c), it can correctly be deduced that (1) the adsorbent layer is already overloaded by the analyte (i.e., again, the analysis is being run in the nonlinear range of the adsorption isotherm), but (2) the lateral interactions (i.e., those of the self-associative type) are negligible among the analyte molecules. The simplest way to approximate this type of the concentration profile is by using the Langmuir isotherm (i.e., the one which has a well-grounded physicochemical explanation).

2.3 PREPARATIVE LAYER CHROMATOGRAPHY AS A PRACTICAL USAGE OF THE NONLINEAR REGION

Due to low aliquots of separated analytes (usually between 10 and 1000 mg) [15], preparative layer chromatography (PLC) can be regarded as one of the most versatile micropreparative isolation techniques available in the arsenal of separation methods. Among those who most frequently utilize PLC, we can find, for example, the organic chemists who work at the microscale level and thus cannot isolate their intermediate and final products with aid of the classical nonpressurized chromatographic columns. Another group are researchers in the field of pharmacognosy, whose task is to isolate pharmacodynamically active constituents from the natural healing materials. PLC can also prove useful for those involved in geochemical and environmental studies, who attempt to isolate trace amounts of analytes often contained in the abundantly available natural matrices. One cannot forget about the applicability of PLC in the life sciences, either, and the list of potential users of this technique is practically endless. An additional and particularly valued feature of PLC (shared in common with the analytical TLC mode) is that, contrary to the advanced column techniques, in adsorption TLC, various different samples of natural origin (e.g., body fluids, liquid environmental samples, etc.), which contain the suspended materials, do not need any extra pretreatment (e.g., of the solid phase extraction [SPE] type), aimed at their removal. In fact, the adsorbent contained in the starting spot (or the starting line) of a typical thin-layer chromatogram purifies the applied sample from insoluble solid particles and also from other contaminants — in this sense acting as a specific inbuilt SPE device, also.

In Chapter 1, we have clearly stated that the contents of this book are going to be devoted exclusively to the capillary flow of the eluents employed in PLC, and we also explained our reasons for making this particular decision. Now let us repeat once again that until now, the specificity of the multiple theoretical (and also practical) aspects of PLC have not been systematically introduced to the planar chromatographic community in form of a separate handbook, even if its procedures were briefly described in several book chapters, e.g., in [16–18]. In order to at least partially fill this undeserved void, we are now going to discuss selected theoretical aspects characteristic of developing chromatograms in the nonlinear adsorption TLC mode.

2.4 LATERAL INTERACTIONS AND THEIR IMPACT ON SEPARATION IN PLC

Lateral interactions — in other words, intermolecular interactions between the molecules of the same analyte or of the two different analytes — depend on the chemical structure of the species involved and more specifically, on the type of functionalities present in their molecules. Simplifying the problem, the analyte molecules can, upon their structure and the resulting ability to form the hydrogen bonds, be divided into the four categories first introduced by Pimentel and McClellan in their famous (and now historical) monography of hydrogen bonding [19]. This very useful classification is given in the aforementioned four categories of compounds.

N: The molecules from this category lack both the "acidic" and the "basic" functionality and, hence, they remain unable to interact through the hydrogen bonds. Aliphatic hydrocarbons are the best examples of the analytes from this particular group of compounds.

A: These molecules are equipped with acidic functionality only and, hence, they can participate in the hydrogen bonds as proton donors. The most representative examples from this group are chloroform ($CHCl_3$) or dichloromethane (CH_2Cl_2).

B: The molecules from this group incorporate electronegative (i.e., basic) heteroatoms (e.g., O, N, S, etc.) and π-electrons, and, hence, they are able to participate in the hydrogen bonds as proton acceptors. Ketones and ethers are among the most representative classes of analytes belonging to this particular group.

AB: The molecules belonging to the last category are equipped with both the acidic and the basic functionalities and, hence, they can interact through the hydrogen bonds both as proton donors and proton acceptors. Acohols and carboxylic acids are among the most representative examples from this group.

Analytes from class N neither self-associate nor participate in the mixed hydrogen bonds. Consequently, they cannot participate in lateral interactions of any kind, either.

Analytes from classes A and B cannot self-associate through the hydrogen bonds and because of this, they cannot participate in the self-associative lateral interactions. However, they can participate in the mixed hydrogen bonds and take part in the mixed lateral interactions. For example, analyte A can laterally interact either with analyte B or analyte AB. In a similar way, analyte B can either interact with analyte A or analyte AB.

Analytes from class AB can self-associate through the hydrogen bonds and, hence, under mild chromatographic conditions, they can participate in the self-associative lateral interactions of the AB … AB type. Moreover, they can participate in the mixed lateral interactions of three different types (AB … A, AB … B, and AB_1 … AB_2).

The enthalpy of the H-bonds among the majority of the organic compounds is relatively low (usually within the range of about 20 kJ per one mol of hydrogen bonds) and therefore they can easily be disrupted. In order to demonstrate the presence of lateral interactions in chromatographic system, low-activity adsorbents are most advisable (i.e., those having relatively low specific surface area, low density of active sites on its surface, and low energy of intermolecular analyte–adsorbent interactions, which obviously compete with lateral interactions). For the same reason, the most convenient experimental demonstration of lateral interactions can be achieved in presence of the low-polar solvents (basically those from the class N; e.g., *n*-hexane, decalin, 1,4-dioxane, etc.) as mobile phases.

The adsorbent most suitable for demonstration of the existence of lateral interactions among the analyte molecules is cellulose. For planar chromatographic purposes, it is available either in the form of ready-made TLC plates precoated with microcrystalline cellulose or as chromatographic paper. The examples that are going to be presented in the subsequent sections of this chapter refer to the chromatographic systems composed of cellulose as a stationary phase and either a single hydrocarbon or ether (e.g., *n*-hexane, *n*-octane, decalin, 1,4-dioxan) as a monocomponent mobile phase. First, the examples from articles [8,20–24] will be presented. These examples focus on lateral interactions among the molecules of a single analyte (i.e., on the self-association through the hydrogen bonds) and on their impact on the concentration profiles of the resulting chromatographic bands. As convenient model analytes, selected alcohols and mono- and dicarboxylic acids were chosen, all of them representing the AB class of analytes, according to Pimentel and McClellan. Shapes of the respective concentration profiles obtained in a nonlinear region of the adsorption isotherm could be approximated with aid of the anti-Langmuir-type adsorption isotherm.

Then, the examples from Reference 23, that focus on retention of the selected binary mixtures of the test analytes (one comprising carboxylic acid and ketone and the other made of alcohol and ketone), chromatographed under the deliberately mild working conditions (microcrystalline cellulose was used as adsorbent and either decalin or *n*-octane as the monocomponent mobile phase) will be discussed. One of the test solutes in each binary mixture (either acid or alcohol) can be viewed as

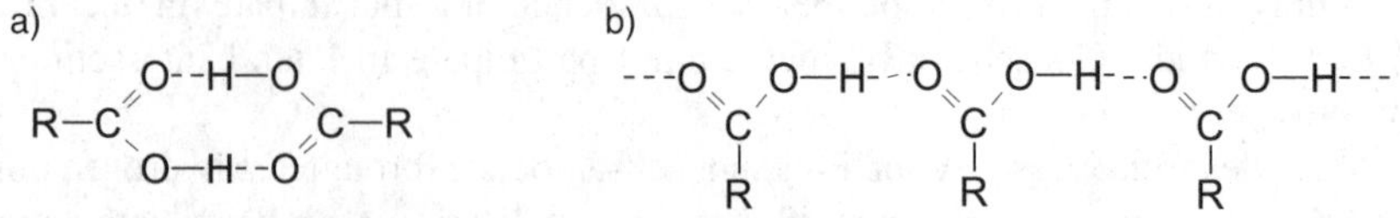

FIGURE 2.5 Schematic representation of hydrogen-bonded self-associative structures of higher fatty acids: (a) cyclic associative dimer and (b) linear associative multimer.

having the double properties of Lewis acid and Lewis base (AB), whereas the other component of each mixture (ketone) is the typical Lewis base (B). Thus, the acid and the alcohol both tend to self-associate by the hydrogen bond and to form mixed associates with the compounds of the A, B, or the AB type. In contrast, ketone cannot self-associate by the hydrogen bonds, but it can participate in the mixed associates either with a Lewis acid (A) or with the compound of the AB type (i.e., either with the carboxylic acid or the aliphatic alcohol).

2.4.1 Self-Associative Lateral Interactions

Self-associative lateral interactions can only occur with the AB-type analytes, chromatographed in sufficiently mild chromatographic conditions. In planar chromatography, this type of lateral interaction was first demonstrated on monocarboxylic fatty acids and α,ω-dicarboxylic acids, chromatographed on microcrystalline cellulose with aid of decalin and 1,4-dioxane as monocomponent eluents, respectively [8,20,23].

2.4.1.1 Higher Fatty Acids

Higher fatty acids can form associative multimers by hydrogen bonding. This is because of the presence of the negatively polarized oxygen atom from the carbonyl group and the positively polarized hydrogen atom from the carboxyl group. Monocarboxylic acids can form cyclic dimers, coupled together by the pair of hydrogen bonds (see Figure 2.5a). It seems, however, that as a result of direct contact of higher fatty acids with an adsorbent, the rings of the predominantly cyclic dimers are forcibly opened (e.g., because of intermolecular interactions of the acidic carboxyl groups with the hydroxyl groups of cellulose), thus considerably shifting the self-associative equilibrium toward the linear associative multimers.

The lengthwise concentration profiles obtained for the higher fatty acids in most cases resembled peaks generated with the aid of the Langmuir-type isotherm of adsorption (see Figure 2.2). Peak profiles like those presented in Figure 2.6 suggesting the presence of lateral interactions [20] were obtained only for the highest concentrations and the shortest aliphatic chain lengths of the acids involved.

2.4.1.2 α,ω-Dicarboxylic Acids

α,ω-Dicarboxylic acids can form linear associative multimers (Figure 2.7) held together by hydrogen bonding (a maximum of four hydrogen bonds can be formed by each molecule of such an acid). This is because of the presence of the two

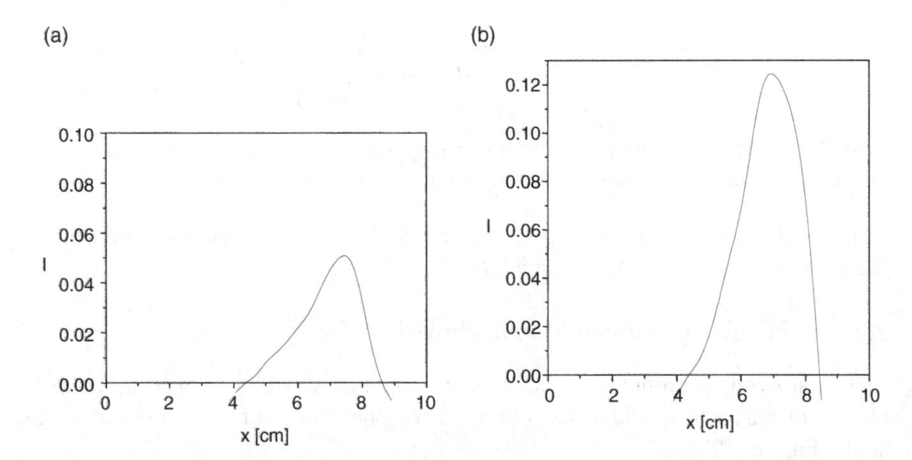

FIGURE 2.6 Concentration profiles of (a) hexadecanoic acid, sample concentration $C = 0.025$ mol l^{-1}, (b) dodecanoic acid, sample concentration $C = 0.1$ mol l^{-1}, obtained on microcrystalline cellulose with decaline as mobile phase [20].

FIGURE 2.7 Schematic representation of the self-association of dicarboxylic acids as a result of intermolecular hydrogen bonds.

negatively polarized oxygen atoms of the carbonyl groups and the two positively polarized hydrogen atoms of the hydroxyl groups.

For α,ω-dicarboxylic acids, the anti-Langmuir concentration profiles are pronounced the most for the compounds with the longest aliphatic chains, as shown in Figure 2.8. From the densitograms obtained, it is apparent that α,ω-dicarboxylic acids are more prone to form anti-Langmuir concentration profiles than monocarboxylic acids. This suggests that lateral interactions are more effective with dicarboxylic acids than with monocarboxylic ones, which seems a natural consequence

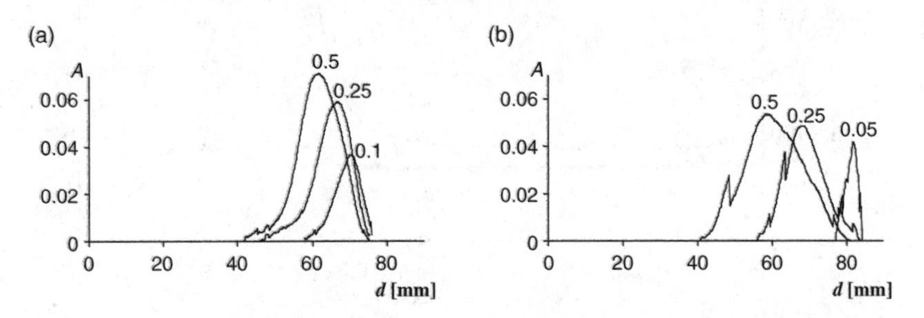

FIGURE 2.8 Concentration profiles of (a) succinic acid (C_4) and (b) suberic acid (C_8), obtained on microcrystalline cellulose with 1,4-dioxan as mobile phase. Sample concentrations were: (a) 0.1, 0.25, and 0.5 mol·l^{-1} and (b) 0.05, 0.25, and 0.5 mol·l^{-1} [8].

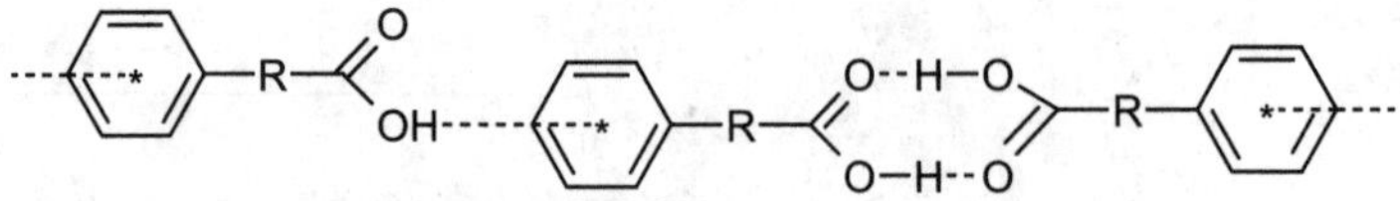

FIGURE 2.9 Mixed linear/cyclic dimers of phenyl-substituted monocarboxylic acids (coupled together through the carboxyl groups and the aromatic π-electrons).

of the ability of the former to give the two hydrogen bonds per each end of the dicarboxylic acid molecule (Figure 2.7).

2.4.1.3 Phenyl-Substituted Monocarboxylic Acids

The most strongly pronounced anti-Langmuir behavior (when comparing the three classes of carboxylic acids) was observed for phenyl-substituted monocarboxylic acids (Figure 2.9).

It is evident that lateral interactions are the strongest with the above presented phenyl-substituted monocarboxylic acids because of at least two factors, namely: (1) the relatively short aliphatic moiety and (2) substitution of the aliphatic chains with phenyl groups. Obviously, the longer the aliphatic chain moiety in a given molecular structure, the greater is the steric hindrance and the more shielded are the respective functionalities. Consequently, the probability of their participation in the hydrogen bonds (i.e., in lateral interactions) is lower. This becomes particularly evident when the higher fatty acids and their respective concentration profiles, indicative of the Langmuir-type isotherms of adsorption are compared with the phenyl-substituted fatty acids and with their substantially different concentration profiles, indicative of the anti-Langmuir-type isotherms of adsorption. The important

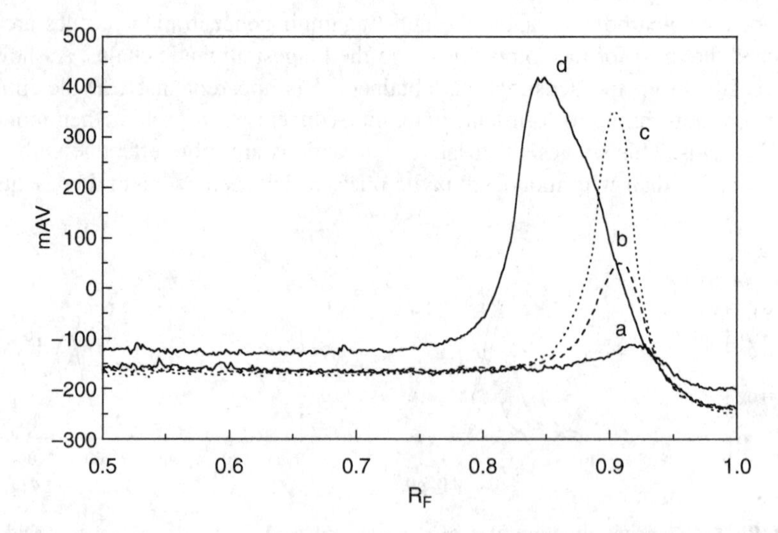

FIGURE 2.10 Concentration profiles of 3-phenylpropionic acid obtained on microcrystalline cellulose with decalin as mobile phase. Sample concentrations were: (a) 0.2, (b) 0.3, (c) 0.4, and (d) 0.5 mol·l⁻¹ [25].

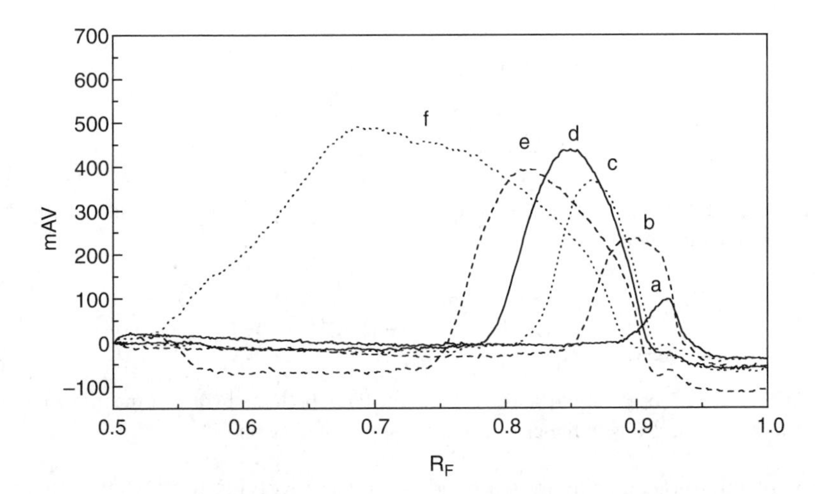

FIGURE 2.11 Concentration profiles of 4-phenylbutyric acid obtained on microcrystalline cellulose with decalin as mobile phase. Sample concentrations were: (a) 0.1, (b) 0.2, (c) 0.3, (d) 0.4, (e) 0.5, and (f) 1.0 mol·l^{-1} [25].

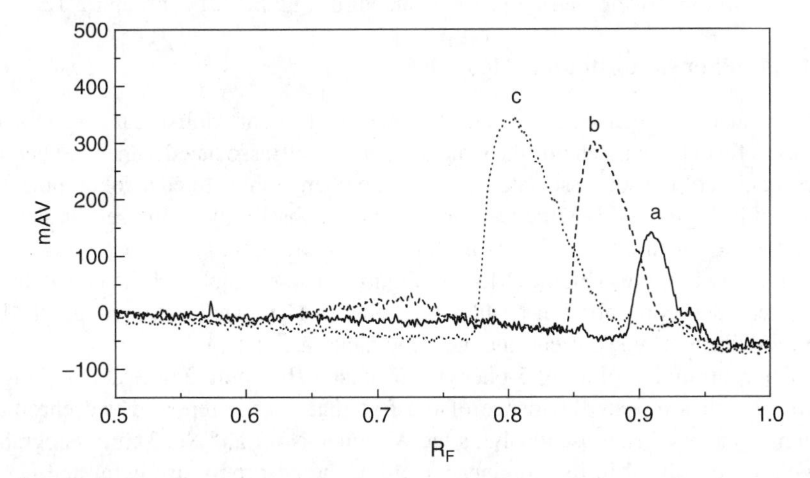

FIGURE 2.12 Concentration profiles of 2-phenylbutyric acid obtained on microcrystalline cellulose with decalin as mobile phase. Sample concentrations were (a) 0.2, (b) 0.3, and (c) 0.5 mol·l^{-1} [25].

structural difference between the phenyl-substituted fatty acids and the other two groups of carboxylic acids discussed in this chapter is their mixed aryl-aliphatic nature. The presence of the flat phenyl group in the structure results in a far less acute steric hindrance than the aliphatic chain and, therefore, it is less effective in preventing the carboxyl functionalities from self-association. The flat phenyl group is, moreover, negatively charged because of its delocalized π-electrons, and therefore also prone to participate in the hydrogen bonds as an electron donor. Summing up, the molecules of the phenyl-substituted fatty acids can participate in two different

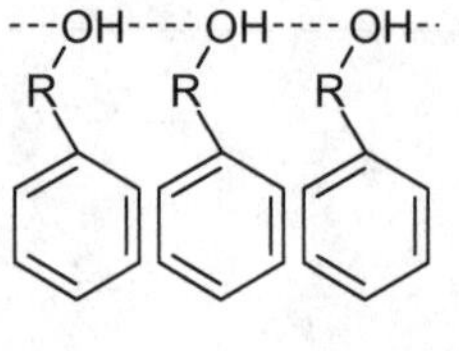

FIGURE 2.13 Schematic presentation of the self-associated chain-like *n*-mer of the phenyl-substituted alcohol.

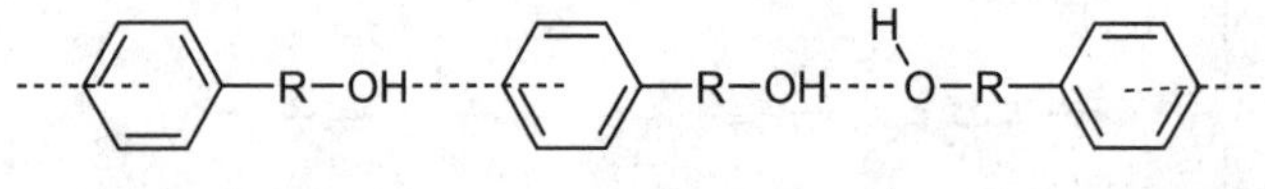

FIGURE 2.14 A possibility to self-associate involving both the hydroxyl functionalities and the aromatic π-electrons in the case of phenyl-substituted alcohols.

kinds of intermolecular hydrogen bonds — those involving the carboxyl functionalities (which result in two hydrogen bonds per acid molecule) and those that involve the aromatic rings (and result in a single hydrogen bond per acid molecule). This dual type of intermolecular interaction is clearly shown in Figure 2.9, and it is also reflected in the densitometric outcome, shown in Figures 2.10 through 2.12.

2.4.1.4 Phenyl-Substituted Alcohols

The manner in which alcohols self-associate is different. Unlike carboxylic acids, alcohols form the long chain-like structures of the self-associated *n*-mers. Schematic presentation of the self-associated chain of the phenyl-substituted alcohol molecules is given in Figure 2.13. Moreover, an alternative possibility of the self-association with the aforementioned alcohols is shown in Figure 2.14.

In the two figures (Figure 2.15 and Figure 2.16), examples of the densitometrically recorded concentration profiles of 5-phenyl-1-pentanol, chromatographed on the two different types of chromatographic paper are shown.

Concentration profiles of 5-phenyl-1-pentanol, shown in Figure 2.15 and Figure 2.16, furnish a repeated evidence of the fact that (1) the reported two chromatographic systems (i.e., those involving the Whatman No. 1 and No. 3 chromatographic paper) were utilized in the nonlinear range of the adsorption isotherm and (2) the alcohol considered is apt for self-association. This is clearly demonstrated by the front tailing of all the concentration profiles and becomes increasingly pronounced with the growing concentration of the alcohol samples.

2.4.2 Mixed-Associative Lateral Interactions

Mixed lateral interactions can occur in the case of the following pairs of molecules: A … B, AB … A, AB … B, and AB_1 … AB_2. Their appearance is even more importunate than that of the self-associative lateral interactions. With the self-associative lateral interactions alone, tailing of chromatographic bands lowers separative performance of a given chromatographic system, whereas with the mixed lateral

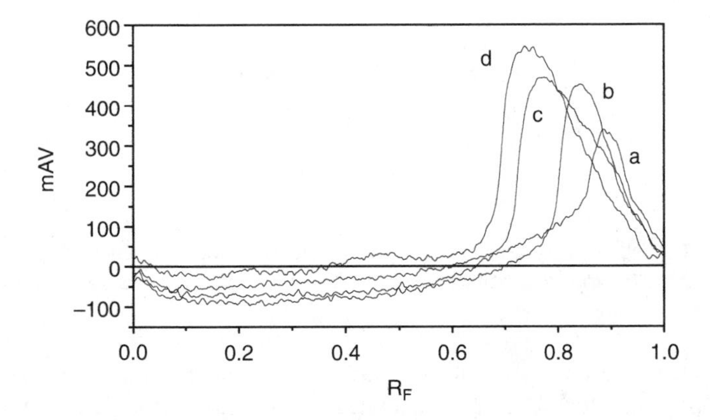

FIGURE 2.15 Concentration profiles of 5-phenyl-1-pentanol, obtained on Whatman No. 3 chromatography paper with *n*-octane as mobile phase. Concentrations of the analyte solutions in 2-propanol were (a) 0.5, (b) 1.0, (c) 1.5, and (d) 2.0 mol l^{-1} [14,25].

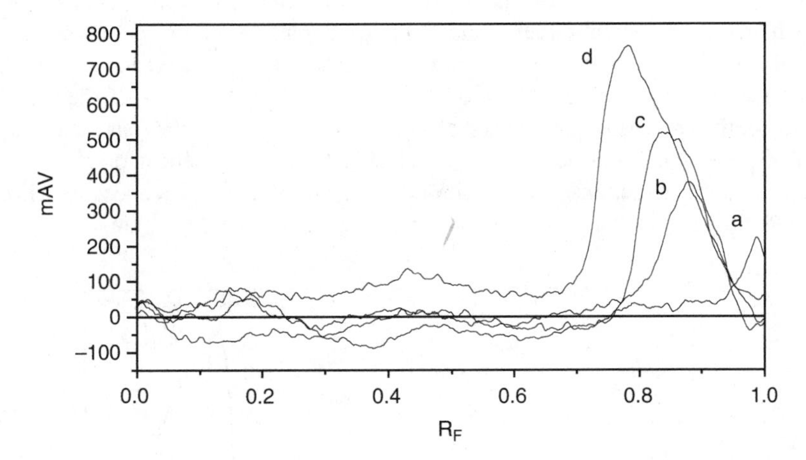

FIGURE 2.16 Concentration profiles of 5-phenyl-1-pentanol, obtained on Whatman No. 1 chromatography paper with *n*-octane as mobile phase. Concentrations of the analyte solutions in 2-propanol were (a) 0.25, (b) 0.50, (c) 0.75, and (d) 1.0 mol l^{-1} [14,25].

interactions, apart from lowering of the system's separative performance, they can even prevent the two different analytes from separation.

Now we are going to demonstrate mixed lateral interactions for the two different AB … B systems (the first composed of acid and ketone and the second composed of alcohol and ketone) and to discuss their impact on separation performance when using mild chromatographic conditions (i.e., the low-active adsorbent and the low-polar eluent). As the binary test systems, we selected the 2-phenylbutyric acid-benzophenone mixture of analytes (i.e., acid and ketone) and the 5-phenyl-1-pentanol-benzophenone mixture of analytes (i.e., alcohol and ketone). Mixed associative dimers composed of the two pairs of analytes are schematically given in Figure 2.17.

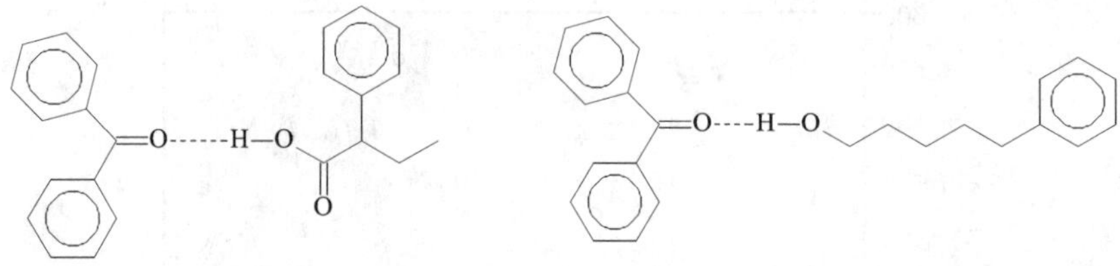

FIGURE 2.17 Schematic presentation of the mixed H-bonded associates of the acid and ketone (2-phenylbutyric acid-benzophenone) and of the alcohol and ketone (5-phenyl-1-pentanol-benzophenone).

Now let us demonstrate selected chromatographic separations in the case of the two aforementioned pairs of analytes, which can participate in the mixed lateral interactions. The persuasive enough pictures are shown in Figure 2.18 and Figure 2.19.

From the examples of the two analytes coupled together by the mixed lateral interactions (see Figure 2.18 and Figure 2.19), it can easily be deduced that in particularly inconvenient circumstances, these interactions can even prevent a pair of analytes from separation (or, in the less acute cases, they can exert another negative effect, leading to partial resolution of the two bands only), in spite of the fact that the respective retention parameters of single analytes under the same chromatographic conditions considerably differ. Hence, from the practical point of view, mixed lateral interactions can be viewed as a harmful circumstance, which can considerably reduce the separation power of a given chromatographic system.

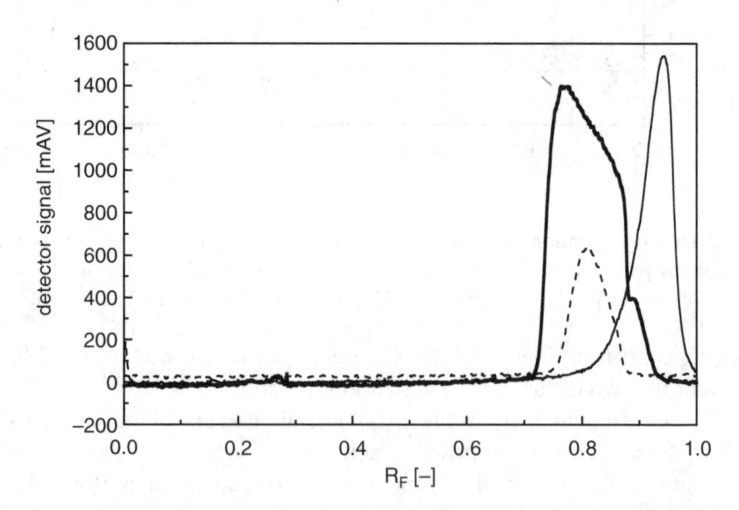

FIGURE 2.18 Comparison of the concentration profiles of 2-phenylbutyric acid (dashed line) and benzophenone (thin solid line) developed as single analytes and as a binary mixture (bold solid line); concentration of 2-phenylbutyric acid in the sample was 1.25 mol l^{-1} and that of benzophenone was 0.10 mol l^{-1}. Microcrystalline cellulose was used as stationary phase and decalin as mobile phase [26].

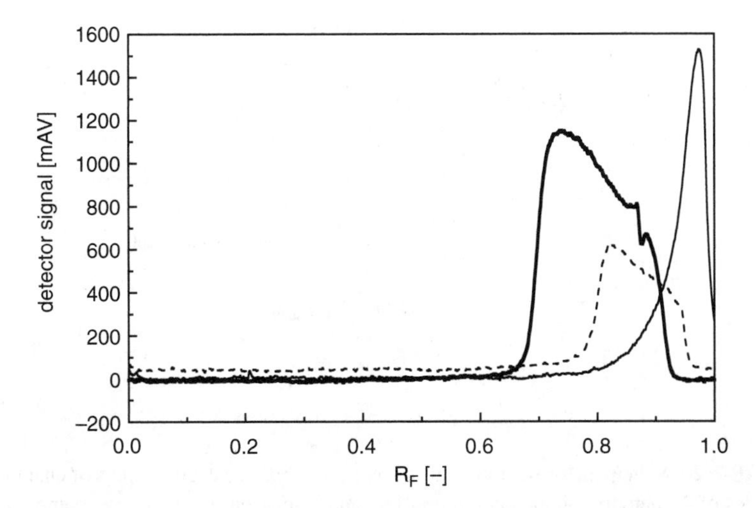

FIGURE 2.19 Comparison of the concentration profiles of 5-phenyl-1-pentanol (dashed line) and benzophenone (thin solid line) developed as single analytes and as a binary mixture (bold solid line); concentration of 5-phenyl-1-pentanol in the sample was 1.50 mol l^{-1} and that of benzophenone was 0.10 mol l^{-1}. Microcrystalline cellulose was used as stationary phase and *n*-octane as mobile phase [26].

2.4.3 Lateral Interactions in the Racemic Mixtures

Initially in Section 2.4, a disadvantageous effect of lateral interactions on separation performance in the nonlinear adsorption TLC mode was discussed adequately. The presented experimental examples of the self- and the mixed-associative lateral interactions were purposely obtained in mild chromatographic conditions and, more specifically, in the systems composed of the low-active adsorbents (either microcrystalline cellulose or chromatographic paper) and of the low-polar monocomponent mobile phases (either hydrocarbons or ether). Undoubtedly, more active adsorbents and more polar eluents intermolecularly interact with the analyte molecules in a more effective manner than the low-active/low-polar ones, more strongly competing with lateral interactions and destroying a considerable proportion of these. However, usage of the more drastic chromatographic conditions is undoubtedly also affected in a similar manner, even if to a lesser extent.

Among the “difficult” (and sometimes referred to as “sensitive”) chromatographic separations, those of enantiomeric antipodes and racemic mixtures are of particularly great importance and of the highest interest. This is because many compounds with a therapeutic effect (and incomparably more often the synthetic species than the natural ones) appear in a clearly defined enantiomeric form and for reasons of safety, need to be isolated from their opposite counterparts. Most pharmacodynamically active compounds are equipped with polar functionalities that make them interact with biological receptors and with the other constituents of a biological environment, and it often happens that these functionalities are of the AB type. In such cases, it can be justly concluded that an almost proverbial difficulty

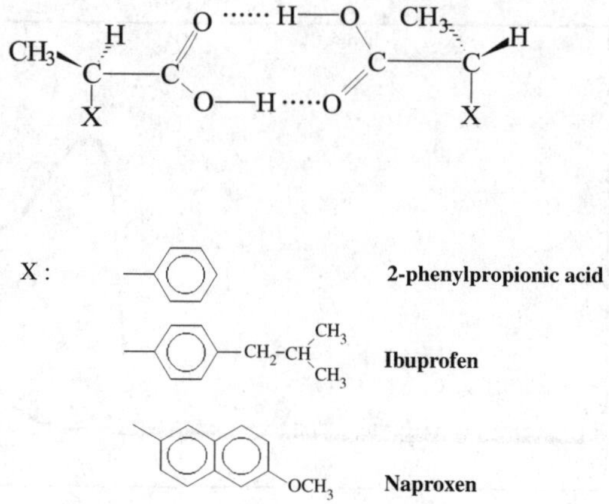

FIGURE 2.20 Schematic presentation of the hydrogen-bonded cyclic dimers of enantiomeric antipodes of 2-phenylpropionic acid, Ibuprofen, and Naproxen (the latter two compounds are drugs from the group of profens).

with chromatographic separation of enantiomeric antipodes can only partially be ascribed to the nature of their elemental composition and molar weight, and a complementary reason can be lateral interactions between the AB_L and AB_D species. For example, compounds derived from 2-phenylpropionic acid (its molecular AB-type structure follows):

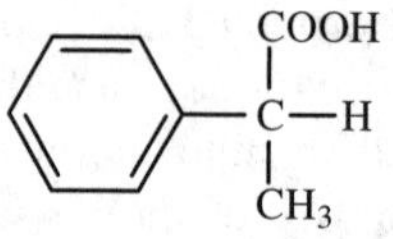

are synthesized as racemic mixtures and several of them are among the most commonly dispensed over-the-counter nonsteroidal antiinflammatory drugs from the group of profens (Ibuprofen being the most popular one). These antipodes can certainly form cyclic dimers, as shown in Figure 2.20.

Needless to add, lateral interactions among the enantiomeric antipodes can very strongly (and, of course, negatively) affect their preparative thin-layer (and also technological column) separation, carried out in the range of the nonlinear isotherm of adsorption.

2.5 THE MEANING OF THE RETARDATION FACTOR (R_F) IN PLC

The traditional method of determination of the numerical values of the retardation factor, R_F, quasi-automatically assumes the following preconditions:

1. Circular (or symmetrically ellipsoidal) chromatographic band shape
2. Gaussian distribution of the mass of the analyte in this band

On the basis of these assumptions, the position of a band on the chromatogram is defined by measuring the distance between the origin and the geometrical center of the band. Despite the considerable imprecision of this definition for skewed (i.e., non-Gaussian) bands, its two features are of great importance:

1. It regards the center of a chromatographic band as the point at which the local concentration of the analyte is the highest.
2. It also regards the center of the chromatographic band as the center of gravity of the mass distribution of the analyte in the band.

For the ideal, i.e., circular or symmetrically ellipsoidal, bands with the Gaussian analyte concentration profiles, the band centers described by assumptions (1) and (2) are, in fact, identical.

For the densitograms recorded from the nonlinear thin-layer chromatograms with non-Gaussian concentration profiles, it can be stated that:

- The numerical value of the R_F coefficient for a given chromatographic band can be determined for the maximum value of the concentration profile of the band (which actually is the point at which local concentration of the analyte is the highest). The R_F coefficient determined according to this definition can be denoted as $R_{F(\max)}$.
- Alternatively, the numerical value of the R_F coefficient can be determined from the center of gravity of the distribution of analyte mass in the band. With nonsymmetrical chromatographic bands, this value cannot be identical with that obtained from the maximum of the analyte's concentration profile. The R_F coefficient determined in this manner can be denoted as $R_{F(\text{int})}$.

To determine the center of gravity of the analyte mass distribution in the chromatographic band, one has to first establish the baseline, then remove the noise from the densitogram, subtract the baseline signal, define the beginning ($i = 0$) and the end ($i = k$) of the chromatographic band, and, finally, calculate the position of its center of gravity using the following relationship:

$$d_{sr.} = \frac{1}{\sum_{i=1}^{k} I\left(\frac{d_i + d_{i-1}}{2}\right) \cdot \left(d_i - d_{i-1}\right)} \cdot \sum_{i=1}^{k} I\left(\frac{d_i + d_{i-1}}{2}\right) \cdot \frac{d_i + d_{i-1}}{2} \cdot \left(d_i - d_{i-1}\right) = \frac{1}{S} \cdot \sum_{i=1}^{k} I\left(\frac{d_i + d_{i-1}}{2}\right) \cdot \frac{d_i + d_{i-1}}{2} \cdot \left(d_i - d_{i-1}\right) \tag{2.19}$$

where S denotes the chromatographic band surface and $I(d_i)$ is the detector signal at a distance d_i.

With a growing implementation of the thin-layer chromatographic laboratories with densitometric scanners and particularly in the case of PLC, it seems quite important to reconsider definition of the R_F coefficient and the ways of its determination. Further, it seems strongly advisable to recommend the retardation factor in its $R_{F(\max)}$ form as a more practical one for the preparative layer chromatographic usage.

2.6 MODELING OF LATERAL INTERACTIONS

In this chapter, we are going to show that using the one- and the two-component multilayer adsorption isotherm models or the models taking into the account lateral interactions among the molecules in the monolayer (discussed in Section 2.1), the overload peak profiles presented in Section 2.4 can be qualitatively modeled.

2.6.1 THE MASS-TRANSFER EQUATIONS

To simulate the empirical concentration profiles, an appropriate mass-transfer model has to be used. One of the simplest models is the model based on the equilibrium-dispersive model, frequently used in column chromatography [1]. It can be given by the following equation:

$$\frac{\partial c_i}{\partial t} + F\frac{\partial q_i}{\partial t} + u\frac{\partial c_i}{\partial x} = D_{a,x}\frac{\partial^2 c_i}{\partial x^2} + D_{a,y}\frac{\partial^2 c_i}{\partial y^2} \tag{2.20}$$

where t is the time, x and y are, respectively, the longitudinal and the perpendicular coordinates of the plate, u is the linear flow rate, F is the phase ratio, $D_{a,x}$ and $D_{a,y}$ are the apparent dispersion coefficients in the directions x and y, and subscript i denotes the ith species.

The exemplary peak profiles, simulated with use of Equation 2.20 for the linear, Langmuir, and the anti-Langmuir isotherms of adsorption are presented in Figure 2.21.

It is clearly visible that longitudinal cross sections of the spots are very similar to the peak profiles shown in Figure 2.1, Figure 2.2, and Figure 2.3 and calculated with the equilibrium-dispersive model (Equation 2.21):

$$\frac{\partial c_i}{\partial t} + F\frac{\partial q_i}{\partial t} + u\frac{\partial c_i}{\partial x} = D_a\frac{\partial^2 c_i}{\partial x^2} \tag{2.21}$$

where $D_a = D_{a,x}$ is the longitudinal apparent dispersion coefficient.

It is quite obvious that dispersion in the direction y (perpendicular to the direction of the development of the chromatogram) cannot change the shape of the longitudinal cross-section of the chromatographic spots. Because of the qualitative nature of our discussion, the dispersion in the direction y was ignored.

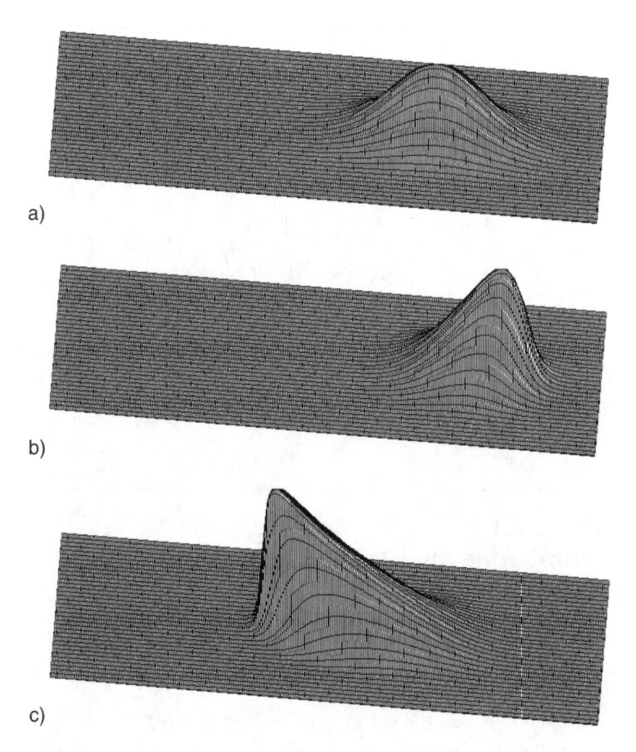

FIGURE 2.21 The exemplary peak profiles simulated with use of Equation 2.20 for (a) the linear, (b) the Langmuir, and (c) the anti-Langmuir isotherm of adsorption.

2.6.2 Modeling of the Self-Associative Lateral Interactions [25]

The most spectacular peak profiles, which suggest self-associative interactions, were obtained for 5-phenyl-1-pentanol on the Whatman No. 1 and No. 3 chromatographic papers (see Figure 2.15 and Figure 2.16). Very similar band profiles can be obtained using the mass-transfer model (Equation 2.21), coupled with the Fowler-Guggenheim isotherm of adsorption (Equation 2.4), or with the multilayer isotherm (Equation 2.7).

In Figure 2.22, the exemplary peak profiles calculated using the Fowler-Guggenheim isotherm are presented. These calculations were performed for the parameter values given in Table 2.1.

To solve Equation 2.21, the initial and the boundary conditions had to be established. It was assumed that the initial concentration and the concentration gradient for $x = L$ were equal to zero. For $x = 0$, the concentration was assumed as equal to the initial concentration, $C = 2$, in the spot in the course of time from $t = 0$ to $t = 0.63$ min.

As it can be seen from the subsequent plots, for the coefficient characterizing interaction energy $\chi = 3$, the theoretical peak profile is very similar to the experimental one. Additional calculations for $\chi = 3$ and the concentrations equal to 2, 1.5, 1, and 0.5 mol l^{-1} resulted in the plots given in Figure 2.23. The obtained plots confirm our

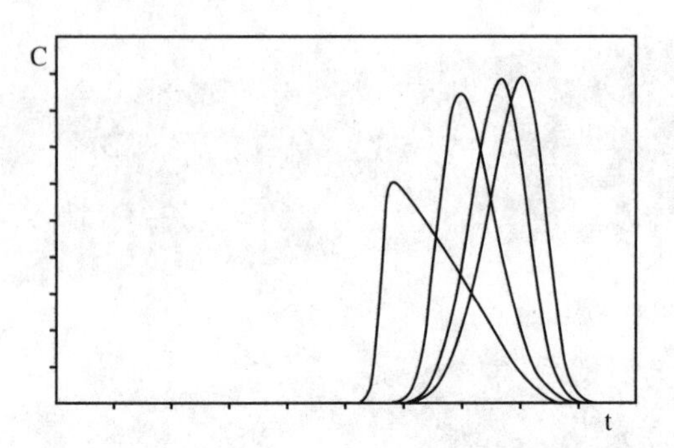

FIGURE 2.22 Peak profiles calculated for the Fowler-Guggenheim model. (1) $\chi = 0$ (equivalent to the Langmuir isotherm), (2) $\chi = 1$, (3) $\chi = 2$, (4) $\chi = 3$ (from the right to the left); $C = 2$ mol l^{-1}.

TABLE 2.1
Terms of the Models Used to Simulate Densitometric Peak Profiles

Magnitude	Value
Analyte migration distance, L	15 cm
Phase ratio, F (assumed value)	0.25
Linear flow rate, u	0.471 cm min^{-1}
Apparent dispersion coefficient, D_a (assumed to be equal to molecular diffusivity)	0.01 cm^2 min^{-1}
Maximum adsorbent capacity, q_s	1 mol l^{-1}
K	1 l mol^{-1}
χ	0, 1, 2, 3

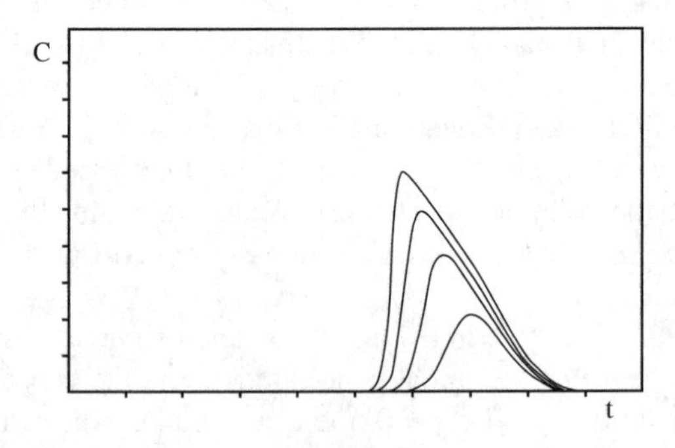

FIGURE 2.23 Peak profiles calculated for the Fowler-Guggenheim isotherm model for χ = 3 and the concentrations of 2, 1.5, 1, and 0.5 mol l^{-1} (peaks from the largest to the smallest, respectively).

suggestion that lateral interactions can be responsible for the observed band profiles. Very similar results are obtained when using the three-layer isotherm model.

TABLE 2.2
Terms of the Models Used to Simulate Densitometric Peak Profiles

Magnitude	Value
Analyte migration distance, L	15 cm
Phase ratio, F (assumed value)	0.25
Linear flow rate, u	0.074 cm min^{-1}
Apparent dispersion coefficient, D_a (assumed to be equal to molecular diffusivity)	0.00033 cm^2 min^{-1}
Maximum adsorbent capacity, q_s	1 mol l^{-1}
Terms of the isotherm represented by Equation 2.10	
Equilibrium constant, K_1	1 l mol^{-1}
Equilibrium constant, K_2	1 l mol^{-1}
Equilibrium constant, K_{11}	2 l mol^{-1}
Equilibrium constant, K_{22}	0 l mol^{-1}
Equilibrium constant, $K_{12} = K_{21}$	2.0 l mol^{-1}
Terms of the isotherm represented by Equation 2.13	
K_1	1 l mol^{-1}
K_2	1 l mol^{-1}
χ_1	2
$\chi_{12} = \chi_{21}$	2
χ_2	0

2.6.3 Modeling of Mixed-Associative Lateral Interactions [26]

The experiments showing the influence of lateral interaction on coelution of the two species were discussed in Subsection 2.4.2. Figure 2.18 and Figure 2.19 give a comparison of single profiles of acid and ketone or of alcohol and ketone with those attained for the binary mixture. Very similar peak profiles can be obtained upon solving Equation 2.21 separately for the alcohol, acid, and ketone with isotherms (Equation 2.4 and Equation 2.7a), and for the binary mixture with the isotherms (Equation 2.9 and Equation 2.10).

Numerical values of the model parameters assumed in Table 2.2, were chosen so as to obtain the best qualitative agreement between the shapes of the experimental and the theoretical peak profiles.

To solve Equation 2.21, the initial and the boundary conditions had to be established. It was assumed that the initial concentration and the concentration gradient for $x = L$ were equal to zero. For $x = 0$, the concentration was assumed to be equal to the initial concentration in the spot in the course of time from $t = 0$ to $t = 4$ min.

The results of calculations for the isotherm models represented by Equation 2.10 and Equation 2.9 are presented in Figure 2.24 and Figure 2.25, respectively.

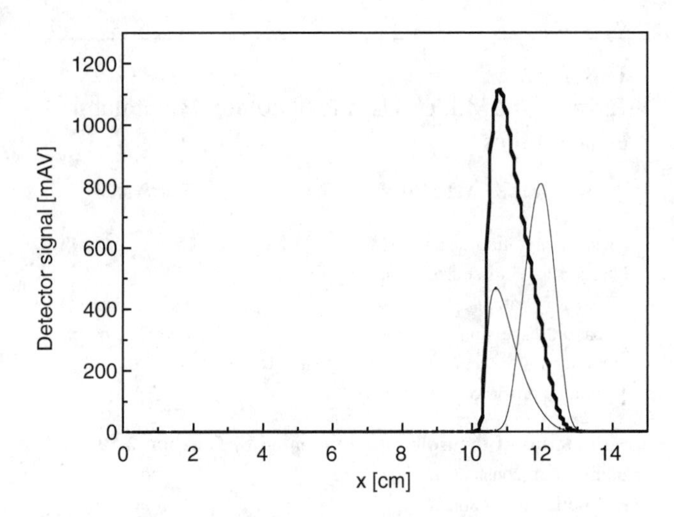

FIGURE 2.24 Calculated signal profiles for the single components (thin lines: the first peak represents 2-phenylbutyric acid and the second peak represents benzophenone) and the mixture (thick line).

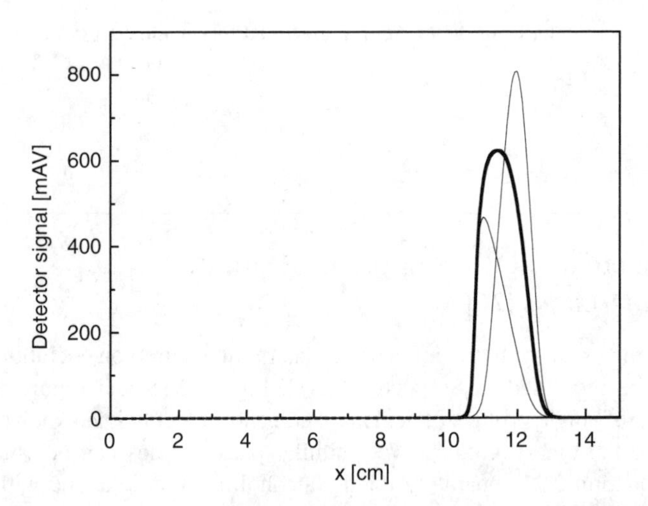

FIGURE 2.25 Calculated signal profiles for the single components (thin lines: the first peak represents 2-phenylbutyric acid and the second peak represents benzophenone) and the mixture (thick line).

As it is apparent from the results obtained, a relatively good qualitative agreement was obtained between the experimental peak profiles and the theoretical profiles simulated for the acid–ketone and for the alcohol–ketone coelution experiments, thus confirming the possibility of the lateral interaction between these species.

2.7 FINAL REMARKS

The main difference between the chromatographic process carried out in the linear and the nonlinear range of the adsorption isotherm is the fact that in the latter case, due to the skewed shapes of the concentration profiles of the analytes involved, separation performance of a chromatographic system considerably drops, i.e., the number of theoretical plates (N) of a chromatographic system indisputably lowers. In these circumstances, all quantitative models, along with semiquantitative and nonquantitative rules, successfully applied to optimization of the linear adsorption TLC show a considerably worse applicability.

In the case of the analytes able to participate in the self-associative lateral interactions (i.e., containing at least one AB functionality in their molecular structure), the negative impact of the interactions exerted on the separation performance depends on the number of the associated monomers per one H-bonded *n*-meric unit, and the higher the number (*n*) of the self-associated analyte monomers in a given aggregate, the more crippled is the separation process.

In the case of two analytes able to participate in the mixed lateral interactions (i.e., able to form the hydrogen bonds of the AB ... A, AB ... B, or AB_1 ... AB_2 type) and chromatographed in mild chromatographic systems (i.e., those composed of a low-active adsorbent and a low-polar mobile phase), mixed lateral interactions can even prevent a given pair of analytes from a successful separation (whereas under the slightly more drastic separation conditions, resolution of a given pair of analytes can be perceptibly worsened, at the least).

Hence, in order to obtain a successful separation of a mixture of the H-bonded analytes, the most recommended are such chromatographic systems in which lateral interactions either are entirely eliminated or manifest themselves to a rather weak extent. Thus, it is evident that in such cases, the most recommended are the chromatographic systems composed of an active enough adsorbent and a polar enough mobile phase.

Finally, an officially updated definition of the retardation factor, R_f, issued by IUPAC is important to the whole field of planar chromatography (the linear and the nonlinear TLC mode included). The importance of such a definition has two reasons. First, it is promoted by the growing access of planar chromatography users for densitometric evaluation of their chromatograms and second, by the vagueness of the present definition in the case of skewed concentration profiles with the samples developed under mass overload conditions.

REFERENCES

1. Guiochon, G., Shirazi, S.G., and Katti, A.M., *Fundamentals of Preparative and Nonlinear Chromatography*, Academic Press, Boston, MA, 1994.
2. Fowler, R.H. and Guggenheim, E.A., *Statistical Thermodynamics*, Cambridge University Press, Cambridge, 1960.
3. Quinones, I. and Guiochon G., *J. Chromatogr. A*, 796, 15–40, 1998.
4. Berezin, G.I. and Kiselev, A.V., *J. Colloid Interface Sci.*, 38, 227–233, 1972.

5. Berezin, G.I., Kiselev, A.V., and Sagatelyan, R.T., *J. Colloid Interface Sci.*, 38, 335–340, 1972.
6. Quinones, I. and Guiochon, G., *Langmuir*, 12, 5433–5443, 1996.
7. Wang, C.H. and Hwang, B.J., *Chem. Eng. Sci.*, 55, 4311–4321, 2000.
8. Kaczmarski, K., Prus, W., Dobosz, C., Bojda, P., and Kowalska, T., *J. Liq. Chromatogr. Relat. Technol.*, 25, 1469–1482, 2002.
9. Martin, A.J.P. and Synge, R.L.M., *Biochem J.*, 35, 1358–1362, 1941.
10. Snyder, L.R., *Principles of Adsorption Chromatography*, Marcel Dekker, New York, 1968.
11. Soczewiński, E., *Anal. Chem.*, 41, 179–182, 1969.
12. Schoenmakers, P., Billiet, H.A.H., Tijssen, R., and De Galan, L., *J. Chromatogr.*, 185, 179–195, 1979.
13. Snyder, L.R., *J. Chromatogr. Sci.*, 16, 223–234, 1978.
14. Kaczmarski, K., Sajewicz, M., Prus, W., and Kowalska, T., Entry in *Encyclopedia of Chromatography*, Cazes, J., Ed., Marcel Dekker, New York, 2004.
15. Nyiredy, Sz., Preparative planar (thin-layer) chromatography, in I.D. Wilson, E.R. Adlard, M. Cooke, and C.F. Poole, Eds., *Encyclopedia of Separation Science*, Academic Press, London, 2000, pp. 888–899.
16. Poole, C.F. and Schuette, S.A., *Contemporary Practice of Chromatography*, Elsevier, Amsterdam, 1984, p. 691.
17. Grinberg, N., *Modern Thin Layer Chromatography*, Marcel Dekker, New York, 1990.
18. Poole, C.F., *Chromatography Today*, Elsevier, Amsterdam, 1992.
19. Pimentel, G.C. and McClellan, A.L., *The Hydrogen Bond*, Freeman, San Francisco, 1960.
20. Prus, W., Kaczmarski, K., Tyrpień, K., Borys, M., and Kowalska, T., *J. Liq. Chromatogr. Relat. Technol.*, 24, 1381–1396, 2001.
21. Sajewicz, M., Pieniak, A., Piętka, R., Kaczmarski, K., and Kowalska, T., *Proc. Int. Symp. Planar Separations Planar Chromatography 2003*, Budapest, Hungary, June 21–23, 2003, pp. 407–414.
22. Sajewicz, M., Pieniak, A., Piętka, R., Kaczmarski, K., and Kowalska, T., *Acta Chromatogr.*, 14, 5–15, 2004.
23. Kaczmarski, K., Sajewicz, M., Pieniak, A., Piętka, R., and Kowalska, *J. Liq. Chromatogr. Relat. Technol.*, 27, 1967–1980, 2004.
24. Sajewicz, M., Pieniak, A., Piętka, R., Kaczmarski, K., and Kowalska, T., *J. Liq. Chromatogr. Relat. Technol.*, 27, 2019–2030, 2004.
25. Pieniak, A., Densitometric investigation of the effect of lateral interactions exerted on retention and separation of analytes in the non-linear variant of adsorption planar chromatography (in Polish). Ph.D. thesis, The University of Silesia, Katowice (Poland), 2006.
26. Pieniak, A., Kaczmarski, K., Sajewicz, M., Zapała, W., Gołebiowska, A., Tomala, R., and Kowalska, T., *Acta Chromatogr.*, 14, 16–36, 2004.

3 Sorbents and Precoated Layers in PLC

Heinz E. Hauck and Michael Schulz

CONTENTS

3.1 SORBENT MATERIALS AND PRECOATED LAYERS FOR STRAIGHT PHASE CHROMATOGRAPHY

Modern planar chromatography is suitable not only for qualitative and quantitative analysis but also for preparative purposes. The separation efficiency of a thin-layer chromatographic system is independent of this intended purpose and is mainly determined by the quality of the stationary phase, that is to say, by the applied coated layer. Therefore, progress in modern planar chromatography can be attributed not only to the development of the efficiency of the instruments but also to a large extent to the availability of high-quality precoated layers. And today, as in the past, bulk sorbents for self production, especially of preparative layer chromatography (PLC) layers, are widely used.

The specific retention of sample molecules to be separated in a straight phase or adsorption chromatographic system is mainly determined by two factors: their interactions with polar surface centers of the solid stationary phase, and by the different sample solubility in the rather nonpolar mobile phase. The most important interactions

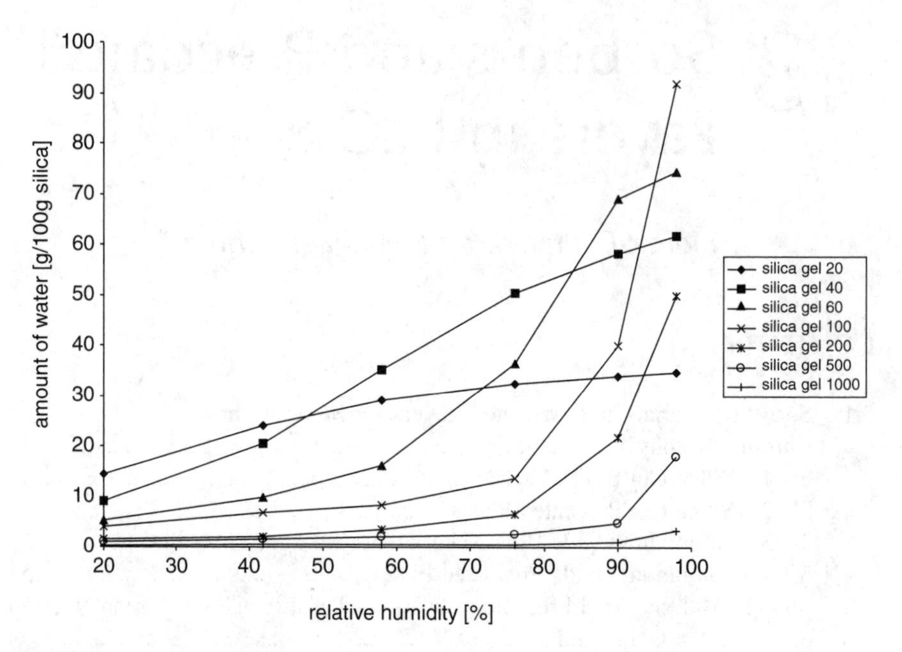

FIGURE 3.1 Water adsorption isotherms of different types of silica gels.

affecting sample retention on the surface of the stationary phase in straight phase chromatography are the hydrogen bondings and induced dipole–dipole interactions.

Because planar chromatography represents, in contrast to column chromatography, an open system, due regard must be given to the influence of the gas phase and especially to the relative humidity of the surrounding atmosphere and its effects on chromatographic behavior. To characterize the capability of silica gel to adsorb water, Figure 3.1 demonstrates the water adsorption isotherms for different silica gel types suitable for chromatographic purposes, whereas Figure 3.2 demonstrates the direct influence of differing water content on the retention behavior of dyestuff molecules in a preparative thin-layer chromatographic system with silica gel as stationary phase.

3.1.1 Silica Gel Bulk Materials

The most frequently used sorbent type for straight phase chromatographic systems in planar chromatography is porous, nonsurface-modified silica gel.

The important physical parameters that describe the chromatographic characteristics of silica gels suitable for straight phase chromatography are as follows:

- Mean hydroxyl group density [1] (about 8 μmol/m^2 independent of silica gel type)
- Specific surface area S_{BET} [2] (between 400 and 800 m^2/g)
- Specific pore volume v_P [3] (between 0.5 and 1.2 ml/g)
- Mean pore diameter (between 40 and 120 Å [4 to 12 nm])
- Mean particle size and particle size distribution (5 to 40 μm)

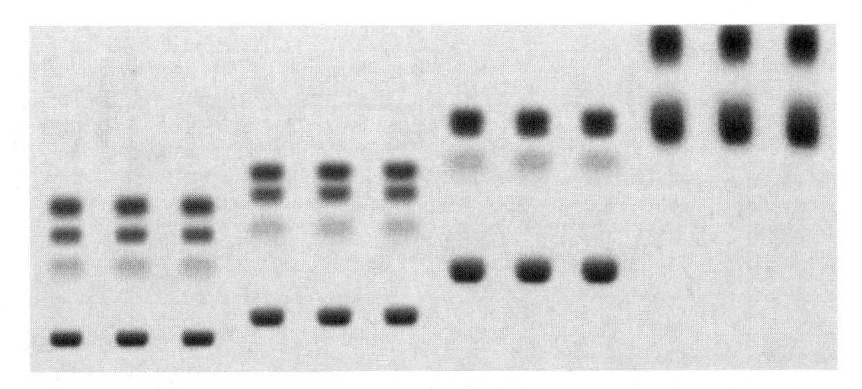

FIGURE 3.2 Separation of lipophilic dyestuffs on PLC plates silica gel 60 depending on different relative humidities. Relative humidities from left to right: 20, 42, 75, and 92%. Mobile phase: toluene.

Table 3.1 lists commercially available bulk silica gels for the manual preparation of PLC layers suitable for straight phase chromatography. These bulk silica gels are produced by different manufacturers, and in some cases they are offered with binding additives and fluorescent indicators. The data summarized in this table are traceable to product information from the concerned manufacturers.

3.1.2 Silica Gel Precoated Plates

More stringent requirements, especially with regard to separation efficiency and reproducibility in preparative planar chromatography also, led to increased application of precoated plates in this field. Figure 3.3 shows a scanning electron micrograph of a cross section through a PLC plate silica gel.

All preparative PLC precoated layers have glass plates as support. They are offered by a variety of manufacturers with or without fluorescent indicators and different binders. The layer thicknesses range uniformly between 0.5 and 2 mm or exhibit a gradient. Furthermore, in some cases the precoated silica gel plates are prescored to avoid cross-contamination from track to track.

To simplify sample application, especially for large sample volumes, without the use of expensive equipment and to obtain good separation efficiencies, preparative precoated plates with so-called concentrating zones or preadsorbent zones have been introduced. These plates consist of two different layer sections with an inert part in front of the surface-active separation part. The mode of operation of such a plate is demonstrated in Figure 3.4a and Figure 3.4b. In these figures, the sequence of the development phases of a separation of dyes on PLC plates silica gel 60 with a concentrating zone is shown. In Figure 3.4a the dye mixture is applied by dipping the concentrating zone into the sample solution, in contrast to Figure 3.4b, which employs a dotlike sample application.

In Table 3.2 the preparative precoated silica gel plates for straight phase PLC from different manufacturers are summarized. Incomplete data are due to incomplete information received from the manufacturers.

TABLE 3.1
Bulk Silicas for Straight Phase Chromatography

Manufacturer	Product	Type of Silica	Pore Size (nm)	Particle Size Distribution (μm)	Mean Particle Size (μm)	S_{BET} (m^2/g)	V_P (ml/g)	pH Value	Binder
Merck Darmstadt	Silica Gel 60 G	60	6	5–40	—	500	0.7	—	Gypsum
Merck Darmstadt	TLC Silica Gel 60 G	60	6	—	15	500	0.7	—	Gypsum
Merck Darmstadt	Silica Gel 60 GF254	60	6	5–40	—	500	0.7	—	Gypsum
Merck Darmstadt	TLC Silica Gel 60 GF254	60	6	—	15	500	0.7	—	Gypsum
Merck Darmstadt	Silica Gel 60 H	60	6	5–40	—	500	0.7	—	No foreign binder
Merck Darmstadt	TLC Silica Gel 60 H	60	6	—	15	500	0.7	—	No foreign binder
Merck Darmstadt	Silica Gel 60 HF254	60	6	5–40	—	—	—	—	No foreign binder
Merck Darmstadt	Silica Gel 60 HF254+366	60	6	5–40	—	—	—	—	No foreign binder
Merck Darmstadt	Silica Gel 60 PF254	60	6	5–40	—	500	0.7	—	—
Merck Darmstadt	Silica Gel 60 PF254+366	60	6	5–40	—	500	0.7	—	—
Merck Darmstadt	Silica Gel PF254 with Gypsum	60	6	5–40	—	500	0.7	—	Gypsum
Macherey-Nagel	Silica G	60	6	2–20	—	500	0.75	—	Gypsum
Macherey-Nagel	Silica G/UV254	60	6	2–20	—	500	0.75	—	Gypsum
Macherey-Nagel	Silica N	60	6	2–20	—	500	0.75	—	Without binder
Macherey-Nagel	Silica N/UV254	60	6	2–20	—	500	0.75	—	Without binder
Macherey-Nagel	Silica G-HR	60	6	3–20	—	500	0.75	—	Gypsum
Macherey-Nagel	Silica N-HR	60	6	3–20	—	500	0.75	—	Without binder
Macherey-Nagel	Silica P/UV254	60	6	5–50	—	500	0.75	—	Organic binder

Macherey-Nagel	Silica P/UV254 with Gypsum	60	6	5–50	—	500	0.75	—	Gypsum
Analtec	Silica Gel G TLC	60	6	—	15	—	—	—	Gypsum
Analtec	Silica GF TLC	60	6	—	15	—	—	—	Gypsum
Analtec	High-Performance Silica Gel	60	6	—	10	—	—	—	—
Analtec	High-Performance Silica Gel F	60	6	—	10	—	—	—	—
Analtec	High-Performance Silica Gel G	60	6	—	10	—	—	—	Gypsum
Analtec	High-Performance Silica Gel GF	60	6	—	10	—	—	—	Gypsum
Analtec	Ergosil surface-treated silica gel	60	6	—	10	—	—	—	—
Scientific Adsorbents Inc.	Silica for TLC	—	—	5–15	—	—	—	—	—
Scientific Adsorbents Inc.	Silica G for TLC	—	—	5–15	—	—	—	—	—
Scientific Adsorbents Inc.	Silica F for TLC	—	—	5–15	—	—	—	—	—
Scientific Adsorbents Inc.	Silica GF for TLC	—	—	5–15	—	—	—	—	—
MP	MP Silica TLC	60	6	5–15	—	500–600	—	6.5–7	—
MP	MP Silica F-TLC	60	6	5–15	—	500–600	—	6.5–7	—
MP	MP Silica G-TLC	60	6	5–15	—	500–600	—	6.5–7	—
MP	MP Silica GF-TLC	60	6	5–15	—	500–600	—	6.5–7	—
J.T. Baker	Silica Gel 7	60	6	5–40	—	—	—	—	—
J.T. Baker	Silica Gel 7G	60	6	5–40	—	—	—	—	Gypsum

FIGURE 3.3 Scanning electron microscope photograph of a PLC plate silica gel 60, layer thickness 0.5 mm.

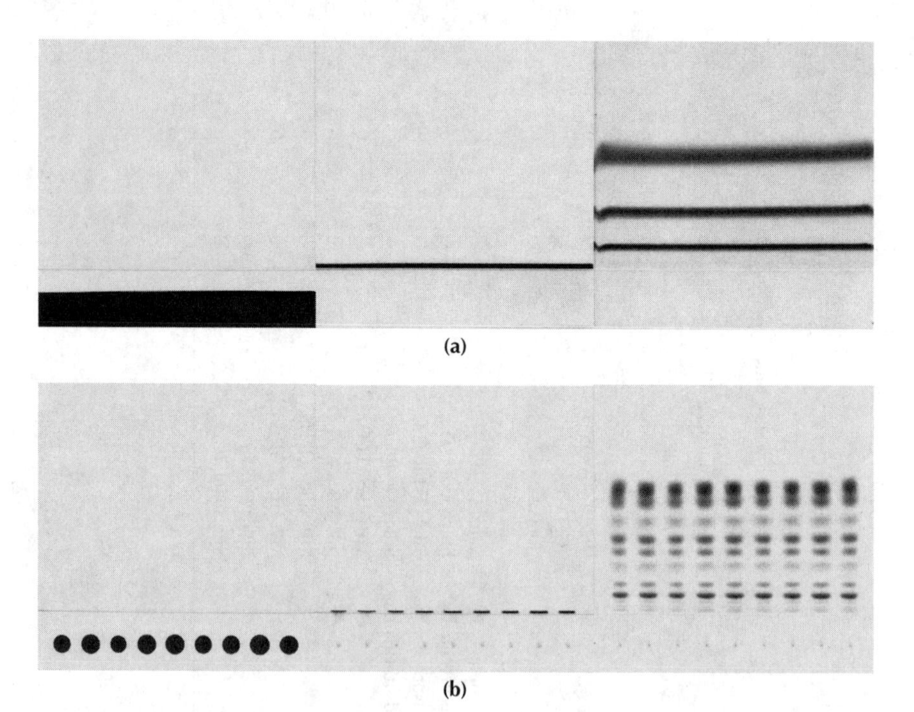

FIGURE 3.4 Stages of the development of PLC plates silica gel 60 with concentrating zones; separation of lipophilic dyestuffs with toluene as the mobile phase. (a) Sample application by dipping in the sample solution, (b) dot-like sample application.

TABLE 3.2
Precoated PLC Glass Plates Silica Gel

Manufacturer	Product	Type of Silica	Pore Size (nm)	Particle Size Distribution (μm)	Excitation Wavelength (nm)	Binder	Layer Thickness (mm)	Format (cm)	Particularities	S_{BET} (m^2/g)	V_P (ml/g)
Merck Darmstadt	Silica Gel 60	60	6	5–40	—	Organic	0.5	20 × 20	—	550	0.8
Merck Darmstadt	Silica Gel 60 F254	60	6	5–40	254	Organic	0.5	20 × 20	—	550	0.8
Merck Darmstadt	Silica Gel 60 F254	60	6	5–40	254	Organic	1	20 × 20	—	550	0.8
Merck Darmstadt	Silica Gel 60	60	6	5–40	—	Organic	2	20 × 20	—	550	0.8
Merck Darmstadt	Silica Gel 60 F254	60	6	5–40	254	Organic	2	20 × 20	—	550	0.8
Merck Darmstadt	Silica Gel 60 F254+366	60	6	5–40	254 + 366	Organic	2	20 × 20	—	550	0.8
Merck Darmstadt	Silica Gel 60 F254 conc.	60	6	5–40	254	Organic	0.5	20 × 20	Concentrating zone 4 × 20 cm	550	0.8
Merck Darmstadt	Silica Gel 60 F254 conc.	60	6	5–40	254	Organic	1	20 × 20	Concentrating zone 4 × 20 cm	550	0.8
Merck Darmstadt	Silica Gel 60 F254 conc.	60	6	5–40	254	Organic	2	20 × 20	Concentrating zone 4 × 20 cm	550	0.8
Macherey-Nagel	Sil G-50	60	6	5–17	—	Organic	0.5	20 × 20	—	500	0.75
Macherey-Nagel	Sil G-50 UV254	60	6	5–17	254	Organic	0.5	20 × 20	—	500	0.75
Macherey-Nagel	Sil G-100	60	6	5–17	—	Organic	1	20 × 20	—	500	0.75
Macherey-Nagel	Sil G-100 UV254	60	6	5–17	254	Organic	1	20 × 20	—	500	0.75
Macherey-Nagel	Sil G-200	60	6	—	—	Organic	2	20 × 20	—	500	0.75
Macherey-Nagel	Sil G-200 UV254	60	6	—	254	Organic	2	20 × 20	—	500	0.75
Whatman	PK5*	150	15	10–12	—	—	0.5	20 × 20	—	—	—
Whatman	PK6F*	60	6	10–12	254	—	0.5	20 × 20	—	—	—
Whatman	PK6F**	60	6	10–12	254	—	1	20 × 20	—	—	—

TABLE 3.2 (Continued)
Precoated PLC Glass Plates Silica Gel

Manufacturer	Product	Type of Silica	Pore Size (nm)	Particle Size Distribution (μm)	Excitation Wavelength (nm)	Binder	Layer Thickness (mm)	Format (cm)	Particularities	S_{BET} (m^2/g)	V_P (ml/g)
Whatman	PK5**	150	15	10–12	—	—	1	20 × 20	—	—	—
Whatman	PK5F*	150	15	10–12	254	—	0.5	20 × 20	—	—	—
Whatman	PK5F**	150	15	10–12	254	—	1	20 × 20	—	—	—
Whatman	PLK5**	150	15	10–12	254	—	1	20 × 20	Concentrating zone	—	—
Whatman	PLK5F**	150	15	10–12	254	—	1	20 × 20	Concentrating zone	—	—
Whatman	PLK5**	150	15	10–12	254	—	1	20 × 20	Concentrating zone	—	—
Whatman	PLK5F**	150	15	10–12	254	—	1	20 × 20	Concentrating zone	—	—
Analtec	Silica Gel G	—	—	—	—	Gypsum	0.5/1/1.5/2	5 × 20/10 × 20/20 × 20/20 × 40	—	—	—
Analtec	Silica Gel G Prep-scored	—	—	—	—	Gypsum	0.25/0.5/1/1.5/2	20 × 20	Scored	—	—
Analtec	Silica Gel GF	—	—	—	254	Gypsum	0.5/1/1.5/2	5 × 20/10 × 20/20 × 20/20 × 40	—	—	—
Analtec	Silica Gel GF Prep-scored	—	—	—	254	Gypsum	0.5/1/1.5/2	20 × 20	Scored	—	—
Analtec	Silica Gel G w/Preadsorbent Zone	—	—	—	—	Gypsum	0.5/1	20 × 20	—	—	—

Analtec	Silica Gel G w/Preadsorbent Zone	—	—	—	—	Gypsum	0.5/1	20 × 20	Scored	—	—
Analtec	Silica Gel G w/Preadsorbent Zone	—	—	—	—	Gypsum	0.5	20 × 20	Channeled	—	—
Analtec	Silica Gel GF w/Preadsorbent Zone	—	—	—	254	Gypsum	0.5/1	20 × 20	—	—	—
Analtec	Silica Gel GF w/Preadsorbent Zone	—	—	—	254	Gypsum	0.5	20 × 20	Scored	—	—
Analtec	Silica Gel GF w/Preadsorbent Zone	—	—	—	254	Gypsum	0.5	20 × 20	Channeled	—	—
Analtec	Silica Gel G UNIPLATE-T Taper Plate	—	—	—	—	Gypsum	Gradient	20 × 20	—	—	—
Analtec	Silica Gel G UNIPLATE-T Taper Plate	—	—	—	—	Gypsum	Gradient	20 × 20	Scored	—	—
Analtec	Silica Gel GF UNIPLATE-T Taper Plate	—	—	—	—	Gypsum	Gradient	20 × 20	—	—	—
Analtec	Silica Gel GF UNIPLATE-T Taper Plate	—	—	—	—	Gypsum	Gradient	20 × 20	Scored	—	—
Analtec	Silica Gel H	—	—	—	—	No foreign binder	0.5	20 × 20	—	—	—

TABLE 3.2 (Continued)
Precoated PLC Glass Plates Silica Gel

Manufacturer	Product	Type of Silica	Pore Size (nm)	Particle Size Distribution (μm)	Excitation Wavelength (nm)	Binder	Layer Thickness (mm)	Format (cm)	Particularities	S_{BET} (m^2/g)	V_P (ml/g)
Analtec	Silica Gel HF	—	—	—	254	No foreign binder	0.5	20 × 20	—	—	—
Analtec	UNISIL Silica Gel G	—	—	—	—	Gypsum	0.5	20 × 20	—	—	—
Analtec	UNISIL Silica Gel GF	—	—	—	254	Gypsum	0.5	20 × 20	—	—	—
Scientific Adsorbents Inc.	Silica Gel — Hard Layer — Organic Binder	—	—	—	—	Organic	0.5/1/2	10 × 20/20 × 20	—	—	—
Scientific Adsorbents Inc.	Silica Gel — Hard Layer — Organic Binder	—	—	—	254	Organic	0.5/1/2	10 × 20/20 × 20	—	—	—
Scientific Adsorbents Inc.	Silica Gel Gypsum	—	—	—	—	Gypsum	0.5/1/2	10 × 20/20 × 20	—	—	—
Scientific Adsorbents Inc.	Silica Gel Gypsum	—	—	—	254	Gypsum	0.5/1/2	10 × 20/20 × 20	—	—	—
MP	MP Silica Plates	—	—	—	254	Organic	0.5	20 × 20	—	—	—
J.T. Baker	Baker Si500F TLC Plate Silica Gel	—	—	—	254	—	0.5	20 × 20	—	—	—
WAKO	Silica 70PF254 Plate	70	7	5–40	254	—	0.7–0.9	20 × 20	—	450	0.8

* = layer thickness of 0.5 mm

** = layer thickness of 1.0 mm

An extensive variety of substance classes have been separated using silica gel as the stationary phase in straight phase adsorption PLC. Some examples are the following:

- Alkaloids [4]
- Antibiotics [5]
- Hydrocarbons [6]
- Lipids [7]
- Mycotoxins [8]
- Natural dyes [9]
- Organometallic compounds [10]
- Pesticides [11]
- Pharmaceutical compounds [12]
- Phenols [13]
- Preservatives [14]
- Steroids [15]

3.1.3 Aluminum Oxide Bulk Materials for Straight Phase PLC

Another fairly important stationary phase in straight phase PLC is aluminum oxide. Comparable with silica gel also in the case of aluminum oxides, hydroxyl groups at the surface of this adsorbent are responsible for the selective retention of sample molecules. The relevant physical parameters for the characterization of aluminum oxides suitable for straight phase PLC are the following:

- Mean hydroxyl group density [16] (about 13 μmol/m^2)
- Specific surface area S_{BET} (between 100 and 350 m^2/g)
- Specific pore volume v_P (between 0.1 and 0.4 ml/g)
- Mean pore diameter (between 40 and 90 Å [4 to 9 nm])
- Mean particle size and particle size distribution (5 to 60 μm)

A special feature of aluminum oxides is the fact that these sorbents are adjusted to different pH values for thin-layer chromatographic purposes. The "acidic" aluminas are in the pH range of 4.0 to 4.5, whereas "neutral" signifies a pH range between 7.0 and 8.0, and "basic" aluminas have a pH value of approximately 9.0 to 10.0.

Table 3.3 lists bulk aluminas for the preparation of homemade PLC plates for straight phase chromatography.

3.1.4 Aluminum Oxide Precoated Plates

To achieve better separation efficiency and reproducibility compared with handmade layers also in the case of aluminum oxides, precoated plates are available for straight phase PLC. They are listed in Table 3.4.

Substance classes that can be separated by straight phase adsorption PLC on aluminum oxide stationary phases are, for example, alkaloids [17].

TABLE 3.3
Bulk Aluminas for Straight Phase Chromatography

Manufacturer	Product	Type of Alumina	Pore Size (nm)	Particle Size Distribution (μm)	S_{BET} (m^2/g)	V_P (ml/g)	pH Value	Binder
Merck Darmstadt	Aluminum oxide 60 G neutral	60	6	5–40	200	0.3	7.5	—
Merck Darmstadt	Aluminum oxide 60 GF254 neutral	60	6	5–40	200	0.3	7.5	—
Macherey-Nagel	Aluminum oxide G	—	—	—	—	—	7.5–8	Gypsum
Macherey-Nagel	Aluminum oxide G/UV254	—	—	—	—	—	7.5–8	Gypsum
Macherey-Nagel	Aluminum oxide N	—	—	—	—	—	9	—
Analtec	Alumina Neutral	—	—	—	—	—	Neutral	—
Analtec	Alumina Basic	—	—	—	—	—	Basic	—
Analtec	Alumina Acid	—	—	—	—	—	Acid	—
Scientific Adsorbents Inc.	Alumina Basic for TLC	—	—	5–15	—	—	Basic	—
Scientific Adsorbents Inc.	Alumina Neutral for TLC	—	—	5–15	—	—	Neutral	—
Scientific Adsorbents Inc.	Alumina Acid for TLC	—	—	5–15	—	—	Acid	—
Scientific Adsorbents Inc.	Alumina with Gypsum for TLC	—	—	5–15	—	—	—	—
MP	MP Alumina B-TLC	—	—	5–25	200	—	Basic	—
MP	MP Alumina N-TLC	—	—	5–25	200	—	Neutral	—
MP	MP Alumina A-TLC	—	—	5–25	200	—	Acid	—
MP	MP Alumina G-TLC	—	—	5–25	200	—	—	—

TABLE 3.4
Precoated PLC Glass Plates Aluminum Oxide

Manufacturer	Product	Type of Alumina	Pore Size (nm)	Particle Size Distribution (μm)	Excitation Wavelength (nm)	Binder	Layer Thickness (mm)	Format (cm)	pH Value	S_{BET} (m^2/g)	V_P (ml/g)
Merck Darmstadt	PLC plates aluminum oxide 60 F254	60	6	5–40	254	Organic	1.5	20 × 20	pH = 9	200	0.3
Merck Darmstadt	PLC plates aluminum oxide 150 F254	150	6	5–40	254	Organic	1.5	20 × 20	pH = 9	70	0.2
Macherey-Nagel	ALOX-100 UV254	60	6	< 60	254	Organic	1	20 × 20	pH = 9	200	—
Scientific Adsorbents Inc.	Alumina Gypsum	—	—	—	—	Gypsum	1	10 × 20/ 20 × 20	—	—	—
Scientific Adsorbents Inc.	Alumina Gypsum	—	—	—	254	Gypsum	1	10 × 20/ 20 × 20	—	—	—

3.2 SORBENT MATERIALS AND PRECOATED LAYERS FOR PARTITION CHROMATOGRAPHY

In contrast to straight phase or adsorption chromatography, partition chromatography involves the separation of sample molecules owing to their different partition coefficients between the liquid stationary and mobile phases. The liquid stationary phase is located in the pores of a sorbent, ideally only acting as a support, making no contribution to the retention of the sample molecules.

In partition chromatographic systems, the selectivity and degree of retention are mainly determined by the compositions of the liquid stationary and mobile phases and by the phase ratio of these two liquids.

Because of the preceding reasons, sorbents for partition chromatography should have rather small specific surface areas in combination with large specific pore volumes.

The loading of the sorbent pores with a liquid stationary phase in PLC can be performed in two different ways:

1. Impregnation prior to the chromatographic development by immersing the PLC plate in a solution of the liquid stationary phase in a suitable solvent, with subsequent evaporation of the solvent. This method results in a homogeneous distribution of the liquid stationary phase all over the PLC layer.
2. *In situ* loading during the chromatographic development. By this method a continuous gradient of the liquid stationary phase is created on the PLC layer in the direction of the mobile phase flow. This is achieved by developing the plate with a suitable multicomponent solvent system.

In contrast to straight phase adsorption chromatography, in the partition chromatographic mode the influence of the gas phase is less important.

3.2.1 Silica Gel

In principle, the bulk silica sorbents and the precoated plates for partition PLC are the same types as those used in straight phase adsorption chromatography (see Table 3.1 and Table 3.2). But there is a trend to apply silica gel types with higher specific pore volumes and lower specific surface areas compared with the sorbents for adsorption chromatography.

The following samples are examples of possible application of silica gel layers in partition PLC:

TABLE 3.5
Kieselguhr Bulk Materials

Manufacturer	Product	Particle Size Distribution (μm)	S_{BET} (m^2/g)	Binder
Merck Darmstadt	Kieselguhr G	5–40	1	Gypsum

- Alkaloids [18]
- Antibiotics [19]
- Antioxidants [20]
- Bile acids [21]
- Carbohydrates [22]
- Naphthalene sulfonic acids [23]
- Peptides [24]
- Pesticides [25]
- Pharmaceutical compounds [26]
- Phenols [27]
- Phospholipids [28]

For partition PLC, a very inert silica bulk material is also offered. This Kieselguhr is a naturally occurring, purified product that is composed of the skeletons of siliceous algae. It is listed in Table 3.5.

Depending on their deposits, Kieselguhrs consist of about 90% silicon dioxide and varying amounts of Al_2O_3, Fe_2O_3, TiO_2, CaO, MgO, Na_2O, K_2O, and various carbonates. The physical parameters for the characterization of Kieselguhrs are as follows:

- Specific surface area S_{BET} (between 1 and 5 m^2g)
- Specific pore volume v_P (between 1 and 3 ml/g)
- Mean pore diameter (between 10,000 and 100,000 Å [1,000 to 10,000 nm])

3.2.2 Aluminum Oxide

Aluminum oxides, similar to silica gels, are available as bulk materials and as precoated plates, to be used not only for straight phase adsorption chromatography, but also for partition PLC (see Table 3.3 and Table 3.4). In particular, the aluminum oxide type 150 (i.e., mean pore diameter 150 Å [15 nm]) is suitable for partition chromatographic purposes.

3.2.3 Cellulose

Both inorganic and organic sorbent materials are suitable for application in partition PLC. The relevant material in this connection is cellulose. Celluloses are natural products with the universal chemical formula $(C_6H_{10}O_5)_x$. These "native" celluloses have a fibrous structure and they need to be ground and purified before use in PLC. Besides native cellulose, "microcrystalline" cellulose can also be used in partition PLC. In this case the cellulose has been recrystallized and is rod-shaped. The specific surface area of celluloses is in the range of about 2 m^2/g.

The bulk celluloses available for the preparation of handmade PLC plates are listed in Table 3.6. The manufacturers, types, and properties of the corresponding PLC precoated plates cellulose are summarized in Table 3.7.

Cellulose layers are predominantly used for the separation of medium-polar substances in PLC such as:

- Alkaloids [29]
- Amino acids [30]
- Phenols [31]

TABLE 3.6
Cellulose Bulk Materials

Manufacturer	Product	Particle Size Distribution (μm)	S_{BET} (m²/g)	Special Features
Merck Darmstadt	Cellulose microcrystalline	<20	—	—
Macherey-Nagel	Cellulose MN 301 (native fibrous)	2–20	1.5	—
Macherey-Nagel	Cellulose MN 301 UV254 (native fibrous)	2–20	1.5	—
Macherey-Nagel	Cellulose 301 HR (native fibrous)	2–20	1.5	Acid washed and defatted
Macherey-Nagel	Cellulose 301 A (native fibrous)	2–20	1.5	For the ^{32}P postlabeling procedure
Macherey-Nagel	Cellulose MN 300 DEAE	—	—	Diethylaminoethyl cell, ion exchanger
Analtec	Avicel Microcrystalline Cellulose	—	—	—
Analtec	Avicel F Microcrystalline Cellulose	—	—	—
J.T. Baker	Cellulose (native fibrous)	2–20	—	—

3.3 PRECOATED LAYERS FOR REVERSED-PHASE CHROMATOGRAPHY

Not only in HPLC, but also in modern thin-layer chromatography, the application of reversed-phase stationary phases becomes increasingly important. The advantage of the hydrophobic layers in comparison with the polar, surface-active stationary phases is the additional selectivity and a reduced likelihood of decomposition of sensitive substances.

Reversed-phase PLC precoated plates are based on silica gel matrices with chemical modifications in such a manner that the accessible polar, hydrophilic silanol groups at the silica gel surface are replaced by nonpolar, hydrophobic alkyl chains via silicon-carbon bonds. For preparative purposes, up to now only PLC precoated RP plates with C-18 modification are available. This abbreviation is often also designated as RP-18, meaning that an octadecyl alkyl chain is chemically bonded to the silica gel surface.

Compared with liquid column chromatography, in PLC there is a certain limitation with respect to the composition of the mobile phase in the case of reversed-phase chromatography. In planar chromatography the flow of the mobile phase is normally induced by capillary forces. A prerequisite for this mechanism is that the surface of the stationary phase be wetted by the mobile phase. This, however, results in a limitation in the maximum possible amount of water applicable in the mobile phase, is dependent on the hydrophobic character of the stationary RP phase. To

TABLE 3.7
Precoated PLC Glass Plates Cellulose

Manufacturer	Product	Type of Cellulose	Particle Size Distribution (μm)	Excitation Wavelength (nm)	Layer Thickness (mm)	Format (cm)	Special Features	S_{BET} (m^2/g)
Macherey-Nagel	CEL 300-50	300	2–20	—	0.5	20 × 20	Native cellulose	1.5
Macherey-Nagel	CEL 300-50 UV254	300	2–20	254	0.5	20 × 20	Native cellulose	1.5
Scientific Adsorbents Inc.	Cellulose	—	—	—	0.5/1	10 × 20/20 × 20	—	—
Scientific Adsorbents Inc.	Cellulose F	—	—	254	0.5/1	10 × 20/20 × 20	—	—

overcome this limitation it is possible to use RP stationary phases with a defined lower degree of modification, to maintain the sorbent's partly hydrophilic character.

3.3.1 Silica Gel RP-18

The silica gel matrix for the RP-18 PLC precoated plates consists either of a type "60" or "150" (indicating the respective mean pore diameter). The manufacturers and the properties of the PLC plates RP-18 available at present are listed in Table 3.8.

The reversed-phase layers in planar chromatography can be used for the separation of a large variety of substances in PLC such as:

- Alkaloids [32]
- Antibiotics [33]
- Antioxidants [34]
- Mycotoxins [35]
- Pesticides [36]
- Pharmaceutical compounds [37]
- Phenols [38]
- Steroids [39]
- Vitamins [40]

3.4 SUMMARY AND PROSPECTS

Planar chromatography has been used for preparative purposes in the past and is still used today. The reasons for the usage of preparative layer chromatography are, first of all, its cost-effectiveness and flexibility.

Until now the application of different types of bulk sorbents used for handmade PLC plates is rather widespread, and in most cases the manufacturers of these materials provide detailed instructions for the preparation of the preparative layers. However, the quality and especially the reproducibility of these handmade plates is frequently rather poor. Due to this, the development of modern TLC and HPTLC, and precoated PLC plates also, becomes increasingly more important.

The most-used stationary phase in PLC is silica gel, with type 60 taking preference. In the future, other sorbents such as the RP materials will also most probably be increasingly used. This will also be true for the case of special PLC plates consisting of layer combinations such as precoated plates with concentrating zones, resulting in simplification of sample application as well as an increase in the efficiency of separation.

REFERENCES

1. Boehm, H.P., *Angew. Chem.*, 78, 617, 1966.
2. Brunauer, S., Emmett, P.H., and Teller, E., *J. Am. Chem. Soc.*, 60, 309, 1938.
3. Unger, K.K., *Porous Silica*, 47, Marcel Dekker, New York, 1990, chap. 5.
4. Waksmundzka-Hajnos, M., Gadzikowska, M., and Hajnos, M.L., *J. Planar Chromatogr.*, 15, 289, 2002.
5. Brown, M.A. and Rajan, S., *J. Agric. Food Chem.*, 34, 470, 1986.
6. Hückelhoven, R., Schuphan, I., Thiede, B., and Schmidt, B., *J. Agric. Food Chem.*, 45, 263, 1997.

TABLE 3.8
Precoated PLC Glass Plates RP-18

Manufacturer	Product	Type of Silica	Pore Size (nm)	Particle Size Distribution (μm)	Excitation Wavelength (nm)	Binder	Layer Thickness (mm)	Format (cm)	Particularities	S_{BET} (m^2/g)	V_p (ml/g)
Merck Darmstadt	PLC plates RP-18 F254s	60	6	5–40	254	Organic	1	20 × 20	—	550	0.8
Macherey-Nagel	RP-18W/ UV254	60	6	5–17	254	Organic	1	20 × 20	Wettable by water	500	0.75
Whatman	PLKC-18F	—	—	10–12	254	—	1	20 × 20	Concentrating zone	—	—

7. Ramadan, M.F. and Mörsel, J.-T., *J. Agric. Food Chem.*, 51, 969, 2003.
8. Stander, M.A, Steyn, P.S., Luebben, A., Miljkovic, A., Mantle, P.G., and Marais, G.J., *J. Agric. Food Chem.*, 48, 1865, 2000.
9. Masuda, T., Toi, Y., Bando, H., Maekawa, T., Takeda, Y., and Yamaguchi, H., *J. Agric. Food Chem.*, 50, 2524, 2002.
10. Singh, N., Mehrotra, M., Rastogv, K., and Srivastava, T., *Analyst*, 110, 71, 1985.
11. Singh, B. and Kulshresta, G., *J. Agric. Food Chem.*, 49, 3728, 2001.
12. Tanaka, R., Aoki, H., Mizota, T., Wada, S.-I., Matsunaga, S., Tokuda, H., and Nishino, H., *Planta Med.*, 66, 163, 2000.
13. Gupta, A.P., Gupta, M.M., and Kumar, S., *J. Liq. Chromatogr.*, 22, 1561, 1999.
14. San, S., Lao, A., Wang, Y., Chin, C.-K., Rosen, R.T., and Ho, C.T., *J. Agric. Food Chem.*, 50, 6318, 2002.
15. Van Boven, M., Daenens, P., Maes, K., and Coklaere, M., *J. Agric. Food Chem.*, 45, 1180, 1997.
16. Combellas, C. and Drochon, B., *Anal. Lett.*, 16, 1647, 1983.
17. El-Imam, Y.M.A., Evans, W.C., and Grout, R.J., *Phytochemistry*, 27, 2181, 1988.
18. Frederich, M., De Pauw, M.-C., Llabres, G., Tits, M., Hayette, M.-P., Brandt, V., Penelle, J., De Mol, P., and Angenot, L., *Planta Med.*, 66, 262, 2000.
19. Beran, M. et al., *J. Chromatogr.*, 558, 265, 1991.
20. Takeoka, G.R. and Dao, L.T., *J. Agric. Food Chem.*, 50, 496, 2002.
21. Dax, C.I. and Müllner, S., *Chromatographia*, 48, 681, 1998.
22. Tarr, F., *Magyar Kémiai Folyórat*, 99, 89, 1993.at
23. Mohau, P., Wickramasinghe, A., and Verma, S., *J. Chromatogr. Sci.*, 31, 216, 1993.
24. Hocart, S.J., Nekola, M.V., and Coy, D.H., *J. Med. Chem.*, 31, 1820, 1988.
25. Zhang, L.Z., Khan, S.U., Akhtar, M.H., and Ivarson, K.C., *J. Agric. Food Chem.*, 32, 1207, 1984.
26. Parimoo, P., Mounisswamy, M., Bharathi, A., and Lakshmi, N., *Indian Drugs*, 31, 211, 1994.
27. Sumaryono, W., Proksch, P., Wray, V., Witte, L., and Hartmann, T., *Planta Med.*, 57, 176, 1991.
28. Utzmann, C.M. and Lederer, M.O., *J. Agric. Food Chem.*, 48, 1000, 2000.
29. Arai, K., Kimura, K., Kushiroda, T., and Yamamoto, Y., *Chem. Pharm. Bull.*, 37, 2937, 1989.
30. Pan, M., Bonness, M.S., and Mabry, T.J., *Biochem. Syst. Ecol.*, 23, 575, 1995.
31. Strack, D., Engel, U., Weissenbock, G., Grotjahn, L., and Wray, V., *Phytochemistry*, 25, 2605, 1986.
32. Houghton, P.J., Latiff, A., and Said, J.M., *Phytochemistry*, 30, 347, 1991.
33. Stout, S.J., Wu, J., Da Cunha, A.R., King, K.G., and Lee, A.U., *J. Agric. Food Chem.*, 39, 386, 1991.
34. Kweon, M.-H., Hwang, H.-J., and Sung, H.-C., *J. Agric. Food Chem.*, 49, 4646, 2001.
35. Capasso, R., Evidente, A., and Vurro, M., *Phytochemistry*, 30, 3945, 1991.
36. Zulalian, J., Stout, S.J., Da Cunha, A.R., Carces, T., and Miller, P., *J. Agric. Food Chem.*, 42, 381, 1994.
37. Bediv, E., Tatli, I.I., Khan, R.A., Zhao, J., Takamatsu, S., Walker, L.A., Goldman, P., and Khan, I.A., *J. Agric. Food Chem.*, 50, 3150, 2002.
38. Fischer, W., Bund, O., and Hauck, H.E., *Fresenius J. Anal. Chem.*, 354, 889, 1996.
39. Hetzel, C.E., Gunatilaka, A.A.L., Kingston, D.G.I., Hofmann, G.A., and Johnson, R.K., *J. Nat. Prod.*, 57, 620, 1994.
40. Khachik, F., Beecher, G.R., and Whittaker, N.F., *J. Agric. Food Chem.*, 34, 603, 1986.

4 Selection and Optimization of the Mobile Phase for PLC

Virginia Coman

CONTENTS

4.1 INTRODUCTION

The most important current problem of planar chromatography is the elaboration of theoretical and experimental methods for predicting the conditions of mixture separation in order to achieve better results. Planar chromatography is an analytical chemistry technique for the separation of mixtures that involves passing of solutes in the mobile phase through the stationary phase. Usually, each component has a

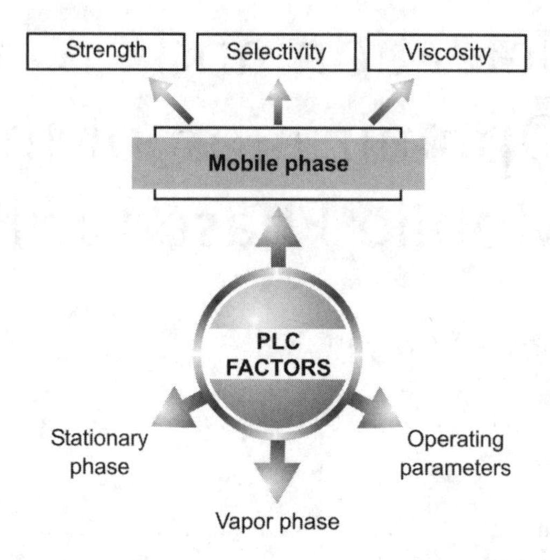

FIGURE 4.1 Important factors in PLC separation.

characteristic separation rate that can be used to identify it and the composition of the original mixture as well. Analytical thin-layer chromatography (TLC) is used to determine the compounds that are in a mixture and their concentrations. Preparative layer chromatography (PLC) may be defined as the TLC of relatively large amounts of material used in order to prepare and isolate quantities of separated compounds for further investigations such as chemical derivatization, structure elucidation (MS, NMR, IR, UV, etc.), and chromatographic standards or biological activity determination [1–4]. PLC is an excellent method for cleaning up synthetic reaction mixtures, natural products, plant extracts, and biotechnological products.

Classical PLC involves migration of a mobile phase by capillary action through a 0.5- to 2-mm layer of adsorbent for separating compounds in amounts of 10 to 1000 mg. This separation method requires a good knowledge of chromatography, the most basic equipment, and simple operational skills. The main aim of PLC is to obtain a maximum yield of separation, not a maximum peak (spot) capacity [3]. The principal factors that may influence a PLC separation [1–4] are shown in Figure 4.1.

As a possible basis for preparative separations, analytical TLC methods are tested first for a quick judgment of the sample, to try to identify unknown compounds, or to optimize a separation before starting larger-scale operations.

Some characteristics, such as the average particle size, the thickness of stationary phase layer, the chamber type, the application of large amounts of samples, the location and detection of the separated compounds, and the removal of interested compounds by elution or extraction, should be taken into account in PLC.

The type of stationary phase, the composition of the mobile phase, the migration distance, the mode of development, and the working temperature may be identical to those of analytical TLC. These procedures have been presented extensively for analytical TLC [5–7] and summarized for PLC [1,2,4,7,8].

The main differences between TLC and PLC are due to the layer thickness and particle size of the stationary phase and the amount of sample applied to the plate.

TABLE 4.1
Comparison between Classical TLC and PLC Methods

Parameter	Classical TLC[a]	Classical PLC[b]
Migration of mobile phase	Capillary action	Capillary action
Stationary phase	Silica gel, alumina, cellulose	Silica gel (mainly), alumina, cellulose
Plate size (cm)	20 × 20	20 × 20; 20 × 40
Volume of stationary phase	Constant	Constant
Particle size (μm)	5–20[b]	5–40
Layer	Precoated	Precoated
Layer thickness (mm)	0.1–0.25	0.5–2.0
Vapor space	Normal chamber Unsaturated or saturated	Normal chamber Saturated 1–2h
Separation mode	Linear	Linear
Development	Ascending Descending	Ascending (frequently) Descending (no advantages)
Separation distance (cm)	10–15	≤18
Time for development (min)	30–200	Longer than TLC
Typical amount of samples	1–5 μl (1ng–10μg)	50–150 mg
Samples per plate	10–20	2–5
R_F value range	0.05–0.95	0.1–0.5[c]
ΔR_F	>0.05	>0.10
Starting spot diameter (mm)	3–6	Streak
Final spot diameter (mm)	6–15	Streak
Resolution	Good	Limited

[a] From Poole, C.F. and Poole, K.S., *Anal. Chem.*, 61, 1257a–1269a, 1989.

[b] From Nyiredy, Sz., Preparative layer chromatography, in *Handbook of Thin-Layer Chromatography,* 3rd ed., revised and expanded, Chromatographic Science Series, 89, Sherma, J. and Fried, B., Eds., Marcel Dekker, New York, 2003, chap. 11.

[c] Soczewiński, E. and Wawrzynowicz, T., Preparative TLC, in *Encyclopedia of Chromatography,* Marcel Dekker, New York, 2001, pp. 660–662.

A comparison between some parameters [1,3,9] that characterize classical TLC and PLC is given in Table 4.1.

For a successful PLC separation, the selection and the optimization of the mobile phase are essential because the separation is generally inferior to that of analytical TLC owing to the larger particle size and particle size distribution of the adsorbent and the overloading of the plate with sample. On thinner layers (0.5 to 1 mm) higher resolution can generally be achieved, whereas on thicker layers (1.5 to 2 mm) resolution is more limited. PLC is suitable for samples containing no more than five compounds [1].

In many cases, the solvent systems determined by analytical TLC are directly applicable to PLC with similar results. A proper mobile phase selected for PLC should have a resolution more than 1.5 in the analytical scale. According to theory, PLC resolution, however, decreases with increasing particle size. Improved separation can be achieved within longer retention times and if the selectivity of the entire system is optimized using another mobile phase. When resolution on thick layers is not so good as on the thinner layers, multiple or stepwise development may be advantageous. Also, spots may be more diffuse on thicker layers [4,7].

4.2 SELECTION OF MOBILE PHASE

Depending on the adsorbent, its activity, and the class of the solute compounds, a wide range of solvents can be used as the mobile phase in TLC, such as single solvents of eluotropic series or solvent mixtures (organic, aqueous, aqueous–organic, and ionic solvents) [7]. The choice of the mobile phase in PLC is similar to that for TLC, and it has a decisive influence on the separation. It is based on eluotropic and isoeluotropic series, depending on the properties of the separated solutes. The PLC choice is also determined by the subsequent recovery of the separated solutes. The less volatile (water, acetic acid, and *n*-butanol) and nonvolatile (buffer solutions and ion association reagents) components should be avoided. In PLC, the mobile phase usually gives R_F values in the lower range of 0.1 to 0.5, because the application of larger sample volumes leads to wide starting zones and increased R_F values. For polar adsorbents, ethyl acetate is recommended as the modifier because it is characterized by good solubility for many nonpolar and moderately polar solutes, rapid equilibrium, and easy evaporation of the separated fractions [3].

Knowledge of the chromatographic process mechanism is very useful in the choice of stationary and mobile phases and of development conditions. In Figure 4.2, the Stahl's triangular scheme for selecting the stationary and mobile phases according to the sample character for determination of proper chromatographic conditions is illustrated. One corner of the central triangle is pointed at the most important property of the sample, the character of the preferred chromatographic system being indicated by the other two corners [10–12].

The separation of different substances of a mixture is one of the most important matters of analytical and preparative chemistry. The most efficient among all the separation methods used in technology and analytical chemistry is the chromatographic one. As it is known, the chromatographic method is based on different

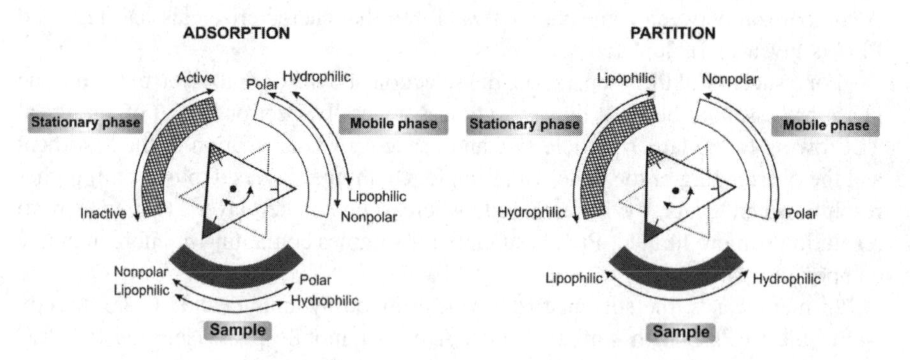

FIGURE 4.2 Stahl triangular scheme to determine proper chromatographic conditions. (Adapted from Hahn-Deinstrop, E., *J. Planar Chromatogr.*, 5, 57–61, 1992. With permission.)

partition of the mixture components between the mobile phase and the stationary one, whereas the components are displaced with different velocity by the mobile phase along the stationary phase. The success of a liquid chromatographic separation depends on the right choice of both the stationary phase and the mobile phase composed of a proper system of solvents [13].

Relative strength of the solvents on polar adsorbents arranged as an eluotropic series in chromatographic elution strength order for pure solvents or mixtures are given in literature [13–16]. The eluotropic series of pure solvents are generally referred to a particular adsorbent.

For preparative separation, the mobile phase can be selected by performing preliminary analytical TLC experiments. In PLC, the chromatographic chamber has to be saturated within 2 h because the development of preparative plates is much slower than the analytical development. In the analytical preassay during the selection of mobile phase composition, the chromatographic chamber must be lined with a sheet of filter paper to obtain a saturated atmosphere with mobile phase vapor. Then, the optimized analytical mobile phase can be transferred unchanged to preparative separations in the saturated developing chamber.

In PLC, an inferior separation is always achieved because the particle size and size distribution of adsorbents for preparative purposes are larger, and the preparative plates are much more overloaded with the separated compounds. This means that an optimized mobile phase is necessary for a successful preparative separation. During the optimization process, the volatility of the individual solvents must be considered so as to avoid several problems in subsequent steps such as elution of the compound from the stationary phase and evaporation of the solvents. It is important to use acetic acid as a component of the preparative mobile phase because of the possibility of chemical degradation during concentration of the isolated compounds [1–4,7,8].

The most important aspects of the selection and optimization of the mobile phase are presented by Touchstone and Dobbins [17], Gocan [13], Siouffi and Abbou [18], Kowalska et al. [19], and Cimpoiu [20].

4.2.1 Requirements for the Mobile Phase

Mobile phases are of a greater variety than the restricted number of stationary phases. Many solvents and their mixtures are used as a mobile phase. The possibility of slight modification of solvent proportions in a mixture permits the increase of mobile phase number and, thus, different results in the component separation of the analyzed sample. That is why the optimum mobile phase selection becomes one of the basic operations for the success of the analysis.

In order to have a better selection, there are the following among the necessary requirements for the mobile phase [15–17]:

- The used solvents must have high purity and low viscosity, strength, and volatility.
- The used solvents must be miscible with each other.
- The samples must be soluble in the mobile phase.

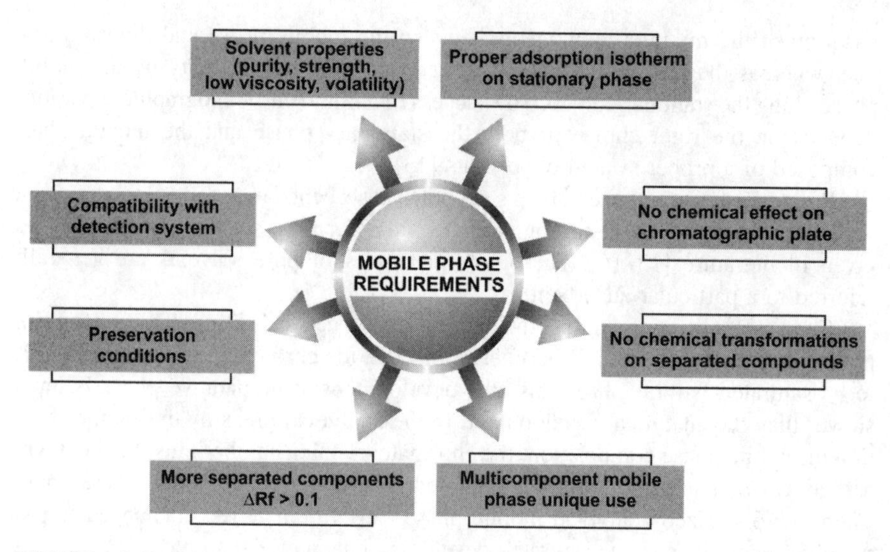

FIGURE 4.3 Requirements for the mobile phase.

- The mobile phase must not chemically affect or dissolve the chromatographic plate (support, adsorbent, and binder).
- The mobile phase should not produce chemical transformations of the separated components because it can modify the chromatographic behavior of the system. Solvents having weak reversible bonds with the solute are recommended.
- The multicomponent mobile phase must not be used repeatedly because the volatility of solvents produces a continuous modification of the solvent ratio, which negatively affects the chromatographic results.
- The recommended mobile phase must assure a proper adsorption isotherm for a given stationary phase. Chapter 2 discusses the adsorption planar chromatography in the nonlinear region.
- For the mobile phase in TLC to separate more components of unknown mixture, ΔR_F difference, must have a minimum value of 0.05, and for a better separation in PLC it must exceed 0.1.
- The mobile phase must be compatible with the detection system and easy to eliminate.
- The reproducibility can be greatly affected by the conditions and the time of mobile phase preservation.

The requirements for mobile phase are presented in Figure 4.3.

4.2.2 Influence of the Stationary Phase

In PLC, polar adsorbents (silica and alumina) and nonaqueous solvents of low viscosity are usually used. Chemically bonded adsorbents (silanized silica) are poorly wettable by aqueous mobile phases and are relatively expensive, thus they are not often applicable [3].

The adsorption isotherm describes the adsorption process in the stationary phase of the solutes from the mobile phase, and it shows the equilibrium concentration of the stationary phase with reference to the concentration of the mobile phase. In TLC, at low concentrations, the adsorption isotherm is linear; the slope changes and the isotherm becomes horizontal when saturation is reached. The best adsorbents are considered to be those with linear adsorption isotherms at relatively high concentrations. Preparative chromatography is based on the nonlinear theory that is presented in detail in Chapter 2. In analytical TLC, linear adsorption isotherms and compact spots are obtained for samples below 1 mg of mixture per 1 g of adsorbent. To increase throughput, PLC is operated under overloaded conditions at 1 mg of mixture per 1 g of adsorbent, or more. The overloading is performed by increasing the concentration or volume, the first being more advantageous [3].

The force driving the mobile phase migration is the decrease in the free energy of the solvent as it enters the porous structure of the layer [6,21,22]. Under these conditions, the velocity at which the solvent front moves is a function of its distance from the solvent entry position. As this distance increases, the velocity declines. Consequently, the mobile phase velocity varies as a function of time and migration distance, and it is established by the system variables.

In TLC, the movement of the solvent front is a function of time described by Equation 4.1:

$$Z_{\mathrm{f}}\left(kt\right)^{1/2} \tag{4.1}$$

where Z_{f} is the migration distance of the solvent front from the point of solvent entry; k, the velocity constant (cm sec^{-1}); and t, the time of contact of the solvent with the adsorbent layer.

The solvent front velocity u_{f} from Equation 4.2 is given by differentiation of Equation 4.1:

$$u_{\mathrm{f}} = k/2Z_{\mathrm{f}} \tag{4.2}$$

The velocity constant k is related to the experimental conditions, as given by Equation 4.3:

$$k = 2K_0 d_{\mathrm{p}}\left(\gamma / \eta\right)\cos\theta \tag{4.3}$$

where K_0 is the layer permeability constant; d_{p}, the average particle diameter; γ, the surface tension; η, the viscosity of mobile phase; and θ, the contact angle between the layer and the mobile phase. K_0 takes into consideration the effect of porosity on the permeability of the layer and the difference between the bulk liquid velocity and the solvent front velocity.

The mobile phase should wet the layer, thus $\cos\theta \neq 0$. In normal phase chromatography, the adsorbent is completely wetted by all solvents and thus $\cos\theta = 1$.

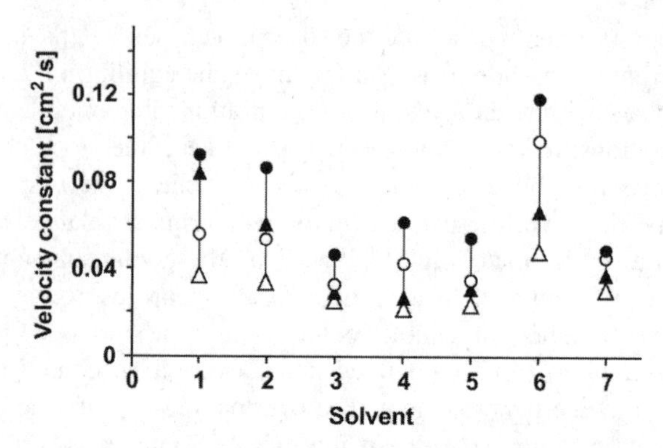

FIGURE 4.4 The influence of layer thickness on mobile phase velocity. Layer thickness: ○, ● = 0.1 mm; △, ▲ = 0.2 mm. Chamber conditions: ○, △ = unsaturated; ●, ▲ = saturated. Solvents: *n*-hexane (1), toluene (2), methylene chloride (3), ethyl acetate (4), chloroform (5), acetonitrile (6), methanol (7). (Data from Poole, S.K. et al., *J. Planar Chromatogr.*, 3, 133–140, 1990.)

In reversed-phase TLC, mobile phases with high content of water do not wet alkyl-bonded silica layers unless partially modified silica plates are used. It is well known that high-viscosity solvents do not generate the same plate number as low-viscosity solvents because of the solute diffusion coefficient and slow mass transfer [21].

The influence of layer thickness on mobile phase velocity and chromatographic properties under different conditions was studied by Poole [22]. The effect of layer thickness on the velocity constant in unsaturated and saturated chambers is given in Figure 4.4. It can be noted that the mobile phase velocity is influenced essentially by the free volume of the layer that has to be filled with the solvent for the mobile phase to advance. This is obviously less for the thinner layer. Although the capacities of layers increase with their thickness, the separation efficiency decreases for thickness above 1.5 mm, thus the optimum PLC layers are those of 0.5- to 1-mm thickness [3]. The influence of layer thickness on chromatographic performance was evaluated from the plot of the height equivalent of a theoretical plate at $R_F = 0.5$ vs. the solvent front migration distance. The principal influence of the layer thickness is to shift the position of the minimum plate height to a greater migration distance of the thinner layer. This is in agreement with the higher mobile phase velocity of thinner layer, which shifts the optimum mobile phase velocity to a greater distance from the solvent level in the chromatographic chamber [23].

For the same adsorbent, different mobile phases can be used according to the aim of chromatographic analysis. Rather than preparing an endless line of chromatographic plates of different thickness, it is easier to change the mobile phase up to the most convenient composition, keeping the same characteristics of the stationary phase. In the case of a hygroscopic adsorbent, the adsorbed water influences its activity.

4.2.3 Properties of Solvents

In order to obtain better separations it is very important to know the bulk physical properties of solvents (viscosity, refractive index, dielectric constant, dipole moment,

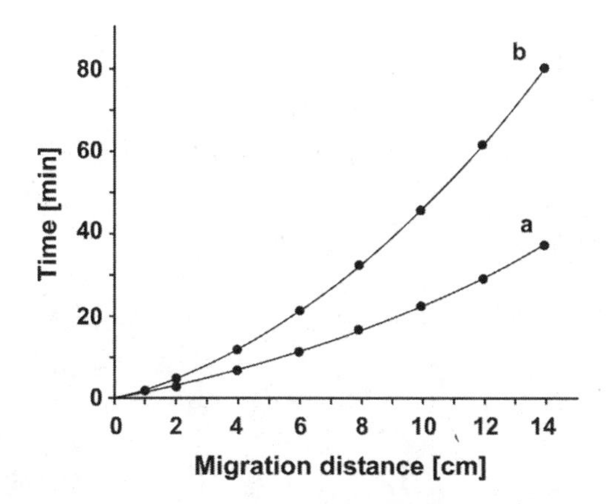

FIGURE 4.5 The influence of solvent viscosity on migration characteristics on preparative plates: (a) Nonviscous solvent (acetonitrile); (b) Viscous solvent (methanol). (Adapted from Botz, L., Nyiredy, S., and Sticher, O., *J. Planar Chromatogr.*, 3, 10–14, 1990. With permission.)

surface tension, etc.), but these are not sufficient to describe the interaction between the solute and solvent during the chromatographic process [13,18,21,24].

The influence of viscosity on migration characteristics on preparative plates of two solvents with different viscosities (acetonitrile, the nonviscous solvent and methanol, the viscous solvent) is presented in Figure 4.5. It can be observed that the solvent with a high flow velocity has a migration time less than the solvent with a low flow velocity, and the migration distance increases as the difference between the migration times of the two solvents increases [25]. In classical PLC, owing to capillary action, the mobile phase velocity is a variable that can be influenced to avoid the high-viscosity solvents during the mobile phase optimization [1].

The physical constants (viscosity η, surface tension γ, and dielectric constant ε) [26] for some common solvents used in chromatography are given in Table 4.2.

One important parameter in the solvent characterization is the solvent polarity, which can be defined in different ways [24]. Polarity is connected with solute–solvent interactions, and these depend on the solute. Thus, there are five types of molecular interactions:

- *Dispersion forces*, the so-called London forces δ_d, are the highest when solute and solvent electrons are polarized. These forces are high when the refractive index values are high. Solvents with high refractive indexes will dissolve solutes with high refractive indexes.
- *Dipole–dipole forces*, the so-called Keesom forces $\delta_{ind}\delta_0$, appear when both the solvent and solute have dipole moments. Strong interactions are produced as a result of dipole alignment. Dipole interactions are determined by the sum of all the dipoles within a molecule.
- *Induction forces*, the so-called Debye forces δ_{ind}, occur in the interaction between a permanent dipole of a solute or a polar solvent and an induced dipole in another compound. They are weak and appear during the analysis of the nonpolar polarized compounds, such as those with multiple

TABLE 4.2
Characteristics of Some Common Solvents Used in Chromatography

Group	Solvent	Mark[a]	P'	x_e	x_d	x_n	$\varepsilon^0_{Al_2O_3}$	$\varepsilon^0_{SiO_2}$	η (cP) 25°C	γ (mN/m) 25°C	ε	π^*		$E_T(30)$	E_T^N
					[42][b]		[31][b]	[44][b]	[26][b]			[51][b]	[53][b]	[50][b]	
	n-Hexane		0.1				0.01		0.300	17.89	1.89	–0.08	–0.11	30.9	0.009
	Cyclohexane		0.2				0.04	0.03	0.894	24.65	2.02	0.00	0.00	30.9	0.006
	Carbon disulfide		0.3				0.15[c]		0.352	31.58	2.63	0.51[d]	0.51	32.8	0.065
	n-Decane		0.4				0.04		0.838	23.37	1.98			31.0	0.009
	Carbon tetrachloride		1.6				0.18	0.11	0.908	26.43	2.24	0.28	0.21	32.4	0.052
	Triethylamine	TEA	1.9	0.56	0.12	0.32	0.48	0.37	0.347	20.22	2.42	0.14	0.09	32.1	0.043
I	Di-*n*-butyl ether	a	2.1	0.44	0.18	0.38			0.637		3.08	0.24	0.18	33.0	0.071
	Diisopropyl ether	b	2.4	0.48	0.14	0.38	0.28			17.27		0.27	0.19	34.1	0.105
	t-Butyl-methyl ether	c	2.7	0.49	0.14	0.37								34.7	0.124
	Diethyl ether	d	2.8	0.53	0.13	0.34	0.38	0.38	0.224	16.65	4.27	0.27	0.24	34.6	0.117
II	*n*-Octanol		3.4	0.56	0.18	0.25			7.288	27.10	10.30			48.1	0.537
	i-Pentanol	a	3.7	0.56	0.19	0.26					13.71			49.0	0.555
	n-Butanol	b	3.9	0.59	0.19	0.25			2.544	24.93	17.84	0.47		49.7	0.586
	i-Propanol	c	3.9	0.55	0.19	0.27	0.82		2.038	20.93	20.18	0.48		48.4	0.546
	n-Propanol	d	4.0	0.54	0.19	0.27			1.945	23.32	20.80	0.52		50.7	0.617
	t-Butanol	e	4.1	0.56	0.20	0.24		0.56	2.096			0.41		43.3	0.389
	Ethanol	f	4.3	0.52	0.19	0.29	0.88		1.074	21.97	25.30	0.54		51.9	0.654
	Methanol	g	5.1	0.48	0.22	0.31	0.95		0.544	22.07	33.00	0.60		55.4	0.762
III	Tetrahydrofuran	a	4.0	0.38	0.20	0.42	0.45		0.456		7.52	0.58	0.55	37.4	0.207
	Quinoline		5.0	0.41	0.23	0.36			0.723	42.59	9.16	0.92	0.93	39.4	0.269
	Diethylene glycol	b	5.2	0.44	0.23	0.33			30.200	44.77	31.82			53.8	0.713
	Pyridine	c	5.3	0.41	0.22	0.36	0.71		0.879	36.56	13.26	0.87	0.87	40.5	0.302

	Triethylene glycol	d	5.6	0.42	0.24	0.34					23.69			52.8	0.682
	Tetramethyl urea		6.0	0.42	0.19	0.39					23.10	0.83		40.9	0.315
	Methyl formamide	e	6.0	0.41	0.23	0.36			1.678		18.48			54.1	0.722
	N,N-dimethyl formamide	f	6.4	0.39	0.21	0.40			0.794		38.25	0.88		43.2	0.386
	N,N-dimethyl acetamide		6.5	0.41	0.20	0.39			1.956			0.88	0.85	42.9	0.377
	Dimethyl sulfoxide	g	7.2	0.39	0.23	0.39	0.62		1.987	42.92	47.24	1.00	1.00	45.1	0.444
IV	Benzyl alcohol	a	5.7	0.40	0.30	0.30			5.474		11.92	0.98		50.4	0.608
	Acetic acid	b	6.0	0.39	0.31	0.30			1.056	27.10	6.20	0.64		51.7	0.648
	Ethylene glycol	c	6.9	0.43	0.29	0.28	1.11		16.1	47.99	41.40	0.92			
	Formamide	d	9.6	0.36	0.33	0.30			3.343	57.03	111.00	0.97		55.8	0.775
V	Dichloromethane	a	3.1	0.29	0.18	0.53	0.42	0.32	0.413	27.20	8.93	0.82	0.73	40.7	0.309
	1,2-Dichloroethane	b	3.5	0.30	0.21	0.49	0.49		0.779	31.86	10.42	0.81	0.73	41.3	0.327
VIa	Ethyl acetate	a	4.4	0.34	0.23	0.43	0.58	0.38	0.423	23.39	6.08	0.55	0.45	38.1	0.228
	Ethyl methyl ketone	b	4.7	0.35	0.22	0.43	0.51								
	Cyclohexanone	c	4.7	0.36	0.22	0.42			2.017	34.57	16.1	0.76	0.68	39.8	0.281
	1,4-Dioxane	d	4.8	0.36	0.24	0.40	0.56[c]	0.49	1.177	32.75	2.22	0.55	0.49	36.0	0.164
	Acetone	e	5.1	0.35	0.23	0.42	0.56	0.47	0.306	23.46	21.01	0.71	0.62	42.2	0.355
	Butyrolactone		6.5	0.34	0.26	0.40						0.87		44.3	0.420
VIb	Benzonitrile	f	4.8	0.31	0.27	0.42	0.44		1.267	38.79	25.90	0.90	0.88	41.5	0.333
	Acetonitrile	f	5.8	0.31	0.27	0.42	0.65[c]	0.50	0.369	28.66	36.64	0.75	0.66	45.6	0.460
	Propylene carbonate	f	6.1	0.31	0.27	0.42						0.83	0.83	46.0	0.472
	Aniline	g	6.3	0.32	0.32	0.36	0.62		3.847	42.12	7.06		1.08	44.3	0.420
VII	Toluene	a	2.4	0.25	0.28	0.47	0.29		0.560	27.93	2.38	0.54	0.49	33.9	0.099
	p-Xylene	b	2.5	0.27	0.28	0.45	0.25		1.197	28.01	2.27	0.43	0.45	33.1	0.074
	Benzene	c	2.7	0.23	0.32	0.45	0.31	0.25	0.604	28.22	2.28	0.59	0.55	34.3	0.111
	Chlorobenzene	d	2.7	0.23	0.33	0.44	0.30		0.753	32.99	5.69	0.71	0.68	36.8	0.188
	Bromobenzene		2.7	0.24	0.33	0.43	0.31		1.074	35.24	5.45	0.79	0.77	36.6	0.182
	Iodobenzene	e	2.8	0.24	0.35	0.41			1.554	38.71	4.59	0.81	0.84	36.2	0.170

TABLE 4.2 (Continued)
Characteristics of Some Common Solvents Used in Chromatography

Group	Solvent	Mark[a]	P'	x_e	x_d	x_n	$\varepsilon^0_{Al_2O_3}$	$\varepsilon^0_{SiO_2}$	η (cP) 25°C	γ (mN/m) 25°C	ε	π*		$E_T(30)$	E_T^N
	Fluorobenzene		3.2	0.24	0.32	0.45			0.550	26.66	5.46	0.62	0.59	37.0	0.194
	Ethoxybenzene		3.3	0.28	0.28	0.44									
	Diphenyl ether		3.4	0.27	0.32	0.41				26.75	3.73	0.66		35.3	0.142
	Anisole	f	3.8	0.27	0.29	0.43			1.056	35.10	4.30	0.73	0.70	37.1	0.198
	Dibenzyl ether		4.1	0.30	0.28	0.42					3.82	0.80	0.80	36.3	0.173
	Nitrobenzene	g	4.4	0.26	0.30	0.44	0.52		1.863	0.40	35.60	1.01	0.86	41.2	0.324
	Nitroethane	h	5.2	0.28	0.29	0.43			0.688	32.13	19.70		0.77	43.6	0.398
	Nitromethane	i	6.0	0.28	0.31	0.40	0.64		0.630	36.53	37.27	0.85	0.75	46.3	0.481
	Chloroform	CF	4.1	0.25	0.41	0.33	0.40	0.26	0.537	26.67	4.81	0.58	0.69	39.1	0.259
VIII	*m*-Cresol	a	7.4	0.38	0.37	0.25			12.9	36.59	12.44			52.4	0.670
	Tetrafluoropropanol	b	8.6	0.34	0.36	0.30								59.4	0.886
	Dodecafluoroheptanol	c	8.8	0.33	0.40	0.27									
	Water	d	10.2	0.37	0.37	0.25			0.890	71.99	80.10	1.09		63.01	1.000

Note: Solvent classification into groups based on solvent polarity (P'), selectivity parameters (x_e proton acceptor, x_d proton donor, x_n dipole interactors) and solvent strength on alumina $\varepsilon^0_{Al_2O_3}$ and on silica gel $\varepsilon^0_{SiO_2}$. Physical constants: viscosity (η), surface tension (γ), dielectric constant (ε). Solvatochromic polarity parameters: π^*, $E_T(30)$ and E_T^N .

[a] Solvent position in Figure 4.9.

[b] Source of data.

[c] Data from Kowalska, T., Kaczmarski, K., and Prus, W., Theory and mechanism of thin-layer chromatography, in *Handbook of Thin-Layer Chromatography,* 3rd ed., revised and expanded, Chromatographic Science Series, 89, Sherma, J. and Fried, B., Eds., Marcel Dekker, New York, 2003, chap. 2.

[d] Data from Kamlet, M.J., Abboud, J.L., and Taft, R.W., *J. Am. Chem. Soc.*, 99, 6027–6038, 1977.

carbon–carbon bonds or heavy atoms. Interaction energy is related to the permanent and induced dipole moments and to the polarizability of two interacting compounds.

- *Hydrogen bonding* δ_h occurs when a proton-acceptor molecule (primary and secondary amines, and sulfoxides) interacts with a proton-donor molecule (alcohols, carboxylic acids, and phenols).
- *Dielectric interactions* take place when ions polarize molecules of a solvent with a high dielectric constant.

The solubility parameter δ_i of a pure solvent defined initially by Hildebrand and Scott based on a thermodynamic model of regular solution theory is given by Equation 4.4 [13]:

$$\delta_i = \left(\Delta E_i^v / V_i\right)^{1/2} \tag{4.4}$$

where ΔE_i^v is the cohesive energy for vaporization per mol of a pure solvent i at zero pressure and V_i is the molar volume of the solvent. The solubility properties can determine the mutual solubility of two liquids if the pure components form an ideal solution.

The theory has been extended to polar solvents by including dispersion δ_d, permanent dipole orientation δ_0, dipole induction δ_{ind}, and hydrogen-bonding interactions δ_h such as acidic δ_a and basic δ_b. In this case the solubility parameter δ_i is given by Equation 4.5:

$$\delta_i = \delta_d^2 + \delta_p^2 + \delta_h^2 \tag{4.5}$$

where $\delta_p^2 = \delta_0^2 + 2\delta_{ind}\delta_d$ describes the polar component and $\delta_h^2 = 2\delta_a\delta_b$, the hydrogen-bonding component.

Liquids without dipole moments (alkanes) have quite low cohesive energy densities, whereas liquids with dipole moments or hydrogen-bonding groups have high cohesive energy densities.

The solubility parameter theory can be also used for the mixed-solvent systems. The total-solubility parameter δ_t is given by the sum of the individual solubility parameters in terms of the volume fractions φ_j in the mixture, according to Equation 4.6:

$$\delta_t = \sum \varphi_j \delta_j \tag{4.6}$$

In fact, the solute retention depends on the solubility parameters of the solute, δ_i, of the mobile phase, δ_m, of the stationary phase, δ_s, and of the phase ratio given by Equation 4.7 [24]:

$$\ln k_i = (V_i / RT)(\delta_s - \delta_m)(\delta_s + \delta_m - 2\delta_i) + \ln n_m / n_s \quad (4.7)$$

Solvent selectivity refers to the ability of a chromatographic system to separate two substances of a mixture. It depends on the chemistry of the adsorbent surface, such as the layer activity and type of chemical modification. The separation power or resolution R_S is given by Equation 4.8 [27]:

$$R_S = 1/4(\alpha - 1)\sqrt{N}\left[k'/(k' + 1)\right] \quad (4.8)$$

where the separation factor α is the ratio of capacity factors k_1/k_2 for the two solutes; N, the theoretical plate number in the adsorbent bed; and k', the average of k_1 and k_2. The capacity factor k' is the equilibrium ratio of total solute in the stationary phase and the total solute in the mobile phase. In Equation 4.8, N reflects the nature of the adsorbent and k' is the average migration velocity of the two solutes and is determined by the mobile phase strength ε^0. The parameter α depends on the adsorbent and the mobile phase composition.

Value of k' can be related to certain properties of the stationary and mobile phases and the solute:

$$\log k' = \log k_p - \alpha' \varepsilon^0 A_s + \Delta \quad (4.9)$$

where k_p is the k' value for pentane as the solvent; α', a function of adsorbent activity; ε^0, the solvent strength parameter; A_s, the area of the adsorbed solute molecule; and Δ, the so-called secondary solvent effect, which can be a function of the adsorbent, solvent, and solute.

Selectivity is largely based on the differences in the interaction between the mobile phase and solute. Different solutes will interact differently with the mobile phase. Therefore, if two solutes cannot be separated with a mobile phase of predetermined strength, they can sometimes be separated using a different mobile phase of the same strength.

4.2.4 Solvent Classification in Eluotropic Series

Several criteria are involved in the classification of solvents used in chromatography. Several authors have defined the solvent relative strength for polar adsorbents in the form of eluotropic series, grouping pure and mixture solvents in the order of their chromatographic elution power. The eluotropic series of pure solvents are generally referred to a particular adsorbent. The solvent strength parameter ε^0 can be defined as the adsorption energy of the solvent per unit area of the standard activity surface. Strength is a single parameter of the solvent's ability to cause migration in chromatography. It is a property of both the stationary phase and the solvent, and it cannot be considered only as a solvent fundamental property. The eluotropic series for various polar adsorbents in terms of the ε^0 parameter are given in the literature [10,13,28].

Snyder [28] has shown that there is a correlation between the ε^0 values for a certain polar adsorbent (silica gel, florisil, and magnesia) and alumina. For example, the ε^0 values for silica, $\varepsilon^0_{SiO_2}$, can be estimated from the data for alumina, $\varepsilon^0_{Al_2O_3}$, by using Equation 4.10:

$$\varepsilon^0_{Al_2O_3} = 1.3\varepsilon^0_{SiO_2} \tag{4.10}$$

Arranging the solvents in separation-strength order, the so-called eluotropic series appeared. This term, introduced by Trappe, was related to the experience with bare silica, where a strong solvent is able to move polar solutes on a polar stationary phase. Later this was improved by the discovery of a direct proportion between the elution strength and the dielectric constant. Because silica is hydrophilic and highly polar, there was a correlation between the eluotropic series and the polarity of a solvent [16,18].

In Figure 4.6, some experimental eluotropic series are given. Some disagreements are observed in the order of the solvents because of the action of some secondary factors producing different solvent behavior [16].

In the mobile phase selection for the separation of compounds on thin silica gel layers, it is necessary to use not only eluotropic series based on the eluting capacity of the solvent but also eluotropic series of compounds established according to their interaction with silica gel.

Another type of eluotropic series experimentally determined on silica gel is based on the polarities of solutes: nonpolar, medium polar, and polar. In Figure 4.7, the eluotropic series for these three groups of solutes are presented. More efficient eluotropic series were obtained by using binary mixture of solvents. Examples are given in Figure 4.8, where the mobile phases are arranged in increasing strength order.

According to Snyder [28], the solvent strength ε^0 is the standard free energy of adsorbed solvent molecules in a standard state, and it is given by Equation 4.11:

$$\varepsilon^0 = \Delta G^0_S / 2.3\, RTA_S \tag{4.11}$$

where ΔG^0_S is the adsorption free energy of solute molecules; R, the Boltzmann constant; T, the absolute temperature; and A_S, the surface that a solvent molecule covers the adsorbent.

The solvent strength is given by the sum of many types of intermolecular interactions according to Equation 4.12 [18]:

$$\varepsilon^0 = C\left[\left(\delta^j_d - \delta^s_d\right)\delta^a_d + \delta^j_0\delta^a_0 + \delta^j_a\delta^a_b + \delta^j_b\delta^a_a\right] \tag{4.12}$$

where C is the constant of the used adsorbent and δ^a_d is the dispersion solubility parameter for the reference solvent S, defined as $\varepsilon^0 = 0$ for *n*-pentane.

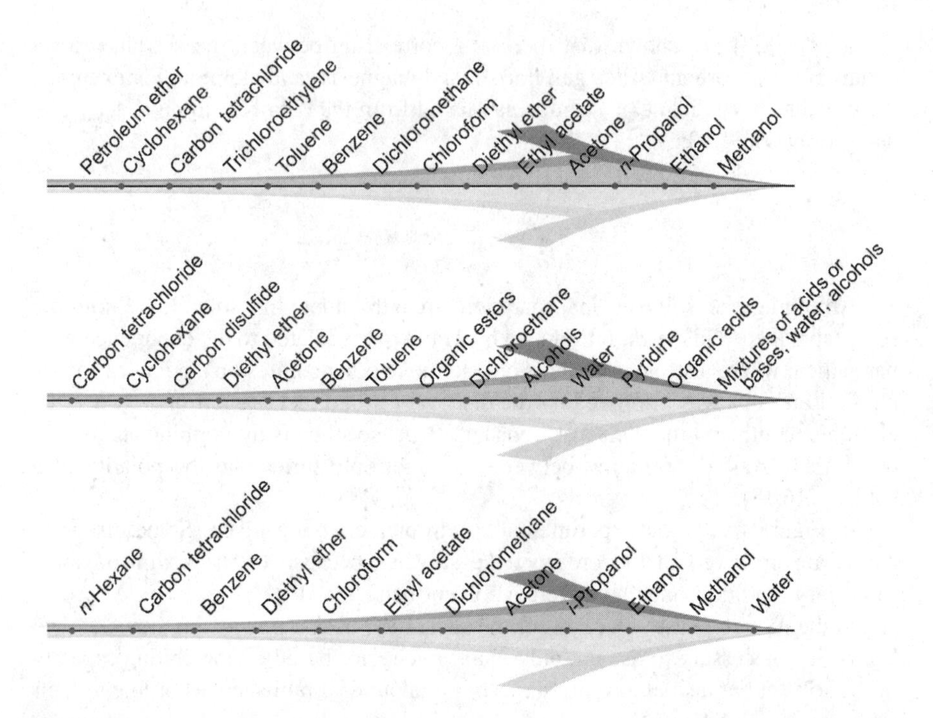

FIGURE 4.6 Eluotropic series of solvents arranged in strength order, experimentally determined on silica gel. (Data from Lederer, E. and Lederer, M., Elution, in *Chromatography (A Review of Principles and Applications)*, Elsevier, Amsterdam, 1957, chap. 4; Pelloni-Tamaş, V. and Johan, F., Developanţi (dizolvanţi şi amestecuri de dizolvanţi folosiţi în cromatografia pe strat subţire (Eluents — solvents and solvent mixtures — used in thin layer chromatography), in *Cromatografia în Strat Subţire (Thin-Layer Chromatography)*, Technical Publishing House, Bucharest, 1971, chap. 4; Bodoga, P., Măruţoiu, C., and Coman, M.-V., Faze mobile (Mobile phases), in *Cromatografia pe Strat Subtire. Analiza Poluanţilor (Thin-Layer Chromatography. Pollutant Analysis)*, Analytical chemistry series, Technical Publishing House, Bucharest, 1995, chap. 3 and references cited therein.)

The solvent strength for a binary mixture ε_{AB} is described by Equation 4.13 [29]:

$$\varepsilon_{AB} = \varepsilon_A^0 + \log\left(X_B 10^{an_b\left(\varepsilon_B^0 - \varepsilon_A^0\right)} + 1 - X_B \right) / \alpha n_b \qquad (4.13)$$

where ε_A^0 and ε_B^0 are the solvent strengths of A and B solvents, respectively, X_B is the mole fraction of solvent B in the mixture, α is the adsorbent activity parameter and n_b is the adsorbent surface area occupied by the molecule of solvent B.

The solvent strength for a ternary mixture was also deduced, and it is given by the Equation 4.14 [29]:

$$\varepsilon_{AC} = \varepsilon_B^0 + \log\left(X_C 10^{an_c\left(\varepsilon_C^0 - \varepsilon_B^0\right)} + X_B \right) / \alpha n_c \qquad (4.14)$$

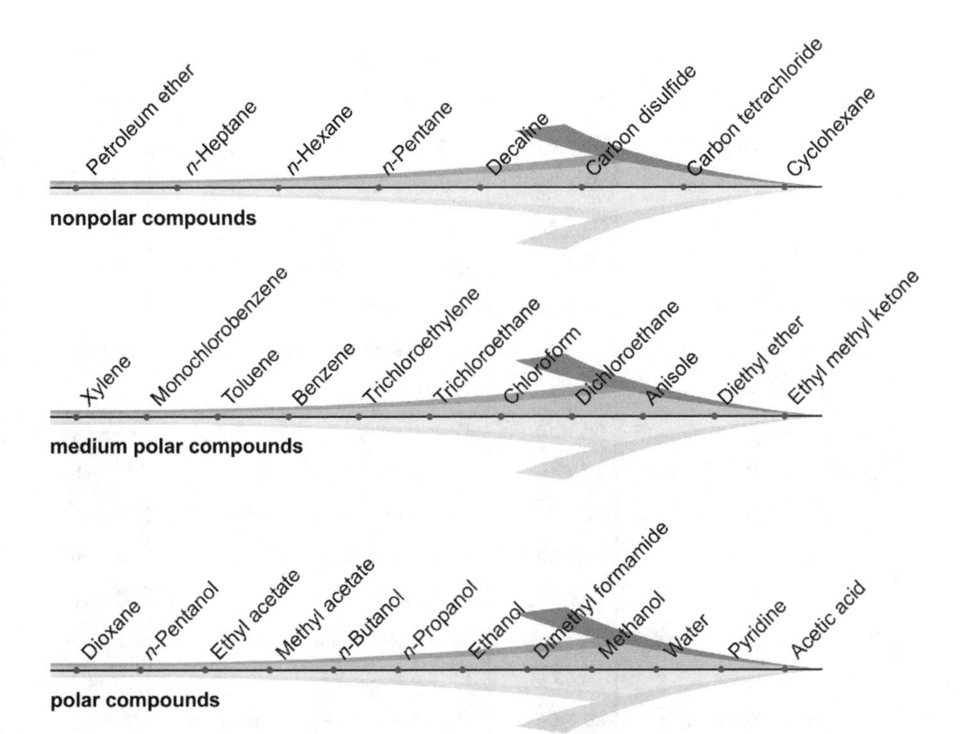

FIGURE 4.7 Extended eluotropic series of solvents arranged in increasing order of R_F values of separated compounds, experimentally determined on silica gel. (Data from Bodoga, P., Măruţoiu, C., and Coman, M.-V., Faze mobile (Mobile phases), in *Cromatografia pe Strat Subţire. Analiza Poluanţilor (Thin-Layer Chromatography. Pollutant Analysis)*, Analytical chemistry series, Technical Publishing House, Bucharest, 1995, chap. 3 and references cited therein.)

Equation 4.13 and Equation 4.14 were tested for a series of mobile phases on alumina [29–31] and silica gel [32]. Two eluotropic series of solvent binary mixtures for alumina ($\alpha = 0.6$) and silica gel ($\alpha = 0.7$) have been calculated by using Equation 4.13, and the obtained data can be used to establish many such series or series of other selectivities [13,28].

Recently, Janjić et al. published some papers [33–36] on the influence of the stationary and mobile phase composition on the solvent strength parameter ε^0 and SP, the system parameter ($SP = \log x_s/x_m$, where x_s and x_m denote the mole fractions of the modifier in the stationary and the mobile phase, respectively) in normal phase and reversed-phase column chromatography. They established a linear dependence between SP and the Snyder's solvent strength parameters ε^0 by performing experiments with binary solvent mixtures on silica and alumina layers.

The separations of some nonionic tensides having biological activity and consisting of ethyleneoxide oligomer mixtures were performed in many different TLC systems (silica and alumina as the stationary phase and single solvent or binary mixtures as the mobile phase). Selectivity was higher on alumina than on the silica layer. Both

Series 1	(v/v)
Chloroform – Diethyl ether	60 + 40
Benzene – Ethyl acetate	50 + 50
Benzene – Diethyl ether	40 + 60
Chloroform – Acetone	85 + 15
Benzene – Methanol	90 + 10
Chloroform – Methanol	99 + 01
Chloroform – Diethyl ether	80 + 20
Cyclohexane – Ethyl acetate	50 + 50
Benzene – Diethyl ether	60 + 40
Benzene – Methanol	95 + 05
Chloroform – Diethyl ether	90 + 10
Benzene – Ethyl acetate	80 + 20
Benzene – Acetone	90 + 10
Chloroform – Acetone	95 + 05
Cyclohexane – Ethyl acetate	80 + 20
Chloroform	100
Benzene – Chloroform	50 + 50
Benzene	100

SERIES 1

Series 2	(v/v)
Dioxane – Water	90+10
Methanol	100
Acetone	100
Dioxane	100
Chloroform – Methanol	80 + 20
Benzene – Acetone	50 + 50
Ethyl acetate – Methanol	99 + 01
Diethyl ether – Dimethyl formamide	99 + 01
Diethyl ether	100
Diethyl ether – Methanol	99 + 01
Benzene – Diethyl ether	10 + 90
n-Butyl acetate – Methanol	99 + 01
Benzene – Ethyl acetate	30 + 70
Chloroform – Acetone	70 + 30
Chloroform – Methanol	95 + 05
Ethyl acetate	100
Cyclohexane – Ethyl acetate	20 + 80
Cyclohexane	100

SERIES 2

FIGURE 4.8 Binary (v/v) eluotropic series of mobile phases experimentally determined on silica gel. The arrows indicate the increase of solvent strength. (Data from Bodoga, P., Măruţoiu, C., and Coman, M.-V., Faze mobile (Mobile phases), in *Cromatografia pe Strat Subţire. Analiza Poluanţilor (Thin-Layer Chromatography. Pollutant Analysis)*, Analytical chemistry series, Technical Publishing House, Bucharest, 1995, chap. 3 and references cited therein.)

strength and selectivity of the mobile phase were influenced mainly by the dielectric constant (≈70%), whereas the effect of refractive index was lower (≈30%) [37].

Solvent selectivity is seen as the factor that distinguishes individual solvents that have solvent strengths suitable for separation. In reality, separations result from the competition between the mobile and stationary phases for solutes based on the differences of all intermolecular interactions with the solute in both phases. Solvents can be organized on selectivity scales that are useful for initial solvent selection, but in a chromatographic separation the properties of the stationary phase must be taken into consideration. Methods that attempt to model chromatographic separation need to consider simultaneously mobile and stationary phase properties [38].

Scott [39] defines the coefficient of distribution K of a solute between the two phases in a chromatographic system by means of the interaction forces in the following way:

$$K = \frac{\text{magnitude of total forces acting on solute in stationary phase}}{\text{magnitude of total forces acting on solute in mobile phase}}$$

$$= \frac{\begin{array}{c}\text{(magnitude of forces between solute and stationary phase)} \times \\ \text{(probability of interaction)}\end{array}}{\begin{array}{c}\text{(magnitude of forces between solute and mobile phase)} \times \\ \text{(probability of interaction)}\end{array}}$$

For a series of n types of interactions, the expression of the distribution coefficient is given by Equation 4.15:

$$K = \left[\sum_{j=1}^{j=n} \varphi_j F_j P_j f_j(T)\right]_s \Bigg/ \left[\sum_{j=1}^{j=n} \varphi_j F_j P_j f_j(T)\right]_m \tag{4.15}$$

where φ is a constant that includes the probability of the contact position determined by the size and geometry of the molecules; F, the magnitude of the respective force between the solute molecule and the phase molecule; P, the probability of molecular interaction, $f(T)$, the thermal energy of the molecule at the time of contact; and the subscripts s and m, the stationary and mobile phases, respectively.

The eluotropic series of solvents were developed to help chromatographers select solvents for multicomponent mobile phases. Solvent strength and eluotropic series are related to the adsorbent activity as a property of a given stationary phase implied in the retention mechanism. The eluotropic series of solvents according to Snyder is related to mono-active-site-type adsorbents. Its imperfection referring to alumina owing to its different types of active sites was studied by Kowalska and Klama [40], who showed that the eluotropic series sequence and solvent strength values are highly dependent on the solute class used and especially the active sites that are occupied in the retention process of a given solute.

4.2.5 Snyder's Solvent Classification

One of the most important problems of planar chromatography is that of the optimization of solvent systems for the separation of mixtures of different samples. An analyst is interested in obtaining the expected result using a minimum number of experiments. Snyder has introduced a new system for solvent classification that permits a logical selection of solvents both in term of polarity indices (P') and selectivity parameters (x_i), proving theoretically the validity of such universal solvent systems [18,38,41,42].

The P' scale of solvent polarity is based on a combination of gas–liquid partition coefficients reported by Rohrschneider [43].

Snyder's papers [41,42] have been really helpful because he gives a classification of a great number of solvents (more than 80) into eight groups for normal phase chromatography based on their properties as proton donors (x_d), or proton acceptors (x_e), and strong dipole interactors (x_n). This classification is the so-called solvent selectivity triangle. The solvents are classified based on their interaction with three prototypical solutes determined by their gas–liquid distribution constants corrected for differences in solvent molecular mass and dispersion interactions. The sum of the three polar distribution constants provides a measure of solvent strength P'. The ratio of the individual polar distribution constants to their sum is a measure of selectivity (x_d, x_e, and x_n). The value of the partition coefficient between the gas phase and the solvents is considered for the test solutes *n*-octane (o), ethanol (e), dioxane (d), and nitromethane (n). Ethanol is used to determine solvent hydrogen-bond basicity (x_e), dioxane hydrogen-bond acidity (x_d), and nitromethane dipole–dipole (x_n) interactions. The interaction coefficients of the solvents are corrected for nonpolar interactions and normalized to the total polarity of the solvent: $x_e + x_d + x_n = 1$. According to Snyder, polarity normalization consequence is a distinction between the solvent strength and solvent selectivity.

Representing each solvent by the three solvent selectivity coordinates and plotting the results on the surface of a triangle, the classification of solvents into eight groups was obtained. Solvents from different groups have different selectivity characteristics and are likely to provide different migration orders. Solvents of the same group are expected to show similar separation properties and are only appropriate for separations.

In Figure 4.9 Snyder's solvent selectivity triangle is presented. The solvents of Table 4.2 are marked in the plot with triangular coordinates for the eight groups.

Solvents grouped in the same region of the triangle will have similar selectivity, whereas solvents from other groups will have different selectivity even if their solvent strengths are similar. Acetone is a solvent characterized by a polarity value of 5.1 (group VI, e). Its x_i values given in Table 4.2 show its polar interaction properties that involve 35% proton-acceptor, 23% proton-donor, and 42% dipole interactions. Solvents near the corners have mainly one kind of selectivity.

The concept of selectivity parameters has a physicochemical relevance, and it is proved experimentally that among solvents with similar functionality there is a great similarity with the selectivity parameters [42]. This fact is very important at the molecular level of the phenomena, and it is the best proof of the predominant role of functionality in intermolecular interactions of the solvent and solute, and the solvent and stationary phase.

In the case of a binary mixture composed of the two solvents A and B, the P' value is a linear sum of the volume fraction (φ) contributions and is given by Equation 4.16:

$$P'_{AB} = \varphi_A P'_A + \varphi_B P'_B \tag{4.16}$$

Optimization of the chromatographic process by Snyder's concept of solvent polarity and selectivity is in fact the optimization of the separation selectivity that

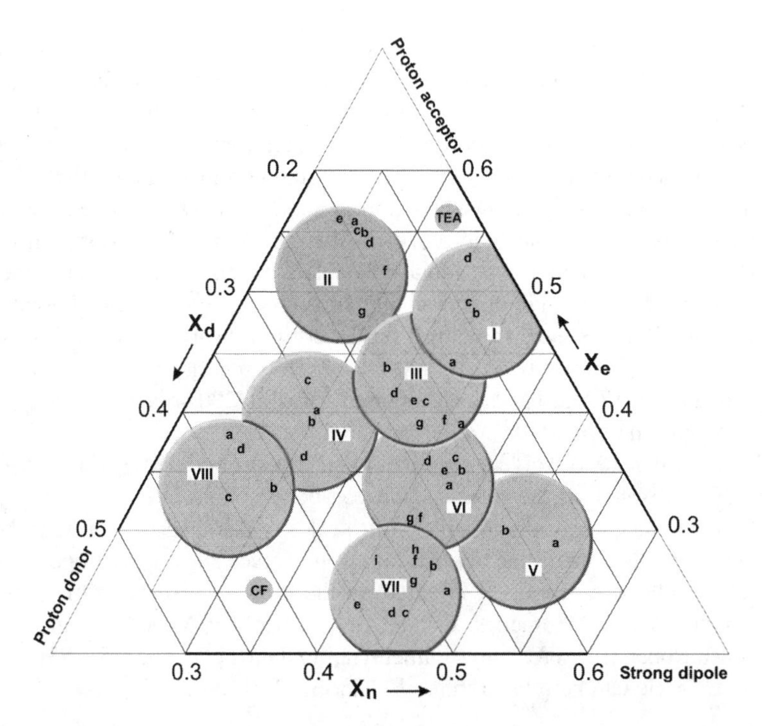

FIGURE 4.9 The Snyder triangle. Every group from Table 4.2 is represented with marks of a, b, c, etc. (Adapted from Snyder, L.R., *J. Chromatogr. Sci.*, 16, 223–234, 1978. With permission.)

can be obtained by using isoeluotropic mixtures of solvents. These mixtures preserve equal elution strength even if they have a composition different from the original mobile phase [19].

4.2.6 Solvatochromic Classification

The dielectric constant and refractive index parameters and different functions of them that describe the reactive field of solvent [45] are insufficient to characterize the solute–solvent interactions. For this reason, some empirical scales of solvent polarity based on either kinetic or spectroscopic measurements have been introduced [46,47]. The solvatochromic classification of solvents is based on spectroscopic measurements. The solvatochromic parameters refer to the properties of a molecule when its nearest neighbors are identical with itself, and they are average values for a number of select solutes and somewhat independent of solute identity.

The *Z scale* developed by Kosower and Mohammad [48,49] is based by the charge-transfer absorption of *N*-ethyl-4-methoxycarbonyl) pyridinium iodide molecule. The wavelength of maximum adsorption (λ_{max} in nm) is measured and the energy (kcal/mol) of this transition becomes the actual polarity measured for a given solvent (Equation 4.17):

$$Z = 28531 / \lambda_{max} \tag{4.17}$$

A disadvantage of the *Z* scale is that in mixtures of very high water content, the position of the charge-transfer absorption peak becomes progressively more difficult to evaluate because of the overlap with the pyridine ring absorption. Thus, the polarity evaluation of binary mixtures of high water content is impossible and this scale is unsuitable for typical reversed phase liquid chromatography (RPLC) use.

The $E_T(30)$ *scale* [46], known as Dimroth and Reichardt's betaine scale, is one of the most used ones, and it is based on the charge-transfer absorption of 2,6-diphenyl-4-(2,4,6-triphenyl-*N*-pyridinio) phenolate molecule. It has one of the largest observed solvatochromic effects of any known organic molecule, amounting to several hundred nanometers in going from a very polar solvent (water, 453 nm) to a nonpolar solvent (diphenyl ether, 810 nm). $E_T(30)$ polarity values are calculated using Equation 4.17 and have been reported for over 200 solvents, as well as for binary solvent mixtures [50].

The $E_T(30)$ scale of solvent polarity is sensitive to both solvent dipolarity and polarizability, as well as the solvent hydrogen-bond donor ability.

The π^* *scale* of solvents (Kamlet–Taft scale) is intended to represent the solute–solvent interactions in the absence of strong forces such as hydrogen-bonding or dipole–dipole interactions. It is based on the π to π^* absorption of a series of seven solutes (six nitroaromatics, such as 4-nitroanisole, *N,N*-diethyl-3-nitroaniline, 1-ethyl-4-nitrobenzene, etc., and 4-dimethylaminobenzophenone) [51–53].

The π^* scale can be related to the Equation 4.18 [46,51]:

$$\pi^* = \left(\nu - \nu_0\right) / 2.232 \tag{4.18}$$

where ν_0 is the frequency (10^{-3} cm^{-1} units) of the maximum of band in cyclohexane.

The π^* value is a measure of solvent dipolarity and polarizability. For nonhaloaliphatic solvents and nonaromatic ones, the π^* parameter is correlated with the permanent dipole moment of the solvent molecule.

In both protonic and nonprotonic solvents, the $E_T(30)$ scale can be related to the π^* and α (hydrogen-bond donor) scales by use of the Equation 4.19 [46,51,54]:

$$E_T(30) = 30.31 + 14.6\pi^* + 16.53\alpha \tag{4.19}$$

The α scale of solvent acidity (hydrogen-bond donor) and the β scale of solvent basicity (hydrogen-bond acceptor) are parameters derived from solvatochromic measurements used in adsorption chromatography [51,54,55].

In Table 4.2 the solvatochromic parameters π^*, $E_T(30)$, and its normalized value E_T^N of common solvents used in TLC are given. Two values for the π^* parameter are mentioned because in time the original solvatochromic values were revised.

The solvatochromic classification of solvents takes into consideration only the polar interactions of the solvents and not their cohesion. The transfer of a solute from one solvent to another occurs with the cancellation of dispersion interactions [38].

The relation between the empirical parameters mentioned in the preceding text and the other parameters described in the literature has been presented by Pytela in a review [56] that summarizes and evaluates the empirical models used for the description of the effects of solvents and solvent mixtures on processes in solutions. The principles of application of the empirical relations and their theoretical basis are analyzed. A survey regarding empirical parameters of individual solvents with respect to the model process and the parameters derived by mathematical–statistical treatment are presented. The section concerning mixed solvents presents empirical models used for the description of the effect of changes in composition of solvent mixtures.

According to the mode of parameter correlation, $E_T(30)$ was introduced in the group of parameters that describe the acidity of solvent and partially its polarity, and π^* in the group of parameters that present the dielectric properties of solvents. Quantitative relations between different parameters of polarity, such as correlation between the π^* scale or $E_T(30)$ and the dielectric constant and refractive index of the solvent or between Z, $E_T(30)$ [57], and π^* have been proposed and verified.

Solvent polarity $E_T(30)$ has been related to the R_F values of some nitro compounds (*o*-, *m*-, *p*-nitroanilines, *o*-nitrophenol, and picric acid) on silica gel G plates. The R_F value of each nitro compound in different solvents was obtained by the linear relationship given by Equation 4.20:

$$R_F = m\left[\log E_T(30) - c\right] \tag{4.20}$$

where m and c are the slope and intercept, respectively [58].

Evaluation of the polarity of a single or a solvent mixture according to the Equation 4.20 is useful when it is difficult to obtain $E_T(30)$ by conventional methodology. This equation shows that the solvent polarity has an important effect during the chromatographic process.

The correlation of Snyder's solvent strength ε^0 with molecular dipolarity and polarizability (π^*) and the hydrogen-bond acidity (α) and the hydrogen-bond basicity (β) solvatochromic parameters for adsorption chromatography can be achieved, although most papers on solvatochromic parameters deal with reversed-phase systems [18].

4.2.7 Preliminary Work for Mobile Phase Selection

The choice of the mobile phase used for the separation of multicomponent mixtures is difficult, but it gives us useful analytical information. It depends on the nature of the compounds to be separated and the material on which the separation is to be carried out (see Figure 4.2). A general rule in making a choice of the mobile phase is to fit the solvent polarity to that of the compounds being separated [15,16,59]. There is a scale of polarities of different classes of compounds as follows: hydrocarbons < ethers < ketones < amines < alcohols < organic acids. In the case of nonpolar compounds, nonpolar solvents and adsorbents with high activity will be chosen. For polar compounds, polar solvents and adsorbents with low activity will be selected. Adsorption chromatography is the most suitable technique for the separation of compounds according to the type and the functionality. The solvents used in PLC are the same as those used for equivalent separations in TLC. In order to obtain the optimum mobile phase, some preliminary work is necessary such as testing by microcircular technique, selection of the solvent character, and selection by selectivity and strength of solvent.

4.2.7.1 Testing by Microcircular Technique

This consists in spotting the separation mixture at intervals on a chromatographic plate and then applying the selected solvents in the center of each spot by means of a thin capillary. The spots migrate radially, and different chromatographic behavior of spotted mixture can be observed [16,59].

The microcircular technique for the mobile phase selection is illustrated in Figure 4.10a. It can be observed that the solvent A3 is the most suitable for nonpolar samples and the solvent B3 is for polar samples.

4.2.7.2 Selection by Solvent Character

Because of the similarities between the separations achieved in PLC and TLC of a certain sample, it is often convenient to try out a variety of solvents from Snyder's groups on thin layers before selecting the most suitable one for use. This is shown in Figure 4.10b. Two or three solvents with better separations are selected, and then different binary or ternary mixtures are tried again till the desired results are obtained [60].

For optimum mobile phase selection we have to consider certain chemical characteristics besides the solvent strength [15,16]. From this point of view, the chromatographic solvents can be divided as follows:

- Inert solvents — they cannot form hydrogen bonds (hydrocarbons)
- Proton donors and acceptors from hydrogen bonds (alcohols, water, amines, organic acids, amides, etc.)
- Proton donors within hydrogen bonds (chloroform)
- Proton acceptors within hydrogen bonds (aldehydes, ketones, ethers, esters, and nitroderivatives)
- Proton donors (acids)
- Proton acceptors (bases)

According to this classification, some practical conclusions can be drawn for the direct relation "solvent–solute":

- Proton-acceptor solvents are preferred at the donor solute separation and *vice versa.*
- For obtaining a much better selectivity, it is necessary to use a mixture of three to five solvents instead of a simple one. This mixture must dissolve all the solutes. In some cases, "universal" mobile phases were obtained, modifying the ratio of solvents in order to have the desired separation.
- The mobile phase can vary according to the developing technique, resulting in fine separation (within the homologous series) or ordinary separation (of group compounds).

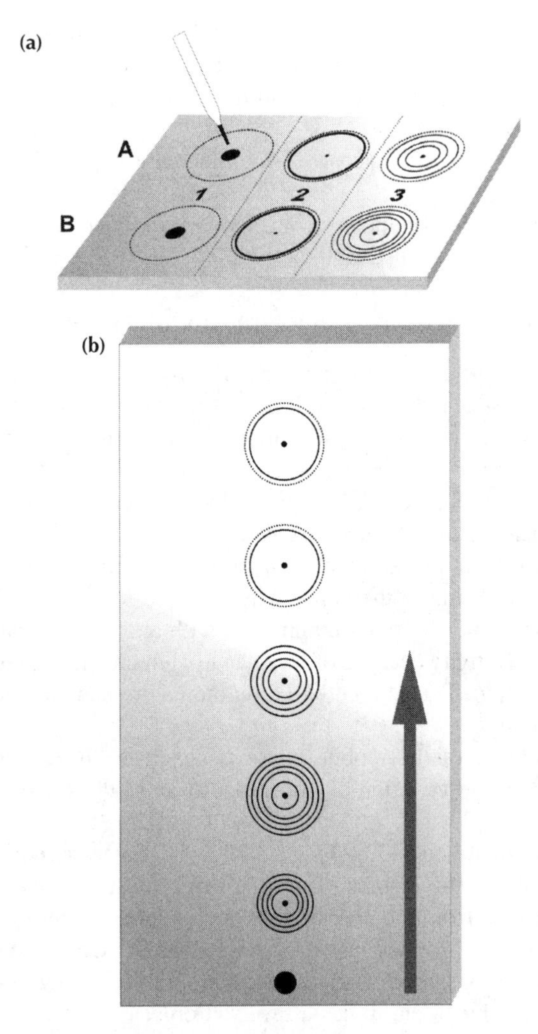

FIGURE 4.10 Mobile phase selection by microcircular technique. a. Sample of known composition: A = nonpolar compound; A1 = *n*-hexane; A2 = acetone; A3 = *n*-hexane–acetone, 60+40, v/v; B = polar compound; B1 = methanol; B2 = water; B3 = methanol–water, 70+30, v/v. b. Sample of unknown composition: testing with solvents of different Snyder's groups and binary solvent mixture.

In the multicomponent mobile phase, there are solvents with precise destinations:

- Training solvents in front of the mobile phase, which have high strength and are added in low quantities
- Breaking solvents, which have low strength and are in large quantities, being the main components
- Solvents for homogeneity, which are used when the first two solvent categories are not soluble

- pH correctors, which are used occasionally for the pH arrangement when some disturbing elements are present in the system
- Activators, which are used to avoid the spot compacting for producing a R_F variation
- Solvents for reducing viscosity, which increase migration velocity

In different chromatographic systems, some types of solvents can play as activator or breaking agents, respectively.

4.2.7.3 Selection by Selectivity and Strength of Solvent

Eluotropic series can be expressed more precisely by the Snyder parameter ε^0, given in Table 4.2 for alumina and silica gel. It is important to take into account eluotropic series and know some rules when solvents are used according to this table [16,17]:

- Increase in the Snyder parameter produces a decrease of the separation and an increase of R_F and reversed.
- The polar solvent proportion in multicomponent mixtures decreases exponentially according to the polarity increase.
- The substitution of a polar component with another one can improve or decrease selectivity because of the change of interactions in the chromatographic system. This can be based on the nature and chemical reactivity of the solute.
- If the mobility of the mobile phase is adequate, the selectivity can be improved by the variation of the composition of the solvent mixture.

The retention of a solute in TLC and PLC is characterized by the R_F value defined as the ratio of the distance from the origin to the center of the separated zone and the distance from the origin to the mobile phase front.

In reversed-phase thin-layer chromatography (RP-TLC), the choice of solvents for the mobile phase is carried out in a reversed order of strength, comparing with the classical TLC, which determines a reversed order of R_F values of compounds. The reversed order of separation assumes that water is the main component of the mobile phase. Aqueous mixtures of some organic solvents (diethyl ether, methanol, acetone, acetonitrile, dioxane, *i*-propanol, etc.) are used with good results.

The separation can be optimized by the alteration of the mobile phase composition. The more the polarity of solutes, the less the content of water of mobile phase must be. If water–alcohol mixture is used as mobile phase, the resolution can be improved by using alcohol with different chain lengths and different water proportions.

If an adequate resolution is not obtained, the mobile phase components should be changed for a better selectivity, preserving the same strength.

Solvent strength determines the R_F value, but not the selectivity. The mobile phase can be established by using the polarity index P' proposed by Snyder. The highest values of P' represent the strongest solute adsorbed in conventional TLC but represent the weakest for the separation in reversed phases. Sometimes aqueous polar mixtures cannot totally wet the chemically bonded layer. For this reason, checking the layer wetting is first recommended. Selectivity can be improved by adding 1 to 20% tetrahydrofuran, dimethyl sulfoxide, or dimethylformamide.

4.3 OPTIMIZATION OF THE MOBILE PHASE

The elaboration of the most efficient chromatographic systems for the optimization of velocity and resolution of the chromatographic process is necessary for solving different analytical problems. The most important factor in the TLC optimization is the mobile phase composition. Taking into consideration the similarity in the retention mechanism between TLC and PLC, the optimized TLC mobile phase can be transferred to the preparative chromatographic system. There are different accepted models and theories for the separation and optimization of chromatographic systems [19,20,61].

4.3.1 Thermodynamic Model

Adsorption, a surface phenomenon, is the basis of many gas or liquid mixture separation and purification methods. It is also the basis of adsorption chromatographic methods used for the analysis of complex mixtures. The knowledge of adsorption mechanisms is useful in choosing the suitable systems providing optimum separation.

Ościk's thermodynamic model is one of the first models describing chromatographic separation in adsorption chromatography [62,63]. His studies on adsorption are based on the thermodynamics of the adsorption process from solution, and the studies establish a relation between certain thermodynamic values and chromatographic parameters. Analyzing adsorption of the multicomponent mobile phases, Ościk obtained an equation for the relation between the data of an adsorption process from multicomponent mobile phases, emphasizing the case for binary mobile phases. A simple equation for R_M values of the solutes relates the adsorption affinity of these adsorbed compound from multicomponent mobile phases to the adsorption affinity of the compound from pure solvents. The relation between the adsorption affinity of a given solute and its R_M value for a multicomponent mobile phase is given by Equation 4.21 [62]:

$$R_{M\Sigma i} = \sum_{i}^{n} \varphi_i R_{Mi} + \sum_{i}^{n-1} \left(\varphi_i^s - \varphi_i\right)\left(\log k_{i,n}^{\infty} + R_{Mi} - R_{Mn}\right) + Y \quad (4.21)$$

where Y is a value that depends on the type of eluted solute and on the mobile phase components. For ideal mobile phases, $Y = 0$, and for the regular mobile phases, its values are not high and can be neglected.

On the basis of this relation, the direct application of TLC adsorption in investigations of the adsorption mechanism from mobile phases is possible [61–65].

For a binary mobile phase consisting of a mixture of solvents 1 and 2, the $R_{M1,2}$ value is given by Equation 4.22 [64]:

$$R_{M1,2} = \varphi_1 \Delta R_{M1,2} + \left(\varphi_1^s - \varphi_1\right)\left(\Delta R_{M1,2} + A_z\right) + R_{M2} + Y \quad (4.22)$$

where $R_{M1,2}$ represents the R_M value of solute z investigated in the binary mobile phase, and R_{M1} and R_{M2} are the R_M values of solute z in pure solvents 1 and 2, respectively; φ_1 and φ_1^s represent the volume (mole) fraction of the more active component of mobile phase (solvent 1 considered) in free state and adsorbed state on the adsorbent surface, respectively; $\Delta R_{M1,2} = R_{M1} - R_{M2}$ is the difference between the R_M values of the eluted compound in pure solvents 1 and 2; and $A_z = \log k_{1,2}^{\infty}$, where $k_{1,2}^{\infty}$ is the hypothetical partition coefficient of solute z between solvent 1 and solvent 2 of the binary mobile phase that represents the molecular interactions of solute z and the mobile phase components.

The R_M value is related to the logarithmic function of R_F value by Equation 4.23 [64,65]:

$$R_M = \log\left[\left(1 - R_F\right)/R_F\right] \tag{4.23}$$

The difference $\varphi_1^s - \varphi_1$ expresses the adsorption of the components of the binary mobile phase. It can be determined experimentally from the adsorption isotherms of the mobile phase components, and it is defined by the distribution function K_1 of the mobile phase components [64]:

$$K_1 = \left[\varphi_1^s\left(1-\varphi_1\right)\right]/\left[\varphi_1\left(1-\varphi_1^s\right)\right] \tag{4.24}$$

$$\varphi_1^s - \varphi_1 = \left[\varphi_1\left(K_1 - 1\right)\left(1-\varphi_1\right)\right]/\left[1+\left(K_1 - 1\right)\varphi_1\right] \tag{4.25}$$

For ideal or regular mixtures as mobile phases, it can be assumed that

$$\Delta R_{M1,2} = -\log K_1 \tag{4.26}$$

According to this relation, the distribution function K_1 can be estimated from the chromatographic data of a solute using pure solvents as the mobile phase. Equation 4.25 shows that the difference $\varphi_1^s - \varphi_1$ for each component of the mobile phase can be calculated without the adsorption isotherm data.

The adsorption mechanism in chromatography on alumina differs from that on silica gel because of the structural differences between these adsorbents. Relationships between the R_M values of solutes and the adsorption data for the mobile phase components on silica gel G and alumina G have been investigated by Różyło [64,65]. The theoretical and experimental results obtained by the relation $R_{M1,2} = f(\varphi_1)$ show a good agreement for the two adsorbents.

Ościk and Chojnacka [63] use TLC adsorption in the investigation of six aromatic hydrocarbons (naphthalene, diphenyl, anthracene, pyrene, chrysene, and acenaphthene) on silica gel G by elution with different binary mobile phases (trichloroethylene–benzene, carbon tetrachloride–benzene, *n*-heptane–trichloroethylene, and *n*-heptane–benzene). Assuming the ideal relationship, Equation 4.26, the $R_{M1,2}$ values were calculated according to Equation 4.22. In the chromatographic systems with trichloroethylene and benzene as the mobile phase the plots are linear, whereas in the system with *n*-heptane as the mobile phase component the relationship is not linear. The differences in the shape of the $R_{M1,2}$ vs. φ_1 plots for aromatic hydrocarbons can be explained by molecular interactions in the adsorption process. The R_M values calculated and measured were in good agreement, which confirms the applicability of Ościk's equation in the examination of interactions between the components of adsorption systems. Analysis of the equation parameters can give important data concerning these interactions.

The A_Z parameter can be calculated from chromatographic data by transforming the fundamental equation (Equation 4.22) and substituting values obtained from chromatograms for concentration 0.3, 0.5, and 0.7 of one of the components, or by transformation of Equation 4.22 to the linear form assuming that

$$A_Z + \Delta R_{M1,2} = C \tag{4.27}$$

Because many adsorbents have heterogeneous surfaces, Equation 4.22 should be modified by introducing of an additional coefficient, resulting in Equation 4.28, so that it can be used in systems employing such adsorbents:

$$R_{M1,2} = \varphi_1^{1/m} \Delta R_{M1,2} + \left(\varphi_1^s - \varphi_1\right)\left(A_Z + \Delta R_{M1,2}\right) + R_{M2} + \Delta H^a \tag{4.28}$$

where m is the heterogeneity parameter with values between 0 and 1, and ΔH^a is the difference between the excess enthalpies of mixing of the surface adsorbed and free (nonadsorbed) solutions if the chemical potential formula of a eluted compound X in the adsorbent is introduced. For ideal and regular solutions, ΔH^a is equal to zero, and for nonideal solutions, it is small and negligible [61]. The Equation 4.28 offers the possibility of using a thermodynamic model for optimization of chromatographic systems, especially those with active mobile phases and heterogeneous adsorbent surfaces.

4.3.2 Snyder–Soczewiński Model

Snyder and Soczewiński created and published, at the same time, another model called the S–S model describing the adsorption chromatographic process [19,61]. This model takes into account the role of the mobile phase in the chromatographic separation of the mixture. It assumes that in the chromatographic system the whole surface of the adsorbent is covered by a monolayer of adsorbed molecules of the mobile phase and of the solute and that the molecules of the mobile phase components occupy sites of identical size. It is supposed that under chromatographic process conditions the solute concentrations are very low, and the adsorption layer consists mainly of molecules of the mobile phase solvents. According to the S–S model, intermolecular interactions are reduced in the mobile phase but only for the chromatographic systems containing mobile phase and solutes of low polarity without hydrogen bonds.

Retention of solute molecules (*X*) is performed by the equilibrium displacement of mobile phase molecules (*M*) from the adsorbed monolayer. There is a competition between solute and solvent molecules for a site on the adsorbent surface in a liquid–solid adsorption process, which is expressed by the Equation 4.29 [13]:

$$X_{\text{liq}} + nM_{\text{ads}} \leftrightarrow X_{\text{ads}} + nM_{\text{liq}} \tag{4.29}$$

where the subscripts "liq" and "ads" denote the solute molecules in free state in the mobile phase and in the adsorbed state, respectively, and *n* is a coefficient that shows the number of mobile phase molecules replaced by a solute molecule at the adsorption moment.

The S–S adsorption model assumes a linear relationship given by Equation 4.30:

$$R_{\text{M}} = R_{\text{M}}^{\text{Y}} - m \log x_{\text{polar}} \tag{4.30}$$

where R_{M}^{Y} is the R_{M} value obtained experimentally for a given solute (*Y*) in pure strong polar modifier; *m*, a coefficient calculated by fitting the experimental data; and x_{polar}, the volume fraction of the polar modifier in the binary mobile phase.

4.3.3 PRISMA Model

The PRISMA model is a system for the optimization of two- to five-component mobile phases, developed by Nyiredy et al. to simplify the optimization process in different planar and column chromatographic systems [66]. This model for the selection of solvents and optimization of the mobile phase was developed first for TLC and high-performance liquid chromatography (HPLC) [38,67].

The model has the advantage that it requires only a simple table containing the polarity index P' and selectivity x_i group for a number of solvents (Table 4.2). The model is based on Snyder's classification of solvents [41,42] according to their characteristics to interact as proton acceptors (x_e), proton donors (x_d), or dipoles (x_n).

The PRISMA model is a three-dimensional geometric design (Figure 4.11) correlating solvent strength (S_T) with mobile phase selectivity. It consists of three parts: an irregular top part, the so-called frustum; a regular middle part; and the basic part, namely the platform. The basic part represents the modifiers that can be added in low and constant concentration (0.1 to 0.3%) to improve the separation and to reduce tailing. The regular part is used for mobile phase optimization of nonpolar and weakly polar compounds in normal phase chromatography. The frustum is used to optimize the separation of polar compounds [20]. The PRISMA model works well for reversed-phase systems too [68]. The length of each edge of the prism (S_A, S_B, and S_C) represents the strength of the pure solvents A, B, and C, respectively [18,20,61,68]. Mobile phase composition is characterized by the solvent strength and the selectivity points (Ps). By cutting the prism at the level of the lowest

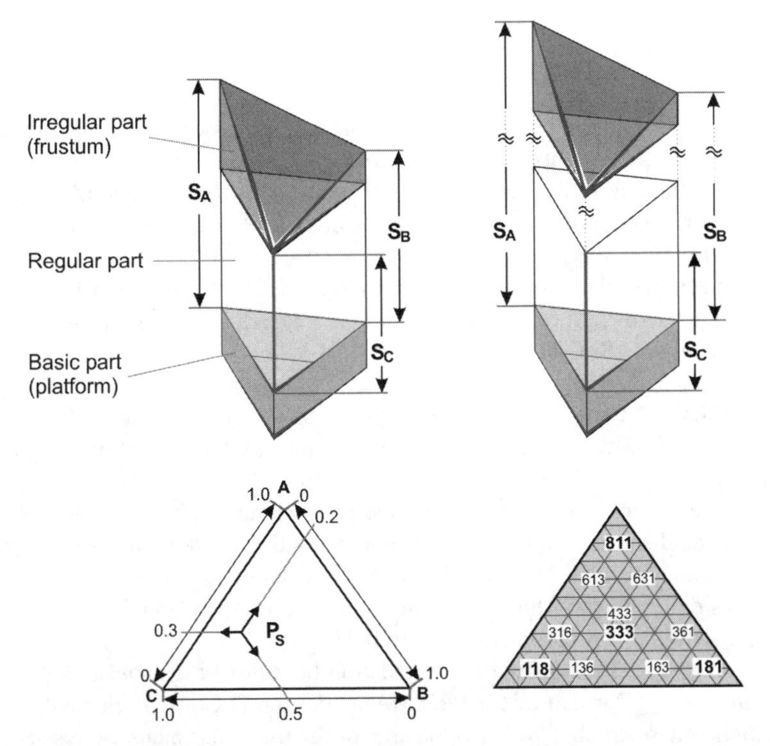

FIGURE 4.11 The PRISMA model. (Adapted from Siouffi, A.-M. and Abbou, M., Optimization of the mobile phase, in *Planar Chromatography, A Retrospective View for the Third Millennium*, Nyiredy, Sz., Ed., Springer Scientific, Budapest, 2001, chap. 3. With permission.)

edge S_C (weakest solvent) along a plane parallel to the platform, two parts, one irregular and the other regular, are obtained.

The top and the planes of the regular prism are parallel equilateral triangles representing weaker solvents diluted with a suitable solvent to adjust the strength. The volume fraction φ of diluted solvents of the mobile phase can be represented by the selectivity points Ps as three digit numbers of which the sum is 10 obtained by multiplying the volume fractions by 10 and arranging them in diminishing solvent strength order (Figure 4.11). The four basic selectivity points are 333, 811, 181, and 118. Optimization starts with the solvent mixture of the center point Ps = 333. In order to have R_F values in the range of 0.2 to 0.8 for all sample compounds, this composition is diluted. In the case of an insufficient separation other selectivity points (631, 361, 163, 136, 316, 613, 433, 343, and 334) are tested.

Horizontal and vertical correlations of hR_F values of nonpolar compounds and the selectivity points at different levels of the solvent strength using saturated TLC systems were given by Nyiredy et al. [18,67] applying the PRISMA model:

$$hR_F = a\left(P_S\right)^2 + b\left(P_S\right) + c \tag{4.31}$$

Mobile phase optimization by the PRISMA system follows the steps [20,38,66,69,70]:

- Selection of basic parameters of chromatographic process (stationary phase, proper pure solvents according to Snyder's classification, and vapor phase)
- Investigation of the chromatographic system using the PRISMA model
- Determination of the optimum composition of the mobile phase using the PRISMA model
- Selection of the development mode and of the chromatographic technique
- Transfer of the optimized TLC mobile phase to other chromatographic techniques (PLC in our case)

The choice of the stationary phase depends on the type of solutes. Silica gel is usually used if there are no special requirements as far as stationary phase is concerned.

Morita et al. [69] optimized the mobile phase composition using the PRISMA model for rapid and economic determination of synthetic red pigments in cosmetics and medicines. The PRISMA model has been effective in combination with a super modified simplex method for facilitating optimization of the mobile phase in high performance thin layer chromatography (HPTLC).

Pelander et al. [71] studied the retardation behavior of cyanobacterial hepatotoxins in the irregular part of the PRISMA model for TLC at 16 selectivity points. The mobile phase combination and the area of the triangular plane were selected in the preassay. The retardation of all the toxins followed the relation for hR_F. The cyanobacterial hepatotoxins behaved predictably in the selected systems in the irregular part of the PRISMA model.

Cimpoiu et al. [72] made a comparative study of the use of the Simplex and PRISMA methods for optimization of the mobile phase used for the separation of a group of drugs (1,4-benzodiazepines). They showed that the optimum mobile phase compositions by using the two methods were very similar, and in the case of polar compounds the composition of the mobile phase could be modified more precisely with the Simplex method than with the PRISMA.

4.3.4 Window Diagram Method

The Window diagram method for the optimization of separation was developed by Laub and Purnell [73], and it has been used both for gas chromatography and HPLC. Recently it is applied in TLC and HPTLC [19,74–76].

Equation 4.32 expresses the difference between the retention parameters ΔR_F of two solutes as a chromatographic response function:

$$\Delta R_F\left[\left(k_2'-k_1'\right)/\left(1+k_2'\right)\left(1+k_1'\right)\right] \tag{4.32}$$

where the capacity factors k_1' and k_2' are given as logarithmic functions of the volume fraction of the polar solvent in the mobile phase system. Plotting ΔR_F vs. the mobile phase composition and taking into consideration all peak pairs, the window diagram is obtained for the identification of the optimum solvent composition [74].

Prus and Kowalska [75] dealt with the optimization of separation quality in adsorption TLC with binary mobile phases of alcohol and hydrocarbons. They used the window diagrams to show the relationships between separation selectivity α and the mobile phase composition (volume fraction x_1 of 2-propanol) that were calculated on the basis of equations derived using Soczewiński and Kowalska approaches for three solute pairs. At the same time, they compared the efficiency of the three different approaches for the optimization of separation selectivity in reversed-phase TLC systems, using RP-2 stationary phase and methanol and water as the binary mobile phase. The window diagrams were performed presenting plots of α vs. volume fraction x_1 derived from the retention models of Snyder, Schoenmakers, and Kowalska [76].

The Window diagram method is seldom used for ternary or quaternary mobile phases because of a large variety of intermolecular interactions that appear in such systems.

The advantage of the window diagram method is that the optimum mobile phase composition can be easily located visually or using a computer.

4.3.5 Computer-Assisted Methods

The experimental results obtained in the laboratory by the researchers can be monitored using computer programs with help of empirical equations or models. Most of the computer-assisted procedures have been developed for HPLC separations and mainly for RPLC, and some of them are commercially available.

Computer-assisted techniques for optimizing HPTLC have been available for a number years and are well presented in the literature [20,77–82]. The strategy of integrating these techniques is to solve each separation problem in the simplest way possible.

Kiridena and Poole [80] demonstrated the use of retention maps for computer-aided-method development for the separation of a mixture of analgesic compounds.

Pelander et al. [81] developed a computer program for optimization of the mobile phase composition in TLC. They used the desirability function technique combined with the PRISMA model to enhance the quality of TLC separation. They applied the statistical models for prediction of retardation and band broadening at different mobile phase compositions; they obtained using the PRISMA method the optimum mobile phase mixtures and a good separation for cyanobacterial hepatotoxins on a normal phase TLC plate and for phenolic compound on reversed-phase layers.

Coman et al. [82] used a new modeling of the chromatographic separation process of some polar (hydroxy benzo[*a*]pyrene derivatives) and nonpolar (benzo[*a*]pyrene, dibenz[*a,h*]anthracene, and chrysene) polycyclic aromatic compounds in the form of third-degree functions. For the selection of the optimum composition of the benzene–acetone–water mobile phase used in the separation of eight polycyclic aromatic compounds on RP-TLC layers, some computer programs in the GW-BASIC language were written.

4.3.6 Chemometric Methods

The optimization of chemical system using mathematical methods has become widespread in the chemical sciences [20,77,83].

The discriminating power (*DP*) of a set of chromatographic systems is defined as the probability of identifying two randomly selected solutes in at least one of the systems. For k chromatographic systems in which N compounds are investigated, the *DP* is given by Equation 4.33 [18]:

$$DP = 1 - 2M / N(N-1) \tag{4.33}$$

where M is the total number of matching pairs.

Principal component analysis (PCA) is a statistical method having as its main purpose the representation in an economic way the location of the objects in a reduced coordinate system where only p axes instead of n axes corresponding to n variables ($p < n$) can usually be used to describe the data set with maximum possible information [20,76,83]:

$$P_j = a_{j1}x_1 + a_{j2}x_2 + ... + a_{jn}x_n \tag{4.34}$$

where x_i represents the original variables and the coefficients a_{j1}, a_{j2}, … , a_{jn} give an indication of the relative importance of the corresponding raw variable in the factor.

Two or three principal components usually provide a good summary of all original variables.

Using this procedure is analogous to finding a set of orthogonal axes that represents the directions of greatest variety in the data. In PCA one considers each row in the data matrix to be a point in multidimensional space with coordinates defined by the values corresponding to the appropriate n columns in the data matrix.

Bota et al. [84] used the PCA method to select the optimum solvent system for TLC separation of seven polycyclic aromatic hydrocarbons. Each solute is treated as a point in a space defined by its retention coordinates along the different solvent composition axes. The PCA method enables the selection of a restricted set of nine available mobile phase systems, and it is a useful graphical tool because scatterplots of loading on planes described by the most important axes will have the effect of separating solvent systems from one other most efficiently.

Betti et al. [85] present a chemometric approach for the analysis of flavonoids and related compounds by planar chromatography.

Soczewiński and Tuzimski [86] performed a chemometric characterization of the R_F values of pesticides for TLC systems.

4.4 CONCLUSION

PLC is a routine practice in many organic synthesis laboratories and is very frequently used. For preparative purposes, the solvent can migrate through the stationary phase by capillary action. Preparative work is carried out in the same way as for TLC. The mobile phase system in PLC is selected after preliminary experiments on analytical plates. Therefore, TLC can be regarded as a pilot method for PLC.

Snyder's classification of solvent properties is important in the selection of the chromatographic conditions and the optimization of the chromatographic processes.

It is important to know the influence of the physicochemical parameters of the mobile phase (dipole moment, dielectric constant, and refractive index) on solvent strength and selectivity. The main interactions in planar chromatography between the molecules of the mobile phases and those of solutes are caused by dispersion forces related to the refractive index, dipole–dipole forces related to the dipole moment, induction forces related to a permanent dipole and an induced one, hydrogen bonding, and dielectric interactions related to the dielectric constant. Solvent strength depends mainly on the dipole moment of the mobile phase, whereas the solvent selectivity depends on the dielectric constant of the mobile phase.

On the basis of Snyder's system for characterization of solvents the PRISMA method for mobile phase optimization has been developed. This system enables the optimization of solvent strength and mobile phase selectivity and also the transfer of the optimized mobile phase to different planar chromatographic techniques, in our case the PLC.

Procedures used vary from trial-and-error methods to more sophisticated approaches including the window diagram, the simplex method, the PRISMA method, chemometric method, or computer-assisted methods. Many of these procedures were originally developed for HPLC and were applied to TLC with appropriate changes in methodology. In the majority of the procedures, a set of solvents is selected as components of the mobile phase and one of the mentioned procedures is then used to optimize their relative proportions. Chemometric methods make possible to choose the minimum number of chromatographic systems needed to perform the best separation.

For practical purposes, it is most suited to combine the analyst's practice with computer assistance, in this context the PRISMA model being very efficient.

All the advantages of these methods for the optimization of the mobile phase by means of preassays in TLC can be exploited at the preparative scale. Finally, the separated zones may be easily removed from PLC plate and eluted in order to isolate quantities of the expected compounds. In PLC selection of the mobile phase, the subsequent recovery of the separated zone should be taken into consideration also.

In its turn, PLC can be used as a pilot technique for column preparative chromatography with the same system of mobile and stationary phases.

REFERENCES

1. Nyiredy, Sz., Preparative layer chromatography, in *Handbook of Thin-Layer Chromatography,* 3rd ed., revised and expanded, Chromatographic Science Series, 89, Sherma, J. and Fried, B., Eds., Marcel Dekker, New York, 2003, chap. 11.

2. Nyiredy, Sz., Possibilities of preparative planar chromatography, in *Planar Chromatography, A Retrospective View for the Third Millennium*, Nyiredy, Sz., Ed., Springer Scientific, Budapest, 2001, chap. 20.
3. Soczewiński, E. and Wawrzynowicz, T., Preparative TLC, in *Encyclopedia of Chromatography*, Marcel Dekker, New York, 2001, pp. 660–662.
4. Touchstone, J.C. and Dobbins, M.F., Preparative thin layer chromatography, in *Practice of Thin Layer Chromatography*, John Wiley & Sons, New York, 1978, chap. 12.
5. Nyiredy, Sz., Ed., *Planar Chromatography, A Retrospective View for the Third Millennium*, Springer Scientific, Budapest, 2001.
6. Geiss, F., *Fundamentals of Thin Layer Chromatography (Planar Chromatography)*, Hüthig, New York, 1987.
7. Zweig, G. and Sherma, J., Eds., Thin layer chromatography, in *Handbook of Chromatography*, II, CRC Press, a Division of the Chemical Rubber Co., Cleveland, 1972, chap. I.V.
8. Grinberg, N., Preparative thin-layer chromatography, in *Modern Thin-Layer Chromatography*, Chromatographic Science Series, 52, Grinberg, N., Ed., Marcel Dekker, New York, 1990, pp. 427–434.
9. Poole, C.F. and Poole, K.S., *Anal. Chem.*, 61, 1257a–1269a, 1989.
10. Liteanu, C. and Gocan, S., Optimizarea procesului cromatografic (Optimization of chromatographic process), in *Cromatografia de lichide cu gradienţi (Liquid Chromatography with Gradients)*, Technical Publishing House, Bucharest, 1976, chap. 5.
11. Hahn-Deinstrop, E., *J. Planar Chromatogr.*, 5, 57–61, 1992.
12. Stahl, E., *Angew. Chem.*, 73, 646–654, 1961.
13. Gocan, S., Mobile phase in thin-layer chromatography, in *Modern Thin-Layer Chromatography*, Grinberg, N., Ed., Chromatographic Science Series, 52, Marcel Dekker, New York, 1990, chap. 3.
14. Lederer, E. and Lederer, M., Elution, in *Chromatography (A Review of Principles and Applications)*, Elsevier, Amsterdam, 1957, chap. 4.
15. Pelloni-Tamaş, V. and Johan, F., Developanţi (dizolvanţi şi amestecuri de dizolvanţi folosiţi în cromatografia pe strat subţire (Eluents — solvents and solvent mixtures — used in thin layer chromatography), in *Cromatografia în Strat Subţire (Thin-Layer Chromatography)*, Technical Publishing House, Bucharest, 1971, chap. 4.
16. Bodoga, P., Măruţoiu, C., and Coman, M.-V., Faze mobile (mobile phases), in *Cromatografia pe Strat Subţire. Analiza Poluanţilor (Thin-Layer Chromatography. Pollutant Analysis)*, Analytical chemistry series, Technical Publishing House, Bucharest, 1995, chap. 3 and references cited therein.
17. Touchstone, J.C. and Dobbins, M.F., The mobile phase, in *Practice of Thin Layer Chromatography*, John Wiley & Sons, New York, 1978, chap. 5.
18. Siouffi, A.-M. and Abbou, M., Optimization of the mobile phase, in *Planar Chromatography, A Retrospective View for the Third Millennium*, Nyiredy, Sz., Ed., Springer Scientific, Budapest, 2001, chap. 3.
19. Kowalska, T., Kaczmarski, K., and Prus, W., Theory and mechanism of thin-layer chromatography, in *Handbook of Thin-Layer Chromatography*, 3rd ed., revised and expanded, Chromatographic Science Series, 89, Sherma, J. and Fried, B., Eds., Marcel Dekker, New York, 2003, chap. 2.
20. Cimpoiu, C., Optimization, in *Handbook of Thin-Layer Chromatography*, 3rd ed., revised and expanded, Chromatographic Science Series, 89, Sherma, J. and Fried, B., Eds., Marcel Dekker, New York, 2003, chap. 3.
21. Guiochon, G. and Siouffi, A., *J. Chromatogr. Sci.*, 16, 598–609, 1978.
22. Poole, C.F., *J. Planar Chromatogr.*, 2, 95–98, 1989.

23. Poole, S.K., Ahmed, H.D., Belay, M.T., Fernando, W.P.N., and Poole, C.F., *J. Planar Chromatogr.*, 3, 133–140, 1990.
24. Héron, S. and Tchapla, A., *Analusis*, 21, 327–347, 1993.
25. Botz, L., Nyiredy, S., and Sticher, O., *J. Planar Chromatogr.*, 3, 10–14, 1990.
26. Lide, D.R. and Frederikse, H.P.R., Eds., Fluid properties, in *CRC Handbook of Chemistry and Physics*, CRC Press, Boca Raton, FL, 1995, chap. 6.
27. Snyder, L.R., *J. Chromatogr.*, 63, 15–44, 1971.
28. Snyder, L.R., *Principles of Adsorption Chromatography*, Marcel Dekker, New York, 1968.
29. Snyder, L.R., *J. Chromatogr.*, 13, 415–434, 1964.
30. Snyder, L.R., *J. Chromatogr.*, 8, 178–200, 1962.
31. Snyder, L.R., *J. Chromatogr.*, 16, 55–88, 1964.
32. Snyder, L.R., *J. Chromatogr.*, 11, 195–227, 1963.
33. Janjić, T.J., Vučković, G., and Ćelap, M.B., *J. Serb. Chem. Soc.*, 65, 725–731, 2000.
34. Janjić, T.J., Vučković, G., and Ćelap, M.B., *J. Serb. Chem. Soc.*, 66, 671–683, 2001.
35. Janjić, T.J., Vučković, G., and Ćelap, M.B., *J. Serb. Chem. Soc.*, 67, 179–186, 2002.
36. Janjić, T.J., Vučkovic, G., and Ćelap, M.B., *J. Serb. Chem. Soc.*, 67, 481–487, 2002.
37. Cserháti, T., *J. Chromatogr. Sci.*, 31, 220–224, 1993.
38. Poole, C.F. and Dias, N.C., *J. Chromatogr. A*, 892, 123–142, 2000.
39. Scott, R.P.W., *J. Chromatogr.*, 122, 35–53, 1976.
40. Kowalska, T. and Klama, B., *J. Planar Chromatogr.*, 10, 353–357, 1997.
41. Snyder, L.R., *J. Chromatogr.*, 92, 223–230, 1974.
42. Snyder, L.R., *J. Chromatogr. Sci.*, 16, 223–234, 1978.
43. Rohrschneider, L., *Anal. Chem.*, 45, 1241–1247, 1973.
44. Snyder, L.R., *J. Chromatogr.*, 25, 274–293, 1966.
45. Kooling, O.W., *J. Phys. Chem.*, 96, 6217–6220, 1992.
46. Johnson, B.P., Khaledi, M.G., and Dorsey, J.G., *Anal. Chem.*, 58, 2354–2365, 1986.
47. Cheong, W.J. and Carr, P.W., *Anal. Chem.*, 61, 1524–1529, 1989.
48. Kosower, E.M., *J. Am. Chem. Soc.*, 80, 3253–3260, 1958.
49. Kosower, E.M. and Mohammad, M., *J. Am. Chem. Soc.*, 90, 3271–3272, 1968.
50. Reichardt, C., *Chem. Rev.*, 94, 2319–2358, 1994.
51. Kamlet, M.J., Abboud, J.L.M., Abraham, M.H., and Taft, R.W., *J. Org. Chem.*, 48, 2877–2887, 1983.
52. Kamlet, M.J., Abboud, J.L., and Taft, R.W., *J. Am. Chem. Soc.*, 99, 6027–6038, 1977.
53. Laurence, C., Nicolet, P., Dalati, M.T., Abboud, J.L.M., and Notario, R., *J. Phys. Chem.*, 98, 5807–5816, 1994.
54. Taft, R.W. and Kamlet, M.J., *J. Am. Chem. Soc.*, 98, 2886–2894, 1976.
55. Kamlet, M.J. and Taft, R.W., *J. Am. Chem. Soc.*, 98, 377–383, 1976.
56. Pytela, O., *Collect. Czech. Chem. Commun.*, 53, 1333–1423, 1987.
57. Bekárek, V. and Juřina, J., *Collect. Czech. Chem. Commun.*, 47, 1060–1068, 1982.
58. Qasimullah, A.A., Andrabi, S.M.A., and Qureshi, P.M., *J. Chromatogr. Sci.*, 34, 376–378, 1996.
59. Abbott, D. and Andrews, R.S., *An Introduction to Chromatography*, Longmans, Green, London, 1965, p. 27.
60. Bauer, K., Gros, L., and Sauer, W., *Thin Layer Chromatography — An Introduction*, Merck booklet, 1991, p. 20.
61. Różyło, J.K. and Siembida, R., *J. Planar Chromatogr.*, 10, 97–107, 1997.
62. Ościk, J. and Chojnacka, G., *J. Chromatogr.*, 93, 167–176, 1974.
63. Ościk, J. and Chojnacka, G., *Chromatographia*, 11, 731–735, 1978.
64. Różyło, J.K., *J. Chromatogr.*, 116, 117–124, 1976.

65. Różyło, J.K., *J. Chromatogr.*, 93, 177–182, 1974.
66. Nyiredy, Sz., Dallenbach-Tölke, K., and Sticher, O., *J. Planar Chromatogr.*, 1, 336–342, 1988.
67. Nyiredy, Sz. and Fatér, Z., *J. Planar Chromatogr.*, 8, 341–345, 1995.
68. Reich, E. and George, T., *J. Planar Chromatogr.*, 10, 273–280, 1997.
69. Morita, K., Koike, S., and Aishima, T., *J. Planar Chromatogr.*, 11, 94–99, 1998.
70. Nyiredy, Sz., Systematic method development in classical and forced-flow planar chromatography, in *Proc. Int. Symp. on Planar Separations, Planar Chromatography 2001*, Nyiredy, Sz., Ed., Research Institute for Medicinal Plants, Budakalász, 2001, pp. 137–148.
71. Pelander, A., Sivonen, K., Ojanperä, I., and Vuorela, H., *J. Planar Chromatogr.*, 10, 434–440, 1997.
72. Cimpoiu, C., Hodişan, T., and Naşcu, H., *J. Planar Chromatogr.*, 10, 195–199, 1997.
73. Laub, R.J. and Purnell, J.H., *J. Chromatogr.*, 122, 71–76, 1975.
74. Walters, F.H. and Deming, S.N., *Anal. Chim. Acta*, 167, 361–367, 1985.
75. Prus, W. and Kowalska, T., *J. Planar Chromatogr.*, 8, 205–215, 1995.
76. Prus, W. and Kowalska, T., *J. Planar Chromatogr.*, 8, 288–291, 1995.
77. Siouffi, A.M. and Phan-Tan-Luu, R., *J. Chromatogr. A*, 892, 75–106, 2000.
78. Nurok, D., Knotts, K.D., Kearns, M.L., Ruterbories, K.J., Uhegbu, C.E., and Alberti, P.C., *J. Planar Chromatogr.*, 5, 350–358, 1992.
79. Wang, Q.-S. and Yan, B.-W., *J. Planar Chromatogr.*, 6, 296–299, 1993.
80. Kiridena, W. and Poole, C.F., *J. Planar Chromatogr.*, 12, 13–25, 1999.
81. Pelander, A., Summanen, J., Yrjönen, T., Haario, H., Ojanperä, I., and Vuorela, H., *J. Planar Chromatogr.*, 12, 365–372, 1999.
82. Coman, V., Măruţoiu, C., and Puiu, S., *J. Chromatogr. A*, 779, 321–328, 1997.
83. Brown, S.D., Blank, T.B., Sum, S.T., and Weyer, L.G., *Anal Chem*, 66, 315R–359R, 1994.
84. Bota, A., Sârbu, C., Măruţoiu, C., and Coman, V., *J. Planar Chromatogr.*, 10, 358–361, 1997.
85. Betti, A., Lodi, G., and Fuzzati, N., *J. Planar Chromatogr.*, 6, 232–237, 1993.
86. Soczewiński, E. and Tuzimski, T., *J. Planar Chromatogr.*, 12, 186–189, 1999.

5 Sample Application and Chromatogram Development

Gertrud E. Morlock

CONTENTS

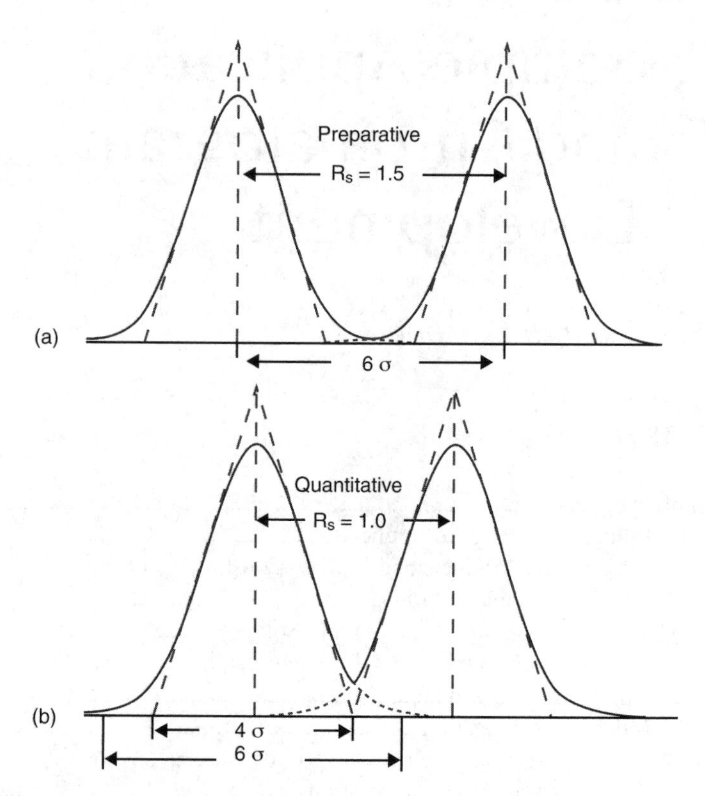

FIGURE 5.1 Required resolution for preparative purposes (R_s = 1.5) in contrast to that for quantitative purposes (R_s = 1.0).

5.1 SAMPLE APPLICATION

The highest separation standards are set for preparative separations because the resolution has to be at least 1.5 (6 σ), ensuring that peaks are separated completely (Figure 5.1). It is based on the fact that the whole substance peak is intended to be recovered to the highest degree of purity. In contrast, for quantitative purposes a resolution of 1.0 (4 σ) is sufficient, because then a peak is pure at its maximum and can be evaluated by peak height.

Because a sample application determines the achievable quality of the chromatographic result, its high-quality performance is a crucial precondition, especially for preparative separations. During a sample application three basic requirements must be met, i.e., the applied sample volume should be accurate with respect to the yield calculated later on, the layer must not be damaged during the application process, and the starting zone should be applied in as narrow a band as possible. Having in mind the two principal ways to transfer a sample onto the plate, i.e., contact spotting and spray-on application, only the latter one fits the purpose of transferring high volumes onto the layer (besides well-trained manual application as a band). For maximal separation power of the chromatographic system, the application zone should be as small as possible in the direction of chromatography. Depending on the volume and quantity of the substance to be applied, a zone width of 5 mm should not be exceeded on preparative layers. Automated application devices should be used, if possible, ensuring a narrower band width, thus improved resolution, and last but not least, ensuring homogeneous distribution of the sample over the entire length of the band as well. The latter aspect is much more relevant for aliquot scanning of bands used for quantitative determinations. However, also in preparative separations, local overloading due to inhomogeneous distribution of the sample can also distort the band applied during subsequent chromatography, as one strives for a compromise between maximum amount and resolution — a compromise that is susceptible to overloading.

For good manufacturing practice, some aspects have to be considered before application that involve the constituents of the sample solution: the property of the solvent used for dissolution, and the concentration of the solution applied onto the layer. It must be clear that the application pattern is completely different for preparative purposes in contrast to analytical separations. Manual application by well-trained analysts is especially helpful for highly concentrated solutions. Benefits of proper instrumentation are shown, and guidance is provided for choosing the proper instrument and crucial parameters that are involved.

5.1.1 Influence on Sample Application

A lot of emphasis is laid on three possible basic interferences during the application process: (1) single constituents of the sample solution crystallize out, (2) the property of the solvent used for dissolution, and (3) the concentration of the solution generating an unacceptably broad application zone. Such interferences could be avoided with proper planning before application.

5.1.1.1 By-Products and Further Constituents of the Sample Solution

In addition to the extracted or synthesized product, the reagents in excess, by-products, and further constituents of the sample solution are factors that have to be taken into account, because they all contribute to the concentration of the application solution. Further constituents, such as catalysts or buffer salts (for adjusting the pH value necessary to enable the intended reaction), can crystallize out during automated application and clog the syringe. Additionally, the sample solution can polymerize during storage, or even during the automated application process, and clog the syringe as well. In both cases, manual application accomplished by a focusing step should be preferred. If the sample crystallizes out at the stainless steel cone of the needle guiding of automated devices, it is recommended first to increase the solvent delivery rate involving less evaporation of the solvent. In most cases this step is successful. Otherwise, the solution has to be diluted or manual application should be performed.

5.1.1.2 Property of the Sample Solvent

Silica gel or alumina is usually used as the stationary phase. For these active phases a solvent of weak elution power is recommended for dissolution of the sample. A

FIGURE 5.2 Aqueous samples can cause breaking up and breaking off of the layer (left side) during application.

solvent of high volatility, which already evaporates during the application process to a great extent, improves the shape of the starting zone as well. Thus, the sample should preferably be dissolved in an apolar, volatile solvent such as *n*-hexane or ether.

Application of large volumes, especially of aqueous samples, can cause washing away of the silica gel layer (Figure 5.2). Special care has to be given to dosage speed, or the band has to be applied as (a rectangular) area, followed by an additional focusing step. Often, partial dissolution in an organic solvent, such as methanol–water (1:1), eases application greatly in contrast to pure aqueous solutions.

These aspects of solvent property similarly apply to precoated impregnated silica gel plates, e.g., by ammonium sulfate, silver nitrate, or magnesium acetate, as well as to microcrystalline cellulose precoated plates. On preparative RP phases, water has the lowest elution power. Therefore, more polar or aqueous solvents should be preferred. In contrast to HPTLC RP-18 layers, on which such aqueous solutions remain as a drop on the surface and are not able to penetrate through the lipophilic layer, on preparative RP phases, pure aqueous application solutions can be applied owing to the minor degree of C18 modification.

5.1.1.3 Concentration of the Solution

Amounts of 10 mg to 1 g are applied as a band onto the layer depending on the layer thickness, which can be between 0.5 and 2 mm for plates used for commercial purposes — layers of self-prepared preparative plates can even be up to 10-mm thick. However, the latter is tricky; qualification and experience are required for proper preparation of such thick layers. For example, 1 ml of a 1% solution is applied onto a 0.5-mm layer, resulting in 10 mg on the plate. In most cases, however, 50 to 150 mg samples are applied onto preparative layers, affording higher application volumes or more concentrated solutions. For the spray-on technique, the concentration is important because the components of the sample should be adsorbed throughout the entire thickness of the layer during application. Highly concentrated solutions, i.e., above 1% considering all substances dissolved, applied by automated spray-on technique, seem to adsorb mainly on the surface (most part of the solvent is evaporated) and generate sparkling spots over the sorbent area nearby (Figure 5.3).

Moreover, such solutions can also cause the sample to crystallize out in the syringe. In this case it is recommended to dilute the sample and apply larger volumes. Of course, a further possibility is the application of highly concentrated solutions manually with a pipette or, if the sample does not crystallize out in the syringe, semiautomated by the Alltech TLC sample streaker.

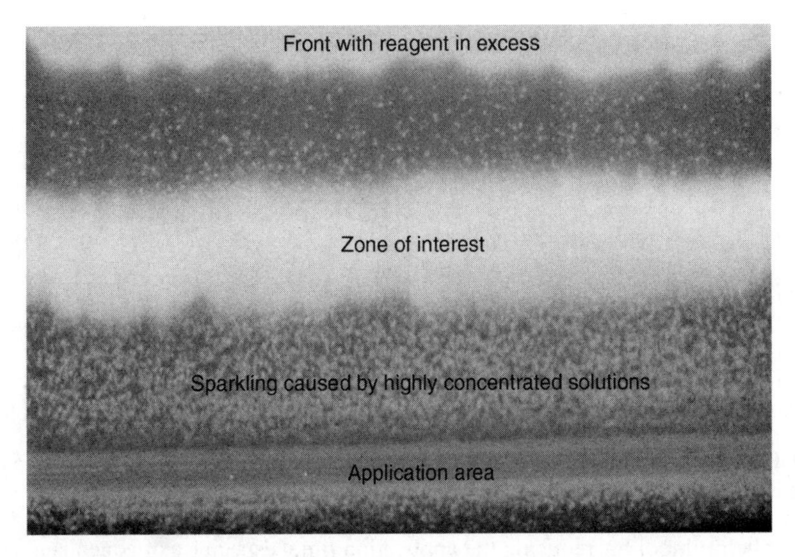

FIGURE 5.3 Sparkling spots over the sorbent area near the application area caused by highly concentrated solutions (in this case 1.5% solution).

5.1.2 Application Pattern

The application pattern for preparative purposes (Figure 5.4) is completely different compared to analytical ones. The analytical application pattern, for example, spots or bands up to 12 mm, is useless in this context. Samples are preferably applied as a narrow band over the whole plate length of 20 cm. The side distance from both edges is about 2 cm. Because of the edge effect, which can cause an uneven band bent up or down at the plate edges, these areas are left free or used for spotwise application of a control standard. Because certain software (CAMAG winCATS, Desaga AS 30 software) calculate the pattern from the distance between the center of application position and the left-hand edge of the plate position in the x-direction, 100 mm has to be edited as first application position for a band length of 160 mm, resulting in an edge distance of 20 mm each.

Large volumes of an aqueous sample should be applied slowly, by a heated spray nozzle or as a rectangular area. Figure 5.5 shows the scheme for an application area of 160-mm length times 10-mm width in contrast to standard 1-mm width. For area application higher dosage speeds can be selected, thus significantly reducing application time. Application by the heated spray nozzle (see automated devices) is a

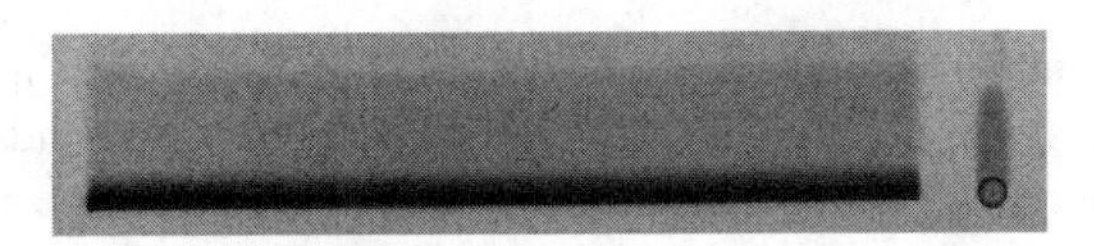

FIGURE 5.4 Typical application pattern for a preparative sample applied as band, if desired with application of a control standard (spot at right side), documentation of the lower part of the plate after chromatography at UV 254 nm.

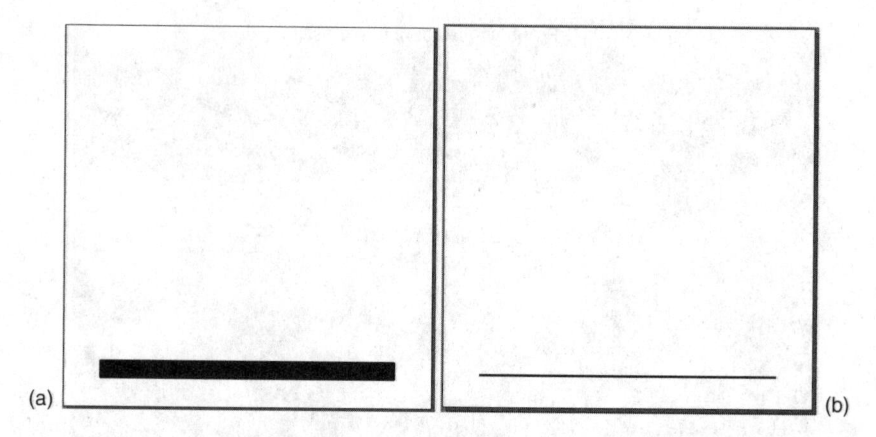

FIGURE 5.5 Application pattern for a large aqueous sample applied as 160 × 10 mm area (left) in contrast to the standard pattern of 160 × 1 mm (right).

further helpful tool for reducing the application time, especially of aqueous nonvolatile sample solutions.

5.1.3 Modes of Application

The plate used for application should be of a high degree of purity. It is recommended to prewash layers and after careful drying to keep them protected by a glass plate cover until chromatography (see Section 5.2.2.1). For best resolution, the application of the sample must be uniform with respect to the concentration and shape of the starting zone. Three main modes of application can be distinguished: (1) manual application, requiring the lowest level of equipment but highest personal qualification, (2) automated devices, generating the narrowest and most homogeneous application zone independent of personal qualification, and (3) solid phase sample application, which may be reasonable in certain cases. It is clearly understood that after application, the starting zone should be dried as soon as possible to avoid further substance distribution. A cold current of air is generally recommended if not otherwise indicated, e.g., for oxidation-prone or volatile samples. Care should be taken that asymmetrical evaporation of the solvent does not occur. Be aware that residual traces of polar application solvents can influence the chromatographic result.

5.1.3.1 Manual Application and Focusing

Practice, skill, and care are preconditions to apply the sample as a band manually using a simple syringe or pipette, especially well-trained hands are required, otherwise an inhomogeneous distribution of the sample over the entire length of the unacceptable band can clearly be seen (Figure 5.6). During the subsequent chromatography process, dissolution of the components in the mobile phase can be a limiting factor, and local overloading can distort the band applied and thus reduce resolution.

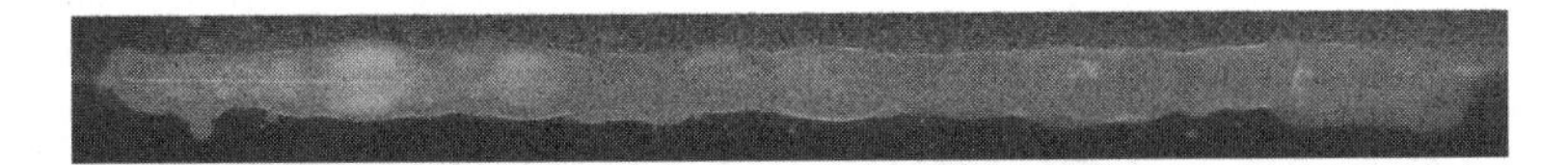

FIGURE 5.6 Example of improper manual application resulting in an inhomogeneous distribution of the sample over the entire length of the unacceptable band, documentation after chromatography at UV 366 nm.

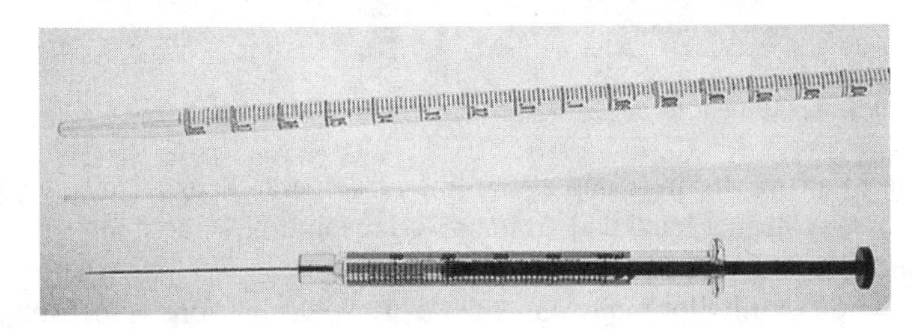

FIGURE 5.7 Equipment for manual application, e.g., 500-μl Hamilton syringe (lower), Pasteur pipette (middle), and 2-ml measuring pipette (upper).

Some equipment for manual application is shown in Figure 5.7. A Hamilton syringe of 500 μl, or a measuring or volumetric pipette of, say, 1 or 2 ml, with Peleus ball, or a Pasteur pipette with an aspirating bulb can be employed.

The pipette or syringe is carefully moved over an imaginary line without touching the layer surface (Figure 5.8). Depending on the volume to be applied, this procedure has to be performed several times so that the lines perfectly overlay each other. For highly polar application solvents, intermediate drying steps keep the starting band narrower. In case of contact with the layer, the use of a Teflon tip on the end of the syringe avoids damage of the layer.

A very helpful tool for manual application can be the employment of layers with a concentrating zone. The so-called concentrating or preadsorbent zone is a small part of the plate that is covered with an inert but highly porous adsorbent such as diatomaceous earth. Various precoated preparative layers with a preadsorbent zone are commercially available. The effect of the concentrating zone is depicted elsewhere in detail (see Chapter 3, Figure 3.4). In brief, the preadsorbent zone serves as a platform for manual application of any desired performance quality. When development starts, soluble components migrate with the mobile phase front and are

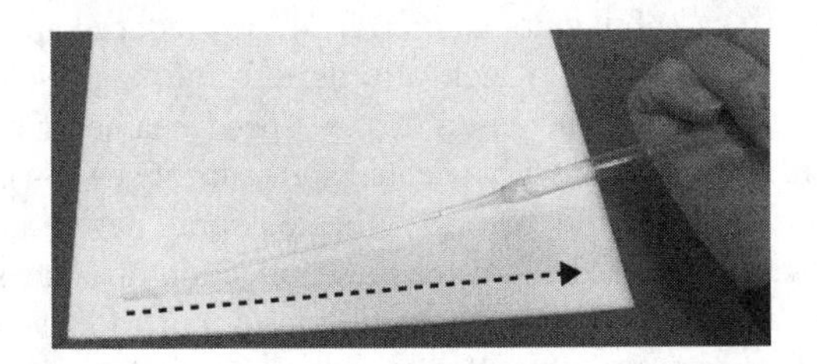

FIGURE 5.8 Manual application by carefully moving the Pasteur pipette over an imaginary line without touching the layer.

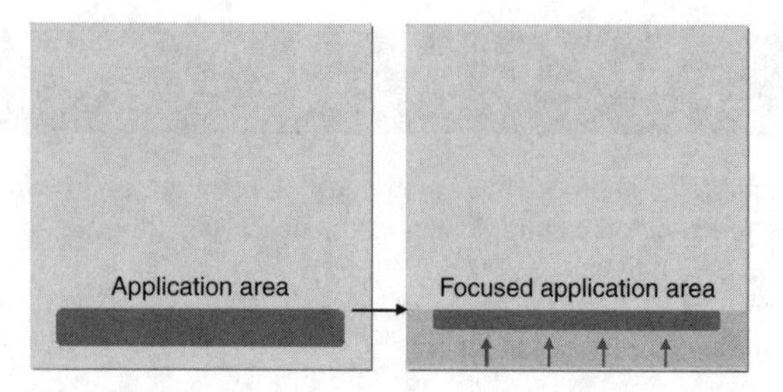

FIGURE 5.9 Improvement of a starting zone by focusing.

focused in a narrow starting band at the line of sorbent change. Thus, such layers enable a narrow starting band and an improved resolution without any employment of instrumentation.

A further possibility for improvement of a given starting zone is focusing. Before chromatography, a broad starting zone can be focused by development of the layer in a solvent of high elution power, i.e., methanol on silica gel layers. The development is stopped when the lower part of the substance zone has just migrated over the upper part (Figure 5.9). Normally, this lasts not longer than a minute. Drying of the plate follows. This focusing technique is also indispensable if the sample has been sprayed on as an area.

5.1.3.2 Automated Application

Availability of modern instrumentation [1] has unlocked the full power of planar chromatography. Especially in the field of preparative chromatography, a good resolution between two zones is of great importance, as shown at the beginning. However, a prerequisite is a compact, narrow starting zone because starting zones sprayed on as narrow bands ensure the highest resolution attainable with any given planar chromatographic system. If high volumes are applied, commercially available spray-on applicators apply bands of less than 3-mm width. Using a controlled stream of nitrogen, the sample is softly sprayed from a syringe onto the layer as a homogeneous band over the whole plate length, thus requiring no direct contact with the layer. Instruments with different levels of automation can be selected. Concerning the final result of generating a narrow starting zone, they are considered to be equal in performance. The decision might just be one of the following such as (1) cost-efficiency (Linomat 5), (2) full automation (ATS4), (3) reduced application time due to the use of a 500-μl syringe, which affords less refilling cycles (TLC sample streaker, Linomat 5), and (4) increased dosage speed enabled by application as a rectangular area in combination with a heated spray nozzle (ATS4). The application of samples as area or the employment of a heated spray nozzle are only available with the last device — a clear benefit for the application of polar sample solvents.

All in all, the application process typically lasts much longer than is usual for quantitative or qualitative purposes. If a sample volume of 1 ml is applied using a dosage speed of 500 nl/sec, which is the fastest dosage speed for Linomat 5, the application lasts about 33 min and 50 min for the fastest dosage speed of 3 sec/μl using AS 30. However, the highest dosage speed can be selected using the ATS4, i.e. 2.5 µl/sec. Thus, 1 ml could be applied in 7 min of pure application time plus time required for filling the syringe nine to ten times, totaling about 15 min. Of course, such a high dosage speed might afford spray-on as area and a subsequent focusing step for more polar application solvents. Regarding high dosage speed using ATS4, a further helpful option is the application of the sample by a heated spray nozzle. Most advantageous of all automated devices, except the Alltech TLC sample streaker, is the self-adjusting plate support. It allows the use of layers differing in thickness without any readjustment to the spray nozzle. This makes changing between conventional TLC and HPTLC layers, on glass plates or sheets and preparative layers, easy. This is a useful feature if several operators employing different layers want to use the same sample application device.

5.1.3.2.1 Semiautomatic Application

Semiautomatic devices suited for preparative purposes are the CAMAG Linomat 5, the Desaga HPTLC applicator AS 30, and the Alltech TLC sample streaker. For all devices, the syringe has to be filled manually with sample solution and rinsed after sample application. Except for the Alltech TLC sample streaker, each of these instruments can be employed either as software-controlled or as a stand-alone device. The former is more convenient for creation, editing, and saving of the application pattern and instrument parameters.

The simplest device for semiautomatic application is the Alltech TLC sample streaker (Figure 5.10). This device is especially designed for preparative layer chromatography. Large sample volumes precisely measured can be applied as a uniform band by a 250- or 500-µl syringe. The band applied is made continuous by forcing the syringe plunger downward as it is shifted along a sloping stainless steel bar. In other words, during forced sloping to the right, the syringe plunger is pressed downward. Thus, the liquid is forced out of the syringe onto the layer, accompanied by a lateral movement. Only as the hand moves the carriage sideward, the liquid is forced out per centimeter of travel, enabling constant application despite irregularities in hand motion. Without scratching the adsorbent surface, the sample is applied as a narrow band of 1-mm width. Time saved by semiautomatic application is

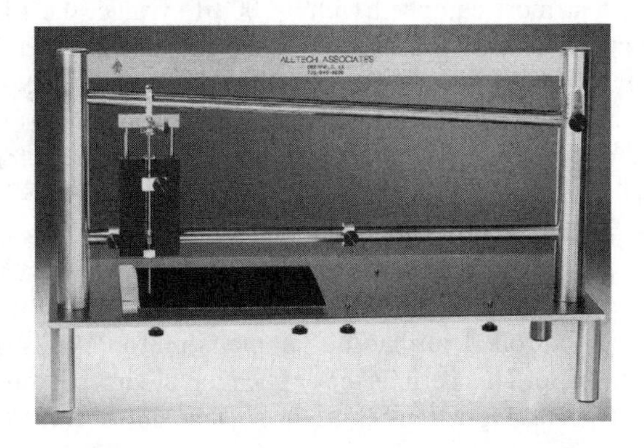

FIGURE 5.10 Alltech TLC sample streaker (semiautomatic).

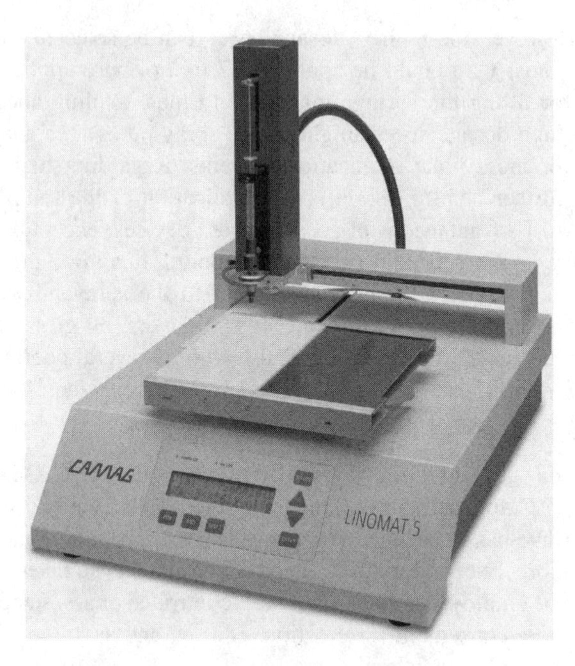

FIGURE 5.11 CAMAG Linomat 5 (semiautomatic).

obviously about 90% compared to manual application. However, adjustments of the device for variations in layer thickness, as well as verification of the guide rod and slope quality, add more time before application. This technique differs from the spray-on technique, which applies the solution as an aerosol. The spray-on technique is used by all other automated devices that are to be discussed later.

The CAMAG Linomat 5 (Figure 5.11) has been designed for the application of samples in the form of narrow bands. This device is perfectly suited for application of large volumes if the 500-μl syringe is employed instead of the 100-μl standard syringe. For sample volumes above 500 μl, the syringe has to be refilled and the application has to be started once more. However, in comparison to other automated devices, which are at most equipped with a 100-μl syringe, the Linomat 5 shows the most efficient application time owing to the availability of a 500-μl syringe, which affords fewer refilling cycles by a factor of five than a 100 μl one.

Particularly for preparative purposes, it is advantageous to start with a compact, narrow application zone because it guarantees optimum resolution for a given planar chromatographic system. Therefore, the instrument uses the spray-on technique for applying samples onto the chromatogram, i.e., the sample liquid is sprayed onto the layer from the tip of the syringe needle by means of compressed air or an inert gas (Figure 5.12). The tip of the needle is thereby about 1 mm above the layer, and the stage movement is controlled so that the sample is uniformly distributed over the entire length of the band. This permits the application of larger sample volumes than is possible with contact sample transfer, as the solvent almost completely evaporates during the process. Even when strongly polar solvents are used (e.g., methanolic or aqueous solutions), the application zones remain compact and narrow. For

FIGURE 5.12 Close-up view of the spray nozzle (CAMAG Linomat 5).

application of large volumes of such polar solvents, however, a very slow application rate is recommended. The slowest selectable rate for Linomat 5 is 30 nl/sec. By the way, such slow rates are not relevant for preparative purposes because application of 1 ml would take about 9 h. Reasonably slow application rates are those starting with 140 nl/sec, affording about 2 h for application. The possibility of applying a kind of manually generated area by just shifting the syringe tower by 1 mm (e.g., during syringe filling) is advantageous. Thus, the application zone can be improved for polar solvents.

Another device applicable for semiautomatic spray-on technique is the Desaga HPTLC applicator AS 30 (Figure 5.13). For application of large volumes, the 100-µl dosage syringe has to be mounted instead of the 10-µl standard syringe. A stepping motor moves the syringe plunger downward, and a gas stream transfers the sample from the needle tip onto the plate. A second stepping motor moves the syringe tower sideways across the plate. During spray-on application, there is no contact with the layer regardless of thickness, because of the advantage of a self-adjusting plate

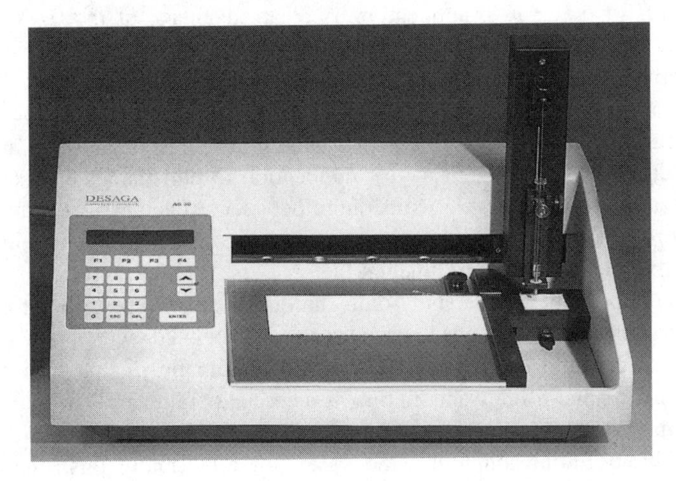

FIGURE 5.13 Desaga HPTLC applicator AS 30 (semiautomatic).

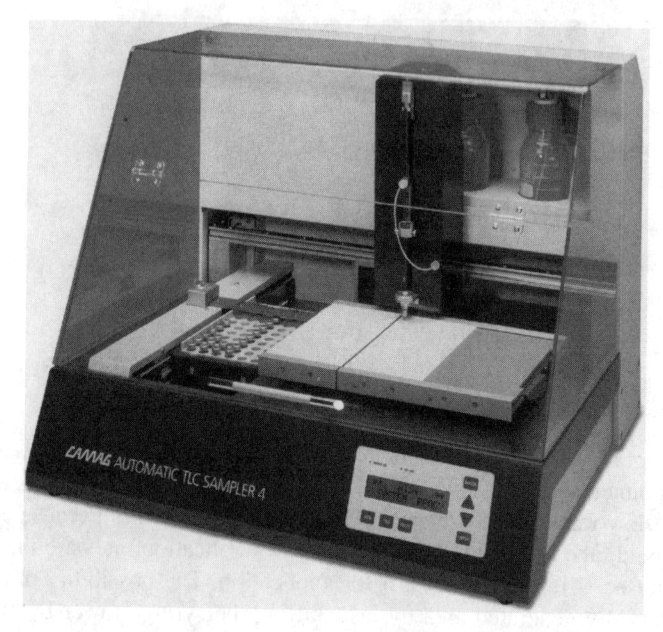

FIGURE 5.14 CAMAG automatic TLC sampler 4 (ATS4, fully automated).

support. In regard to nonvolatile polar samples solvents, the slowest AS 30 delivery rate for application is 120 sec/μl. However, such a slow application rate is not reasonable for preparative purposes, as discussed earlier.

5.1.3.2.2 Fully Automated Application

Fully automated devices ease the tedious application of high volumes (for example, 2 ml), for which the syringe has to be refilled several times. There are two commercially available instruments with this capability, but only one perfectly fulfills the requirements of preparative application. A clear advantage of the CAMAG ATS4 (automatic TLC sampler 4, Figure 5.14) is the spray-on as rectangular area, especially useful for the application of nonvolatile, large sample volumes. The same applies to a special option for faster application, especially of nonvolatile, large sample volumes, by a heated spray nozzle (Figure 5.15). Application of an aqueous extract reduces the time necessary for application to half the time by heating the spray nozzle to 60°C. A nozzle temperature between 30 and 60°C can be selected. Furthermore, the ATS4 applies substances under a cover that better protects the mostly active plate from environmental factors.

The same principle of the spray-on technique as already described for Linomat 5 applies to this device. However, special care has to be taken with single instrument parameters regarding loss of sample owing to automatic rinsing operations. For example, the syringe-filling vacuum time is a crucial parameter. During the selected syringe-filling vacuum time, the side port of the syringe is opened in order to suck the air bubble or initial sample into the waste bottle. Normally, this is an important feature of quantitative planar chromatography, because the initial part of the sample withdrawn into the syringe might partly be dissolved in the rinsing solvent with

FIGURE 5.15 Heated spray nozzle of CAMAG ATS4.

which the syringe is still filled in the region not used by the sample volume. Thus, an accurate concentration to be applied onto the layer is guaranteed if the syringe vacuum time is selected (e.g., as 4 sec). But for preparative purposes, when the sample is valuable and none should be lost, 0 sec has to be selected.

Further, in parameter selection, it is recommended to choose the feature "empty syringe before filling with next sample" to avoid mixing with the rinsing solvent when the sample is withdrawn from the syringe. Small instrument details incorrectly chosen can falsify the yield calculated later on, or result in loss of costly sample. For application of the whole sample volume, i.e., to minimize the vial dead volume, the needle can be adjusted closer to the bottom of the vial, but special care should be taken with that parameter because wrong adjustment may destroy the needle.

If the sample volume to be applied exceeds 1 ml, which is the maximal editable application volume, the overspotting tool can be used. With this tool, a second (up to 1 ml) or additional volume can be applied from the same vial onto the same zone. A limiting factor is the supply of syringe volumes not exceeding 100 µl, thus enabling automated refilling of the syringe more often than by using Linomat 5. However, as the ATS4 application is fully automated and offers the highest dosage rate for application, it is just an aspect of time and not personnel labor.

A second fully automated device, the HPTLC applicator AS 30 (described earlier), can be employed in connection with a sampling device. Automated refilling of the syringe is performed by editing a volume factor, e.g., 10 for application of 10 times 100 µl. This device can be recommended if loss of sample is not relevant (e.g., owing to automatic rinsing operations that afford at least 70 µl dead volume for a minimal 20-cm tube connection). However, the fully automatic mode is not recommended for valuable samples. Sample volume still present in the Teflon tube between the sampler and AS 30 syringe will be wasted and lost because this operation cannot be circumvented by the user.

5.1.3.3 Solid Phase Sample Application (SPSA)

An alternative to spray-on in the form of a rectangular area, the solid phase sample application (SPSA) is suited for application of especially large, nonvolatile sample volumes on preparative layers [2]. Therefore, the sample is dissolved in a suitable

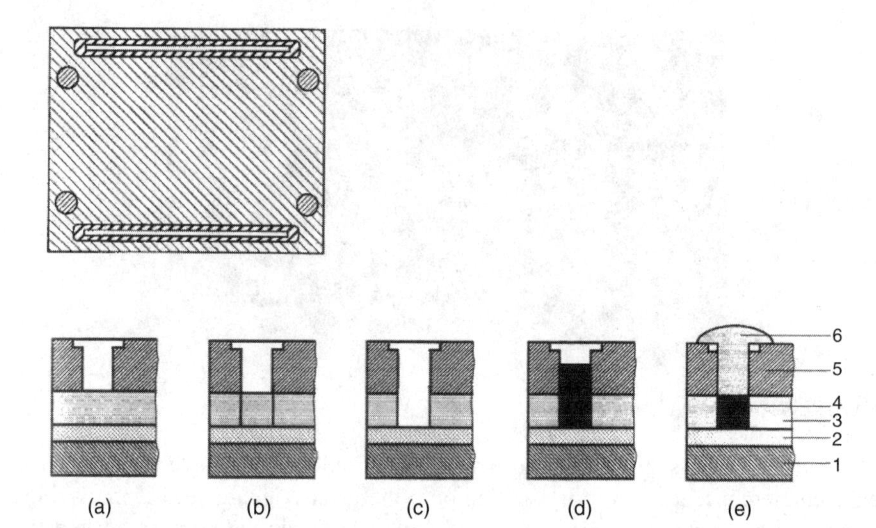

FIGURE 5.16 Template scheme (top view) for solid phase sample application (SPSA) and process of performance (cross section of steps a to e): 1 — base of the device, 2 — glass plate, 3 — adsorbent layer, 4 — sample, 5 — top of the device, 6 — plunger to compress. Step a: Template placed onto the preparative plate; Step b: Marking by means of a thin needle; Step c: Scraped out channel on the preparative plate; Step d: Filling in of the prepared mixture of sample and deactivated adsorbent; Step e: Compression by means of a plunger. (From Botz, L., Nyiredy, Sz., and Sticher, O., *J. Planar Chromatogr.*, 3, 10–14, 1990. With permission.)

solvent and mixed with five to ten times its weight of a deactivated inert adsorbent, e.g., kieselguhr or RP-18. This mixture is carefully dried by rotary evaporation. Then, the dry mixture is filled in the specially prepared channel of a preparative layer. For preparation of the channel, the layer is covered by a hard aluminum template for marking by means of a thin needle (Figure 5.16a and Figure 5.16b). The template is equipped with two channels of 190 × 5 mm, i.e., one for the 1-mm layers that can be refilled with 0.5-g sample inclusive of deactivated adsorbent, and another for the 2-mm layers and 1 g for refilling. Thus up to 100-mg (1-mm layer) and 200-mg (2-mm layer) samples can be applied in maximum amounts, which is suitable for most preparative cases. Then, carefully scraping out of the original adsorbent follows to form a regular channel profile (Figure 5.16c). The prepared dry mixture of sample and deactivated adsorbent is filled into the channel and pressed evenly with a plunger (Figure 5.16d and Figure 5.16e). The optimum contact between the compressed adsorbent and the stationary phase prevents the material from falling out of the channel when the plate is placed vertically into the chromatographic trough chamber. No special care is needed in handling of these layers except RP-18 as deactivated support adsorbent. In this case the channel might be better covered by a small thin glass strip of 190 × 5 mm to avoid a minor part or small amount falling out.

However, the sample might not be sensitive to the evaporation step. Nevertheless, it is a cost-effective alternative for obtaining a good distribution of nonvolatile, large sample volumes across the whole plate length and, thus, for improving resolution if no spray-on equipment is available.

5.2 CHROMATOGRAM DEVELOPMENT

This chapter is devoted to classical development of preparative layers by capillary action, which is widely used as an economical, routine method for the isolation of 10 to 150 mg pure amount of substance. Forced-flow techniques by external pressure such as overpressured layer chromatography (OPLC) or by centrifugal forces, e.g., rotation planar chromatography (RPC), are discussed elsewhere [3,4]. The main difference is the nature of mobile phase migration and the separation of higher amounts, i.e., OPLC up to 300 mg and RPC up to 500 mg (the latter on plates with 4-mm layer thickness instead of 2 mm). Though better resolution might be obtained with forced-flow techniques, extensive equipment is a precondition, and further improvement is necessary regarding some aspects.

5.2.1 Scale-Up

The particle size and size distribution of adsorbents for preparative purposes are higher and wider, respectively, compared to analytical ones. In addition, the adsorbent layer is much thicker and effectively overloaded with the compounds. These items make resolution difficult, which must even be better than for quantitative separations as discussed in Section 5.1. These facts necessitate an excellent and superior strategy to find the best separation, i.e., the mobile phase with the best selectivity (see Chapter 4). It was also shown that plates with a thickness gradient, called Uniplate-T taper plate [5], could improve resolution in the lower-hR_F range.

Separations on analytical plates are significantly more rapid than on preparative layers, and the best solvent system can preliminarily and most effectively be selected by their performance on such thinner plates. Besides others, the selection of the proper mobile phase can easily be performed using the CAMAG HPTLC Vario Chamber and analytical TLC plates sized 10 × 10 cm. Because of six channels, six different solvents or solvent ratios can be tested in one run (Figure 5.17 and Figure 5.18). This kind of special chamber is very time saving. The best analytical separation system is then transferred to a trough chamber system still on analytical plates with a layer thickness of 0.25 mm. The resolution should be checked for various high amounts on the plate applied as 10-mm bands to find out the unacceptable overloading level. The maximum amount of sample that still shows an acceptable resolution is then transferred to, e.g., a 2-mm preparative layer (Figure 5.19). Here, a plate thickness that is higher by a factor of 8 and a 16-times-longer band (160-mm band) are considered. For this example, the maximal amount is multiplied by 128 (16 × 8) and applied as a 160-mm band onto a preparative layer of 2-mm thickness. Typically, amounts of 50 to 150 mg are separated on preparative layers. Regarding the best mobile phase system, it can be transferred unchanged to preparative layers; however, special care must be taken to use the same degree of chamber saturation and similar chamber geometry.

5.2.2 Prewashing of Preparative Layers

For preparative purposes, it is recommended to prewash the plates. Silica gel or alumina layers are mostly employed, and such active sorbents adsorb not only water

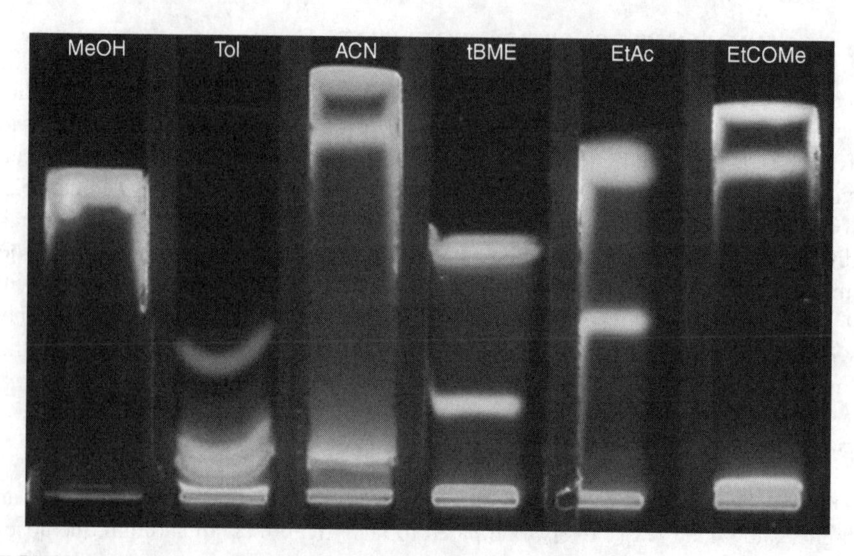

FIGURE 5.17 Analytical preselection of a suitable selectivity of the mobile phase on silica gel using the CAMAG HPTLC Vario Chamber.

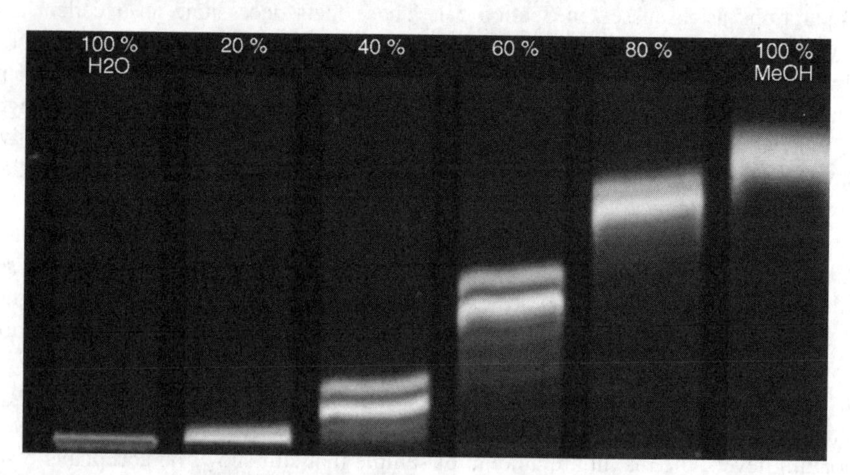

FIGURE 5.18 Analytical preselection of the ratio of two solvents on RP-18 phases using the CAMAG HPTLC Vario Chamber.

to become equilibrated with the humidity but also impurities from the surrounding air. Such impurities are often visible as dark interference fronts along the whole plate containing fluorescence indicator (UV 254 nm) and numbered according to the fronts seen, i.e., β-, y-front, etc. Also, soluble binder components dissolved by usage of special polar mobile phase solvents or contaminants, i.e., plasticizer from the shrink-wrapped package, can cause such interfering secondary fronts. The number and position of the interfering fronts can vary, depending on the composition of the mobile phase. At significant polarity changes of the mobile phase mixture, such interferences are predominantly fixed and visible. Prewashing can be performed for a set of plates if proper storage is guaranteed and repeated contamination prevented.

FIGURE 5.19 Scale up to preparative separation on layers of 2-mm thickness using solvent of channel 5 (EtAc, ethyl acetate) in Figure 5.17.

Solvents for prewashing should be of high purity and eluting power to be able to desorb impurities from the layer; otherwise prewashing is useless. The following techniques for prewashing of layers can be employed.

5.2.2.1 Techniques of Prewashing

To get a clean plate background and to avoid contamination of the synthesized or extracted compound by an impurified plate, immersion into the prewashing solvent overnight (at least 8 h) is an effective method. A second mode of prewashing is previous chromatography with the prewashing solvent up to 5 mm beneath the upper edge of the plate but at least 1 cm longer than the migration distance of the subsequent chromatography. This technique is also called blank chromatography [6], in which impurities of a blank plate are concentrated at the upper edge of the plate. Note that it is important to mark the direction of prewashing on the top right-hand side of the prewashed plates because for chromatography, the same direction has to be used (Figure 5.20, Reference 7).

There is no universally agreed technique for prewashing. In the literature only a few studies are described regarding the benefits of prewashing [8–10]. A study on silica gel plates compared the two modes, and the authors considered methanol-based immersion for several minutes as a more efficient prewashing technique than prewashing by single development [9]. However, a combined technique of both modes as follows was deemed best: first, single development and then, after an intermediate drying step, immersion for 5 min in the same solvent (methanol) [10]. Immersion times are reported to lie between 1 to 7 min [9] and at least 8 h — the latter surely yields a better cleaning effect. Different solvents for prewashing of normal phases are described in the literature, i.e., methanol [11–14], methanol containing ammonia [15], methanol mixed with other solvents, such as dichloromethane (1:1) [16] or with chloroform (2:1) [17], or also acetone [18,19] as well as sequential prewashing with chloroform–methanol (2:1) followed by acetone [20].

FIGURE 5.20 Direction of prewashing should be marked. (FromHahn-Deinstrop, E., *Applied Thin-Layer Chromatography — Best Practice and Avoidance of Mistakes*, Wiley-VCH, Weinheim, Germany, 2000. With permission.)

A study compared several solvents and mixtures with each other, and according to investigations it was demonstrated that methanol was superior to other solvents and mixtures [8]. If prewashing with methanol is not satisfactory, e.g., if a lipophilic mobile phase is to be used later on, a further possibility is prewashing with the developing solvent, e.g., *n*-hexane–ethyl acetate (7:3) [21], however, for constancy of the layer property, acids or bases should be kept out of the mixture. A very pure solvent of high eluting power is generally recommended for the respective adsorbent; thus, methanol for silica gel or alumina phases and a lipophilic one for reversed phases are recommended. It is self-evident that the degree of purity of the prewashing solvent should be as high as possible, at least HPLC, AMD, or spectroscopy grade.

Effective prewashing can be performed for a set of plates; however, proper storage must be guaranteed to avoid any recontamination. A special simultaneous separating chamber (Desaga) of the same dimensions as the standard chambers but with vertical grooves in the transverse sides can hold up to five plates. This chamber is effectively employed for prewashing of plates if filled with a suitable solvent.

This chamber can also be used for storage of plates. Alternatively, the Alltech Multiple Plate Development Rack (Teflon or anodized aluminum) can be used to hold six plates for transfer into and out of the standard tanks.

Finally, an elaborate, but very effective prewashing procedure used in routine analytical trace analysis [22] can be adapted for preparative layers. Several plates can easily be handled simultaneously if placed in a special rack. The whole rack with the set of plates and a respective number of pure glass plates, used for covering prewashed plates later on, can be placed and immersed into the prewashing solvent in a size-adapted glass box. After cleaning a set of three plates and three cover plates per liter solvent by immersion for at least 8 h (overnight), the solvent has to be discharged or disposed. To avoid any interference whatsoever, two prewashing steps are subsequently performed by immersion, the second step with a completely fresh solvent. Prewashed plates are dried and covered by the precleaned pure glass plates by clamping a Teflon strip (PTFE sealing band) around the side edges of both glass plates. Glue and additives from normal adhesive tapes cause recontamination. During application of the sample it is recommended to keep the residual part of the layer still covered by the glass plate just shifted upward by about 3 cm. Not until chromatography is the prewashed layer delivered from the glass plate (to protect the adsorbent from recontamination).

5.2.2.2 Drying and Storage of Prewashed Plates

Drying of prewashed plates is mostly performed in a very clean cabinet desiccator or on a TLC plate heater flushed with purified nitrogen gas under a self-made cover. The cabinet desiccator should exclusively be used for the purpose of plate drying to avoid recontamination. In handy racks up to 10 plates (made of light alloy by Desaga or CAMAG) or 21 plates (made of anodized aluminum, polished steel, and PTFE by Baron) can easily be transported, placed, and dried vertically in the desiccator so that the moisture can easily escape upward. The vertical position of the plates in the rack promotes air circulation through convection. Typically, drying is performed at temperatures above the boiling point of the prewashing solvent for a definite time period. For example, methanol and adsorbed water from the ambient air are evaporated at temperatures of 120°C for 0.5 to 2 h depending on layer thickness (0.5 to 2 mm). To just get rid of adsorbed methanol, lower temperatures of about 80°C are sufficient. A further possibility for thinner preparative layers (0.5 to 1 mm) is drying in a desiccator that is evacuated up to 1 mbar by a vacuum pump overnight and flushed with purified nitrogen gas for opening [22]. The whole space-economizing rack used for prewashing and drying of plates can be placed and stored in a special large-sized desiccator (Desaga).

After cooling down of the plate in a protected atmosphere or in a desiccator, the prewashed dried plate should be covered by a clean glass plate during application or for long-term storage in a protected atmosphere. For example, a specially designed drying chamber (Figure 5.21) ensures protection against dust and dampness when storing TLC plates [10]. It consists of a glass chamber with a polypropylene cover as used in laboratories and can be provided with drying racks for TLC plates (made of anodized aluminum, polished steel, and PTFE) able to handle up to 21 plates.

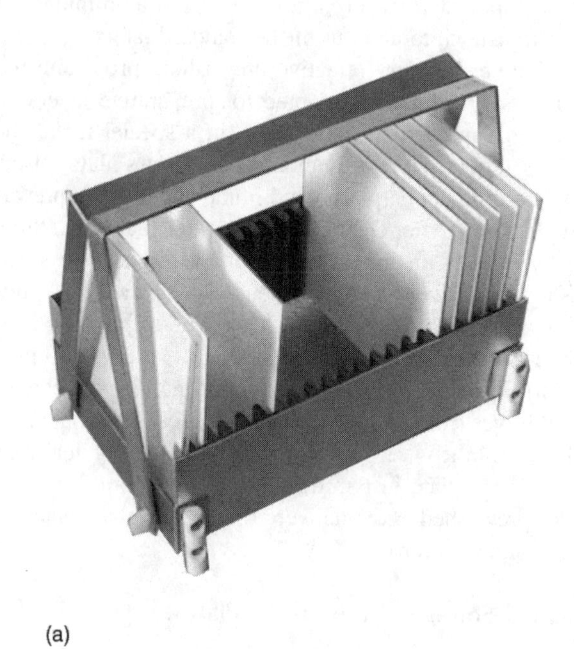

(a)

(b)

FIGURE 5.21 (a) Baron drying rack and (b) drying chamber for TLC plates.

The presence of legs and side protection made of PTFE allow insertion of the hot drying rack into desiccators. The air is filtered and circulates through an integrated 12-V fan; humidity is indicated through a hygrometer. Standard desiccator drying agents can be employed for this device as well.

5.2.2.3 Road Map to Your Own Procedure

Proper treatment of plates begins with opening of the plate packet and ends with documentation. The rule that plates should be handled on the outer edges using only the fingertips of both hands and should preferably be transported in TLC racks is obvious. Proper handling of the whole prewashing step is a precaution against recontamination. Often, little care is taken for drying and storage of prewashed plates. For a simple start regarding active adsorbents on the plate, previous development with methanol (HPLC grade) is most cost-effective and useful in getting rid of interfering β-fronts. If results are not satisfactory, immersion overnight can be employed. It is essential to establish a clean drying and storage facility, otherwise all previous effort would be nullified. During the application process prewashed plates should be covered by a clean glass plate (described earlier) because active adsorbents such as silica gel or aluminum oxide strongly adsorb impurities from ambient air.

5.2.3 Development

Regarding development, normal silica gel layers are tried first if not otherwise indicated. Selection of the best developing solvent has to be performed systematically, and its preparation should be proper. Various modes of development and separation chambers, both of which can influence the separation, have to be considered for development, with focus on the classical method of chromatogram development. Aspects of the chamber climate are helpful as background information and can improve resolution.

5.2.3.1 Handling of the Mobile Phase

Emphasis has to be on choice and proper handling of the mobile phase. In Chapter 4 different approaches for mobile phase selection are discussed. General hints for selection are the avoidance of the following:

- Problematic solvent compositions such as mixtures containing ether because of its very low boiling point and vapor pressure, which cause problems, especially during summer.
- Chlorinated hydrocarbons, which are more or less treated as hazardous solvents.
- Mixtures forming two phases *a priori*, or at a given miscibility gap, or around the latter, because small temperature or volume changes can cause demixing.
- Major differences in polarity in mobile phase mixtures if secondary fronts interfere.

- Solvents of high viscosity. A reduced viscosity of the mixture results in an increased migration rate, less diffusion and, thus, a focused band. For example, it is advantageous to select methanol instead of propanol because of its lower viscosity.

Developing solvents should be composed of as few solvents as possible; however, they mostly consist of 2 to 4 solvent components. It is recommended that developing solvents be prepared fresh every time and to use them only once. Some clues for handling of mobile phase mixtures are as follows:

- Solvent compositions that are critical in regard to reproducibility could be ones that contain alcohols and acids because of the capability to form esters, or ones composed of esters owing to saponification possibilities. Such mixtures should always be prepared fresh to ensure reproducible results.
- For preparation of solvent mixtures, the single solvents should be measured out separately and then properly mixed in a vessel with a ground glass stopper (avoidance of volume contraction and measuring errors).
- To save costs, only the mobile phase volume intended for use should be calculated and prepared. Typically, the lower end of the plate should be immersed several millimeters, i.e., just about 4 mm, for preparative purposes.

But even when single solvents are employed as developing solvents, attention has to be given to purity grade, manufacturer, batch, and stabilizers contained.

5.2.3.2 Separation Techniques

There is great flexibility in planar chromatography owing to the multiplicity of separation techniques [23]. However, just a few development techniques are employed for preparative purposes (Figure 5.22). Besides forced-flow techniques, such as overpressured layer chromatography (OPLC, forced flow by external pressure) or rotation planar chromatography (RPC, forced flow by centrifugal forces), radial developments such as the latter and single, or even multiple, linear developments can be employed. Forced-flow techniques are discussed in detail elsewhere [3,4] and are not covered in this chapter.

Single linear developments are mostly employed in the vertical mode. The applicability of the horizontal mode is discussed in Chapter 6. For circular and anticircular developments, the movement of the mobile phase is two-dimensional; however, from the standpoint of sample separation it is a one-dimensional technique. Circular developments result in higher hR_F values compared to linear ones under the same conditions, and compounds are better resolved in the lower-hR_F range. The same effect is noticed on plates with a layer thickness gradient (see Section 5.2.1). On the other hand, using anticircular developments, compounds are better resolved in the upper-hR_F range.

Multiple developments can be classified into uniform multiple developments (UMD, also called unidimensional multiple developments) for methods in which the migration distance and mobile phase composition are constant and incremental multiple development (IMD) with mostly ascending migration distances. However,

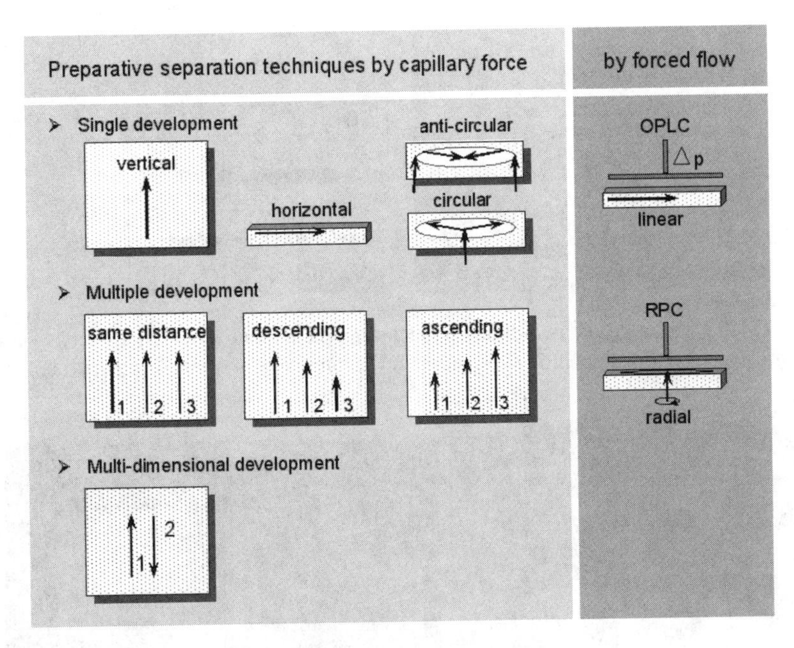

FIGURE 5.22 Multiplicity of preparative separation techniques.

a descending migration distance is preferred to improve the resolution of two compounds in the lower-hR_F range, which does not require the complete migration distance again. Multiple developments with the same mobile phase can be employed to improve resolution to a minor extent. Different mobile phases can be employed as step gradients to enable the separation. The so-called gradient multiple development (GMD) implies constant migration distances, whereas the migration distance differs in bivariate multiple developments (BMD) [24].

Two-dimensional developments are only reasonable in the opposite direction (180°). For example, the starting band is applied in the upper part of the layer, and for the first chromatography the trough is filled up with solvent just beneath the starting band (Figure 5.23). Using a mobile phase of weak solvent strength, only lipophilic interferences are eluted. After drying, the upper part (stripe) of the plate carrying the eluted interferences is cut off, the residual plate is turned 180°, and developed with the mobile phase in the usual manner.

5.2.3.3 Separation Chambers

Different TLC chambers exist [25], and their given vapor space can generally influence the chromatographic result depending on the chamber form and type [26]. In classical preparative layer chromatography, the most frequently used chamber type is the flat-bottom (plain ground flange edge) or the twin-trough chamber 20 × 20 cm (Figure 5.24). Heavy lids made of glass ensure a gastight seal. To ensure a tight seal for an optimal equilibration atmosphere, a grease seal should be avoided to prevent contamination and lipophilic interferences in the chromatogram. A better way is to heavily load the light lids using a lead ring. The TLC chambers mentioned

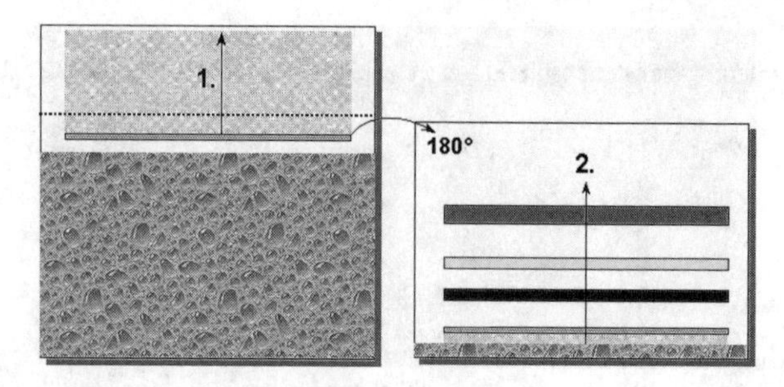

FIGURE 5.23 Example of a two-dimensional development (plate turned by 180°), first development for removal of lipophilic substances and second development for separation.

FIGURE 5.24 Developing chambers 20 × 20 cm for classical preparative purpose of some manufacturers (left to right): CAMAG twin-trough chamber and flat-bottom chamber, Desaga standard separating chamber, Analtech chamber.

earlier can be used for the simultaneous separation of maximum of two preparative plates if resolution is not affected by the narrower vapor phase in the front region of the plate, and if increased disruption of the saturated chamber climate is not too detrimental by placing two plates into the chamber. The employment of horizontal chambers is discussed in detail elsewhere (see Chapter 6). Generally, 40 to 100 ml of mobile phase is used for development of preparative layers, depending on migration distance and layer thickness. The plate should lean against the side wall and be positioned in a perpendicular fashion. The angle between the plate and bottom of the tank affects the rate of development as well as the shape of the spots [27]. It is obvious that the mobile phase flow increases as the angle decreases toward the horizontal line (until it is 0° in the horizontal developing mode). Generally, an angle of 75° is recommended for optimum development. The migration distance should be optimized. Often, a longer migration time and distance do not imply a better resolution because of increased diffusion effects. The maximum migration distance is 18 cm and is limited by the nature of capillary flow. For reproducible separations, temperature and humidity should be documented.

Especially for temperature-dependent partition processes, besides temperature-influenced humidity of the vapor phase, a thermostat device is recommended for development. The Baron TLC thermobox 200 and Desaga TLC thermobox respectively (Figure 5.25) can be adapted to suit current development chambers using a variable covering mask. A Peltier element with integrated temperature control

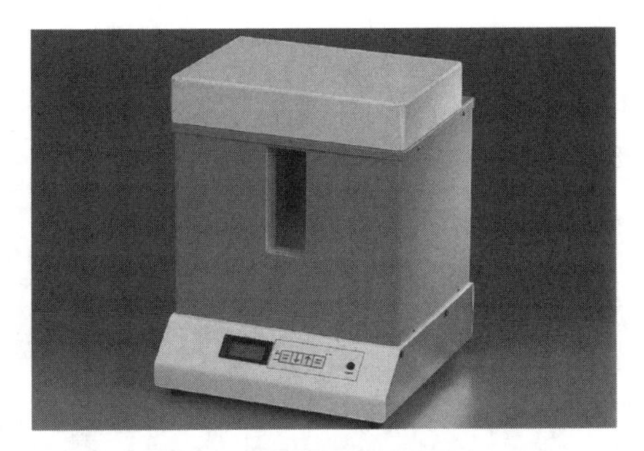

FIGURE 5.25 Automatic development chamber Desaga TLC thermobox 200.

provides the desired temperature. The range of temperature control is at least 10°C below and at least 20°C above the surrounding temperature (±0.5°C). Air circulation is integrated for even temperature distribution. Because of an insulating window and an illuminated interior, the development process can be observed.

A further automated developing chamber, the CAMAG ADC2, is regrettably limited to plate dimensions of maximum 20 × 10 cm and thus is not suitable for preparative purposes. However, the Baron TLC-Mat (Figure 5.26, also obtainable from Desaga [28,29]) can be supplied with a larger polypropylene solvent trough and thus be adapted for preparative plates. The device works with a sensor for front recognition and exactly indicates the development time. Automated development is carried out in a light-and-air-protected chamber and stopped after a defined time by removal of the solvent in the trough. After chromatography, the solvent vapor is removed in the chamber by an integrated blower.

For a protected development of light-sensitive substances, a specially designed cover, the Baron Protective Cover against Light for Development Chambers, provides auxiliaries. For various chamber sizes of different manufacturers, adjusted models are available. As a special option, a connection for protective gas can also be installed.

5.2.3.4 Drying of Developed Plates

To avoid continuing diffusion after development, the plate must immediately be dried to get rid of the mobile phase solvents. Therefore, a cold stream of air from a simple hair dryer (however, involving danger of recontamination) or a plate heater at low temperatures (e.g., 35°C) flushed with nitrogen gas under a self-made cover is generally recommended. Typically, the drying step lasts about 15 to 40 min depending on layer thickness and volatility of the solvent. Another device for drying of plates up to 20 × 20 cm is the Baron cold air dryer (Figure 5.27). Within the cold air dryer, the plates can protectively be freed from the mobile phase after development. By means of a filter, fans extract the air at a rate of 200 m^3/h, and exhaust air exits at the back [28]. Recently a further TLC Dryer [30,31] was developed which removes the solvent vapors from the upper layer of the sorbent by a laminar air or inert gas flow and a constant temperature gradient in the chromatographic flow direction. Besides less diffusion of analytic components due to accelerated drying,

FIGURE 5.26 Automatic development chamber Baron TLC-Mat adjustable for preparative purposes.

a non-homogeneous in-depth distribution of compounds inside the sorbent is eliminated. However, the latter aspect has more importance for quantitative HPTLC. The device can dry up to four plates (in separate compartments) in parallel. Generally, such drying devices should be placed beneath a fume hood. Special care has to be given to thermally labile or oxidation-prone sample components. Besides desiccators evacuated for plate drying (see Section 5.2.2.2), the Alltech Controlled Atmosphere Chamber also provides an inert atmosphere for drying developed plates, which is ideal for applications with sensitive samples to protect them from decomposition. It eliminates solvent vapors in the laboratory when used with vacuum. Generally, scraping out the band of interest should follow as soon as possible, because prolonged substance contact with the active layer might lead to decomposition.

5.2.4 Influencing Factors — Chamber Climate

One of the most crucial influencing factors in planar chromatography is the vapor space and the interactions involved. The fact that the gas phase is present, in addition to stationary and mobile phases, makes planar chromatography different from other chromatographic techniques. Owing to the characteristic of an open system the stationary, mobile, and vapor phases interact with each other until they all are in equilibrium. This equilibrium is much faster obtained if chamber saturation is employed. This is the reason for differences in separation quality when saturated and unsaturated chambers are used. However, the humidity of the ambient air can also influence the activity of the layer and, thus, separation. Especially during sample application, the equilibrium between layer activity and relative humidity of the

FIGURE 5.27 Baron cold air dryer for drying of plates up to 20 × 20 cm.

ambient air is established within minutes [32], thus making the need for a previous activation step of the layer questionable. It is recommended that a conditioning step be performed for adjustment of the activity of the layer or an alkaline or acidic conditioning of the layer just before chromatography. For special purposes, development with a reduced vapor phase in so-called sandwich chambers (S-chambers) can be advantageous for separation.

5.2.4.1 Unsaturated Chamber

Driven by capillary action the mobile phase moves up the preparative layer until the desired migration distance is reached, and chromatography is interrupted by drying the plate. However, multiple effects can influence the separation on active adsorbents provided the chamber is closed and reasonably tight. Mostly mobile phase mixtures are used for development of plates, implying evaporation predominantly of the most volatile solvent in the mobile phase mixture at the trough bottom (Figure 5.28, effect 1) until equilibrium is established. Depending on the vapor pressure of the individual components, the composition of the vapor phase can differ significantly from that of the mobile phase. Driven by capillary action, the solvent mixture moves up the layer so that the most polar component of the mixture is adsorbed predominantly to the active parts of the adsorbent, leading to a ratio change of the solvent mixture (Figure 5.28, effect 2) during chromatography. Under certain conditions (involvement of impurities, see Section 5.2.2), such ratio changes can cause the formation of secondary fronts that are also very prominently present in the sandwich configuration. Simultaneously to effect 1, the part of the layer that is already wetted with the mobile phase interacts with the gas phase. The less polar components especially are given off in the gas phase (Figure 5.28, effect 3). While still dry, the stationary phase adsorbs molecules from the gas phase whereby, the polar components especially will be withdrawn from the gas phase and loaded on to the adsorbent

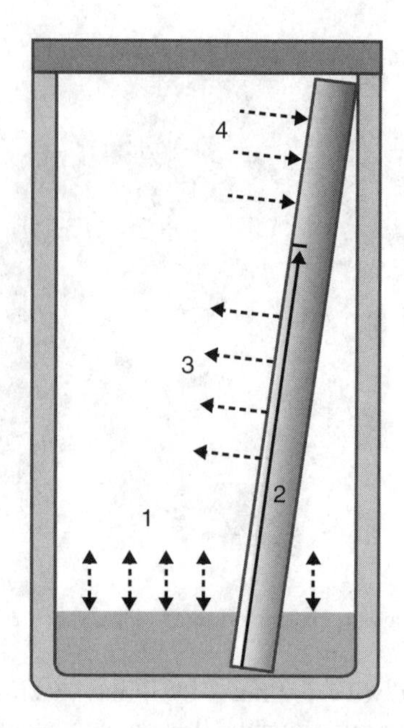

FIGURE 5.28 Influencing factors of chamber climate in a trough chamber: 1 — evaporation of low volatile components of the mobile phase from the trough bottom until saturation, 2 — sorptive saturation, 3 — evaporation from the layer, 4 — adsorption of the vapor phase.

(Figure 5.28, effect 4). This process approaches an equilibrium called adsorptive saturation.

Owing to the ratio change, evaporation effect, adsorptive saturation, and other factors, which can be considered a kind of uncontrolled gradient, it is not correct to call such developments isocratic. Only the composition of the developing solvent at the time when it is placed into the chamber is known, whereas the mobile phase changes during migration. However, the mentioned effects are reduced on midpolar or apolar stationary phases as well as if single solvents are used as mobile phase, if a sandwich configuration is chosen or by the employment of chamber saturation.

5.2.4.2 Chamber Saturation

Chamber saturation is recommended for better reproducibility of the separation — especially if multicomponent mobile phase mixtures are composed of solvents differing in volatility or polarity to a great extent. Moreover, chamber saturation can improve resolution of two components or reduce the formation of secondary fronts. For chamber saturation, the large tank sides are lined with a sheet of filter paper 20 × 20 cm each. During the filling of the mobile phase into the chamber, it is poured onto the filter, which is then completely wetted and soaked by the mobile phase. Note that the wet filter paper is dipped into the mobile phase at the trough bottom. The prepared closed tank will become saturated within 15 to 30 min depending on the volatility of the solvent components (without wetted filter paper it needs more

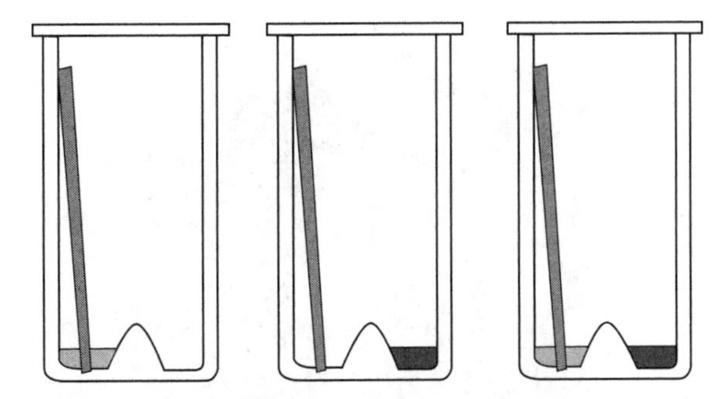

FIGURE 5.29 Modes of conditioning in a CAMAG twin-trough chamber: chromatography as in a flat-bottom chamber; however, less solvent consumption and previous uninterrupted conditioning with the mobile phase (left), preconditioning mode with a different solvent (middle), layer conditioning during chromatography with a solvent differing from mobile phase (right).

than 2 h). Then the plate is placed in such a manner that it interrupts the chamber climate as little as possible, i.e., the lid is moved slightly to the back as required.

5.2.4.3 Conditioning of the Layer Just before Chromatography

A flat-bottom chamber implies opening the lid for insertion of the plate or mobile phase. The CAMAG twin-trough chamber [33,34], on the other hand, has a glass ridge along its base that divides the tank into two troughs and thus allows the user a conditioning of the layer without interruption of the chamber climate. By placing the plate in one trough and the mobile phase in the other one (Figure 5.29, left trough), controlled conditioning of the layer with the mobile phase is possible in a wetted filter-paper-lined twin-trough chamber for a defined time period. Then, development is started by tilting the chamber to transfer solvent into the trough in which the plate is still placed. Thus, an uninterrupted conditioning of the layer with the mobile phase is possible. Preconditioning with a different solvent such as an acidic or alkaline solution is also possible (Figure 5.29, trough in the middle). This mode is preferred for humidity control of the layer if a saturated salt solution or a diluted sulfuric acid solution is filled in the other trough. After a defined conditioning time, the mobile phase is poured along the trough wall with as little interruption as possible, so that the lid is moved slightly to the back and the glass plate is tilted to the required extent. During chromatography, the specific layer conditioning can be maintained with a solvent differing from the mobile phase (Figure 5.29, right trough).

Two further devices are advantageous for reproducible chamber saturation and uninterrupted conditioning of the layer. Plates can be time-controlled and adjusted to the specific chamber atmosphere using the Baron conditioner headpiece (Figure 5.30). Upon completion of the preselected conditioning time (maximum 60 min) the plate is automatically lowered into the mobile phase. Further handling is continued in the usual manner. The conditioner headpiece can be adapted to standard chambers. In combination with a twin-trough chamber, an uninterrupted preconditioning with a different solvent is possible.

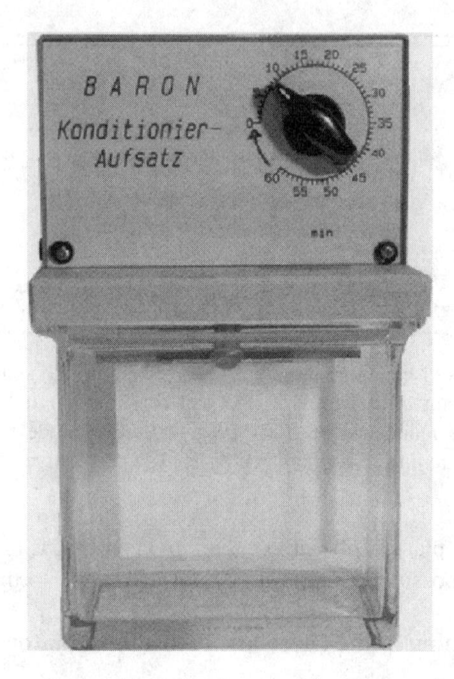

FIGURE 5.30 Baron conditioner headpiece in use with a Desaga chamber 20 × 20 cm.

A second device for reproducible chamber saturation is the Alltech Latch-Lid equilibration apparatus. The modified lid has two retractable glass hooks passing through Teflon-threaded collars, which are used to raise and lower the rack inside the closed chamber. Thus, the plate placed in the rack can be conditioned and thereafter lowered into the mobile phase.

In connection with conditioning of layers and chamber saturation, the virtual front has to be mentioned. During chromatography, components of the mobile phase that have been loaded onto the layer via the gas phase are pushed ahead of the true but invisible solvent front and generate a visible virtual front in front of the real one. Stopping chromatography at the same migration distance as that used for unsaturated or sandwich chambers, the hR_F values are generally lower owing to the visible virtual front that has been referred to stop chromatography [32].

5.2.4.4 Sandwich Configuration

The discussed effects, such as evaporation and adsorptive saturation, are prevented by placing a counter plate at a distance of one or a few millimeters from the chromatographic layer. The development with such a reduced vapor phase in the so-called sandwich chambers (S-chambers) can improve the separation. The glass-backed 20 × 20 cm plate forms one wall of the chamber with the adsorbent facing inward. A glass plate with spacers, called counter plates, is clamped to this plate and forms the other wall of the chamber (Figure 5.31, left [32]).

Vapor space in the chamber is minimal and eliminates the need for presaturation. However, for specific mobile phase mixtures (e.g., composed of chloroform and

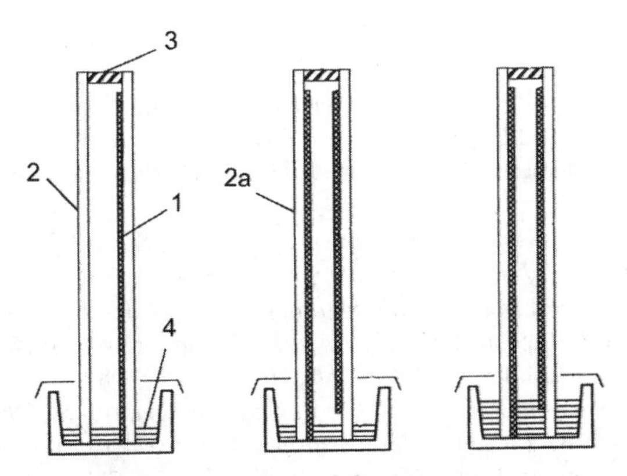

FIGURE 5.31 Unsaturated (left) and saturated S-chamber showing saturation process (middle) and subsequent development (right): 1 — plate, 2 — counter plate, 2a — counter plate wetted with mobile phase, 3 — spacer, 4 — solvent trough.

ethanol in which the apolar component has a heavier specific vapor), the compensation of the density gradient causes a wavy convection and thus an irregular β-front movement. In such cases, it is recommended to employ a saturated sandwich configuration. Therefore, the counter plate is lined with a wetted filter paper or consists of a wetted chromatographic plate. A lower strip is scraped off the chromatographic plate to prevent immersion into the solvent mixture during saturation. After a defined saturation period (Figure 5.31, middle), development follows by filling up the solvent trough with the mobile phase (Figure 5.31, right). The formation of secondary fronts, which are prominently observed for unsaturated sandwich developments, can also be reduced in the saturated sandwich mode. Further, saturated sandwich developments can be performed in horizontal chambers (see Chapter 6).

REFERENCES

1. Omori, T., Modern sample application methods, in *Planar Chromatography — A Retrospective View for the Third Millennium*, Nyiredy, Sz., Ed., Springer Scientific, Budapest, 2001, p. 120.
2. Botz, L., Nyiredy, Sz., and Sticher, O., *J. Planar Chromatogr.*, 3, 10–14, 1990.
3. Nyiredy, Sz., Ed., Possibilities of preparative planar chromatography, in *Planar Chromatography — A Retrospective View for the Third Millennium*, Springer Scientific, Budapest, 2001, p. 386.
4. Nyiredy, Sz., Preparative layer chromatography, in *Handbook of Thin-Layer Chromatography*, Sherma, J. and Fried., B., Eds., Marcel Dekker, New York, Basel, 2003, p. 307.
5. Alltech Associates Inc., 2051 Waukegan Road, Deerfield, IL 60015, USA, www.alltechweb.com.
6. Hahn-Deinstrop, E., *J. Planar Chromatogr.*, 6, 313–318, 1993.
7. Hahn-Deinstrop, E., *Applied Thin-Layer Chromatography — Best Practice and Avoidance of Mistakes*, Wiley-VCH, Weinheim, Germany, 2000.

8. Maxwell, R.J., Yeisley, S.W., and Unruh, J., *J. Liq. Chromatogr.*, 13, 2001–2011, 1990.
9. Maxwell, R.J. and Unruh, J., *J. Planar Chromatogr.*, 5, 35–40, 1992.
10. Maxwell, R.J. and Lightfield, A.R., *J. Planar Chromatogr.*, 12, 109–113, 1999.
11. Huf, F.A., Harberts, J.C.M., and Span, M.A., *Pharmaceutisch Weekblad,* 114, 660–663, 1979.
12. Meyyanathan, S.N. and Suresh, B., *J. Chromatogr. Sci.*, 43, 73–75, 2005.
13. Bodart, P. et al., *J. Planar Chromatogr.*, 11, 38–42, 1998.
14. Morlock, G. et al., *CAMAG Bibliography Service (CBS),* 93, 14–15, 2004.
15. Argekar, A.P., Kapadia, S.U., and Raj, S.V., *J. Planar Chromatogr.*, 9, 208–211, 1996.
16. Sherma, J. and Miller, R.L., *Am. Lab.*, 16, 126–127, 1984.
17. Agarwal, N.L. and Mital, R.L., *J. Inst. Chemists (India)*, 47, 239–241, 1975.
18. Thielemann, H., *Fresenius' Zeitschrift fuer Analytische Chemie*, 264, 32, 1973.
19. Wuthold, K. et al., *J. Planar Chromatogr.*, 16, 15–18, 2003.
20. Hack, M.H. et al., *J. Planar Chromatogr.*, 15, 396–403, 2002.
21. Erdelyi, B., Birincsik, L., and Szabo, A., *J. Planar Chromatogr.*, 16, 246–248, 2003.
22. Burger, K., Laborinterne Arbeitsvorschrift: Dünnschicht-Chromatographie (DC) — Vorbehandlung und Lagerung der DC-Platten, February 4, 1998, pp. 1–3.
23. Kovar, K.-A. and Morlock, G., Dünnschicht-Chromatographie, in *Untersuchungsmethoden in der Chemie,* Naumer H. and Heller, W., Eds., Stuttgart, 1997, p. 57.
24. Nyiredy, Sz., Ed., Multidimensional planar chromatography, in *Planar Chromatography — A Retrospective View for the Third Millennium*, Springer Scientific, Budapest, 2001, p. 110.
25. Dzido, T., Modern TLC chambers, in *Planar Chromatography — A Retrospective View for the Third Millennium,* Nyiredy, Sz., Ed., Springer Scientific, Budapest, 2001, p. 68.
26. Geiss, F., Schlitt, H., and Klose, A., *Zeitschrift fuer Analytische Chemie*, 213, 331–346, 1965.
27. Abbott, D.C. et al., *Analyst*, 89, 480–488, 1964.
28. Lothar Baron, Laborgeräte, Insel Reichenau, Germany, www.baron-lab.de.
29. Sarstedt (Desaga), Nümbrecht, Germany, www.sarstedt.com.
30. Prosek, M., Golc-Wondra, A., Vovk, I., and Zmitek, J., Controlled drying in quantitative TLC, *Proceedings of the International Symposium on Planar Separations*, Siofok, 77–92, 2005.
31. TLC Dryer type P4, Iskra PIO d.o.o., Trubarjeva c. 5, SLO 8310 Šentjernej, Slovenia.
32. Geiss, F., *Fundamentals of Thin-Layer Chromatography,* Hüthig Verlag, Heidelberg, 1987.
33. Jänchen, D., *J. Chromatogr.*, 33, 195–198, 1968.
34. CAMAG, Muttenz, Switzerland, www.camag.com.

6 On Methodical Possibilities of the Horizontal Chambers in PLC

Tadeusz H. Dzido and Beata Polak

CONTENTS

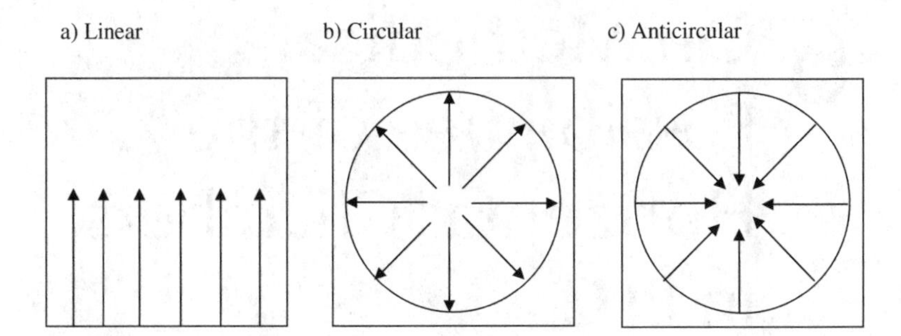

FIGURE 6.1 Different modes of chromatogram development.

6.1 INTRODUCTION — MODES OF CHROMATOGRAM DEVELOPMENT (LINEAR, CIRCULAR, AND ANTICIRCULAR)

There are two different modes of development of planar chromatograms applied in the practice of preparative layer chromatography (PLC): linear development and radial development. The most popular mode is linear development, in which the mobile phase moves from one edge of the chromatographic plate to the opposite edge, with the adsorbent layer and mobile phase being of constant width (Figure 6.1a). The radial development is broken down into two more modes: circular and anticircular development. If the mobile phase is delivered to the center of the chromatographic plate, or close to it, and moves toward periphery of the adsorbent layer, a circular mobile phase front is formed (Figure 6.1b); this mode of chromatogram development is known as circular development. Mobile phase movement in the opposite direction is observed when the anticircular mode of chromatogram development is applied, from the periphery of the adsorbent layer to its center (Figure 6.1c).

6.2 CHAMBER TYPES

The preceding modes of chromatogram development can be employed both in preparative planar chromatography and analytical thin-layer chromatography (TLC). Two main types of chambers are available for use in planar chromatography: N-chambers (normal, conventional chambers) and S-chambers (sandwich chambers). N-chambers are made as rectangular or cylindrical glass vessels and are the most popular chambers in chromatographic practice. This is probably owing to their low price and very simple operation. Some early models of sandwich chambers were vertical. Nowadays, S-chambers are usually offered by manufacturers as horizontal sandwich chambers. Some sandwich chambers can operate in both modes, i.e., with vapor phase (as normal chambers) and without vapor phase (as sandwich chambers). The name horizontal indicates that the chromatographic plate is horizontally positioned in this chamber. This type of chamber has gained popularity in chromatographic practice owing to many different methodical possibilities of performing chromatogram development and very low solvent consumption. Chromatogram development in horizontal chambers is somewhat faster in comparison to the vertical configuration. This means that in the horizontal chamber, the efficiency of the chromatographic system should be higher because of the value of the mobile phase flow rate, which should then be closer to the optimal value. Some authors [1] have reported shape distortion of the spots during development using the horizontal mode. They recommended that the angle of the plate should be 75°. However, this statement should be verified for contemporary manufactured stationary phases.

6.2.1 Horizontal Chambers for Linear Development

One of the first horizontal chambers described in the literature was constructed by Brenner and Niederwieser [2]. In this chamber the chromatographic plate was supplied with solvent by means of a wick (Figure 6.2). This mode of solvent delivery to the adsorbent layer was applied in the H-separating chamber from Desaga [3], Vario KS-chamber [4,5] from CAMAG (Muttenz, Switzerland), and sequence-TLC device by Bunčak (Scilab) [6]. Whereas the first chamber has been mainly applied for analytical separations, the last two were used for analytical and preparative separations as well. A cross section of the sequence-TLC chamber is shown in Figure 6.3. The chromatographic plate with the adsorbent layer facedown is held by the frame (2) and can be moved over the support (1), which is equipped with a mobile phase trough in the middle. The development of chromatogram is started by putting the magnet holder (3) on the cover plate (5), which attracts the iron-cored wick to the adsorbent layer. Then, the solvent from the trough is supplied to the adsorbent

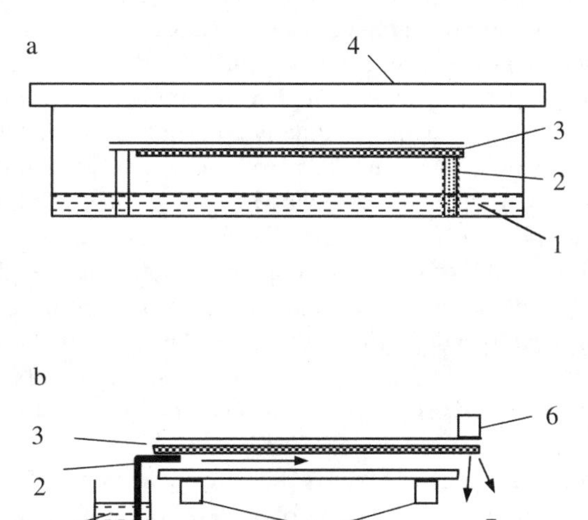

FIGURE 6.2 Brenner and Niederwieser S-chamber: (a) normal development, (b) continuous development; 1 — mobile phase, 2 — wick, 3 — chromatographic plate with adsorbent layer face up, 4 — cover plate, 5 — support, 6 — heater, 7 — evaporation of the mobile phase. (From Brenner, M. and Niederwieser, A., *Experientia* 17, 237–238, 1961.)

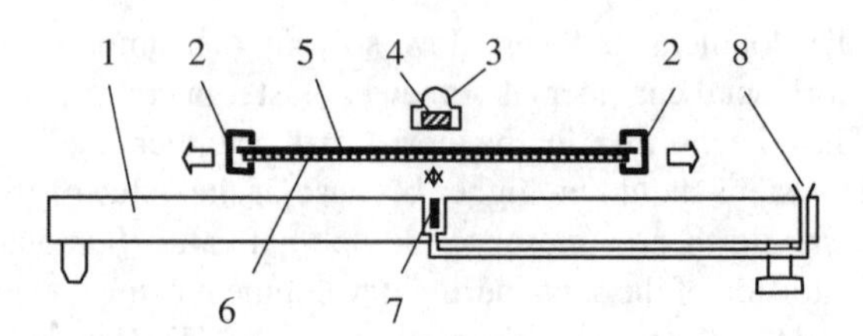

FIGURE 6.3 Sequence developing chamber by Bunčak: 1 — support with mobile phase source (reservoir), 2 — holding frame, 3 — magnet holder, 4 — magnet, 5 — cover plate, 6 — TLC plate, 7 — wick with iron core, 8 — solvent entry. (From Bunčak, P., *GIT Fachz. Lab. (Suppl., Chromatographie)*, G-I-T-Verlag, Darmstadt, 3–8, 1982.)

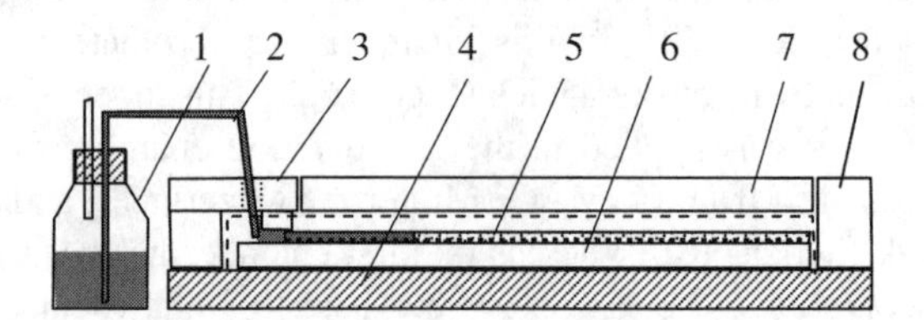

FIGURE 6.4 Horizontal ES chamber: 1 — mobile phase reservoir, 2 — capillary siphon, 3 — small cover plate with distributor plate, 4 — base plate, 5 — adsorbent layer, 6 — carrier plate, 7 — main cover plate, 8 — additional cover plate used to close the chamber. (From Soczewiński, E., *Planar Chromatography,* Vol. 1, Kaiser, R.E., Ed., Huethig, Heidelberg, 1986, pp. 79–117.)

layer (6). The solvent entry position on the adsorbent layer should be previously chosen by sliding the chromatographic plate on skids.

The construction of the original horizontal chamber was reported by Soczewiński [7], who applied an external eluent reservoir equipped with a glass or stainless steel tube (capillary siphon) to deliver the solvent to the chromatographic plate with the adsorbent layer faceup. The cross section of the chamber is presented in Figure 6.4. The tip of the capillary siphon (2) is positioned on the beginning part of the chromatographic plate (6) and under the glass distributor (3), which distributes solvent to the whole width of the chromatographic plate. The distribution plate (3) can be moved forward and onward by a few millimeters according to the direction of chromatogram development, which allows for even distribution of the mobile phase to the chromatographic plate at the beginning stage of chromatogram development. It also allows breakup of the development process at any time. This chamber was named ES (equilibrium sandwich) chamber.

Modification of the ES chamber was reported by Rumiński [8], who applied a glass rod (3) to obtain even distribution of solvent to the adsorbent layer (Figure 6.5). This modification was employed for preparative planar chromatography, especially with plates of large width. Preparative chromatograms can be developed even on 40-cm-wide plates.

The next modification of the ES chamber was reported by Su et al. [9], who introduced a new eluent distributor (3) into the chamber (Figure 6.6). The distributor in the chamber can be moved to the desired position along the chromatographic

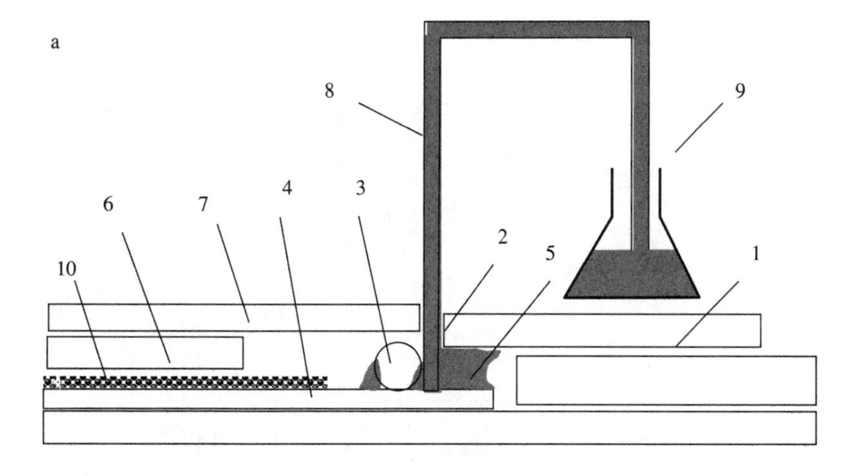

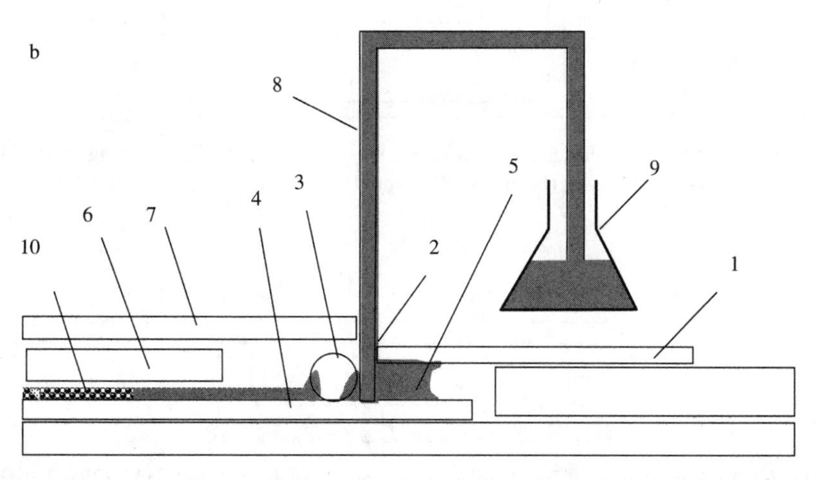

FIGURE 6.5 Cross section of ES chamber modified by Rumiński: 1 — distribution plate, 2 — V-shaped cuts, 3 — distribution rod, 4 — carrier plate, 5 — slit, 6 — cover plate, 7 — tightening plate, 8 — delivery tubes, 9 — external container; a — filling-up mode, b — developing mode. (From Rumiński, J.K., *Chem. Anal. (Warsaw)* 33, 479–481, 1988.)

plate. An improved distributor (funnel distributor) for this chamber is described in the following paper [10] (Figure 6.7). The characteristic feature of the chamber is that the solvent entry position on the adsorbent layer can be changed. This makes the chamber suitable for special modes of chromatogram development, similar to the sequence-TLC device by Bunčak.

The next generation of chambers comprises the horizontal DS chambers (Chromdes, Lublin, Poland) manufactured for plate dimensions ranging from 5 × 10 cm to 20 × 20 cm, depending on chamber type [11,12]. These chambers can

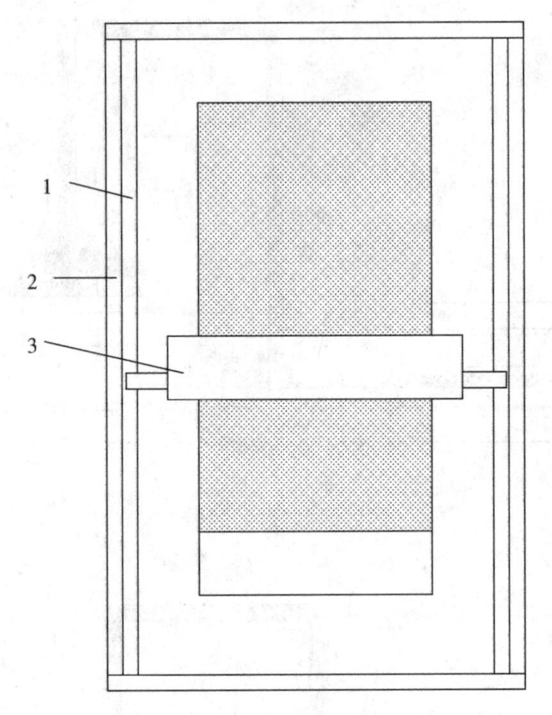

FIGURE 6.6 Top view of ES chamber modified by Wang et al.: 1 — supporting plate, 2 — spacing plate, 3 — distributor. (From Su, P., Wang, D., and Lan, M., *J. Planar. Chromatogr.* 14, 203–207, 2001.)

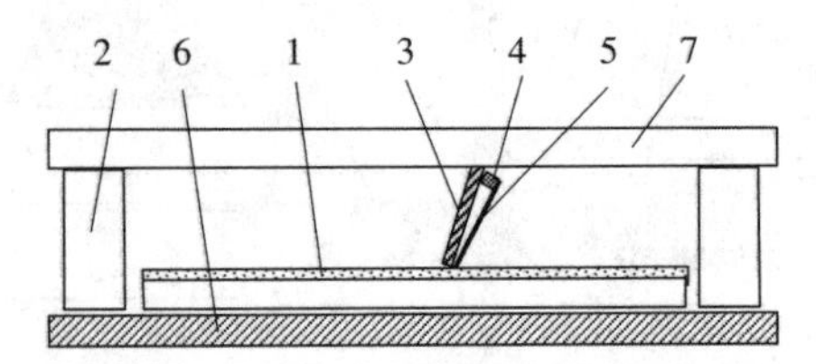

FIGURE 6.7 Cross section of ES chamber with funnel distributor (modified by Wang et al.): 1 — spacing plate, 2 — base plate, 3 — distributor, 4 — glue, 5 — slide, 6 — thin-layer plate, 7 — cover plate. (From Lan, M., Wang, D., and Han, J., *J. Planar. Chromatogr.* 16, 402–404, 2003.)

operate in normal and sandwich mode as well. The cross section of the DS-II chamber type is shown in Figure 6.8. The adsorbent layer on the plate is supplied with eluent from the shallow reservoir, which is positioned approximately on the same level as the chromatographic plate. The eluent reservoir is covered with a glass plate, which forms a flat capillary with the bottom of the eluent reservoir. Any solvent introduced into the reservoir forms a vertical meniscus, which is a very important advantage of the chamber: during chromatogram development, the meniscus of eluent in the reservoir moves in the direction of the chromatographic plate (Figure 6.8b). This feature makes the chamber very economical. It minimizes the

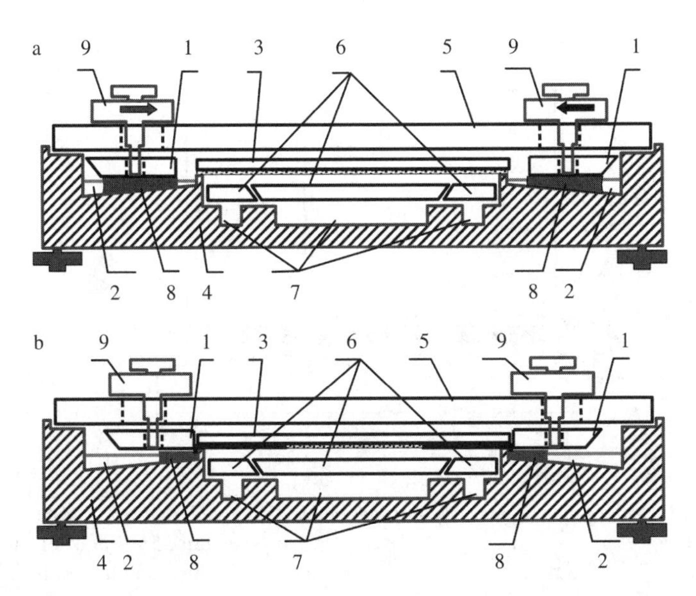

FIGURE 6.8 Horizontal DS-II chamber: (a) before development, (b) during development; 1 — cover plate of the mobile phase reservoir, 2 — mobile phase reservoir, 3 — chromatographic plate facedown, 4 — body of the chamber, 5 — main cover plate, 6 — cover plates (removable) of the troughs for vapor saturation, 7 — troughs for saturation solvent, 8 — mobile phase, 9 — mobile phase distributor.

consumption of solvent; eluent can be exhausted from the reservoir even to the last drops. In addition, the DS chamber enables saturation of the layer with eluent vapors by insertion of some drops of eluent onto the bottom of troughs lined with blotting paper.

A horizontal developing chamber is also manufactured by CAMAG for plates 10 × 10 cm and 20 × 10 cm [13]. Its application is reported in the literature for analytical separations. Therefore, it is not described here in detail.

An interesting construction of the horizontal chamber was recently proposed by Nyiredy [14] (Figure 6.9). It is a fully online chromatograph in which a chromatographic plate is applied instead of the column. The adsorbent layer is covered with a quartz plate so the chromatographic system has no vapor phase. Solvent is delivered to the chromatographic plate from the eluent reservoir by a wick (filter paper), and its movement is caused by capillary forces. The end part of the plate is exposed, and the mobile phase can evaporate, leading to continuous chromatogram development. The sample is injected with a microsyringe using a special injection block. Diode array spectrophotometer is applied to simultaneous band detection. In the next version of the chamber, Nyiredy [15] introduced a pumping system for mobile phase delivery to the chromatographic plate instead of the eluent reservoir with wick. The pumping system can provide isocratic or gradient mode of solvent delivery to the chromatoplate. The mobile phase is still moved through the chromatographic plate by capillary forces. The chromatographic plate is equipped with a special thin channel scratched in the adsorbent layer to ensure even distribution of the mobile phase (Figure 6.10).

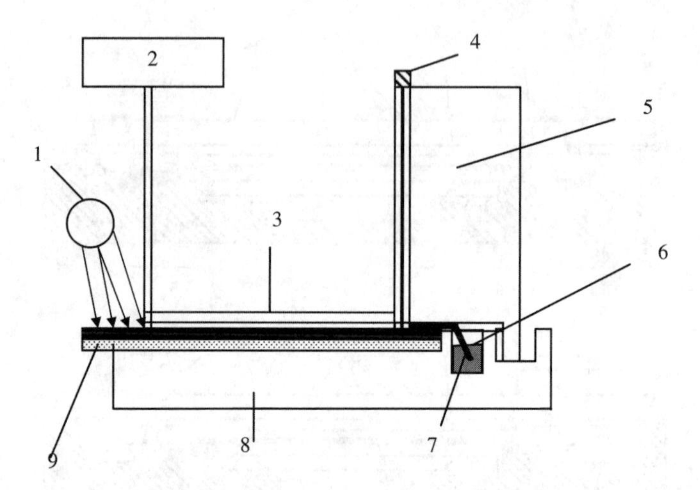

FIGURE 6.9 Horizontal chamber for fully online HPTLC by Nyiredy: 1 — evaporator, 2 — diode-array detector, 3 — quartz glass cover plate, 4 — septum, 5 — injector block, 6 — mobile phase, 7 — filter paper, 8 — Teflon chamber, 9 — chromatoplate. (From Nyiredy, Sz., *J. Planar Chromatogr.* 15, 454–457, 2002.)

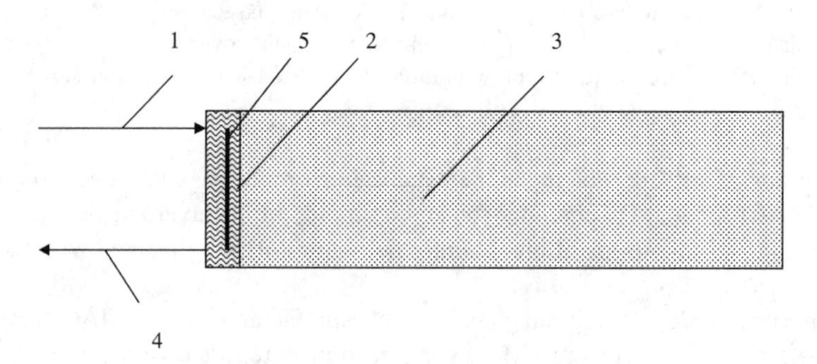

FIGURE 6.10 The cover of the prepared chromatoplate for gradient development in a fully online horizontal chamber: 1 — solvent system inlet, 2 — Silcoflon cover sheet, 3 — chromatoplate, 4 — solvent system outlet, 5 — channel for solvent system. (From Nyiredy, Sz. and Benkö, A., *Proceedings of the International Symposium on Planar Separations, Planar Chromatography 2004*, Nyiredy, Sz., Ed., Research Institute for Medicinal Plants, Budakalász, 2004, pp. 55–60.)

6.2.2 Horizontal Chambers for Radial Development

There are two main types of chamber for radial development, those for circular and anticircular development. However, the mobile phase driven by capillary forces in the radial mode of planar chromatogram development for preparative separations is rarely applied in laboratory practice. At the beginning stage of planar chromatography, this mode of preparative planar chromatogram development was realized with chambers that were designed for analytical separations. The simple chamber for circular development can be set up with a petri dish containing solvent and a wick used to transport the mobile phase to the center of the chromatographic plate [16].

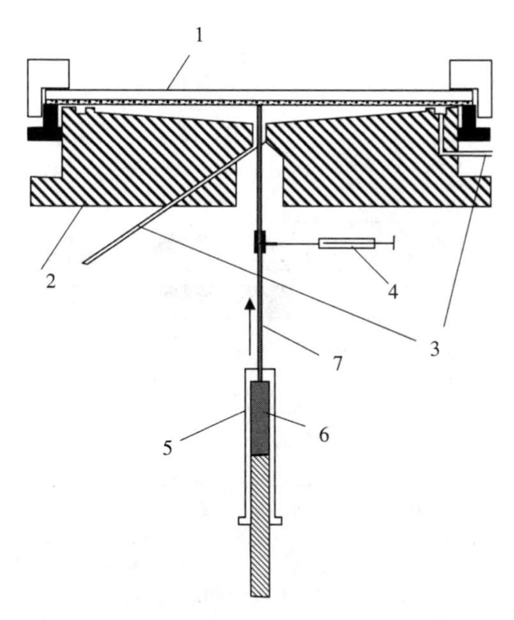

FIGURE 6.11 Cross-sectional view of the circular U–chamber (CAMAG): 1 — chromatographic plate; 2 — body of the chamber; 3 — inlet or outlet for parallel or counter gas flow, to remove vaporized mobile phase, to dry or moisten (impregnate) the plate; 4 — syringe for sample injection; 5 — dosage syringe to maintain the flow of mobile phase; 6 — mobile phase; 7 — capillary. (From Kaiser, R.E., *HPTLC High Performance Thin-Layer Chromatography*, Zlatkis, A. and Kaiser, R.E., Eds., Elsevier, Institute of Chromatography, Amsterdam, Bad Dürkheim, 1977, pp. 73–84.)

Another construction in which the mobile phase is delivered to the center of the chromatographic plate with the syringe connected to the stepping motor was reported by Kaiser [17] and manufactured by CAMAG (Figure 6.11). The sample to be separated can be spotted on the plate in its center or 1 to 2 cm around its center.

A special chamber for preparative chromatogram development was described by Botz et al. [18] and Nyiredy [19] (Figure 6.12). A 20 × 20 cm chromatographic plate with adsorbent layer faceup is placed on the aluminum holder. The stainless steel reservoir is located in the central part of the plate covering the scratched surface of the stationary phase up to a diameter of 2 to 3 cm. Special elastic sealing is located between the adsorbent layer and the stainless steel reservoir of eluent. A magnet is located under the chromatographic plate. It creates a magnetic field that helps to hold the reservoir in an appropriate place. The chamber can operate with and without the presence of vapor phase, which is demonstrated on the left and right side of Figure 6.12, respectively. Development starts when the eluent reservoir is filled up with appropriate solvent from the second reservoir. Development of chromatogram can be stopped by turning off a tap of the second reservoir.

Application of anticircular development to preparative planar chromatography is not popular in spite of the possibility of obtaining a good resolution of mixture components, especially of higher R_F values. Delivery of the mobile phase to the

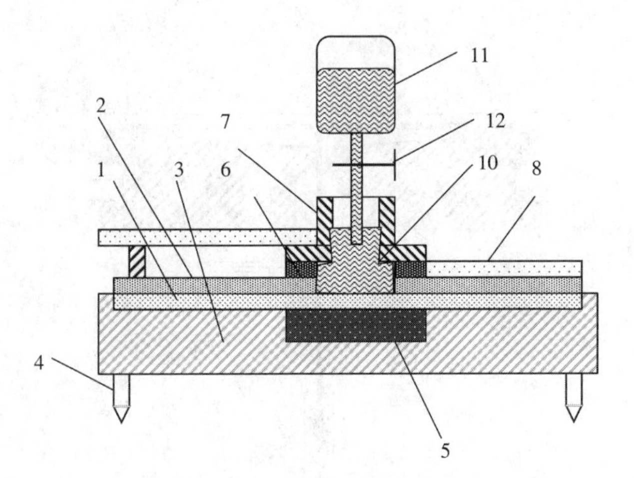

FIGURE 6.12 Chamber for preparative circular planar chromatography: 1 — glass plate, 2 — stationary phase, 3 — aluminum holder, 4 — legs, 5 — magnet, 6 — sealing ring, 7 — stainless steel mobile phase reservoir, 8 — glass cover plate, 9 — metal ring, 10 — mobile phase, 11 — second mobile phase reservoir, 12 — tap. (From Botz, L., Nyiredy, S., and Sticher, O., *J. Planar Chromatogr.* 3, 401–406, 1990; Nyiredy, S., *Handbook of Thin-Layer Chromatography,* 3rd ed., Sherma J. and Fried, B., Eds., Marcel Dekker, New York, Basel, 2003, pp. 307–337.)

adsorbent layer is more complicated in this mode of chromatogram development. In the simplest version, it is accomplished by delivery of the mobile phase to the periphery of the adsorbent layer with wick [20], and in more sophisticated constructions as in the CAMAG anticircular U-chamber with cylindrical capillary [21]. Studer and Traitler designed a special device (adapted U-chamber from CAMAG) for preparative planar chromatography for anticircular separations on 20 × 20 cm chromatoplates [22].

6.3 MODES OF HORIZONTAL CHAMBER APPLICATION TO PLC

6.3.1 Linear Development

6.3.1.1 Isocratic Linear Development

Isocratic linear development is the most popular mode of chromatogram development in analytical and preparative planar chromatography. It can be easily performed in horizontal chambers of all types. The mobile phase in the reservoir is brought into contact with the adsorbent layer, and then the movement of the eluent front takes place. Chromatogram development is stopped when the mobile phase front reaches the desired position. Usually 20 × 20 cm and 10 × 20 cm plates are applied for preparative separations, and this makes the migration distance equal to about 18 cm. Due to the fact that the migration distance varies with time according to the equation $Z_t = ct^{1/2}$ (Z_t, c, and t are the distance of the solvent front traveled, constant, and migration time, respectively), the development of preparative planar chromatograms usually takes a lot of time.

6.3.1.2 Continuous Isocratic Development

In the conventional mode of chromatogram development, the chromatographic plate is closed in the developing chamber. The development is finished when the eluent front reaches the end of the chromatographic plate. However, the development can proceed further if some part of the plate extends out of the chamber; then the eluent evaporates from the end of the plate. This evaporation proceeds at a constant rate and is responsible for the constant flow of the mobile phase. To enhance the efficiency of evaporation, a blower or heating block can be applied to the exposed part of the chromatographic plate. In Figure 6.13, the cross section of a DS chamber application for continuous chromatogram development is shown; also compare with Figure 6.8. Under these conditions, the planar chromatogram development is more similar to the column chromatography mode than to the conventional development. In the case

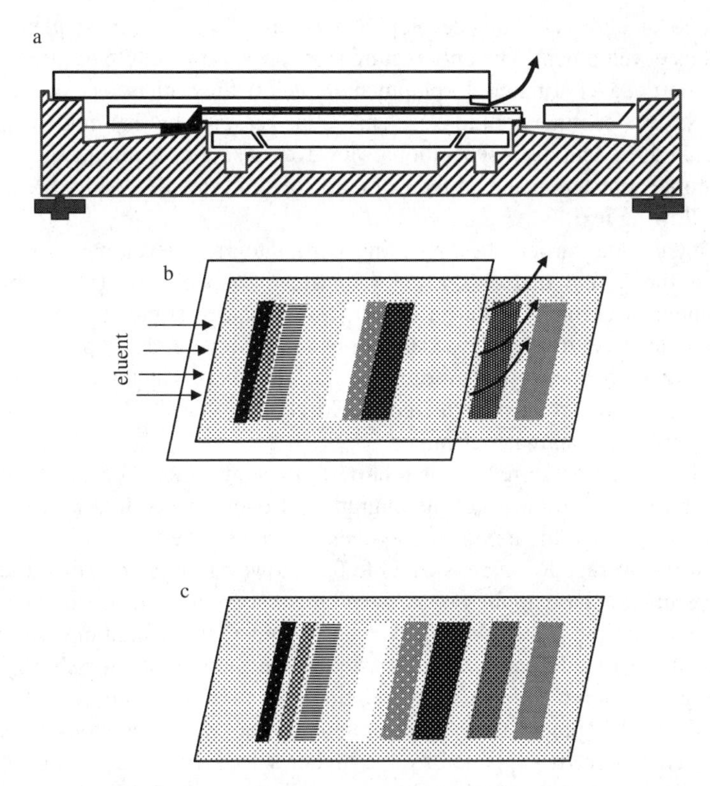

FIGURE 6.13 Schematic demonstration of horizontal DS chamber application for continuous development: (a) cross section of DS chamber during continuous development, (b) part of the plate with bands of lower retention exposed but with bands of higher retention covered to enable further development, (c) final chromatogram.

of incomplete separation of the components of lower R_F values, some increase of separation can be obtained when using this mode. In Figure 6.13b and Figure 6.13c this procedure is schematically demonstrated; the chromatogram development has proceeded to the end (the front of the mobile phase has reached the end of chromatographic plate), and mixture components of higher R_F value are well separated as opposed to those of lower values (Figure 6.13b). In this situation continuous development should be performed. The end part of the chromatographic plate, which comprises the bands of good resolution, needs to be exposed as demonstrated in Figure 6.13b. The components of lower R_F values can migrate through a longer distance, which usually leads to improvements in their separation (Figure 6.13c). If necessary, a larger part of the chromatographic plate can be exposed in the next stage of continuous development to obtain an improvement of separation of components of even higher retention than the one located on the exposed part of the chromatographic plate.

6.3.1.3 Short Bed–Continuous Development

This mode of chromatogram development is, in principle, almost identical with continuous development. The only feature that varies is the length of the developing path. In short bed–continuous development (SB/CD), this path is very short, typically equal to several centimeters [23–25]. This is the reason why this mode is preferentially applied for analytical separations. However, a similar technique is applied for zonal sample application and online extraction of solid samples, which are described in the following text.

As mentioned earlier, the preceding chromatogram development on the full length of the 20-cm plate takes a lot of time. To overcome this problem, the development of chromatogram on a short distance with simultaneous evaporation of the mobile phase from the exposed part of the chromatographic plate can be very conveniently performed by means of horizontal chambers. The mode was introduced by Perry [23] and further popularized by Soczewiński et al. [24,26], using a horizontal equilibrium sandwich chamber.

In this mode, a better resolution relative to the conventional development can be obtained for similar distances of the migration of components. It is well known that the best resolution of the mixture component can be reached in conventional development if the average R_F value is equal to 0.3. However, in the continuous development, the mobile phase applied is of lower eluent strength, e.g., eluent strength that enables reaching the average value $R_F = 0.05$. Under such conditions, several void volumes of the mobile phase should pass through the chromatographic system. If the average migration distance of the component mixture is similar to that of the conventional development, then the resolution obtained with continuous development is better. This effect is explained by the higher selectivity of the chromatographic system with mobile phase of lower eluent strength and by the better kinetic properties of the chromatographic system. At lower eluent strength, molecules of the component mixture spend more time in the stationary phase, and the flow rate of the mobile phase is higher (closer to optimal value) under the condition determined by the efficiency of solvent evaporation from the exposed part of the plate.

The principle of the mode is demonstrated in Figure 6.14 [26]. Several void volumes pass throughout the short chromatographic plate bed, e.g., of 4 cm in length instead of 20 cm in length. Then the mobile phase flow rate is much higher than that at the end of the 20-cm-long plate. However, the flow rate of the mobile phase depends on the efficiency of solvent evaporation. If a more volatile solvent is used, then the flow rate of the mobile phase can be higher. It is often necessary to increase

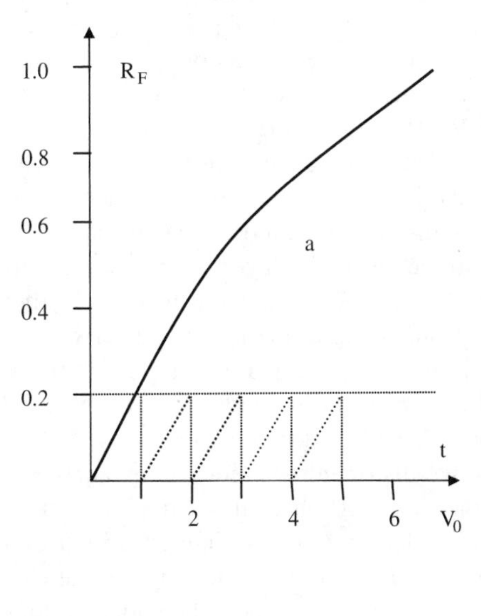

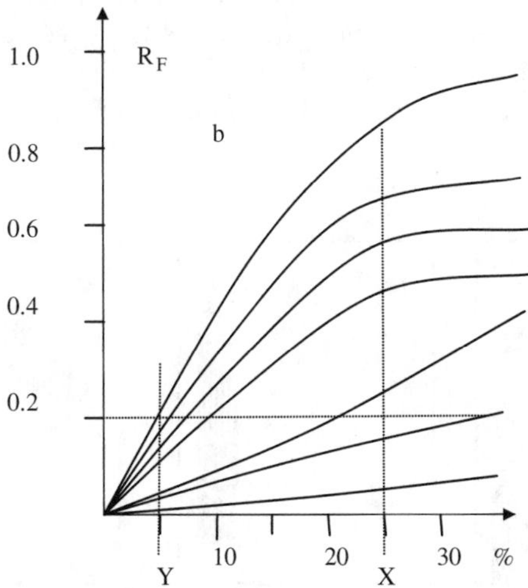

FIGURE 6.14 (a) Principle of SBCD, elution with five interstitial volumes on 4-cm distance (5 × 4 cm) is faster than single development on 20-cm distance (thick line), (b) R_F values of sample components plotted as a function of modifier concentration. Optimal concentration (Y) for SBCD (5 × 4 cm) is lower than for development on the full distance of 20 cm (X). (From Soczewiński, E., *Chromatographic Methods Planar Chromatography,* Vol. 1, Ed., Kaiser R.E., Dr. Alfred Huetig Verlag, Heidelberg, Basel, New York, 1986, pp. 79–117.)

the efficiency of evaporation of the mobile phase from the end of the short plate by the application of a heater or blower.

6.3.1.4 Two-Dimensional Separations

Horizontal chambers can be easily used for two-dimensional separations in the usual way. The only problem seems to be the size of the sample. In a conventional two-dimensional separation used for analytical purposes, the size of the sample is small. The quantity of the sample can be considerably increased when using a spray-on technique with an automatic applicator. Soczewiński and Wawrzynowicz have proposed a simple mode to enhance the size of the sample mixture with the ES horizontal chamber [27]. The principle of this mode is presented in Figure 6.15. In Figure 6.4, two distributing plates have been introduced instead of one in the ES chamber: one short, 1-cm-long distributor for sample application, and the other long for solvent delivery to the adsorbent layer. During sample application as shown in Figure 6.15a, the mobile phase is delivered to the adsorbent layer using a long distribution plate, and the sample mixture is delivered using a small distributor. Then the sample band is evenly formed during the procedure of zonal application according to frontal chromatography mode (see Section 6.3.3.3). After the sample application is performed, both distributors are fed with the appropriate eluent, or the cover plate with two distributors is replaced by another one with a long single distributor. If after the development of the chromatogram the sample is not satisfactory resolved, e.g., as demonstrated in Figure 6.15b, then the second eluent with a different selectivity is delivered to another edge of the chromatographic plate to develop the chromatogram in the direction perpendicular to the former one (Figure 6.15c). After the first development is performed the bands can be rectangular in shape. If necessary, before the second development the bands can be flattened by the preconcentration technique (described in Section 6.3.3.1).

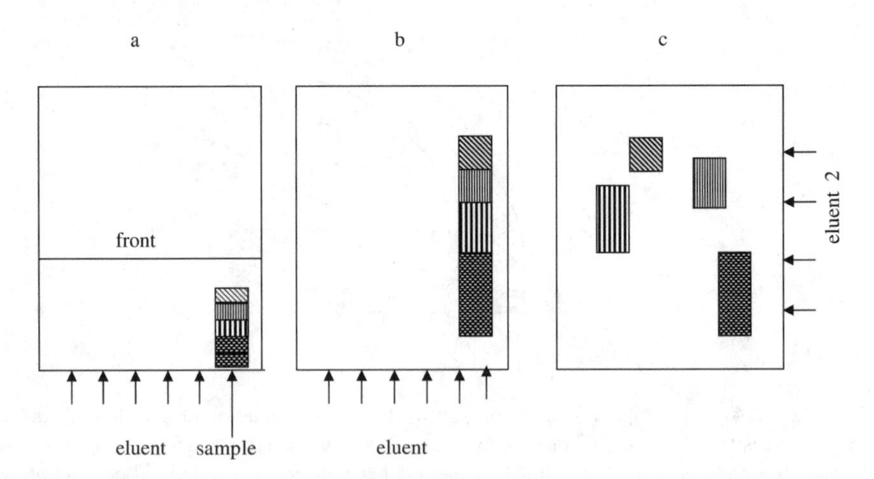

FIGURE 6.15 Two-dimensional micropreparative separation. (a) Formation of rectangular starting band from the short distributor, eluent delivered parallel from the distributor, (b) development in the first direction, (c) development in the second direction.

6.3.1.5 Multiple Development

Multiple development is the mode in which a chromatogram is developed several times on the same plate and each step of the development follows the complete evaporation of the mobile phase from the chromatographic plate of the previous development. Multiple development can be performed in the followimg ways [28]: unidimensional multiple development (UMD), incremental multiple development (IMD), gradient multiple development (GMD), and bivariant multiple development (BMD). In UMD each step of chromatogram development is performed with the same mobile phase and the same migration distance of the eluent front. The same mobile phase, but with an increasing development distance in each subsequent step, is applied in IMD mode. GMD mode applies the same development distance but a different composition of the mobile phase in each step. The most versatile mode of multiple development is BMD, in which the composition and development distance vary in each step of chromatogram development. These modes of chromatogram development are mainly applied for analytical separations owing to very good efficiency, which is comparable to HPLC. The sophisticated device to be used in this mode is manufactured by CAMAG and named AMD2. All modes of multiple development can be easily performed using chambers for automatic development, which are manufactured by some firms. However, these device are relatively expensive. Typical horizontal chambers for preparative planar chromatography should be considered for application in multiple development in spite of more manual operations in comparison to automatic chromatogram development. Especially, horizontal DS chamber could be considered for application in multiple development including preparative separations. This chamber can be easily operated owing to its convenient maintenance, including cleaning the eluent reservoir. For the separation of a more complicated sample mixture, computer simulation could be used to enhance the efficiency of the optimization procedure. However, its applications are still reported for analytical separations [29,30].

All these modes of multiple chromatogram development are mainly applied in analytical separation; however, there are some examples of preparative planar chromatography [31,32].

6.3.1.6 Gradient (Stepwise and Continuous) Development

Sample mixtures containing components of various polarity cannot be often completely separated in one run using isocratic development. The thin-layer chromatogram is then composed of bands of medium retention (R_F values in the range of 0.1 to 0.8), of lower retention ($0.8 < R_F < 1$), and of higher retention ($0.0 < R_F < 0.1$). In this situation bands of lower and higher retention are not well separated, and they are located close to the eluent front and start line respectively. This separation problem is well known as the *general elution problem* and is solved in most cases by gradient elution. In TLC, gradient elution can be realized by applying multiple development (especially BMD) and stepwise or continuous gradient development. The stepwise mode can be easily implemented using horizontal chambers (DS and ES chambers). The principle of the mode is then based on the introduction of mobile phase fractions of increasing eluent strength

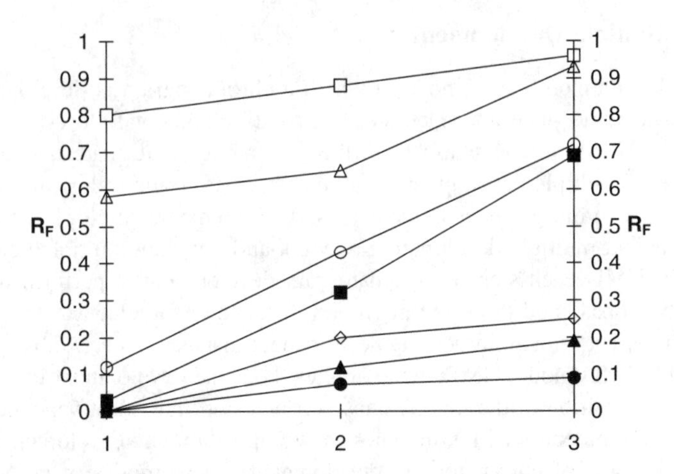

FIGURE 6.16 R_F values of aromatic amines obtained on silica gel plate: 1, 3 — isocratic development with 5 and 50% solutions of methyl ethyl ketone in cyclohexane, respectively, 2 — two-stepwise gradient development with both solvents; open squares, *N, N*-dimethylaniline, open triangles, *N*-ethylaniline, open circles, aniline, diamonds, 2-phenylenediamine, filled squares, 3-phenylenediamine, filled triangles, 4-phenylenediamine, filled circles, 3-aminopyridine. (From Soczewiński, E. and Czapińska, K., *J. Chromatogr.* 168, 230–233, 1979.)

following one after another in a series into the eluent reservoir in the DS chamber or under the distributor plate in the ES chamber [26,29,30]. The most important requirement that should be fulfilled during chromatogram development in this case is that each eluent fraction should be introduced into the reservoir when previous one has been completely exhausted. It was reported that even a two-step gradient can considerably increase the separation of the component mixture [33]. In Figure 6.16, the retention data are presented for three chromatographic systems. Two systems are isocratic with low and high eluent strength, and as can be seen on the illustration, complete separation cannot be obtained in either chromatogram. However, the best separation selectivity was obtained on the third chromatogram, in which the two-step gradient was applied. The stepwise gradient was also applied to preparative separations [34]. In Figure 6.17, preparative chromatograms of Azulan extract are presented with isocratic (Figure 6.17a) and six-step gradient elution (Figure 6.17b), with the concentration of ethyl acetate varying from 10 to 40% in chloroform. As can be seen on the illustration, the separation obtained with the stepwise gradient was considerably improved in comparison to isocratic elution. More bands represented by higher peaks were produced in the gradient chromatogram.

More sophisticated equipment is necessary for continuous gradient elution in a horizontal chamber. Soczewiński and Matysik [35] have proposed a miniaturized gradient generator with two vessels connected with elastic PTFE tubing and filled up with spontaneously mixing solvents. This gradient generator is combined with the ES chamber. However, the device is more suitable for narrow plates (e.g., 5 cm wide). Wider plates do not produce a uniform gradient profile across their area.

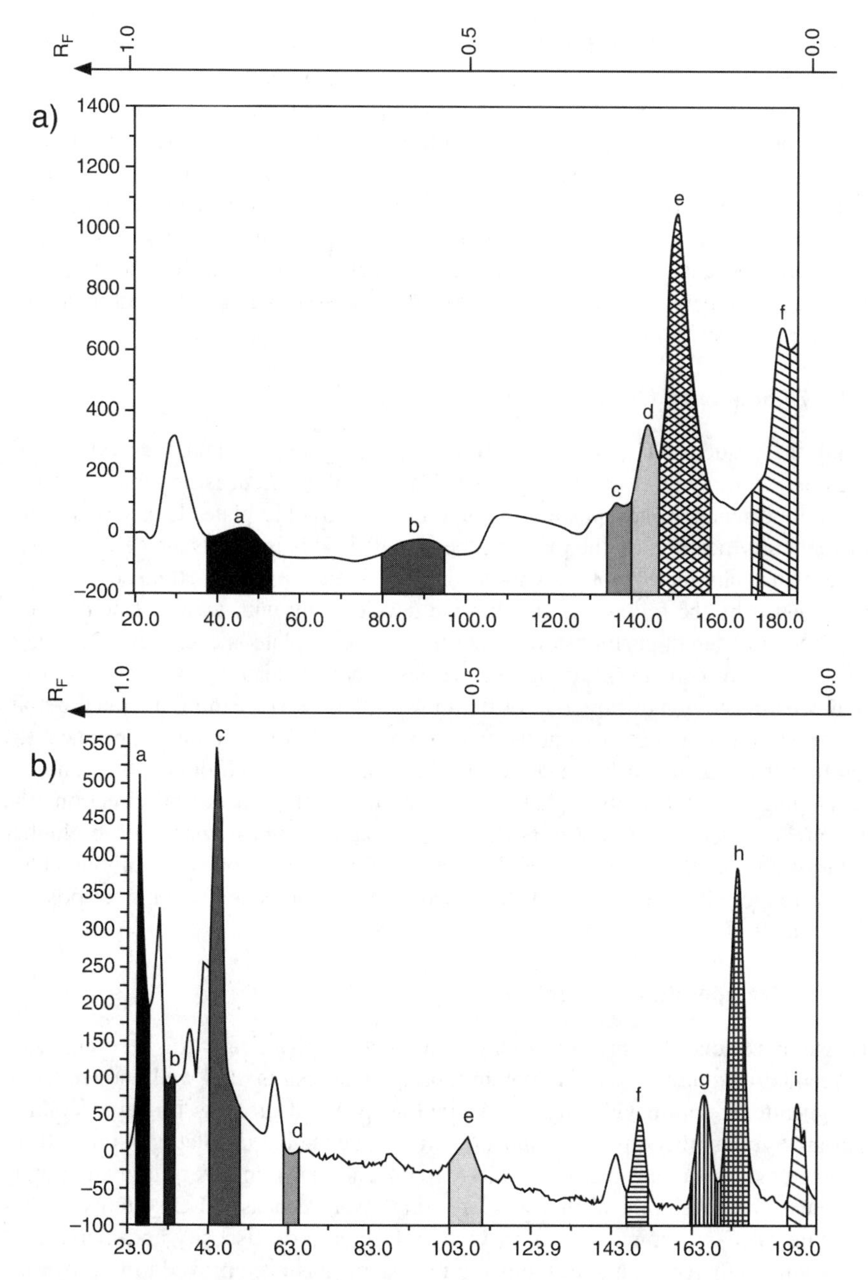

FIGURE 6.17 Densitograms of Azulan extract scanned at 410 nm divided into fractions a to i; (a) isocratic development, ethyl acetate in chloroform (1:5), (b) stepwise gradient development, 10 to 40% v/v of ethyl acetate in chloroform. (From Matysik. G., Soczewiński, E., and Polak, B., *Chromatographia,* 39, 497–504, 1994.)

The device for continuous gradient elution in horizontal chamber described by Nyiredy [15] and presented in the preceding text, (Figure 6.10) seems to be a very interesting solution both for analytical and preparative applications.

Continuous gradient elution can be also easily created by the saturation of the mobile phase during the development in the horizontal DS chamber with another solvent vapor, whose drops are placed on the blotting paper lining the trough bottom of the chamber [36]. High reproducibility of this mode is difficult to obtain and the gradient range is restricted. But for some separations, this mode could be considered because of very good selectivity that can be obtained, especially for mixture components showing various properties of proton–donor and proton–acceptor interactions [36].

6.3.1.7 Sequence Development

Using the sequence-TLC device of Bunčak [6] (Figure 6.3), and the modified ES chamber of Wang et al. [9,10] (Figure 6.6 and Figure 6.7), it is possible to deliver the mobile phase in any position to the chromatographic plate. It means that the solvent entry position on the plate can be changed. This is an advantage for preparative planar chromatography owing to the increased separation efficiency that can be achieved by the following: (1) multiple development with various eluents being supplied to different positions on the chromatographic plate in each step, (2) changing the solvent entry position during the development, leading to increase in efficiency owing to higher flow rate of the mobile phase, (3) cleaning the plate before the development, (4) separating the trace components from the bulk substance, (5) spotting the mixture to be separated in the middle of the chromatoplate, and (6) developing with the mobile phase of lower eluent strength in one direction (the mixture components of lower polarity are separated and components of higher polarity stay on the start line), and developing after evaporation of the mobile phase with stronger eluent in the opposite direction (then components of higher polarity are separated).

6.3.1.8 Temperature Control

Temperature control in preparative layer chromatography is rare. It is recommended to keep temperature constant to obtain reproducible results. It is well known that at a higher temperature viscosity of the mobile phase decreases. In liquid column chromatography this effect is responsible for the reduction of inlet pressure and for the increase of mobile phase velocity in conventional thin-layer chromatography systems. The development time was halved at reversed-phase TLC systems at temperature 58°C in comparison to 15°C [37]. Decrease of viscosity and increase of diffusion coefficient in higher-temperature systems has been proved to improve the separation efficiency by increasing mass transfer of the solute to the stationary phase. However, this advantage is not so obvious for TLC systems. In a TLC system at elevated temperature, stronger evaporation of solvent takes place. This can lead to the formation of bubbles of solvent vapor and dissolved air. On the other hand, higher temperature leads to solubility increase of the sample, which is responsible for increase in capacity of the chromatographic system. This property is very important for preparative separations — samples of higher size can be separated. Taking into account the preceding discussion, an adaptation of the horizontal DS chamber to preparative planar chromatography with temperature control was performed [37]. This chamber enables precise temperature control of the chromatographic system because the chromatographic plate is located between two heating coils connected to the circulating thermostat. Figure 6.18 shows preparative chromatograms of three dye mixtures obtained at temperatures of 20, 30, and 40°C [38]. As can be seen, the separation is poor if the development is performed at 20°C (Figure 6.18a). Temperature increase to 30°C leads to the improvement of resolution (Figure 6.18b). However, the best resolution is obtained when development is performed at 40°C (Figure 6.18c) — all three bands are well resolved at this temperature. The explanation of the effect is connected to the capacity increase of the chromatographic system and retention decrease of the mixture components (longer migration distance of the bands leads to better resolution) at higher temperature.

6.3.2 Radial Development

Radial development of a chromatogram can be performed as circular and anticircular development. The chambers for these development modes are described in the preceding section.

6.3.2.1 Circular Development

The sample mixture in this mode can be applied on the chromatographic plate at its center or close to it. In the former case, the sample components form ring-like zones. An example of such a chromatogram is presented in Figure 6.19 [39]. The chromatogram was obtained with a circular U-chamber from CAMAG (Figure 6.11), which can be used for preparative and analytical separations.

The chamber for circular development presented by Botz et al. [18] and modified by Nyiredy [19] (Figure 6.12) is suitable for preparative planar chromatogram development with mobile phase movement driven by capillary action. Various sample mixtures (solid or liquid) can be applied on the chromatographic plate. Examples of sample application modes are presented in Figure 6.20 [18]. The recommended application mode of the liquid mixture is by the injection of sample solution into the hole of the adsorbent layer located in the center of the chromatographic plate. Then the adsorbent layer is wetted by the sample solution until it is completely absorbed. The remnants of this solution can be transferred to the adsorbent layer by some portion of the mobile phase, which is injected into the center of the chromatographic plate. As seen in Figure 6.20a, the sample forms narrow rings. It is possible to obtain a narrow starting zone by hand operation if the operator has experience with this procedure. These modes of sample applications can be used for preconcentration of sample mixture and for sample extraction from the solid state, e.g., powdered plant material (Figure 6.20b). The solid sample mixture is placed in the hole located in the center of the adsorbent layer. The appropriate solvent should be used to extract the components of interest. In this procedure, preconcentration with

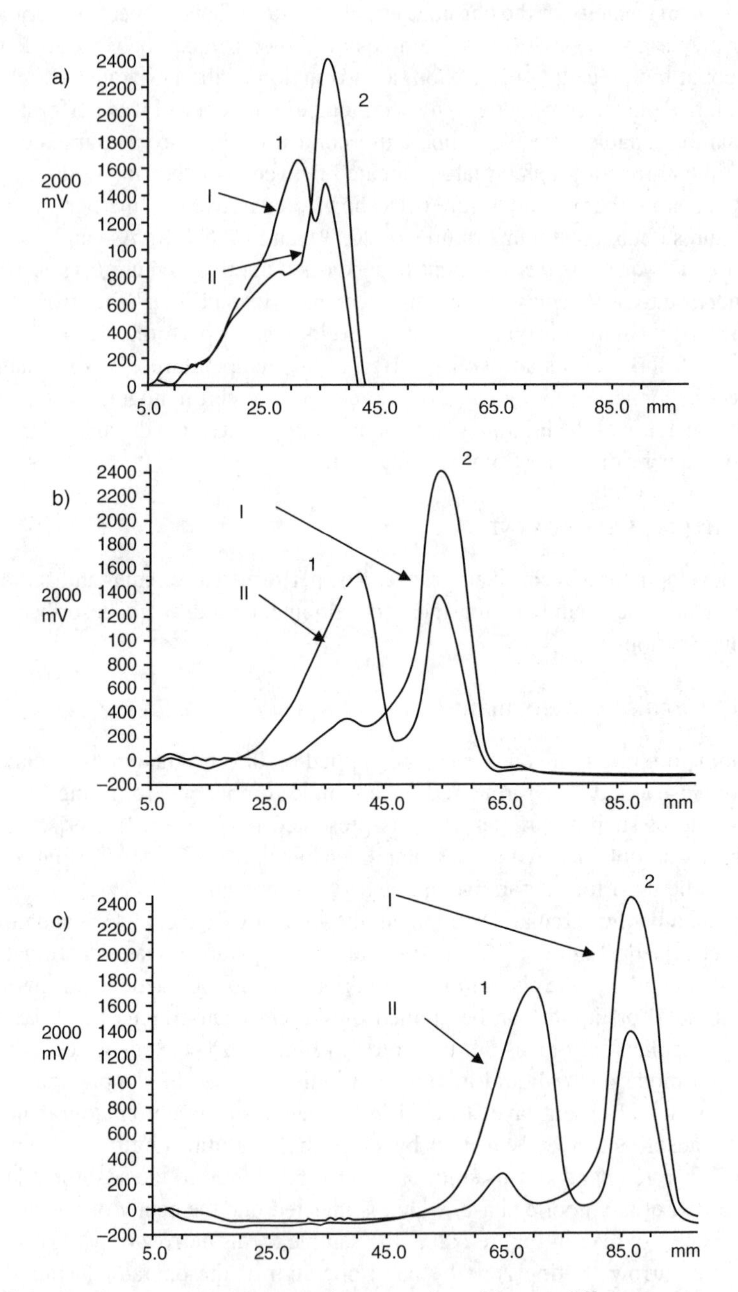

FIGURE 6.18 Chromatogram of preparative separation of dyes at different temperatures; (a) 20°C, (b) 30°C, (c) 40°C. (From Dzido, T.H., Gołkiewicz, W., and Piłat, J.K., *J. Planar Chromatogr.* 15, 258–262, 2002.)

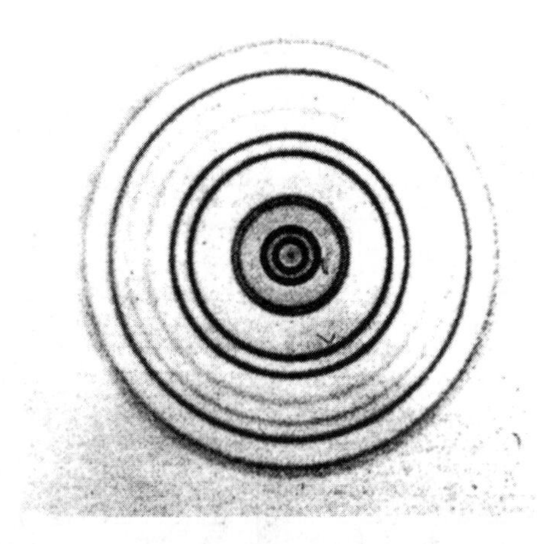

FIGURE 6.19 Circular chromatography of dyes on precoated silica gel high-performance TLC plate; lipophilic dyes, mobile phase: hexane–chloroform–NH_3, 70:30. (From Ripphahn, J. and Halpaap, H., *HPTLC High Performance Thin Layer Chromatography,* Zlatkis, A. and Kaiser, R. E., Eds., Elsevier, Amsterdam, 1977, pp. 189–221.)

continuous extraction of powdered material is advantageous to enhance extraction efficiency. The chamber enables chromatogram development with a sample applied into the scrapped channel in the adsorbent layer, too (Figure 6.20c). The obtained quality of separation of the test dye mixture with circular development was considerably higher than that using linear ascending development performed in a twin-trough chamber. Even linear development using plates with a preconcentration zone produced lower separation quality in comparison to circular development (Figure 6.21) [18]. An additional advantage obtained with circular development in this chamber was a higher increase of separation quality with development distance in comparison to linear development in a twin-trough chamber (Figure 6.22) [18]. The authors advise that separation using circular development with the chamber described has advantages compared to separation efficiency obtained in linear development.

However, the advantages are with respect to chromatogram development in vertical N-chamber (twin-trough chamber). The separation quality was not compared with that using horizontal mode of linear development. The authors have reported one disadvantage of the mode: recovering of bands of interest can be performed only by scraping the adsorbent from the plate after development. This is more complicated than in the case of linear chromatogram development using rectangular plates.

6.3.2.2 Anticircular Development

Anticircular development is very rarely applied in planar chromatographic practice for preparative separation. This mode of separation was introduced by Kaiser [40]. Studer and Traitler adapted the anticircular U-chamber from CAMAG to preparative separations on 20 × 20 cm plates [22], as mentioned earlier. The sample mixture was spotted at the circumference of the plate, and the mobile phase was moved from the circumference to the center of the plate. The bands obtained are elongated, but their diameter measured perpendicular to the direction of mobile phase movement

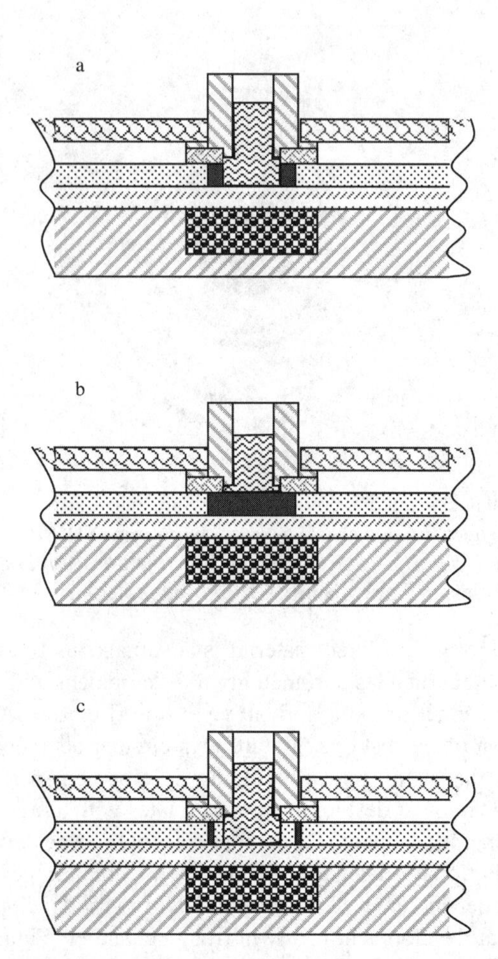

FIGURE 6.20 Sample application using preparative circular planar chromatographic device (the black zones symbolize the applied sample): (a) liquid phase sample application, (b) solid phase application into the center hole of the plate, (c) solid phase sample application into a scrapped channel. (From Botz, L., Nyiredy, Sz., and Sticher, O., *J. Planar Chromatogr.* 3, 401–406, 1990.)

is not increased, as it is in linear or circular development. This mode when applied for preparative separation is characterized by some advantages, according to Studer and Traitler: (1) a larger amount of sample (about twice as much) can be loaded comparing to the situation when linear 20 × 20 cm plates are used, (2) samples can be collected by the "through-flow" technique, and they can be taken at a certain time, (3) the plate can be properly washed because of "through-flow" technique, and (4) the separation can occur approximately five times faster than in linear separation (with the same solvent system) and five times less solvent is used. However, the mode has not gained much popularity in laboratory practice, probably because more sophisticated equipment is necessary to perform the separation.

Issaq [41] has proposed using conventional chambers to perform anticircular development. In this mode, a commercially available chromatographic plate is

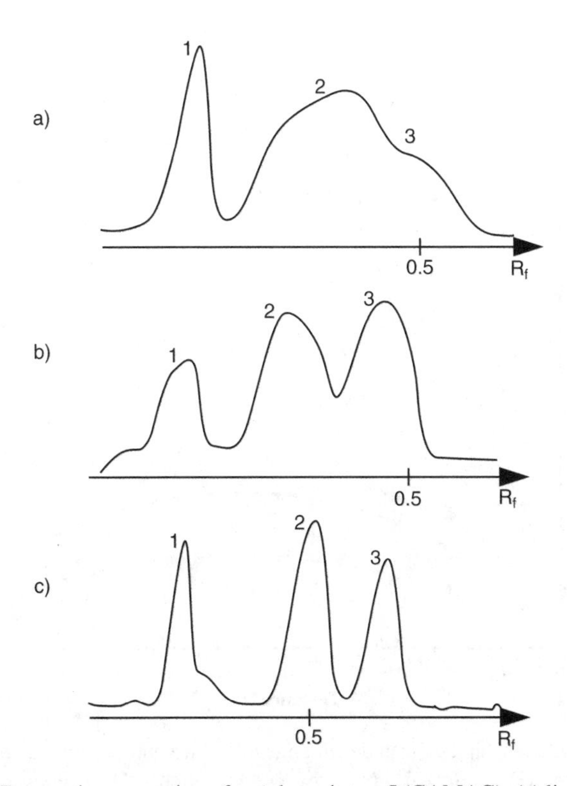

FIGURE 6.21 Preparative separation of test dye mixture I (CAMAG): (a) linear development using preparative plate without concentrating zone (development time: 83 min), (b) linear development using preparative plate with concentrating zone (development time: 76 min), (c) circular development using preparative plate (development time: 38 min). 1 = oracet blue, 2 = oracet red, 3 = butter yellow; detection: VIS at λ = 500 nm. (From Botz, L., Nyiredy, Sz., and Sticher, O., *J. Planar Chromatogr.* 3, 401–406, 1990.)

divided into triangular plates, and the sample (or samples) is spotted along the base of a triangular plate. Wetting of the mobile phase is started when the base of the triangular chromatographic plate is contacted with solvent. This means that all kinds of developing chambers (N-chambers and S-chambers) can be easily used in this mode of chromatogram development. The bands on the plate after preparative chromatogram development are narrower than the original bands on the start line of the plate, depending on their migration distance Figure 6.23 [41]. This means that the bands are more concentrated, require less solvent for development, and less solvent to elute from the plate as well.

6.3.3 Sample Treatment

6.3.3.1 Preconcentration

This procedure is concerned with narrowing the sample zones on the start line on the chromatographic plate. In analytical separation the preconcentration procedure is applied to increase the efficiency and decrease detection limit. In preparative

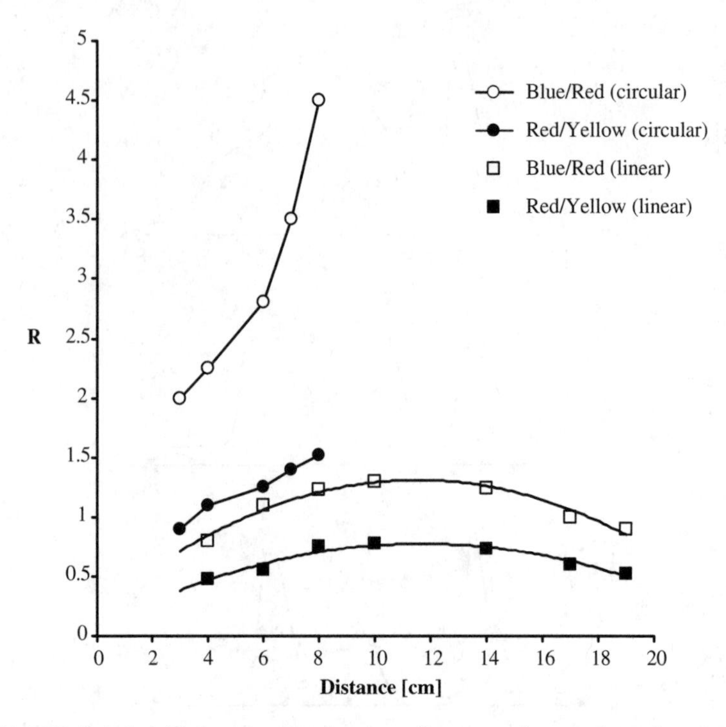

FIGURE 6.22 Relationship between development distance and resolution using circular and linear development. (From Botz, L., Nyiredy, Sz., and Sticher, O., *J. Planar Chromatogr.* 3, 401–406, 1990.)

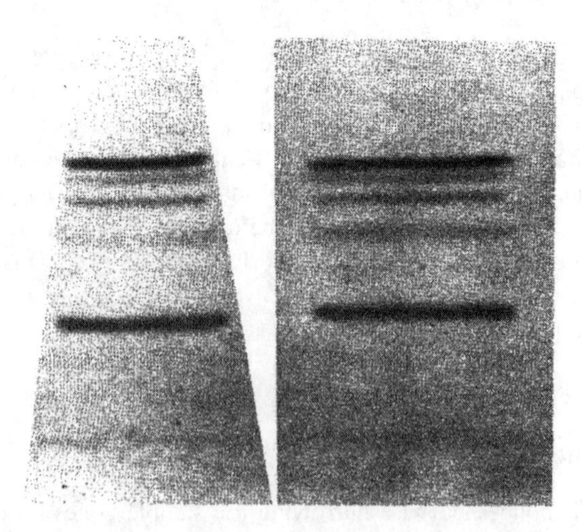

FIGURE 6.23 Comparison of the separation of streaks of dye on triangular and rectangular 5 × 20 cm plates. (From Issaq, H.J., *J. Liq. Chromatogr.* 3, 789–796, 1980.)

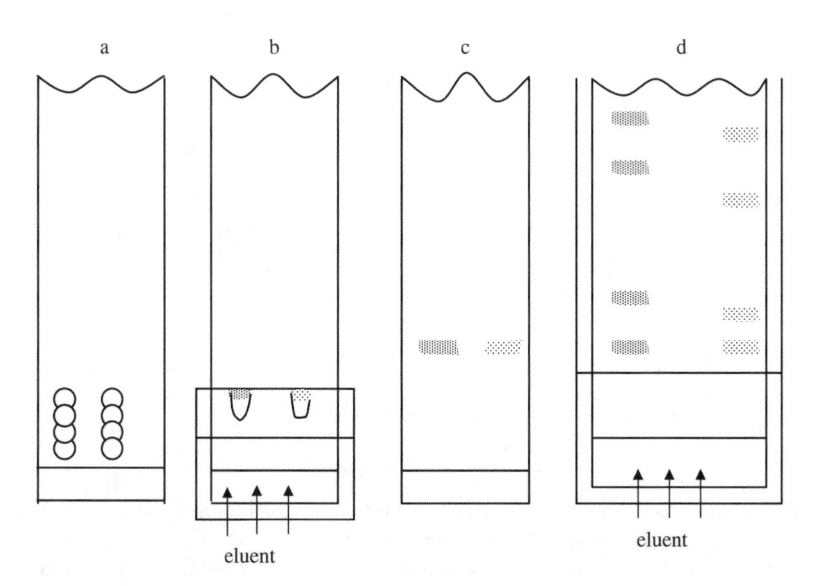

FIGURE 6.24 Online preconcentration of samples on thin layer: (a) spotting the sample solutions in series along the layer, (b) elution with strong, volatile solvent, and concentration beyond the additional short cover plate, (c) evaporation of solvent, (d) elution in a closed chamber. (From Soczewiński, E., *Chromatographic Methods Planar Chromatography,* Vol. 1, Ed., Kaiser R.E., Dr. Alfred Huetig Verlag, Heidelberg, Basel, New York, 1986, pp. 79–117.)

planar chromatography this procedure is used to increase the sample size and efficiency of the separating system. The preconcentration procedure is very similar to continuous development performed in a horizontal chamber. In the first stage, the samples to be separated are spotted along the layer in series 2 to 3 cm long, as shown in Figure 6.24 [26]. The adsorbent layer is then covered with a narrow plate, and volatile solvent is delivered to the adsorbent layer (Figure 6.24a). The sample is eluted in the adsorbent layer and swept beyond the narrow cover plate, forming sharp starting zones that appear owing to solvent evaporation. Evaporation of solvent can be accelerated by application of a blower with air or nitrogen. In the next stage, the cover plate is removed so that the solvent can evaporate. When this happens, the chamber is covered with a lid. Then the development of the chromatogram is performed with a suitable eluent. The preconcentration procedure can be easily performed with ES and DS chambers. Nyiredy's device for fully online TLC can be applied to this procedure, too [15].

The preconcentration procedure can be helpful for elution of the sample component from the chromatographic plate after the separation process. This operation is shown in Figure 6.25 [26]. After the development, the chromatographic plate is placed under UV light to localize separated zones of the component mixture. The zones of interest are separated from the remaining adsorbent area by scraping off the adsorbent below and above them. The plate is then placed in the chamber and developed in the direction perpendicular to the former one. The zone is moved to the side (right side of the plate in Figure 6.25) of the chromatographic plate, forming a narrow zone that can be easily eluted from the chromatographic plate.

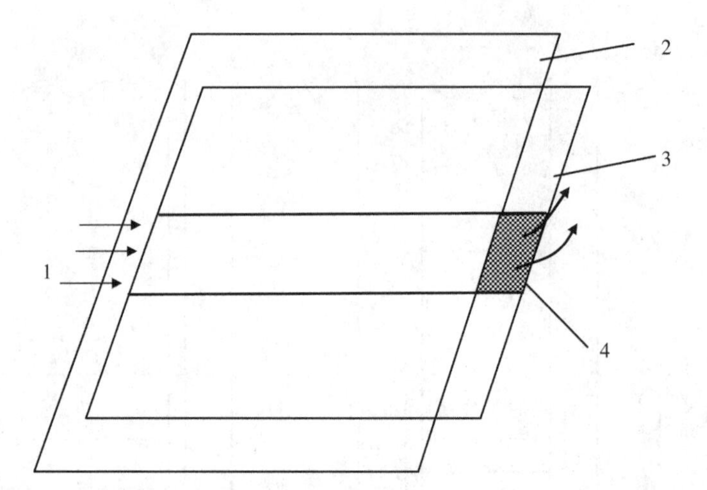

FIGURE 6.25 Concentration of a separated zone in the chamber by elution at a right angle and evaporation of solvent beyond the cover plate: 1 — eluent, 2 — cover plate, 3 — chromatographic plate, 4 — sample compound. (From Soczewiński, E., *Chromatographic Methods Planar Chromatography,* Vol. 1, Ed., Kaiser R.E., Dr. Alfred Huetig Verlag, Heidelberg, Basel, New York, 1986, pp. 79–117.)

6.3.3.2 Online Extraction

The procedure described earlier for sample preconcentration can be easily extended for the online extraction of solid samples, e.g., powdered plant materials. Horizontal configuration of the chromatographic plate in the chamber facilitates this procedure, because the sample to be extracted is then placed on a carrier plate at the beginning part of the adsorbent layer (or in the scrapped channel of the adsorbent layer), which should be directed upward [15,26]. The chamber is covered with a narrow plate, and the development is started with a suitable extracting solvent. In some cases, it is advantageous to put the narrow plate directly on the adsorbent layer to press the sample to be extracted. Extracted components are preconcentrated on the adsorbent layer at the end of the narrow plate, as shown in Figure 6.26 [15].

After extraction and evaporation of the extracting solvent, the chamber is covered with the main lid, and the chromatogram is developed employing the typical procedure. A few small samples or one large sample can be simultaneously extracted using this mode [26].

6.3.3.3 Band Application

For preparative separation, it is necessary to apply large quantities of sample on the chromatographic plate. Using the typical procedure for sample application, the starting zone should form a band. This can be achieved by spotting a series of spots or streaking the solution on the start line (Figure 6.27). An autosampler with a spray-on technique is a convenient way of obtaining a uniform band of sample mixture on the starting line. However, the price of this device is relatively high. The two first modes often form an unwanted complex starting band with components of lower

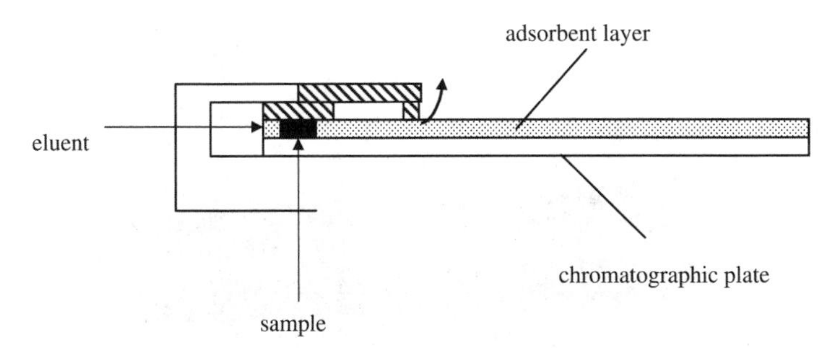

FIGURE 6.26 Principle online extraction in a horizontal chamber. (From Nyiredy, Sz. and Benkö, A., *Proceedings of the International Symposium on Planar Separations, Planar Chromatography 2004*, Nyiredy, Sz., Ed., Research Institute for Medicinal Plants, Budakalász, 2004, pp. 55–60; Soczewiński, E., *Chromatographic Methods Planar Chromatography,* Vol. 1, Ed., Kaiser R.E., Dr. Alfred Huetig Verlag, Heidelberg, Basel, New York, 1986, pp. 79–117.)

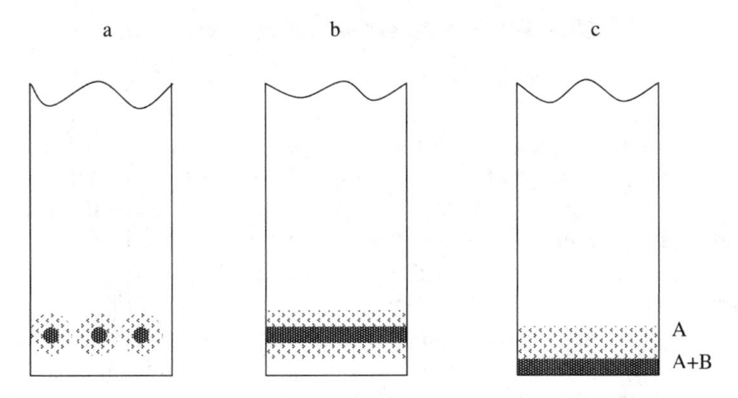

FIGURE 6.27 Starting zones for large sample size: (a) spotting, (b) streaking, (c) delivered with the horizontal DS chamber (from eluent reservoir) — partial separation due to frontal chromatography. (From Soczewiński, E., *Chromatographic Methods Planar Chromatography,* Vol. 1, Ed., Kaiser R.E., Dr. Alfred Huetig Verlag, Heidelberg, Basel, New York, 1986, pp. 79–117; Soczewiński, E. and Wawrzynowicz, T., *J. Chromatogr.* 218, 729–732, 1981.)

retention on the periphery of the band and of higher retention inside the band, especially if a large volume of sample mixture is applied to the plate (Figure 6.27a and Figure 6.27b). This effect can be responsible for decrease of efficiency and is very important, especially if a more complicated sample mixture is to be separated.

Some horizontal chamber types (DS and ES chambers) allow this inconvenience to be circumvented [42,43]. In case of the DS chamber, the sample mixture is introduced into the eluent reservoir, and the chromatogram is developed with this sample solution. This means that the sample solution is introduced into the adsorbent layer from the edge of the chromatographic plate. In this stage of sample application, the process proceeds according to the frontal chromatography principle. The component with the lowest adsorption energy migrates first. As the result of this band application, the component mixture is partly resolved, which is advantageous for performing further separation procedures (development of the chromatogram). In

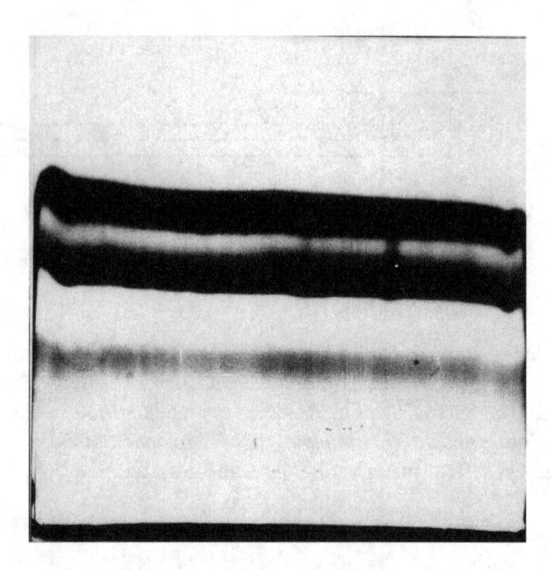

FIGURE 6.28 Preparative chromatogram of dyes obtained after band application.

Figure 6.27c, this procedure is illustrated by showing the band of the mixture throughout the whole width of the chromatographic plate [26,27]. In the next stage, the development of the chromatogram proceeds. It is performed by the introduction of an appropriate developing solvent into the eluent reservoir. It is recommended that at first a small quantity of the eluent be introduced into the reservoir to wash the reservoir walls and eliminate remnants of the sample solution. An example of separation of the sample mixture using this procedure is shown in Figure 6.28. Even a few milliliters of the sample solution can be applied with this mode. However, it depends on the sample mixture.

Soczewiński gives some practical hints to effectively apply preparative chromatography with the zone applied as a band with the ES chamber [26]. These remarks are valid in regard to other chambers as well. For example, for lower eluent strength of the mobile phase, it is recommended that application of the sample be preceded by wetting the layer with some portion of eluent to eliminate solvent demixing. However, if the chamber allows for the saturation of the adsorbent layer — as the DS chamber does — this operation may not be necessary. The next hint is to dissolve the sample in the eluent. It helps avoid disturbances in the composition of the mobile phase. On the other hand, it is known that sample should be dissolved in a solvent of lower eluent strength. In such a circumstance, the band is formed as a more compact zone. This occurs as a result of stronger adsorption of mixture components. This procedure should be applied for separation of more complicated sample mixtures, e.g., plant extract. In the case of simple mixtures, as those obtained in the synthesis of organic compounds, the sample applied as a band can be considerably overloaded. Band application can lead to the increase of separation efficiency owing to preliminary resolution of the sample mixture obtained during the application of the sample. The principle of this mode is shown in Figure 6.29, in which a two-component sample mixture is applied to the chromatographic plate from the eluent

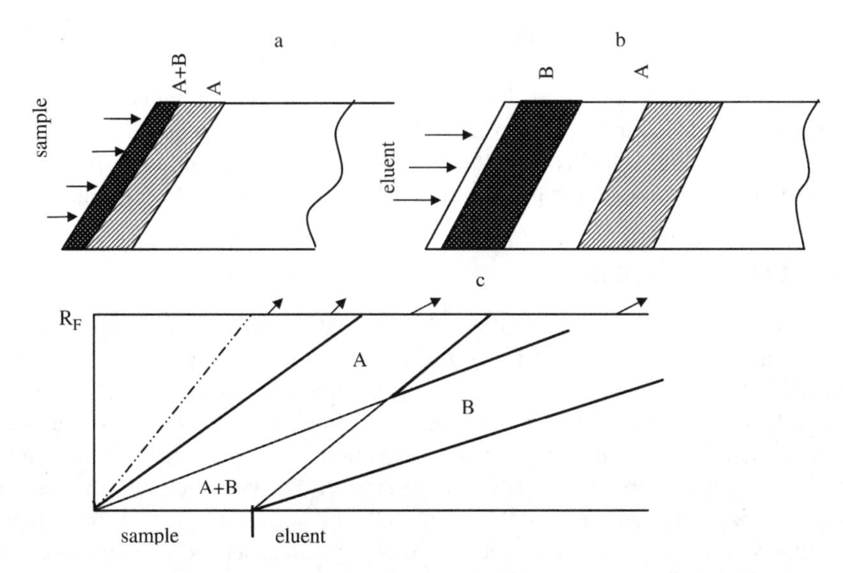

FIGURE 6.29 (a) Delivery of large sample volume with the distributor, (b) subsequent elution, (c) graphical representation of formation of zones in the frontal and elution chromatography process. (From Soczewiński, E., *Chromatographic Methods Planar Chromatography,* Vol. 1, Ed., Kaiser R.E., Dr. Alfred Huetig Verlag, Heidelberg, Basel, New York, 1986, pp. 79–117.)

reservoir. Figure 6.29a shows the application of the sample in overloaded condition and Figure 6.29b and Figure 6.29c show the development of the chromatogram (very wide bands are formed).

As was reported by Soczewiński, a three-component mixture containing 5 mg of each of the ingredients can be completely separated using a 90 × 100 × 0.5 mm layer of silica [26]. This indicates that the capacity of the chromatographic system can be considerably increased by the application of wider and thicker layers of adsorbent on the chromatoplate [44].

6.4 REGENERATION OF THE LAYERS

If a chromatographic plate is not treated with reagents, as it is in case of visualization of the bands, and the layer is not destroyed by scraping the bands to elute separated components, then the following question arises: why not apply this chromatographic plate again, as a column in the HPLC technique? In some laboratories, this approach should be taken into account owing to the reduction of laboratory costs if many ready-for-use plates are applied.

Contemporary evaluation of the chromatographic plate is often performed by the localization of spots in UV light or densitometry scanning. After this procedure is performed, the adsorbent layer structure is not destroyed. The problem is how to remove the separated component mixture from the plate to make it ready to use for the next separation.

Most horizontal chambers enable convenient and economical regeneration of the layers. The procedure is as follows: (1) press the blotting or chromatographic paper with a glass strip to the end of the chromatographic plate using clips, (2) insert the chromatographic plate into the horizontal chamber so that it contacts its edge (which is opposite to that of the blotting paper) with solvent of high eluent strength (e.g., ethyl acetate, acetone, ethanol), (3) allow the solvent to wet the whole adsorbent layer and enter eluted bands from the adsorbent layer to the blotting paper, and (4) remove the chromatographic plate from the chamber and leave for drying.

6.5 CONCLUSIONS

In spite of the ongoing automation of analytical procedures, the application of horizontal chambers is still of interest in contemporary laboratory practice. Many features of the chambers such as simple and convenient operation, and especially versatile methodical possibilities (gradient elution, multiple development, two-dimensional development, sample preconcentration, band application), have resulted in their popularity growing in laboratory practice. A very important advantage of some horizontal chambers is their low solvent consumption (in some types, extremely low), which leads to decrease of operation costs and reduction of environmental pollution (large quantities of solvents remaining after developing process in conventional chambers must be utilized). The next important factor is the price of the chambers. Horizontal chambers are much cheaper in comparison to more complicated devices with a greater amount of automation. Planar chromatography is still automated in a restricted range. Only single operations such as scanning, chromatogram development, and sample application are automated. The proceeding factors imply that the interest in horizontal chambers for analytical and preparative separations will continue.

REFERENCES

1. Abbot, D.C., Egan, H., Hammond, E.W., and Thomson, J., *Analyst* 89, 480–488, 1964.
2. Brenner, M. and Niederwieser, A., *Experientia* 17, 237–238, 1961.
3. Kraus, L., *Concise Practical Book of Thin-Layer Chromatography*, Desaga, Heidelberg, 1993.
4. Geiss, F. and Schlitt, H., *Chromatographia* 1, 392–402, 1968.
5. Geiss, F., Schlitt, H., and Klose, A., *Z. Anal. Chem.* 213, 331–346, 1965.
6. Bunčak, P., *GIT Fachz.Lab. (Suppl., Chromatographie)*, G-I-T-Verlag, Darmstadt, 3–8, 1982.
7. Soczewiński, E., *Planar Chromatography*, Vol. 1, Kaiser, R.E., Ed., Huethig, Heidelberg, 1986, pp. 79–117.
8. Rumiński, J.K., *Chem. Anal. (Warsaw)* 33, 479–481, 1988.
9. Su, P., Wang, D., and Lan, M., *J. Planar. Chromatogr.* 14, 203–207, 2001.
10. Lan. M., Wang, D., and Han, J., *J. Planar. Chromatogr.* 16, 402–404, 2003.
11. Dzido, T.H. and Soczewiński, E., *J. Chromatogr.* 516, 461–466, 1990.
12. Dzido, T.H., *J. Planar Chromatogr.* 6, 78–80, 1993.
13. Jaenchen, D.E., *Handbook of Thin Layer Chromatography*, 2nd ed., Sherma, J. and Fried, B., Eds., Marcel Dekker, New York, 1996, pp. 129–148.
14. Nyiredy, Sz., *J. Planar Chromatogr.* 15, 454–457, 2002.

15. Nyiredy, Sz. and Benkö, A., *Proceedings of the International Symposium on Planar Separations, Planar Chromatography 2004*, Nyiredy, Sz., Ed., Research Institute for Medicinal Plants, Budakalász, 2004, pp. 55–60.
16. Blome, J., *HPTLC High Performance Thin-Layer Chromatography*, Zlatkis, A. and Kaiser, R.E., Eds., Elsevier, Institute of Chromatography, Amsterdam, Bad Dürkheim,1977, pp. 51–71.
17. Kaiser, R.E., *HPTLC High Performance Thin-Layer Chromatography*, Zlatkis, A. and Kaiser, R.E., Eds., Elsevier, Institute of Chromatography, Amsterdam, Bad Dürkheim, 1977, pp. 73–84.
18. Botz, L., Nyiredy, Sz., and Sticher, O., *J. Planar Chromatogr.* 3, 401–406, 1990.
19. Nyiredy, Sz., *Handbook of Thin-Layer Chromatography,* 3rd ed., Sherma J. and Fried, B., Eds., Marcel Dekker, New York, Basel, 2003, pp. 307–337.
20. De Deyne, V.J.R. and Vetters, A.F., *J. Chromatogr.* 103, 177–179, 1975.
21. Bauer, K., Gros, L., and Sauer, W., *Thin Layer Chromatography — An Introduction*, Hüthig, Heidelberg, 1991.
22. Studer, A. and Traitler, H., *J. High Resol. Chromatogr. Chromatogr. Commun.* 9, 218–223, 1986.
23. Perry, J.A., *J. Chromatogr.* 165, 117–140, 1979.
24. Matysik, G., Soczewiński, E., and Matyska, M., *Farmacja Polska* 39, 331–334, 1983.
25. Lee, Y.K. and Zlatkis, K., *Advances in Thin Layer Chromatography, Clinical and Environmental Applications,* Touchstone J.C., Ed., Wiley-Interscience, New York, 1982.
26. Soczewiński, E., *Chromatographic Methods Planar Chromatography,* Vol. 1, Ed., Kaiser R.E., Dr. Alfred Huetig Verlag, Heidelberg, Basel, New York, 1986, 79–117.
27. Soczewiński, E. and Wawrzynowicz, T., *J. Chromatogr.* 218, 729–732, 1981.
28. Szabady, B., *Planar Chromatography, A Retrospective View for the Third Millennium*, Nyiredy, Sz., Ed., Springer, 2001, pp. 88–102.
29. Gołkiewicz, W., *Handbook of Thin-Layer Chromatography,* 3rd ed., Revised and Expanded, Sherma, J. and Fried, B., Eds., Marcel Dekker, New York, Basel, 2003, pp. 153–173.
30. Markowski, W., *Encyclopedia of Chromatography*, Cazes, J., Ed., Marcel Dekker, New York, Basel, 702–717, 2005.
31. Waksmundzka-Hajnos, M., Gadzikowska, M., and Gołkiewicz, W., *J. Planar Chromatogr.* 13, 205–209, 2000.
32. Wawrzynowicz, T., Czapińska, K. L., and Markowski, W., *J. Planar Chromatogr.* 11, 388–393, 1998.
33. Soczewiński, E. and Czapińska, K., *J. Chromatogr.* 168, 230–233, 1979.
34. Matysik, G., Soczewinski, E., and Polak, B., *Chromatographia,* 39, 497–504, 1994.
35. Soczewiński, E. and Matysik, G., *J. Liq. Chromatogr.* 8, 1225–1238,1985.
36. Dzido, T.H. and Polak, B., *J. Planar Chromatogr.* 6, 378–381, 1993.
37. Dzido, T.H., *J. Planar Chromatogr.* 14, 237–245, 2001.
38. Dzido, T.H., Gołkiewicz, W., and Piłat, J.K., *J. Planar Chromatogr.* 15, 258–262, 2002.
39. Ripphahn, J. and Halpaap, H., *HPTLC High Performance Thin Layer Chromatography,* Zlatkis, A. and Kaiser, R.E., Eds., Elsevier, Amsterdam, 1977, pp. 189–221.
40. Kaiser, R. E., *J. High Resol. Chromatogr. Chromatogr. Commun.* 3, 164–168, 1978.
41. Issaq, H.J., *J. Liq. Chromatogr.* 3, 789–796, 1980.
42. Soczewiński, E. and Maciejewicz, W., *J. Chromatogr.* 176, 247–254, 1979.
43. Soczewiński, E., Kuczmierczyk, J., and Psionka, B., *J. Liq. Chromatogr.* 3, 1829–1834, 1980.
44. Soczewiński, E., Wawrzynowicz, T., and Czapińska, K.L., *Chromatographia,* 20, 223–228, 1985.

7 Location of Separated Zones by Use of Visualization Reagents, UV Absorbance on Layers Containing a Fluorescent Indicator, and Densitometry

Bernd Spangenberg

CONTENTS

7.1 INTRODUCTION

Preparative thin-layer chromatography (TLC) is an offline technique because sample application, chromatographic development, and plate assessment are carried out stepwise using various devices. But, in any case, the detection is done by the use of light. The plate is illuminated by incoming light (incident light) and evaluated by the detection of the reflected light. The most convenient way to detect separated

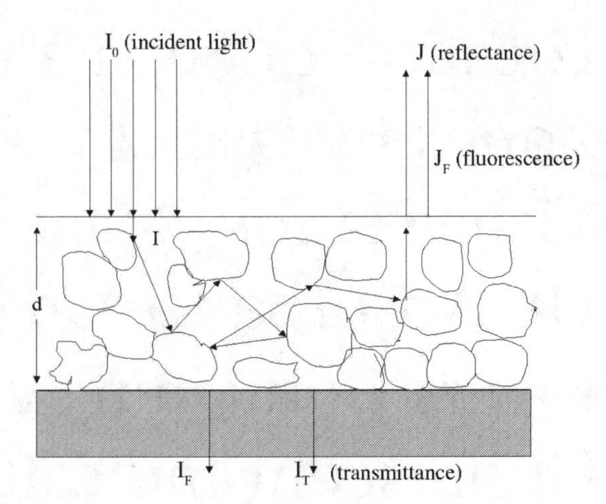

FIGURE 7.1 Interactions of an adsorbent and scattering TLC layer with incident light.

compounds on the plate is visual examination. Unfortunately, this works only in the spectral range from 400 to 800 nm. *In situ* densitometry offers a way of measuring the optical density of the separated spots directly on the plate in the spectral range from 200 to 1100 nm [1–3]. Therefore, a scanner is recommended.

7.2 THE KUBELKA–MUNK THEORY OF REMISSION

During measurements, the illuminating light is either absorbed or reflected by the plate and the reflected light provides the desired information of how much light is absorbed by the sample. The diffuse reflected light is called reflectance, denoted by J. Usually different processes occur during plate illumination. Figure 7.1 shows an overview.

Figure 7.1 schematically shows the interactions of an adsorbent and scattering TLC layer with the incident light.

Complex processes are involved in transmittance and reflectance of scattered radiation, which are theoretically described by Schuster [4]. In an ideal scattering medium all fluxes of light can be summed up as components of two vectors. Vector I stands for the light flux in the direction of the incident light, and the vector J describes the light intensity in the antiparallel direction. With k, the absorption coefficient, and s, the scattering coefficient, the two Schuster equations are as follows:

$$-\frac{\delta I}{\delta x} = -2(k+s)I + 2sJ \tag{7.1}$$

$$\frac{\delta J}{\delta x} = -2(k+s)J + 2sI \tag{7.2}$$

Both equations describe the differential reduction of the two light fluxes (I and J) from top to bottom (I) and from bottom to top (J).

If a sample is illuminated by a parallel light flux with the intensity I_0 and there is no scattering in the sample (s = 0) and no fluorescence, the incoming light is reduced in intensity and leaves the sample as transmitted light I_T. The reduction over a distance d can be calculated from the two Schuster equations as:

$$\ln \frac{I_0}{I_T} = 2kd \tag{7.3}$$

This simple relationship between incident and transmitted light is well known as the Boguert–Lambert–Beer law. This expression renders positive values for $I_T < I_0$. In case of scattering material like TLC plates, a part of the scattered light is emitted as reflectance J from the plate surface to the top. For the first approximation of a parallel incident light beam with the intensity I_0, some radiation may be scattered inside the layer and some radiation may be absorbed either by the sample or by the layer itself. According to the Schuster equations and with the abbreviation R (the diffuse reflectance of an infinitely thick layer),

$$R_\infty = \frac{J}{I_0} \tag{7.4}$$

a linear relationship between the absorption coefficient and the light reduction is given by the so-called Kubelka–Munk equation [5–8]:

$$\frac{(1 - R_\infty)^2}{R_\infty} = \frac{k}{s} \tag{7.5}$$

If the reflected light intensity of a clean plate (J_0) is combined with the reflected value (or the spectrum) of the TLC spot (J) as

$$R = \frac{J}{J_0} \tag{7.6}$$

the Kubelka–Munk equation shows the following form [9]:

$$KM = \frac{(1 - R)^2}{2R} = \frac{a_m}{s} m \tag{7.7}$$

The reflected light intensity of the plain plate J_0 can easily be taken at a surface area free of any other compounds. In this equation, m indicates the mass of the absorbing molecules in the layer. The proportionality factor is called the sample absorption coefficient, a_m. In this expression, the absorption of the TLC layer is taken into consideration by use of J_0 instead of I_0.

7.2.1 Direct Detection in Absorption

The evaluation of thin-layer plates by optical methods is based on a differential measurement of light emerging from the sample-free and the sample-containing zones of the plate. In the UV wavelengths range, only TLC scanners can measure this light loss. Two different scanner systems are on the market. To evaluate a TLC plate, in general, a monowavelength scanner is used, because this type of scanner is, at present, the dominant type worldwide. This system was introduced by H. Jork in the year 1966 [10]. Monowavelength scanners illuminate the plate with nearly monochromatic light. A monowavelength scanner has to illuminate the plate sequentially with light of different wavelengths in order to scan an absorption spectrum directly from a sample spot or to measure chromatograms (in TLC densitograms) at different wavelengths. In general, the reflected light from a clean TLC plate (J_0) is combined with the reflected light of a TLC track (J). Equation 7.8 is generally used for evaluations. This expression is similar to the Boguert–Lambert–Beer Equation 7.3 and describes a functional relation between the logarithm expression and the sample mass.

$$-\ln R = f(a_m m) \tag{7.8}$$

The expression is called "absorbance" and has a positive value if more light is absorbed from a substance spot than from a clean TLC plate surface. Usually, there is no linear relationship between the logarithmic signal and the sample mass in the layer.

In contrast to monowavelength scanners, diode-array scanners use white light for illumination purposes. Diode-array scanners register TLC plate spectra and densitograms at different wavelengths directly from the plate. This kind of scanner was first introduced in the year 2000 [11]. For evaluation purposes, the logarithmic expression of the Boguert–Lambert–Beer law (Equation 7.8) or the Kubelka–Munk equation (Equation 7.7) can be used, because both expressions transform a decreasing reflected light intensity from a sample spot into an increasing measurement signal. In other words, both equations give positive values for reflected light intensities smaller than J_0 [12].

Figure 7.2 shows the separation of a spinach leaf extract on a silica gel high-performance thin-layer chromatography (HPTLC) plate, extracted with acetone at room temperature and separated with *n*-heptane and 1-propanol (9:1, v/v) as the mobile phase. The densitogram is measured at 305 nm by the use of a diode-array scanner (Tidas TLC 2010 diode-array scanner, J&M company, Aalen, Germany) and is calculated with Equation 7.7. Equation 7.8 would result in a similar plot. In this densitogram, all plate areas emitting lesser light intensities than the reference area are plotted as positive signals. The spinach extract was simply applied bandwise at

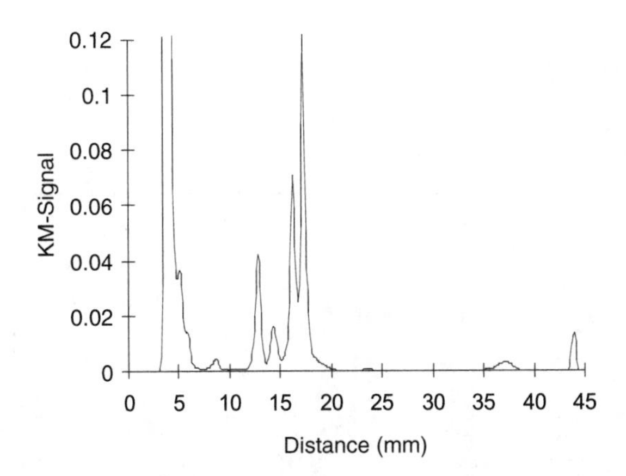

FIGURE 7.2 Kubelka–Munk densitogram of a spinach leaf extract taken at 305 nm. The two peaks at 16-mm and 17-mm distances are due to chlorophyll b and chlorophyll a.

5-mm distance without further pretreatment, and therefore a large signal at the point of application can be seen. The front signal is visible at 44-mm distance. Four weak signals seen as yellow bands on the plate are distinguishable in the Kubelka–Munk densitogram as peaks at 8-, 13-, 14.5-, and 37-mm distances. These peaks are from the compounds neoxanthin, violaxanthin, lutein, and carotene. The two larger peaks at 16- and 17-mm distances are due to chlorophyll b and chlorophyll a. The disadvantage of monowavelength scanners is that the information from reflected light is measured at a single wavelength. Figure 7.2 includes no sample light absorptions at wavelengths other than 305 nm. Additional measurements are recommended for further information.

In contrast, a diode-array scanner offers much more information. Figure 7.3 shows the same sample as a Kubelka–Munk contour plot, measured with a diode-array device in the wavelength range from 200 to 600 nm. This contour plot combines 450 single spectra covering a distance of 45 mm on the plate. From this contour plot, 512 single densitograms are extractable. In Figure 7.3 the spectrum of the peak at 13 mm (violaxanthin) is plotted on the left-hand side. The spectral information can be beneficially used for the identification of separated sample spots. The four yellow spots show absorptions between 250 and 450 nm, whereas the two chlorophylls have additional absorption signals beyond 550 nm, which indicate their original green color.

7.2.2 Direct Detection in Fluorescence

The term fluorescence is used for a transformation of absorbed light into a light (J_F) of lower energy. The extent of this transformation is described by the quantum yield factor q. Light intensity absorbed by the sample can be calculated from the light intensity reflected from the clean plate surface (J_0) minus the light intensity (J) reflected from a sample spot [13].

$$J_F(\lambda_F) = q\left[J_0(\lambda_A) - J(\lambda_A)\right] = qJ_0(1-R) \tag{7.9}$$

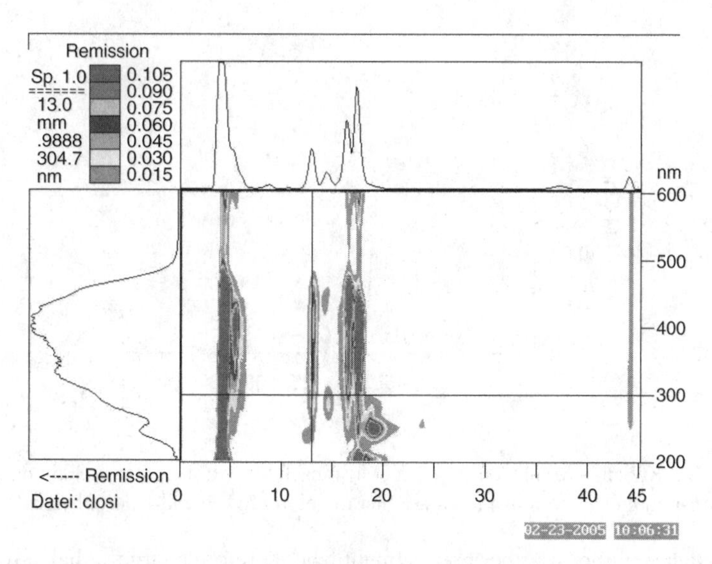

FIGURE 7.3 Kubelka–Munk contour plot of a spinach leaf extract, taken at 305 nm. The two signals at 16- and 17-mm distances are due to chlorophyll b and chlorophyll a, which show absorptions beyond 550 nm.

In general, the wavelengths of the fluorescent light are shifted to higher wavelengths (λ_F) in relation to those of the absorbed light (λ_A).

Fluorescent measurements are done by directly measuring the emitted fluorescence intensity. A crucial advantage of fluorescence spectrometry is that the measurement signal increases with increasing illumination power. For densitometric measurements, the TLC plate is usually illuminated by a mercury lamp showing a strong excitation at 356 nm. By use of a cutoff filter at 400 nm to suppress the excitation wavelengths, monowavelength scanner can measure all fluorescent emissions in the visible range very sensitively. If a deuterium lamp is used for illumination purposes, a diode-array scanner can register fluorescence and absorptions simultaneously, because the wavelength-shifted fluorescence adds additional light intensity to the reflected light of the TLC plate [12]. This additional light intensity (a difference between J and J_0) results in a positive Kubelka–Munk signal, because the nominator in Equation 7.7 is squared. A cutoff filter is not recommended because a deuterium lamp emits mainly in the range from 260 to 380 nm, whereas fluorescent light is usually emitted above 400 nm. In Figure 7.4, the pyrene absorption spectrum and the Kubelka–Munk spectrum are shown, measured without a cutoff filter by use of an diode-array scanner. Pyrene shows structured absorptions in the range from 200 to 350 nm and strong fluorescent signals above 370 nm. Both spectra have been extracted from the same set of measurement data by use of Equation 7.7 and Equation 7.8.

Figure 7.4 shows the absorption spectrum of pyrene, evaluated with Equation 7.8 (above), and the Kubelka–Munk spectrum of pyrene (below), evaluated by use of Equation 7.7.

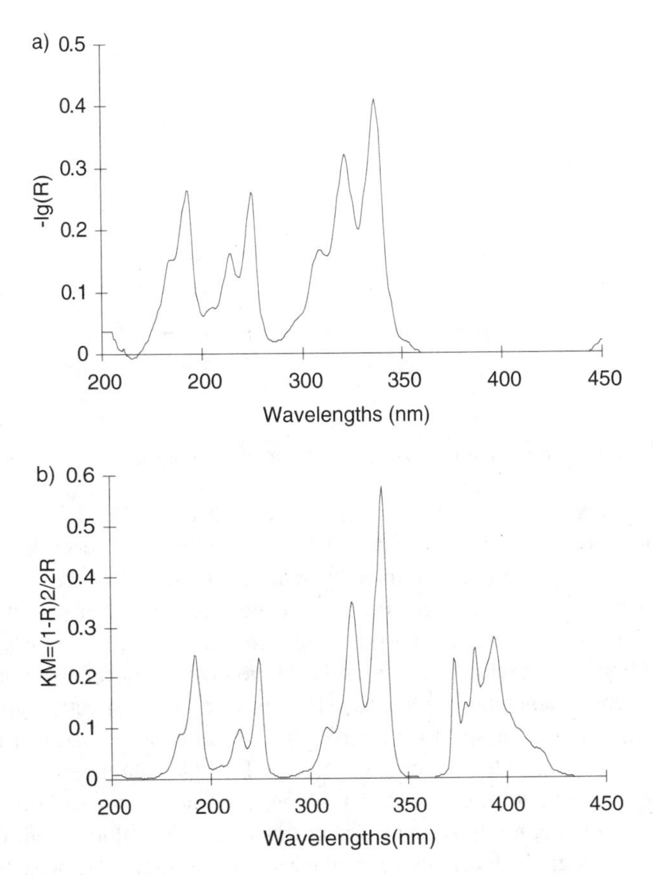

FIGURE 7.4 Absorption spectrum of pyrene, evaluated with Equation 7.8 (above), and the Kubelka–Munk spectrum of pyrene (below), evaluated by use of Equation 7.7.

It is well known that fluorescence from an RP-18 phase is much brighter than from a silica gel plate, because the coating of RP-18 material blocks nonradiative deactivation of the activated sample molecules. By spraying a TLC plate with a viscous liquid, e.g., paraffin oil dissolved in hexane (20 to 67%), the fluorescence of a sample can be tremendously enhanced. The mechanism behind fluorescence enhancement is to keep molecules at a distance either from the stationary layer or from other sample molecules [14]. Therefore, not only paraffin oil, but a number of different molecules show this enhancement effect.

A scanner is not necessarily recommended for the identification of fluorescent substances, because the fluorescent light is mainly emitted in the visible region (vis-region). In this case, it is possible to use the eyes as a detection system.

7.2.3 Fluorescence Quenching

For the detection of UV-absorbing substances that show no fluorescence, TLC plates are very often prepared with a fluorescence indicator. Commonly, an inorganic dye

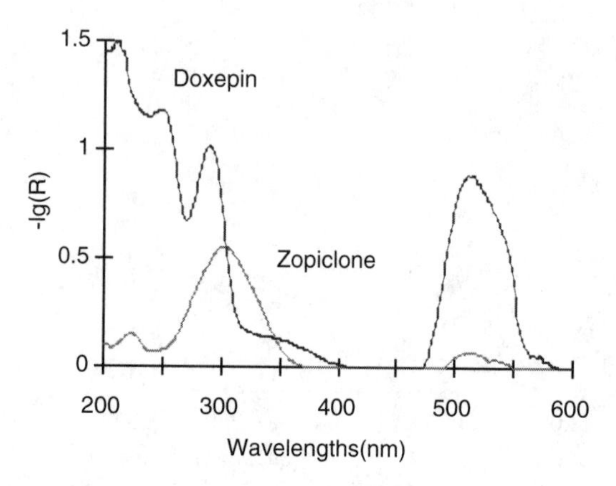

FIGURE 7.5 Absorption spectra of doxepin and zopiclone evaluated with Equation 7.8.

(manganese-activated zinc silicate) is used. This dye absorbs light at 254 nm, showing a green fluorescence at ~520 nm. Sample molecules in the layer cover the fluorescent dye and inhibit its light absorption. In comparison to an uncovered area, sample spots or sample bands show lower light intensity in the vis-region because the covered fluorescent dye cannot transform absorbed light into a fluorescence emission. Black zones on a fluorescent background will indicate the position of the components. The term "fluorescence quenching" is often used for this decrease of reflected light intensity.

Figure 7.5 shows the spectra of doxepin and zopiclone (two pharmaceutical substances). Both spectra have been taken from a TLC plate containing a fluorescent dye. Doxepin shows a strong absorption at 254 nm, and in its remission spectrum an additional large signal appears at 520 nm caused by the fluorescent dye in the TLC layer. This signal represents the light loss of the sample spot at 520 nm in comparison to a sample-free zone. Zopiclone, in contrast, shows nearly no absorption at 254 nm. Therefore, the fluorescence inhibition signal at 520 nm is very weak. Obviously, zopiclone covers the fluorescent dye insufficiently, and this underlines the fact that only compounds with a strong absorption at 254 nm will show fluorescence inhibition.

Figure 7.5 shows the absorption spectra of doxepin and zopiclone evaluated with Equation 7.8.

However, the sensitivity of this detection method is lower than the sensitivity of reflection measurements.

7.3 PLATE DYEING

7.3.1 NONDESTRUCTIVE DYEING (BY IODINE VAPOR AND WATER)

The separation of uncolored samples is usually done on TLC plates containing a fluorescent dye so as to use the fluorescence-quenching effect for sample location. If such plates are not available or if samples show no quenching effect, two universal reagents can help. If the TLC plate is simply sprayed with water, sample spots are very often not translucent but white. Clearer zones can be obtained sometimes by first saturating the TLC plate with water and then letting the plate dry until the zones are distinctly visible [15]. In this way, sample spots can be identified without sample destruction.

Another reversal location method is to expose the TLC plate to iodine vapour in a closed chamber that contains some iodine crystals. Iodine is lipophilic and accumulates in lipophilic sample spots, showing a brown color on a pale yellow-brown background. The same result is obtained by spraying with an iodine solution (250 mg iodine dissolved in 100 ml of heptane). In nearly all cases, this iodine accumulation is totally reversible without altering the sample, because outside the closed chamber iodine evaporates quickly from the plate. Caution should be taken with this iodine treatment in the case of unsaturated compounds because iodine vapor can react with double bonds [16].

7.3.2 Fluorescence and pH Reagents

Coloring reactions are well known for nearly all kinds of substances, and in the literature hundreds of these staining reactions are listed [17–20]. The disadvantage of plate staining is that odorous compounds or hazardous chemicals have to be used. Its advantage is that a coloring reaction is done simply by spraying the plate or parts of the plate. In analytical TLC, an additional plate heating is often recommended, but these reactions can be ruled out in preparative TLC because plate heating can alter or even destroy the sample. Therefore, heating should be avoided at any stage of the chromatographic process and staining agents are recommended for reacting at ambient temperatures. This greatly reduces the available number of reagents.

For safety purposes, manual spraying should be carried out in a laboratory fume hood or a special TLC spray cabinet, wearing protective spectacles and laboratory gloves. The spray should be applied from a distance of about 15 cm with a uniform up-and-down motion. A commercially available spray system (DESAGA ChromaJet DS 20, Desaga, Heidelberg, Germany) sprays computer-controlled zones of desired extent with various reagents. Automatic spraying needs less chemicals and reduces the exposition of the laboratory staff.

Figure 7.6 shows the computer-controlled spray system ChromaJet DS 20 (with permission of Desaga).

If destructive reagents are necessary, a vertical channel at 1-cm distance to the edge of the plate should be scraped in the layer. Afterwards the plate is covered by a sheet of paper, leaving the small stripe at the edge uncovered. Spraying at the uncovered stripe reveals the spot location without diffusion of the reagent into the plate center [3]. Several such stripes can be sprayed successively to test different reaction types. Stripe spraying on each side of the plate serves as a good guide to locate even sloping sample bands. Nevertheless, universal reagents should be preferred as they show coloring reactions with nearly all kinds of sample molecules.

Well known is the use of fluorescent dyes as unspecific staining reagents [17–20] such as rhodamine B (100 mg dissolved in 100 ml of ethanol), rhodamine 6G (50 mg dissolved in 100 ml of ethanol), 8-(anilino) naphthalene-1-sulfonic acid (100 mg dissolved in 100 ml of water), or berberine (10 mg dissolved in 100 ml of ethanol). The combination of 100 mg of rhodamine B and 35 mg of 2′,7′-dichlorofluorescein dissolved in 20 ml of ether, 70 ml of ethanol (95%), and 16 ml of water

FIGURE 7.6 A computer-controlled spray system (with permission of Desaga, Heidelberg, Germany).

showing excellent contrast for black-and-white or colored photography has been published [20]. Especially, lipids sprayed with such dyes show bright fluorescent zones on more-or-less dark backgrounds. During illumination, all these fluorescent dyes transform absorbed energy into light emissions. Contact with the TLC layer opens a radiationless pathway that competes with the fluorescent emission. If sample molecules such as lipids block the layer contact, the fluorescence is distinctively brighter in comparison to the sample-free zones. This process is identical to the fluorescent enhancement effect described for paraffin oil.

The pH indicator shows the acid or basic properties of sample molecules. Commonly used for acid indicating are solutions of bromocresol green (20 mg dissolved in 10 ml of ethanol combined with 1 ml of 0.1-molar aqueous NaOH) or bromophenol blue (20 mg dissolved in 10 ml of ethanol, pH-adjusted with 0.1-molar NaOH or 0.2% aqueous citric acid). In the presence of acids, 2,6-dichloroindophenol (40 mg dissolved in 100 ml of ethanol) changes the color from blue to red. The fluorescent dye acridine orange (20 mg dissolved in 100 ml of ethanol) changes pH-dependently the color of its fluorescence from yellow-green to yellow.

7.3.3 Universal Detection Reagents (Charring)

If heating is necessary to perform a coloring reaction, a part of the TLC plate must be cut off for reagent dipping or spraying. This detached part can be heat treated without destroying the main sample share. Charring is often the only way of sample staining for molecules showing no reactive groups. Charring breaks down the original components into other visible compounds or, in the extreme, into pure carbon. This decomposition process of organic samples mainly result in black to brown zones on a white background. A TLC plate with an inorganic binder is recommended to avoid a black background. Some of the following charring reagents produce fluorescent zones at lower temperatures (80 to 120°C) before final charring occurs at higher temperatures (above 160°C).

Sulfuric acid in combination with $MnCl_2$ is often used as an universal reagent. Two-hundred mg of $MnCl_2 * 4H_2O$ have to be dissolved in 30 ml of water and must be subsequently diluted with 30 ml of methanol. Finally, 2 ml of concentrated H_2SO_4 must be added carefully. After spraying, 15 min of heating at 120°C is necessary to complete the reaction. Organic compounds form brown spots on a white background [17].

The combination of copper salts and sulfuric acid is another widely used charring reagent. For preparation, 10 ml of concentrated H_2SO_4 has to be mixed carefully with 90 ml of acetic anhydride. Three grams of copper-II-acetate is dissolved in 100 ml of 8% phosphoric acid. After plate dipping or plate spraying, 15 min of heating at 125°C is recommended [17].

Phosphomolybdic acid oxidizes most organic compounds under formation of a blue-grey dye. For reagent preparation, 250 mg of phosphomolybdic acid must be dissolved in 50 ml of methanol. The plate must be heated to 120°C until the predominantly blue spots are visible [17].

Vanillin reacts with many compounds, forming dark-colored zones. For reagent preparation, 1 g of vanillin (4-hydroxy-3-methoxybenzaldehyde) should be dissolved in 25 ml of ethanol, which is subsequently diluted with 25 ml of water and 35 ml of concentrated H_3PO_4. A variation of this reagent is to mix 500 mg of vanillin, dissolved in 85 ml of methanol, 10 ml of acetic acid, and 5 ml of concentrated H_2SO_4. After spraying, the plate must be kept for 5 to 15 min at 120 to 160°C. A reagent with similar properties can be obtained by mixing 50 mg of 4-dimethylaminobenzaldehyde with 1 ml of concentrated H_2SO_4 dissolved in 100 ml of ethanol [17–20]. This reagent is known as *van Urks reagent,* whereas mixing a solution of 50 mg of 4-dimethylaminobenzaldehyde in 50 ml of methanol with 10 ml of concentrated HCl is known as *Ehrlich reagent* [17]. Similar reactions can also be performed using anisaldehyde (4-methoxybenzaldehyde), which reacts with sugars and glycosides, too. To prepare *Ekkert reagent*, dissolve anisaldehyde (0.5 ml) in a mixture of 85 ml of methanol and 10 ml of acetic acid, carefully adding 8 ml of concentrated H_2SO_4.

7.3.4 Selective Detection Reagents

In the literature, a large number of more-or-less selective detection reagents are published [1,3,17–20]. The presented selection of reagents cover reactions that can be easily performed. The reagents remain stable for days when stored in a refrigerator.

The *Dragendorff reagent* is available in different compositions, and it mainly stains nitrogen-containing compounds, producing colored zones on a white background. For solution a: 1.7 g of basic bismuth nitrate and 20 g of tartaric acid are dissolved in 80 ml of water. For solution b: 16 g of potassium iodide are dissolved in 40 ml of water. The final spray reagent is mixed from solution a and b in the same ratio, and 5 ml of this solution is added to a solution of 10 g of tartaric acid in 50 ml of water [17–20].

Iodoplatinate reagent stains all organic compounds containing nitrogen. Thioles, thioether, or sulfoxides give reactions as well. For reagent preparation, solution a (5% platinic chloride in water) and solution b (10% aqueous KI) have to be mixed in amounts of 5 ml and 45 ml and diluted with 100 ml of water. The addition of concentrated HCl (1 to 10 parts of the spray solution) will often increase the sensitivity [18].

Sulfur-containing samples show colored spots when sprayed with 2,6-dibromoquinone-4-chlorimide (*Gibbs reagent*). For preparation, 2 g of this compound is dissolved in 100 ml of acetic acid or ethanol. Heating to 110°C is necessary to give a reaction. This reagent also creates colored zones when samples contain phenols. For reactions with phenols, only the less-reactive 2,6-dichloroquinone-4-chlorimide can be used under the same conditions.

Tetracyanoethylene (500 mg TCNE dissolved in 100 ml of acetonitrile or ethylene acetate) reacts with aromatic compounds, forming colored zones on a slightly yellow background. In some cases, heating to 100°C is necessary to complete the reaction.

Antimony chloride, such as 4% $SbCl_3$ in 50-ml $CHCl_3$ or 50-ml glacial acid, forms various colors characteristic of the compounds showing carbon double bonds (*Carr–Price reagent*).

Ninhydrin (1,2,3-indantrione, 100 mg dissolved in 50 ml of methanol) transforms all samples containing molecules with NH_2 groups, such as amino acids, peptides, or amines, into red- or purple-colored products. To perform a reaction, at least 5 to 10 min heating at 120°C is necessary.

Compounds with hydroxy groups, such as flavonoids, sugars, or hydroxy acids, often react with 1% diphenyl boric acid β-aminoethylester in methanol or ethanol. Various colors are formed by many natural products (*Neu reagent*).

Aldehydes and ketones can be located as orange or yellow zones after reaction with 2,4-dinitrophenylhydrazine hydrochloride (200 mg dissolved in 2 ml of aqueous HCl). In some cases, heating to 100°C is necessary.

In the year 1994, T. Takao et al. published a staining reaction to identify radical-scavenging activity [21]. Takao et al. used the compound 1,1-diphenyl-2-picrylhydrazyl (DPPH), which changes its color, in the presence of antioxidants such as ascorbic acid or rutin, from blue-purple to colorless or yellow. Forty mg of DPPH has to be solved in 20 ml of acetone or methanol. The reaction is complete immediately.

REFERENCES

1. Nyiredy, Sz., Ed., *Planar Chromatography. A Retrospective View for the Third Millennium*, 1st ed., Springer, Budapest, 2001.
2. Sherma, J. and Fried, B., Eds., *Handbook of Thin-Layer Chromatography*, 3rd ed., Marcel Dekker, New York, 2003.

3. Sherma, J. and Fried, B., Preparative thin layer chromatography, in *Preparative Liquid Chromatography,* Journal of Chromatography Library, Bidlingmeyer, B.A., Ed., Vol. 38, Elsevier, Amsterdam, 105–127, 1987.
4. Schuster, A., *Astrophys. J.*, 21, 1–22, 1905.
5. Geiss, F., *Fundamentals of Thin Layer Chromatography*, Hüthig-Verlag, Heidelberg, 1987.
6. Kubelka, P. and Munk, F., *Z. Tech. Physik*, 11a, 593–601, 1931.
7. Kubelka, P. and Munk, F., *J. Opt. Soc. Am.,* 38, 448, 1067, 1948.
8. Kortüm, G., *Reflextionsspektroskopie*, Springer, Heidelberg, 1969.
9. Spangenberg, B., Post, P., and Ebel, S., *J. Planar Chromatogr. Mod. TLC,* 15, 11–18, 2002.
10. Jork, H., *Z., Anal. Chem.,* 221, 17–33, 1966.
11. Spangenberg, B. and Klein, K.-F., *J. Chromatogr. A*, 898, 265–269, 2000.
12. Spangenberg, B. and Klein, K.-F., *J. Planar Chromatogr. Mod. TLC*, 14, 260–265, 2001.
13. Spangenberg, B. and Weyandt-Spangenberg, M., *J. Planar Chromatogr. Mod. TLC*, 17, 164–168, 2004.
14. Spangenberg, B., Lorenz, K., and Nasterlack, St., *J. Planar Chromatogr. Mod. TLC*, 16, 331–337, 2003.
15. Touchstone, J.C., Perspectives, in *Modern Thin-Layer Chromatography,* Chromatographic Sciences Series, Vol. 52, Grinberg, N., Ed., Marcel Dekker, New York and Basel, 1990, pp. 465–480.
16. Nichaman, Z., Sweeley, C.C., Oldham, N.N., and Olson, R.E., *J. Lipid Res.*, 4, 484–485, 1963.
17. Jork, H, Funk, W., Fischer, W., and Wimmer, H., *Thin Layer Chromatography,* Vol. 1a and Vol. 1b, VCH Verlagsgesellschaft, Weinheim, 1990 and 1993.
18. Touchstone, J.C., *Practice of Thin Layer Chromatography*, 3rd ed., John Wiley & Sons, New York, 1992, pp. 143–183.
19. Stahl, E., Ed., *Thin-Layer Chromatography*, 2nd ed., Springer-Verlag, Berlin, 1969.
20. Barret, G.C., Nondestructive detection methods in paper and thin layer chromatography, in *Advances in Chromatography*, Vol. 11, Marcel Dekker, New York, 1974, pp. 145–179.
21. Takao, T., Kitatani, F., Watanabe, N., Yagi, A., and Sakata, K., *Biosci. Biotech. Biochem.*, 51, 1780–1783, 1994.

8 Additional Detection Methods and Removal of Zones from the Layer

Joseph Sherma

CONTENTS

8.1 INTRODUCTION

As explained in Chapter 1, classical preparative layer chromatography (PLC) involves flow of the mobile phase by capillary action. The method uses relatively basic equipment and is not expensive.

PLC is used for separations of 2 to 5 mg of sample on thin-layer chromatography (TLC) plates (0.25-mm layer thickness) or high-performance TLC (HPTLC) plates (0.1-mm thickness). In these instances, the method is termed micropreparative TLC. The isolation of one to five compounds in amounts ranging from 5 to 1000 mg is carried out on thicker layers. PLC is performed for isolation of compounds to be used in other tasks, i.e., further identification by various analytical methods, such as ultraviolet (UV) solution spectrometry [1] or gas chromatography/mass spectrometry (GC/MS) [2], obtaining analytical standards, or investigations of chemical or biological properties [3].

The purity of recovered compounds depends on the purity of all materials used in the PLC process, such as the solvents, and the cleanliness of the tank, sample containers, etc. Plates stored in cardboard boxes or plates with polymer binders exposed to light and air will become contaminated. Prewashing of plates by development with the mobile phase, methanol–dichloromethane (1:1), or 1% acetic acid or 1% ammonium hydroxide in diethyl ether (depending on whether the subsequent mobile phase is acidic or basic) will clean the layer. The prewashed plates are vacuum-dried and stored in a vacuum desiccator prior to use to keep them clean.

PLC is normally performed on homemade or commercially precoated 20 × 20 cm or 20 × 40 cm plates containing layers that are 0.5 to 2.0 mm (500 to 2000 μm) thick with 5 to 40 μm particle size (average about 25 μm). Preparative layers are usually formulated with gypsum ($CaSO_4$ hydrate) binder (termed a G-plate), which leads to a "soft" layer that can be removed by scraping more easily for compound recovery (see following text) than if an organic binder of the type that is typical in analytical layers was used. The plates may also contain a preadsorbent strip composed of diatomaceous earth or inert silica along the lower edge for more convenient sample application. Plates coated with plain silica gel adsorbent have been mostly used for PLC, but alumina layers are sometimes also used for normal phase adsorption PLC, silver-ion-impregnated silica gel for separations according to the degree of compound unsaturation [4], cellulose for normal phase liquid–liquid partition PLC, and silica gel layers chemically bonded with C-2 (ethyl) or C-18 (octadecyl) groups for reversed-phase separations. Resolution is improved on thin preparative layers (0.5 to 1 mm), but capacity increases on thicker layers (sample capacity increases roughly with the square root of the layer thickness [5]). The Uniplate-T preparative plate (Analtech, Newark, DE) has a 0.7 mm thick concentration zone and a tapered silica gel G or GF layer gradually increasing from 0.3 mm at the bottom to 1.7 mm at the top; during development, a negative mobile phase velocity gradient is formed that keeps each component focused as a narrow band, leading to a great reduction in zone elongation and overlapping and improved resolution [5,6].

Samples are applied as a narrow streak across the layer, manually with a pipet or using a commercial application instrument, such as the automated CAMAG (Wilmington, NC) Linomat 5 and TLC Sampler 4, or the Analtech manual sample streaker. The sample is usually a 5 to 10% (w/v) solution in a solvent that should be as weak (nonpolar for silica gel) and volatile as possible. Samples applied to a preadsorbent zone are focused by the mobile phase migration into a narrow band at the interface with the separation layer. Online application of large volumes of samples is possible with certain horizontal chambers provided with distributors (CAMAG linear chamber or DS chamber, Chromdes, Lublin, Poland); when the mixture contains a few components and the selectivity of the system is high, large volumes of samples can be introduced from the edge of the layer so that components are already partly separated in the application stage (frontal chromatography mode) and, because of mutual displacement effects, may be fully separated during development with high yield [7].

Linear development is carried out in the ascending or horizontal mode in a saturated (lined with thick filter paper) or unsaturated [8] rectangular chamber

(N-tank) or sandwich-type chamber with single or multiple development using a mobile phase that has been established previously to be optimum for the required separation. Preliminary analytical TLC experiments can be used to aid the selection of PLC mobile phases; a difference of greater than 0.1 R_F for a separation in analytical TLC usually will lead to good separations in a comparable PLC system [9]. Marked improvement of separation efficiency in the separation of complex samples can be obtained by stepwise gradient elution [7]. A circular PLC chamber has been described that offers special advantages for separating compounds with low R_F values [6].

After the preparative plate has been developed and the mobile phase removed, the separated compounds must be located prior to their recovery from the plate. Mobile phase removal after development should not involve heating if the compound of interest is thermally labile; place the plate in a desiccator for vacuum drying.

Naturally colored compounds can be detected simply by viewing in daylight (white light). Naturally fluorescent compounds can be detected by viewing under UV light (usually 366 nm). A layer with 254 or 366 nm fluorescent phosphor or indicator (usually marked "F" or "UV" on commercial plates) will show the separated substances when examined under the respective UV lamps if the compounds absorb at or near these wavelengths. This detection approach is termed *fluorescence quenching* or *fluorescence inhibition*. Fluorescent phosphors impregnated into layers on PLC plates, such as zinc silicate, are generally insoluble in the solvents used to elute compounds (except when strong acids are used, e.g., HCl) and should not add significant contaminants.

Chapter 7 of this book discusses direct detection by color, fluorescence, and fluorescence quenching by eye and by using a densitometer. Densitometry is valuable for detecting UV-absorbing zones that are not visible on plates without a fluorescent indicator. It can also be applied for identification of unknown zones on PLC plates by comparison of their *in situ* spectra with the spectra of known standards chromatographed on the same plate or a library of computer-stored reference spectra. In addition, densitometry can be used to document the results of PLC by recording a densitogram of the bands (slit-scanning densitometer) or an image of the whole layer (videodensitometer, digital camera densitometer, or office flatbed scanner densitometer), and it can define the exact locations of zone boundaries that may not be clearly ascertained by viewing with the eyes, thereby allowing more accurate zone removal. Examples of the use of densitometry in PLC are the identification of Chinese drugs on silica gel by their densitometric fluorescence spectra [10] and monitoring separations of taxoids on silica gel F and silanized silica gel F layers by scanning at 366, 254, and 230 nm [11].

Chapter 7 describes nondestructive (reversible) detection of substances by using iodine or water. Iodine can detect a wide variety of chemical classes as brown to yellow-brown zones on a preparative layer after exposure to its vapor or dipping into a 1% iodine in chloroform solution. Another general, nondestructive technique for detection of hydrophobic compounds in PLC uses water as a spray reagent [12–14].

Finally, Chapter 7 discusses the use of selected fluorescence and pH indicator detection reagents, universal charring methods, and selective derivatization reagents that produce colored zones.

8.2 ADDITIONAL DETECTION METHODS

8.2.1 Other Detection Reagents

The following are additional detection reagents not mentioned in Chapter 7, with specific references to their use in PLC: 0.1% or 0.2% methanolic 2′,7′-dichlorofluorescein for location of fatty acid isomers and derivatives from milk and oils on a Ag^+-impregnated silica gel layer [15,16], $FeCl_3$ reagent for isoprenylated xanthones on silica gel [17], fluorescein reagent for cholesterol oxidation products on silica gel [18], bromine vapor and densitometry for fatty acids and triglycerides from papaya seed oil on argentation silica gel [19], and orcinol reagent for serum glycolipids on silica gel [20].

8.2.2 Hot Wire Decomposition

Using an 8-in. Nichrome wire strip connected with springs to a variac, organic decomposition can be carried out in a very fine line to locate zones. The resulting carbon is small in quantity and will not interfere in subsequent elutions.

8.2.3 Detection of Radioactive Zones

Compounds that are radioactive can be located on a preparative layer by contact film autoradiography, electronic autoradiography, and storage phosphor screen imaging [21–23]. These methods differ in terms of factors such as simplicity, speed, sensitivity, and resolution, and the method of choice depends on the available equipment, reagents, and instrumentation. All are nondestructive, and the detected compounds can be recovered without change for later studies.

8.2.3.1 Contact Film Autoradiography

Contact film autoradiography involves exposure of the radiolabeled zones on a developed and dried chromatogram with photographic (x-ray) film, leading to the appearance of dark zones on the developed film. When properly exposed, the resolution of the zones on the film may be comparable with that on the original chromatogram and as high as any other radioactivity detection method, and the methodology is simple. The major disadvantage of the method is the long exposure times required (hours to weeks, depending on the type of isotope and radioactivity level). The three major exposure methods are direct exposure (autoradiography, useful for all beta emitters), direct exposure with an intensifying screen placed behind the film (improves the detection efficiency for gamma-emitting and high-energy beta-emitting isotopes), and fluorographic exposure (fluorography; for detection of lower-energy isotopes such as tritium [24], the layer is coated or impregnated with a scintillator followed by direct exposure to x-ray film).

Examples of PLC with autoradiography detection include the published studies on ^{3}H labeling of 1-^{3}H-PAF-aceter [25]; diazinon and related compounds from plant material [26]; metabolic fate of triamcinolone acetonide in laboratory animals [27]; synthesis of 4-S-cysteaminyl-[U-^{14}C]phenol antimelanoma agent [28]; radiolabeled isomers of hexachlorocyclohexane [29]; photolysis of imazapyr herbicide in aqueous media [30]; and absorption, excretion, metabolism, and residues in tissues of rats treated with ^{14}C-labeled pendimethalin (Prowl) herbicide [31].

8.2.3.2 Electronic Autoradiography

Early instrumental detection of radioactive compounds on preparative layers often involved use of a methane gas-flow counter [32]. Today, radioactive compounds can be detected after PLC by use of digital autoradiography (DAR) using a Berthold digital autoradiograph (EG&G Berthold, Wilbad, Germany) [33]. This instrument has a 20 × 20 cm sensitive area that contains a 600 × 600 wire grid, the multiwire proportional chamber (MWPC). Measurements are made by using a 9:1 mixture of argon–methane as the counting gas bubbled through methylal at 2.8°C and a flow rate of 5 ml/min. The positive high-voltage potentials used for ^{3}H- and ^{14}C-labeled compounds are 2040 and 1200 V, respectively, signal analysis is achieved by measuring 5 × 360,000 detector cells/sec, and DAR measurement time (run time) is optimized on the basis of the amount of radioactivity applied to the plate [33]. TLC-DAR has been rated as a good method for preparative applications [34].

8.2.3.3 Storage Phosphor Screen Imaging

A phosphor screen can be used like an autoradiography film to detect radioactive zones on PLC plates. The available screens are sensitive to x-rays and beta and gamma emissions from isotopes such as ^{14}C, ^{3}H, ^{35}S, ^{125}I, ^{131}I, ^{32}P, and ^{33}P. The screen captures the latent images produced by the ionizing radiation, and during reading, as a result of red-laser-induced stimulation, blue light is emitted from the screen in proportion to the amount of radioactivity in the zone. The resulting digital image allows for detection over a wide sensitivity range. TLC with phosphor imaging detection was described as a highly rated method for use in preparative applications [21].

8.2.4 Biological Detection

Micropreparative TLC and PLC are used quite often for the purification and isolation of biologically active compounds [e.g., 35], and several detection methods are based on the biological activity of such compounds. As with destructive or nonreversible reagents that impart color, UV absorbance, or fluorescence to compounds to allow their detection, small spots of the sample should be applied to the outer side margins of the plate when using biological detection reagents. The spots in the margins are then sprayed with the requisite reagent, after covering or masking the major streaked portion of the plate with glass, cardboard, or foil and scribing vertical channels between the streaks and spots. The desired substances are then visible on each side of the plate and serve as a guide for scribing horizontal lines across the center of the plate to outline the areas that contain most of the compounds of interest. An alternative procedure is to streak the sample across the entire layer and to use the outer edges of the streaked sample outside the mask as guide areas to be sprayed. Analtech sells Prep-Scored silica gel G and GF plates that are prescored 2.5 cm from each side edge; the sample streak is extended onto the scored areas, and

FIGURE 8.1 Prep-Scored plate with scraped zone. (Photograph supplied by Analtech, Newark, DE.)

following development the side portions are snapped away from the center portion of the plate (Figure 8.1). After detection, the edge strips are placed next to the center section to indicate the areas for collection. These plates must be used for reagents applied by dipping, and they eliminate the chance of sprayed reagent wicking into the center region of the plate and also any undesirable effects that a heating step needed for the detection procedure may cause. The aforementioned procedures can be used with any PLC plate if a glass cutter is available for vertical scoring after development. Heating of preparative plates during the chromatography and detection procedures can lead to decomposition of the compounds of interest and should be avoided so that they can be recovered in an unaltered state.

Cholinesterase-inhibiting pesticides (e.g., organophosphate and carbamate pesticides) are detected by dipping the developed chromatogram in a solution of the enzyme cholinesterase followed by incubation for a short period. Then the plate is dipped in a substrate solution, e.g., 1-naphthyl acetate/fast blue salt B. In the presence of the active enzyme, 1-naphthyl acetate is hydrolyzed to 1-naphthol and acetic acid, and the 1-naphthol is coupled with fast blue salt B to form a violet-blue azo dye. The enzyme is inhibited by the pesticide zones, so the enzyme–substrate reaction does not occur; pesticides are, therefore, detected as colorless zones on a violet-blue background [36].

Direct bioautography involves application of a suspension of a suitable microorganism growing in a broth to a developed plate. Incubation in a humid atmosphere enables growth of bacteria, and dehydrogenases from the living microorganisms convert tetrazolium salt into the corresponding intensely colored formazin. Antibacterial compounds appear as clear spots against a colored background [37]. A simple contact bioautography detection was used in the isolation of an antimicrobial bromoditerpene from marine alga [38], and seeded agar overlay bioautography was used with PLC in the study of the antimicrobial activity of extracts of some *Bignoniaceae* from Malaysia [39]. Zone location by bioautography can also be carried out on a filter paper print of a preparative chromatogram; this method was used during the PLC of the neutral macrolide antibiotic, neutramycin, from an extract of a mash filtrate [40].

The use of "bioactivity-guided PLC" was reported in a study of hyperatomarin, an antibacterial prenylated phloroglucinol from *Hypericum atomarium* ssp. *degenii* [41]. Crude diethyl ether–petroleum ether (1:2) extracts of plant material (about 50 mg/plate) were developed on 2-mm silica gel plates with toluene–ethyl acetate (95:5) mobile phase. In the bioassay, several test microorganisms (e.g., *Staphylococcus aureus* ATCC 25923) were used, antibacterial activity was determined by the agar dilution method using Mueller-Hinton medium, and the antibiotic erythromycin was used as a positive control.

Immunostaining was used to detect compounds such as steroidal alkaloid glycosides after chromatographic separation on a silica gel plate, transfer to a polyvinylidene difluoride (PVDF) membrane, and treatment of the membrane with sodium periodate solution followed by bovine serum albumin (BSA) solution, resulting in a solasodine–BSA conjugate. The PVDF membrane was then immersed in antisolamargine MAb, followed by peroxidase-labeled secondary MAb. When the substrate and H_2O_2 were added, clear blue spots appeared [42]. Protocols for the immunological detection of glycosphingolipids with monoclonal antibodies on layers were reviewed [43]. Examples of published applications of PLC with immunostaining detection are structural studies of gangliosides from the YAC-1 mouse lymphoma cell line [44] and purification, characterization, and serological properties of a glycolipid antigen reactive with a serovar-specific monoclonal antibody against *Leptospira interrogans* serovar canicola [45].

Bioluminescence can be used for specific detection of separated bioactive compounds on layers (BioTLC) [46]. After development and drying the mobile phase by evaporation, the layer is coated with microorganisms by immersion of the plate. Single bioactive substances in multicomponent samples are located as zones of differing luminescence. The choice of the luminescent cells determines the specificity of detection. A specific example is the use of the marine bacterium *Vibrio fischeri* with the BioTLC format. The bioluminescence of the bacteria cells on the layer is reduced by toxic substances, which are detected as dark zones on a fluorescent background. BioTLC kits are available from ChromaDex, Inc. (Santa Ana, CA).

8.3 REMOVAL OF SUBSTANCES FROM THE LAYER

The material located in the scribed area of the plate is recovered by removing the adsorbent zone, eluting the substance from the adsorbent with a suitable solvent, and separating the residual adsorbent. The final step involves concentrating the eluate, usually by evaporation; this should be done at as low a temperature as possible and in an inert gas stream, such as nitrogen, so that the recovered compound is not decomposed or otherwise altered. After concentration, the compound may be recrystallized from an appropriate solvent or the solution used for further analyses or studies.

The outlined areas of the layer containing the substances of interest are scraped off cleanly down to the glass (or aluminum or plastic) backing on the plate with a spatula, scalpel, or razor blade (Figure 8.1). An area about 10% larger than the visualized sample area should be scraped to compensate for three-dimensional development of the zone in the layer. The loosened adsorbent is transferred to a sheet of glassine weighing paper and poured into a small glass vial or tube having a solvent-resistant cap. If a centrifuge is available, the container can be a 15-ml centrifuge tube. One or two drops of water can be added to displace the component from the adsorbent sites. Wetting the layer by spraying to dampness with some ethanol or the PLC mobile phase can reduce the loss of fine adsorbent particles by flaking and blowing during the scraping and transfer steps.

Various apparatuses have been described for removal of the layer material. For example, a simple, inexpensive micropreparative system for ng-to-mg amounts of compounds involved removal of thin lines of layer material with a fast-moving drill followed by elution by siphoning eluent through a specially formed sintered glass at one end of the zone or spot and collecting the eluent at the other with a piece of filter paper carton, which was then extracted by soaking and centrifugation [47].

The organic solvent used to elute the compound must be adequately strong (polar for the adsorbent silica gel) and a good solvent for the component. Absolute methanol should be avoided as a single solvent because silica gel itself and some of its common impurities (Fe, Na, SO_4) are soluble in this solvent and will contaminate the isolated material. Solvent containing less than 30% methanol is recommended, or ethanol, acetone, chloroform, dichloromethane, or the mobile phase originally used for PLC are other frequently used choices for solute recovery. Water is not recommended because it is so difficult to remove by evaporation during the concentration step (removal by lyophilization is necessary). A formula that has been used to calculate the volume of solvent needed when the PLC mobile phase is chosen for elution is:

$$\text{Solvent volume} = (1.0 - R_F)\,(10)\,(\text{volume of the scrapings})$$

Example: For a compound with a $R_F = 0.2$ and a scraping volume = 0.5 ml, the predicted elution volume is (1.0 – 0.2) (0.5) (10) = 4.0 ml of solvent.

The selected solvent is added to the container with the scrapings, and the tube is shaken mechanically or by hand or is vortex-mixed for up to 5 min. The contents of the vial are allowed to settle by gravity or the tube is centrifuged, and the supernatant liquid carefully removed by pipet or decantation. More complete removal of the adsorbent can be obtained by membrane filtration, centrifugation, or the use of a Soxhlet apparatus [48]. A second portion of solvent can be added to the scrapings to repeat the extraction, and the two solutions combined. If it is necessary to definitely exclude all silica gel particles from the solution, a filtering medium should be used that is capable of retaining at least 5-µm particles (most PLC layers contain silica gel particles ranging from 5 to 40 µm). If the filtrate is still cloudy owing to the presence of microscopic silica gel particles, taking the sample to dryness and redissolving may leave the silica gel undissolved and yield a clear solution of the compound of interest.

An alternative approach [5] is to add enough water to cover the adsorbent material and to extract the aqueous suspension several times with an immiscible

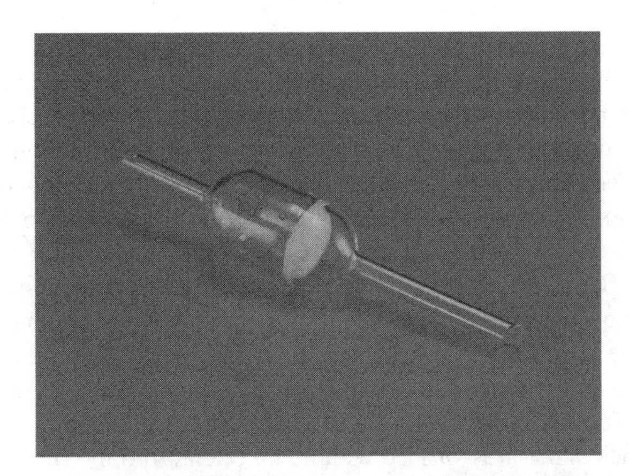

FIGURE 8.2 Sample recovery tube for recovery of samples from PLC plates (Catalog No. 416400-0023). The approximate bulb capacity is 25 to 30 ml, overall length 180 mm, and fritted disc porosity coarse. (Photograph supplied by Kimble/Kontes, Vineland, NJ.)

organic solvent. Before solvent evaporation, colloidal silica is removed by filtration through a membrane filter.

Manual collection of the scrapings into a small column instead of a vial allows direct elution of the separated compound. As an example, Jozwiak et al. [49] scraped bands from 0.5-mm silica gel PLC plates into 10 mm ID × 120 mm length columns with funnel-shaped inlets and narrow outlets, closed with glass wool; adsorbed alkaloids were eluted with acetone followed by methanol.

Instead of scraping and manual collection of the adsorbent, the band can be sucked off the plate with a "vacuum-cleaner"-type apparatus. Dekker [50] described an apparatus for the isolation of compounds from layers by elution and direct Millipore filtration, and Platt [51] designed a zone collector that used vacuum to transfer separated zones from layers directly to vials for liquid scintillation counting of radioactivity.

Sample recovery tubes are available commercially for the removal and recovery of adsorbent from a preparative layer. An example is shown in Figure 8.2. A similar sample recovery tube is available from Bodman (Aston, PA; Catalog No. GSRT-2). With these tubes, the scrapings are directly recovered from the plate using suction, and elution is accomplished with an appropriate solvent under vacuum while the adsorbent is retained on the disc.

Vacuum collectors of this type may not be suitable for sensitive substances, because the adsorbent containing the desired substance is in constant contact with a stream of air and oxidation may occur [6]. In this case, the scraped adsorbent can be placed manually in the empty tube and then extracted with solvent with the aid of vacuum.

A homemade combination scraper-collector used in my laboratory can be made from a Pasteur pipet (229 mm × 7 mm OD; Fisher Scientific Co., Pittsburgh, PA; Catalog No. 13-678-6B) [52]. The pipet is cut with a file 60 mm from the top and 65 mm from the tip to produce a pipet that is approximately 100 mm long, which is then plugged with glass wool. The pipet tip is attached to a vacuum pump or aspirator, and the top is used to scrape the zone on the layer containing the compound of interest. Scraping and vacuum collection are thus accomplished simultaneously. The pipet is removed from the vacuum and used as a column. A collection vial is covered with aluminum foil, and the column (pipet) is pushed through the foil cover into the vial. The elution solvent is allowed to percolate through the column into the vial. If the glass wool plug is sufficiently tight (but not too tight to restrict gravity flow of the solvent), most or all of the adsorbent will be retained in the column.

Wing and BeMiller [48] removed carbohydrate bands from a silica gel H preparative plate by utilizing a piece of 25-mm OD glass tubing that had a 6-mm opening for the adsorbent to enter and a glass wool wad to prevent it from being sucked through to the vacuum line, which was attached to the other end of the glass tubing. Chloroform (20 ml) was added to the adsorbent in the tubing, the slurry was filtered, and the residue was washed with 20 ml of chloroform. The chloroform was removed under vacuum and the residue redissolved in chloroform to remove the final traces of silica gel.

Recovery of the desired substance should be carried out as quickly as possible because the longer it is in contact with the adsorbent, the more likely it is that decomposition will occur.

8.4 RECENT APPLICATIONS OF PLC

PLC on thick layers for the purpose of purification and isolation of separated compounds dates back to the late 1950s, but the method is far from obsolete because it is still being applied currently with the same basic methodology in many research areas. The following is a selection of recent studies involving PLC that were chosen to illustrate the diversity of possible applications: purification of the sesquiterpene lactone anthecotulide [53], persistence and metabolism of impazapyr in four typical Chinese soils [54], separation and characterization of lignin compounds from walnut shell oil [55], preparation of both antipodes of enantiopure inherently chiral calix (4) crown derivatives [56], kinetic and chemical assessment of the UV/H_2O_2 treatment of antiepileptic drug carbamazepine [57], diterpenes from the aerial parts of *Salvia candelabrum* and their protective effects against lipid peroxidation [58], effect of the *N*-acetylglycine side chain on the antimicrobial activity of xanthostatin [59], inner molecular order and fluidity gradient of phospholipid bilayers after lipoperoxidation [60], the tryptophan metabolism of *Malassezia furfur* [61], phenolic acids in yacon leaves and tubers [62], characterization of side-chain oxidation products of sitosterol and campesterol [63], antioxidant activity of extracts of defatted seeds of niger (*Guizotia abyssinica*) [64], isolation and quantitative analysis of cryptotanshinone, an active quinoid terpene formed in callus of *Salvia miltiorrhiza* BUNGE [65], biosynthesis of tetrahydrofuranyl fatty acids from linoleic acid by Clavibacter sp ALA2 [66], ruthenium complexes of 2-[(4-(arylamino)phenyl)azo]pyridine [67], isolation of saponins with viral entry inhibitory activity [68], isolation and purification of coumarins from *Peucedanum tauricum* Bieb. fruits [69], atropisomerism in monopyrroles [70], alkaloids from *Boophane disticha* with affinity to the serotonin transporter in rat brain [71], *in vivo* activity of released cell wall lipids of *Mycobacterium bovis* bacillus Calmette-Guerin [72], purification of the sesquiterpene lactone anthecotulide [53], and fractionation of a pesticide mixture by micropreparative TLC with zonal application and horizontal development [73].

Other applications of PLC are presented in Chapters 9 to 16 of this book.

REFERENCES

1. Gyeresi, A., Gergely, M., and Vamos, J., *J. Planar Chromatogr.–Mod. TLC* 13, 296–300, 2000.
2. Liu, Q.-T. and Kinderlerer, J.L., *J. Chromatogr. A* 855, 617–624, 1999.
3. Waksmundzka-Hajnos, M. and Wawrzynowicz, T., *J. Liq. Chromatogr. Relat. Technol.* 25, 2351–2386, 2002.
4. Jham, G.N., Velikova, R., Nikolova-Damyavova, N., Rabelo, S.C., Teixeira da Silva, J.C., Alessandra de Paula Souza, K., Valente, V.M.M., and Cecon, P.R., *Food Res. Int.* 38, 121–126, 2005.
5. Poole, C.F., *The Essence of Chromatography*, Elsevier, New York, 2003, pp. 848–850.
6. Nyiredy, Sz., Preparative layer chromatography, in *Handbook of Thin Layer Chromatography*, 3rd ed., Sherma, J. and Fried, B., Eds., Marcel Dekker, New York, 2003, pp. 307–338.
7. Soczewinski, E. and Wawrzynowicz, T. Preparative TLC, in *Encyclopedia of Chromatography*, Cazes, J., Ed., Marcel Dekker, New York, 2001. pp. 660–662.
8. Nyiredy, Sz., *J. AOAC Int.* 84, 1219–1231, 2001.
9. Nyiredy, Sz., Ed., Possibilities of preparative layer chromatography, in *Planar Chromatography*, Springer Scientific, Budapest, Hungary, 2001. pp. 386–409.
10. Sun, W., Sha, Z., and Yang, J., *Zhongguo Zhongyao Zazhi* 24, 619–622, 1993.
11. Hajnos, M.L., Glowniak, K., Waksmundzka-Hajnos, M., and Kogut, P., *J. Planar Chromatogr.–Mod. TLC* 14, 119–125, 2001.
12. Gritter, R.J. and Albers, R.J., *J. Chromatogr.* 9, 392, 1962.
13. Tate, M.E. and Bishop, C.T., *Can. J. Chem.* 40, 1043–1048, 1962.
14. Chen, C.H., Chang, C., Nowotney, A.M., and Nowotney, A., *Anal. Biochem.* 63, 183–194, 1975.
15. Cruz-Hernandez, C., Deng, Z., Zhou, J., Hill, A.R., Yurawecz, M.P., Delmonte, P., Mossoba, M.M., Dugan, M.E.R., and Kramer, J.K.G., *J. AOAC Int.* 87, 545–562, 2004.
16. Ratnayake, W.M.N., *J. AOAC Int.* 87, 523–539, 2004.
17. Hano, Y., Matsumoto, Y., Shinohara, K., Sun, J.Y., and Nomura, T., *Planta Med.* 57, 172–175, 1991.
18. Caboni, M.F., Costa, J., Rodriguez-Estrade, M.F., and Lercker, G., *Chromatographia* 46, 151–155, 1997.
19. Nguyen, H. and Ndjiiska, H., *Fat. Sci. Technol.* 97, 20–23, 1995.
20. Ilinov, P., Katzarowa, E., Dimov, S., and Zaprianova, E.T., *J. Liq. Chromatogr. Relat. Technol.* 20, 1149–1157, 1997.
21. Hazai, I. and Klebovich, I. Thin layer radiochromatography, in *Handbook of Thin Layer Chromatography*, 3rd ed., Sherma, J. and Fried, B., Eds., Marcel Dekker, New York, 2003, pp. 339–360.
22. Stultz, C.L.M., Sullivan, M.T., Macher, B.A., and Stack, R.J., *Anal. Biochem.* 219, 61–70, 1994.
23. Reichert, W.L., Stein, J.E., French, B., Goodwin, P., and Varanasi, U., *Carcinogenesis* 13, 1475–1479, 1992.
24. Sheppard, H. and Tsien, W.H., *Anal. Chem.* 35, 1992, 1963.

25. Morgat, J.L., Roy, J., Wichrowski, B., Michel, E., Heymans, F., and Godfroid, J.J., *Agents Actions*, 12, 705–706, 1982.
26. Gilmore, D.R. and Cortes, A., *J. Chromatogr.* 21, 148–149, 1966.
27. Gordon, S. and Morrison, J., *Steroids* 32, 25–35, 1978.
28. Somayaji, V.V., Wiebe, L.I., and Jimbow, K., *NucCompact* 20, 158–159, 1989.
29. Grundey, U. and Kraus, P., *J. Chromatogr.* 117, 242–244, 1976.
30. Mallipudi, N.M., Stout, S.J., Dacunha, A.R., and Lee, A.H., *J. Agric. Food Chem.* 39, 412–417, 1991.
31. Zulalian, J., *J. Agric. Food Chem.* 38, 1743–1754, 1990.
32. Schulze, P.E. and Wenzel, M., *Angew. Chem.* 74, 777–779, 1962.
33. Klebovich, I., Application of planar chromatography and digital autoradiography in metabolism research, in *Planar Chromatography*, Nyiredy, Sz., Ed., Springer Scientific, Budapest, Hungary, 2001, pp. 293–311.
34. Szunyog, J., Mincsovics, E., Hazai, I., and Klebovich, I., *J. Planar Chromatogr.–Mod. TLC* 11, 25–29, 1998.
35. Hajnos, M.L., Glowniak, K., Waksmundzka-Hajnos, M., and Piasecks, S., *Chromatographia* 56, S91–S94, 2002.
36. Morlock, G. and Kovar, K.-A., Detection, identification, and documentation, in *Handbook of Thin Layer Chromatography*, 3rd ed., Sherma, J. and Fried, B., Eds., Marcel Dekker, New York, 2003, pp. 207–238.
37. Botz, L., Nagy, S., and Kocsis, B. Detection of microbiologically active compounds, in *Planar Chromatography*, Nyiredy, Sz., Ed., Springer Scientific, Budapest, Hungary, 2001. pp. 489–516.
38. Caccamese, S., Cascio, O., and Compagnini, A., *J. Chromatogr.* 478, 255–258, 1989.
39. Houghton, P.J., Mat Ali, R., and Azizol, M., *Pharm. Pharmacol. Lett.* 7, 96–98, 1997.
40. Lefemine, D.V. and Hausmann, W.K., *Antimicrob. Agents Chemother.* (1961-70), 134–137, 1963.
41. Savikin-Fodulovic, K., Aljancic, I., Vajs, V., Menkovic, N., Macura, S., Gojgic, G., and Milosavljevic, S., *J. Nat. Prod.* 66, 1236–1238, 2003.
42. Tanaka, H., Putalin, W., Tsuzaki, C., and Shoyama, Y., *FEBS Lett.* 404, 279–282, 1997.
43. Ishikawa, D. and Taki, T., *Methods Enzymol.* 312, 157–159, 2000.
44. Muething, J., Peter-Katalinic, J., Hanisch, F.G., and Neumann, U., *Glycoconj. J.* 8, 414–423, 1991.
45. Ono, E., Takase, H., Naiki, M., and Yanagawa, R., *J. Gen. Microbiol.* 133, 1329–1336, 1987.
46. Kreiss, W., Eberz, G., and Weisemann, C., *CAMAG Bibliography Service (CBS)* 88, 12–13, 2002.
47. Laeufer, K., Lehmann, J., Petry, S., Scheuring, M., and Schmidt-Schuchardt, M., *J. Chromatogr. A* 684, 370–373, 1994.
48. Wing, R.E. and BeMiller, J.N., Preparative thin layer chromatography, in *Methods in Carbohydrate Chemistry*, Whistler, R.L. and BeMiller, J.N., Eds., Academic Press, New York, 1972, pp. 60–64.
49. Joswiak, G., Wawrzynowicz, T., and Waksmundzka-Hajnos, M., *J. Planar Chromatogr.–Mod. TLC* 13, 447–451, 2000.
50. Dekker, D., *J. Chromatogr.* 168, 508–511, 1979.
51. Platt, S.G., *Anal. Biochem.* 91, 357–360, 1978.
52. Fried, B. and Sherma, J., *Thin Layer Chromatography: Techniques and Applications*, 4th ed., Marcel Dekker, New York, 1999, pp. 235–248.

53. Meyer, A., Zimmermann, S., Hempel, B., and Imming, P., *J. Nat. Prod.* 68, 432–434, 2005.
54. Wang, X., Wang, H., and Fan, D., *J. Environ. Anal. Chem.* 85, 99–109, 2005.
55. Mathias, E.V. and Halkar, U.P., J. *Anal. Appl. Pyrol.* 71, 515–524, 2004.
56. Cao, Y.D., Luo, J., Zheng, Q.Y., Chen, C.F., Wang, M.X., and Huang, Z.T., *J. Org. Chem.* 69, 206–208, 2004.
57. Vogna, D., Marotta, R., Andreozzi, R., Napolitano, A., and d'Ischia, M., *Chemosphere* 54, 497–505, 2004.
58. Janicsak, G., Hohmann, J., Zupko, I., Forgo, P., Redel, D., Falkay, G., and Mathe, I., *Planta Med.* 69, 1156–1159, 2003.
59. Kim, S.K., Ubukata, M., and Isono, K.J., *Microbiol. Biotechnol.* 13, 998–1000, 2003.
60. Megli, F.M. and Sabatini, K., *Chem. Phys. Lipids* 125, 161–172, 2003.
61. Mayser, P., Stapelkamp, H., Kramer, H.J., Podobinska, M., Wallbott, W., Irlinger, B., and Steglich, W., *Antonie Van Leeuwenhoek J. Gen. Mol. Microbiol.* 84, 185–191, 2003.
62. Simonovska, B., Vovk, I., Andrensek, S., Valentova, K., and Ulrichova, J., *J. Chromatogr. A* 1016, 89–98, 2003.
63. Johnsson, L. and Dutta, P.C., *J. Am. Oil Chem. Soc.* 80, 767–776, 2003.
64. Shahidi, F., Desilva. C., and Amarowicz, R., *J. Am. Oil Chem. Soc.* 80, 845–848, 2003.
65. Wu, C.T., Mulabagal, V., Nalawade, S.M., Chen, C.L., Yang, T.F., and Tsay, H.S., *Biol. Pharm. Bull.* 26, 845–848, 2003.
66. Hosokawa, M., Hou, C.T., Weisleder, D., and Brown, W., *J. Am. Oil. Chem. Soc.* 80, 145–149, 2003.
67. Das, C., Saha, A., Hung, C.H., Lee, G.H., Peng, S.M., and Goswami, S., *Inorg. Chem.* 42, 198–204, 2003.
68. Gosse, B.K., Gnabre, J.N., Ito, Y., and Huang, R.C., *J. Liq. Chromatogr. Relat. Technol.* 25, 3199–3211, 2002.
69. Glowniak, K., Bartnik, M., Mroczek, T., Zabza, A., and Wierzejska, A., *J. Planar Chromatogr.–Mod. TLC* 15, 94–100, 2002.
70. Boiadjiev, S.E. and Lightner, D.A., *Tetrahedron-Assymetry* 13, 1721–1732, 2002.
71. Sandager, M., Nielsen, N.D., Stafford, G.I., van Staden, J., and Jaeger, A.K., *J. Ethnopharmacol.* 98, 367–370, 2005.
72. Geisel, R.E., Sakamoto, K., Russell, D.G., and Rhoades, E.R., *J. Immunol.* 174, 5007–5015, 2005.
73. Tuzimski, T., *J. Planar Chromatogr.–Mod. TLC* 18, 39–43, 2005.

Section II

9 Medical Applications of PLC

Jan Błądek and Anna Szymańczyk

CONTENTS

9.1 INTRODUCTION

PLC studies in the field of medicine cover the processes of separation, purification, and determination of xenobiotics or endogenous substances present in body fluids and tissues. Such research is carried out with the following aims:

1. Studies of a network of metabolic pathways and their disorders. A metabolic disorder occurs when there is a blockage in one of the pathways because the enzyme required to catalyze the reaction is malfunctioning or missing. This results in an accumulation of metabolites on one side of the block and a deficiency of essential chemicals on the other. In many metabolic disorders, early diagnosis may prevent permanent functional damage or death.
2. Pharmacokinetic and toxicological studies of xenobiotics. For example, monitoring of therapeutic drugs is necessary in order to improve patient outcomes by adjusting the drug dose to obtain a certain serum concentration and also to predict those individuals who might develop an adverse drug reaction or toxicity. Toxicological studies are usually of a screening character; they are carried out in order to detect hard intoxication and drug abuse. This class of research also includes environmental medicine. A wide variety of xenobiotics, such as polycyclic aromatic hydrocarbons or dioxins, are analyzed in human body fluids for the investigation of environmental and occupational exposure.

Medical analyses are extremely complex and include a number of steps from sample collection to the final report of the results. Intermediate steps of such analyses (sample pretreatment) are the most tedious and time consuming, and their main aim consists of adequate reduction of matrix interference and the isolation and concentration of analytes [1]. Of course, there is no single ideal technique for biological sample preparation; there are several (dialysis, hydrolysis, ultrafiltration, dilution, liquid–liquid or solid–phase extraction, etc.), each with its own advantages and disadvantages. Within this group of methods, PLC can be singled out as the most advantageous, a fact that justifies its wide use in medical analyses.

9.2 BIOLOGICAL MATRICES AND THEIR COMPONENTS

The applications of PLC in medicine continue to grow because of the ease of operation and increased selectivity with many new stationary phases, as well as the interfacing of automation and robotics, which makes sample preparation extremely straightforward. When compared with other techniques of separation and purification of analytes, PLC is characterized by lower costs, less rigorous sample preparation, and the ability to analyze multiple samples simultaneously. At least two properties distinguishing the use of PLC in medicine can be identified: first, there is the aforementioned diversity of aims for which separation may be carried out. Lipids, enzymes, amino acids, and bile acids are substances of especially high research and diagnostic value; considerable attention is also devoted to pharmacokinetic and toxicological studies of drugs. Second, there is the diversity of matrices from which those substances of clinical interest are to be separated. At present, studies of highest importance include analyses of blood, urine, and tissue; analyses of saliva or sputum, hair, human milk, or breath are not as common.

9.2.1 Short Characteristics of the Matrices

Biological matrices are very complex; apart from the analytes, they usually contain proteins, salts, acids, bases, and various organic compounds. Therefore, effective sample preparation must include particulate cleanup to provide the component of interest in a solution, free from interfering matrix elements, and in an appropriate concentration.

The subject of blood analyses is plasma or serum; in some cases (especially in postmortem analyses), whole blood analyses have to be performed as well. Plasma is a liquid obtained after centrifuging an unclothed sample of blood (after prior treatment with an anticlotting agent), i.e., it is a fraction of blood obtained after the separation of red blood cells. Plasma contains approximately 90% water, 8% proteins, 1% organic acids, and 1% salts. A blood sample without an anticoagulant undergoes very rapid coagulation, and the liquid obtained as a result of its centrifugation (serum) contains 35 to 45 mg/ml albumins, 30 to 35 mg/ml globulins, 4 to 7 mg/ml lipids, 7 mg/ml salts, and 1.34 to 2.0 mg/ml carbohydrates. As opposed to plasma, serum does not contain fibrinogen and certain clotting factors. Urine is one of the most commonly studied biological matrices used for drug screening, monitoring workplace exposure to chemicals, and for forensic purposes. It is a universal means of excretion of both parent xenobiotics and their metabolites. The determination of analytes in urine is convenient as sampling is noninvasive. It is relatively easy to collect and is a moderately complex matrix, which contains both organic and inorganic constituents. Tissue samples can originate from the muscle, liver, myocardium, kidney cortex, cerebellum, and brain stem. Usually, they are prepared before extraction by a stepwise process that leads to a disruption of the gross architecture of the sample. This modifies the physical state of the sample and provides the extraction medium with a greater surface area per unit mass.

As with urine, saliva (sputum) is easy to collect. The levels of protein and lipids in saliva or sputum are low (compared to blood samples). These matrices are viscous, which is why extraction efficiency of xenobiotics amounts to only 5 to 9%. By acidifying the samples, extraction efficiencies are improved as the samples are clarified, and proteinaceous material and cellular debris are precipitated and removed. Some xenobiotics and their metabolites are expressed in hair. Hair is an ideal matrix for extraction of analytes to nonpolar phases, especially when the parent xenobiotics are extensively metabolized and often nondetectable in other tissues (parent molecules of xenobiotics are usually less polar than metabolites). Hair is a popular target for forensic purposes and to monitor drug compliance and abuse. Human milk may be an indicator of exposure of a newborn to compounds to which the mother has been previously exposed. The main components of human milk are water (88%), proteins (3%), lipids (3%), and carbohydrates in the form of lactose (6%). At present, increasing attention is devoted to the determination of xenobiotics in breath. This matrix, however, contains only volatile substances, whose analysis is not related to PLC applications.

9.2.2 Substances of Medical Interest

9.2.2.1 Amino Acids and Proteins

Amino acids are organic acids containing an amino group. The naturally occurring amino acids are α-amino acids (the NH_2 group is attached to the carbon atom next to the –COOH group), the one exception being β-alanine. The separations and analyses of amino acids in urine, blood, and tissues are crucial because an early estimation of free amino acids may prevent neurological damage and mental retardation in young infants with inborn errors of amino acid metabolism. Significant meaning in clinical chemistry is also ascribed to analyses of other amino acids; e.g., separation of N-terminal amino acids is applied in the research of peptide structures. Proteins are another large group of compounds of great importance to the structure and functioning of living matter. These substances may serve as structural elements, enzymes, hormones, oxygen carriers, antibodies, and many other important types of compounds. They are compounds of large molecular weight and contain carbon, hydrogen, oxygen, nitrogen, and, with a few exceptions, also sulphur. Proteins may be classified as simple, which contain only amino acids, or conjugated, which contain amino acids and nonamino-acid substances such as nucleic acids, carbohydrates, lipids, metals, or phosphoric acids. Amino acids are the fundamental structural units of protein and are the end result of the complete hydrolysis of these substances.

9.2.2.2 Carbohydrates

A class of naturally occurring compounds of carbon, hydrogen, and oxygen, in which the latter elements are in the same proportion, or in nearly the same proportion, as in water. Mono-, di-, tri-, oligo-, and polysaccharides, ketose, tri-, tetra-, pento-, and hexose, as well as reducing and nonreducing sugars, have a great importance in the life sciences. Several diseases are accompanied by increased elimination of various groups of sugars in the urine and feces.

9.2.2.3 Lipids

They are important constituents of all living organisms. They occur in a free or bound form, and as lipoproteins they comprise complex mixtures of different classes of compounds. Seven different classes of lipids, including fatty acids, phospholipids, cholesteryl ester, cholesterol, triglycerides, diglycerides, and monoglycerides, are isolated and separated. Fatty acids and phospholipids are both anionic, cholesterol and cholesteryl esters are extremely nonpolar, and mono-, di-, and triglycerides all contain polar ester and hydroxyl groups. In clinical chemistry, lipids are classified according to the number of hydrolytic products per mole. Simple (or neutral) lipids release two types of products (fatty acids and glycerol). Complex (or polar) lipids give three or more products, such as fatty acids, glycerol, phosphoric acid, and an organic base. A natural lipid mixture comprises different classes of simple and complex lipids. Lipids play a vital role in virtually all aspects of biological life. Disturbances in the lipid metabolism of the organism lead to various disorders and malfunctions. For example, glycolipids have been reported to be associated with differentiation, development, and organogenesis. These glycolipids have an active role in the formation of cataracts. The glycolipid biosynthesis pathway is initiated by the glucosyltransferase-catalyzed synthesis of glucosylceramide. Nonesterified fatty acids ("free" fatty acids, which occur in blood predominantly in association with albumins) play a significant role in the physiological control of carbohydrate metabolism. Serum lipid fatty acid compositions have been determined in diabetes, coronary artery disease, as well as renal disease. The gangliosides (a family of sialic acid containing glycosphingolipids) are localized mainly on the outer surface of plasma membranes. Gangliosides are antigenic and are involved in immunological mechanisms of certain autoimmune diseases.

9.2.2.4 Bile Acids

These are steroids with an alcoholic group and five-carbon-atom side chain terminating in a carboxyl group. The naturally occurring bile acids do not occur free; there are linked to the amino acids taurine and glycine. They are found in the serum, gastrointestinal fluid, liver biopsies, urine, feces, and hepatic and gall bladder bile. The bile acids are amphiphilic molecules and have several structural characteristics, such as a C_{24} carboxyl group, β-linked A and B rings of the steroid backbone, a C_3 hydroxyl group, further hydroxyl groups at C_6, C_7, or C_{12} only, and conjugation of the C_{24} carboxyl moiety with taurine or glycine only. Separation of these substances and their conjugates from biological matrices is primarily the subject of diagnostics. Much attention is also devoted to analytical procedures for the determination of naturally occurring bile acid patterns (important in diagnosis). A number of inherited metabolic diseases (e.g., Zellweger syndrome, neonatal, infantile Refsum disease, and peroxisomal bifunctional protein deficiency) in which the normal conversion of cholesterol to the primary C_{24} bile acids is disrupted have been characterized.

9.2.2.5 Drugs

They are usually classified on the basis of their designated use (antibiotics, cardioactive drugs, antioxidants, antidepressants, etc.), but a different criterion has been introduced for analytical purposes, namely, the ionic state of drugs at various pH values. On the basis of this criterion, the following types are distinguished: basic, acidic, neutral, and amphoteric drugs. Drugs containing an amine group may accept a proton at low pH levels and become cations. Most drugs (over 80%) contain nitrogen, which means that they are cations (at an appropriate pH). Drugs that contain an acidic functional group, such as a carboxylic acid, are capable of donating a proton at pH above 5 and becoming an anion. Drugs that contain neither acidic nor basic functional groups are neutral drugs; they are not ionic at any pH. The fourth category comprises drugs that contain both acidic and basic functional groups; they are called amphoteric drugs. Amphoteric drugs may be either cationic, anionic, or zwitterions (both positively and negatively charged at the same time), depending on the pH. The object of investigation of drugs in clinical chemistry is assimilation and excretion, harmfulness, metabolism, etc.

9.2.2.6 Biogenic Amines

These are four monoamines synthesized and secreted within many mammalian tissues, including various regions in the brain, sympathetic nervous system, enlero-chromaffin cells of the digestive tract, and adrenal medulla. These biogenic amines (indoleamine and catecholamines — dopamine, norepinephrine, and epinephrine) are synthesized within the cell from their precursor amino acids and have been associated with many physiological and behavioral functions in animals and humans.

9.2.2.7 Nucleic Acids

They are polynucleotides in which the nucleotides are linked by phosphate bridges. They are usually found chemically bound to proteins to form nucleoproteins. Nucleic acids are subdivided into (1) ribonucleic acids (RNA), containing the sugar D-ribose and (2) desoxyribose nucleic acids (DNA), containing the sugar D-desoxyribose. These compounds are found in all living cells. *Nucleosides* are compounds obtained during partial decomposition (hydrolysis) of nucleic acids, and they contain a purine or pyrimidine base linked to either D-ribose, forming ribosides, or D-desoxyribose, forming desoxyribosides (e.g., adenosine, cytidine, guanosine, or uridine). *Nucleotides* are the fundamental units of nucleic acids. The nucleotides found in nucleic acids are phosphate monoesters of nucleosides (e.g., adenylic acids, guanylic acids, or uridylic acids); some of them are important coenzymes. The term nucleotides is also used for compounds not found in nucleic acids but containing substances other than purines and pyrimidines. These nucleotides are modified vitamins and function as coenzymes (examples are riboflavin phosphate, flavin adenine nucleotide, diphosphopyridine nucleotide, or coenzyme A).

9.2.2.8 Enzymes

They are chemical substances (e.g., amylase, lipase, trypsin, oxidase, and urease) occurring in all living cells, which are the necessary starting materials for carrying out the chemical reactions of biological processes. Enzymes cause hydrolysis, coagulation, oxidization, etc. Enzyme activity depends on their concentration and the substrate concentration, pH, and temperature. Enzymes metabolize a vast array of clinically, physiologically, and toxicologically important compounds, including drugs, steroids, and carcinogens. For example, cytochrome P450 (a super family of hemoproteins) catalyzes the oxidative as well as reductive metabolism of a wide variety of compounds of both xenobiotic and endogenous origin.

9.3 SEPARATION AND DESORPTION TECHNIQUES

The basic aim of PLC applications in clinical chemistry, apart from the recovery of standards of endogenous substances, consists of structural identification of isolated (without further separation) substances of relatively high purity. Therefore, the majority of works devoted to this topic pertain to semipreparative separation. Obtaining low amounts of analytes, achieved by coupling TLC with modern analytical methods, is sufficient for this purpose. Semipreparative PLC is virtually a specific property of biomedical chemistry. This means that the separation techniques and visualization methods applied in analytical TLC differ only slightly from those used in PLC [2,3]. The differences between the two methods emerge only during the stage of removal and recovery of the substances of biomedical interest from the layer.

9.3.1 Specificity of Chromatographic Systems

Nonmodified silica gel is used most commonly for the separation of substances of medical interest. The separation is based on the interactions (hydrogen bonding, van der Waals' forces, and ionic bonding) between the molecules of drugs, lipids, bile acids, etc., and the silica gel. Alumina has similar properties but is rarely used. Successful separation of endogenous substances, drugs, or their metabolites can also be achieved using physically or chemically modified silica gel.

Preparative, nonmodified silica gel plates are used in the processes of recovery from matrices and purification of both xenobiotics and endogenous substances. Commercially prepared plates of silica gel have uniformity of thickness and stability of the silica gel layer, which determines the reproducibility of separations. Scientific literature offers works aimed at determining whether various commercial plates differ in their effect on the recovery of biogenic substances. For example, Sowa and Subbaiah [4] studied the recovery of unsaturated fatty acids. Five commercial brands of silica gel plates were compared in this study, which was carried out using phosphatidylcholines (PC) — 17:0–17:0, 16:0–18:1, 16:0–18:2, and 16:0–20:4, cholesteryl esters (CE) — 17:0, 18:1, 18:2, and 20:4, and free fatty acids (FFA) — 17:0, 18:1, 18:2, and 20:4. The PC samples were recovered using a mixture of chloroform:methanol:water (65:25:4, v/v/v). FFA and CE were chromatographed separately using the following mobile phase: hexane:diethylether:acetic acid (70:30:1, v/v). Following visualization, the areas corresponding to the standards were scraped; analytes were methylated (directly on the plate or after elution) and determined by gas chromatography (GC). The proposed solution was used for the evaluation of the effectiveness of recovery of different unsaturated fatty acids present in PC of human plasma. The studies showed that some commercial TLC plates are not suitable for the direct methylation of fatty acids in the presence of silica gel after the TLC separation of the parent lipid classes. There is a selective loss of unsaturated fatty acids in presence of J.T. Baker and Whatman plates, apparently because of oxidative degradation that occurs during the methylation step. No loss of unsaturated fatty acids occurs even on these plates if the lipids are first eluted from the silica gel.

Ravandi [5] and coworkers examined oxidative lipid destruction in various biological systems. Milligram quantities of core aldehydes (choline — PC and ethanolamine — PE phosphatides) were obtained from unsaturated triacylglycerols (TG) and glycerophospholipids by first preparing the ozonides. The ozonides were isolated by PLC and reduced to the corresponding core aldehydes with triphenylphosphine in chloroform. The separation conditions were the following — the stationary phase was silica gel H and the mobile phase was heptane–isopropyl ether–acetic acid (60:40:4, v/v.). The aldehyde bands were located by spraying the plate with the Schiff reagent, which gave a purple color. Other PLC bands were identified by spraying the plate with 2,7-dichlorofluorescein and viewing the plate under UV light. The aldehydes were recovered from the plate by scraping the gel and extracting it with chloroform. The PC ozonides were similarly reduced, but the reduction products were purified by PLC using the polar solvent system chloroform–methanol–acetic acid–water (75:45:12:6, v/v). The phospholipid aldehydes were recovered by extraction of the gel with chloroform–methanol–water (65:25:4, v/v). The recovered substances were resolved and characterized by liquid chromatography with online mass spectrometry.

Physical modification of adsorbents depends on their impregnation with various additives not miscible with the mobile phase (additives are adsorbed on the support). The most significant nonpolar modifier is crude paraffin; main nonorganic modifiers include silver nitrate, boric acid, mercuric acetate, or urea. For example, the incorporation of 5 to 20% $AgNO_3$ into the silica gel permits effective complexing of π double bonds of analytes with the silver ions. The chemical modification of silica gels consists of adjoining varied functional groups to the chromatographic bed (usually silica gel). This process eliminates the stripping of these groups by the mobile phase. In clinical chemistry, nonpolar phases are of the highest practical significance. The retention data obtained on these phases are usually comparable to those obtained earlier on plates impregnated with paraffin.

In medical analyses, silver ion chromatography is mainly used for the separation of lipid classes into molecular types depending on the number, configuration, and the position of the double bond in the fatty acid moieties. This adsorbent retards the migration of the polyunsaturated fatty acids more than the oligounsaturated fatty acids, without materially affecting the migration of the saturated fatty acids. Among the works confirming the correctness of separations on modified silica gels, the work by Rezanka [6] is of particular interest. A mixture of fatty acid methyl esters from cod-liver oil transesterified by MeONa in MeOH was separated a by two-dimensional elution on silica gel plates impregnated with urea (first dimension) and silver nitrate (second dimension). The slurry was prepared by mixing silica gel with a 10% urea solution or a 10% solution of $AgNO_3$. A 0.5-mm layer of silica gel with urea was put on a glass plate, and the free surface was overlayered with silica gel with $AgNO_3$. The plate was dried again and activated for 1 h at 105°C. The lower, right-hand corner of the plate was loaded with 250 to 500 μg of the sample, and the plate was developed with butyl acetate. After the chromatographic run was finished, the plate was dried, turned around by 90° and developed with a mixture of hexane–diethyl ether–methanol (90:10:1, v/v) in which the polar solvent had to be present to disrupt the urea complex. The chromatogram was sprayed with a 0.1% ethanol solution of 2′,7′-dichlorofluorescein and the corresponding spots marked under UV light. The silica gel was scraped and extracted with a mixture of hexane–diethyl ether (1:1, v/v). The fatty acid methyl esters were converted into oxazolines and analyzed by GC/MS (Figure 9.1).

Numerous endogenous substances and commercially available pharmaceuticals are racemic mixtures. Therefore, it is an important problem of clinical chemistry to develop methods for resolution of enantiomers and for establishing enantiomeric purity, because these substances exhibit different biological and physiological

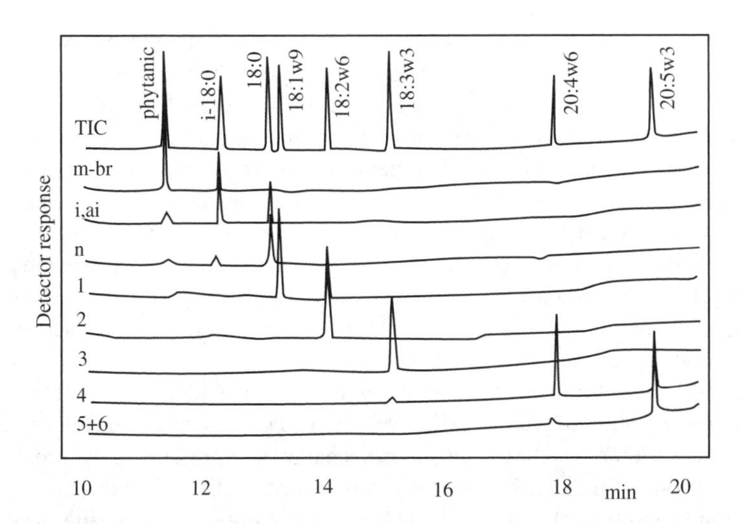

FIGURE 9.1 GC/MS of total methyl esters of fatty acids standards and single fractions after PLC. Abbreviations: first number (18 or 20) = number of carbon atoms in the chain; second number (0 to 5) = number of double bounds; n = normal chain; i =isoacid; ai = anteisoacid; m-br =multi-branched acid. (From Rezanka, T., *J. Chromatogr. A*, 727, 147–152, 1996. With permission.)

behavior. In the case of PLC, the said tasks are realized through the use of chiral plates or through chiral mobile phase additives. Silica gel impregnated with (–)-brucine, L-proline–Cu(II) complex, and L-arginine–Cu-(II) complex have been used for the resolution of D, L-amino acids; L-ascorbic acid or (+)-tartaric acid have been used as impregnators for the resolution of enantiomeric PTH amino acids. Aboul-Enein and coworkers [7] have also examined the enantiomeric resolution of some 2-arylpropionic acids using L-(–)-serine-impregnated silica as the stationary phase. The enantiomeric resolution of certain 2-arylpropionic acids was achieved on silica gel plates impregnated with optically pure L-(–)-serine as a chiral selector. The mobile phase enabling successful resolution of (±)-ibuproxam and (±)-ketoprofen was acetonitrile–methanol–water in the ratio 16:4:0.5, v/v, and for (±)-tiaprofenic acid, the same was used in the ratio 16:3:0.5, v/v. The effect of concentration, temperature, and pH of the impregnating chiral selector on resolution has been studied. The procedure was applied successfully to resolve commercial ampoules of ketoprofen dosage formulation.

Of course, preparative or semipreparative separation of substances of clinical interest is realized not only on silica gels or impregnated silica gels, but favorable results are also obtained on different stationary phases (e.g., cellulose or ion exchangers).

Celluloses (native or microcrystalline) are organic sorbents. They have a low specific surface area and are applied mainly in partition chromatography, especially for the separation of relatively polar compounds. Works on the topic include those by Whitton and coworkers [8], who examined biosynthetic pathways for the formation of taurine in vertebrates. Taurine and its precursor amino acids were extracted from tissues, and the purified supernatant was spotted onto cellulose plates. The chromatograms were developed in butanol–acetic acid–water (11:6:3, v/v) and after drying were sprayed with a solution of 1% ninhydrin in acetone and heated for 5 min at 100°C. On both sides of each plate, a set of standards of the compounds to be quantified was run to locate bands containing taurine and its precursors. The bands containing cysteic acid and hypotaurine were then scraped from the plate and transferred into scintillation vials and counted. To establish the purity of bands after development in one dimension, some plates were developed in a second dimension using methanol–pyridine–water (25:1:5, v/v) as the solvent. In this manner tissue concentrations of precursors of taurine and taurine itself were estimated in haemolymph, nervous tissue, and muscle; the information thus obtained was used in the evaluation of taurine biosynthesis.

Ion-exchange chromatography (IEC) is a technique widely used for the fractionation and purification of proteins. IEC uses charged groups on the surface of a protein to bind to an insoluble matrix of an opposite charge. The protein dipolar ion displaces the counterions of the matrix functional groups and will itself be displaced with an increasing proportion of counterion. This is usually done by increasing the concentration of ions in the elution buffer. The most important parameters in IEC are the choice of ion-exchange matrix and the initial conditions, including buffer type, pH, and ionic strength. Detailed information on this topic can be found in the works of Luo et al. [9].

Organic or inorganic solvent and even solutes of a strong acid or base, ion-pairing agents, ion-exchanger agents, etc., can be used in TLC as a mobile phase. It is also worth stressing that the mobile phase is evaporated after development, and it does not interfere with the solute spots during the visualization process. A number of mixtures with various solvent percentages can be used in clinical applications, with their compositions depending on the nature of separated analytes. For example [10], the four mixtures (1) ethyl acetate–methanol–30% ammonia (85:10:15, v/v), (2) cyclohexane–toluene–diethyl amine (65:25:10. v/v), (3) ethyl acetate–chloroform (1:1, v/v), and (4) acetone have been proposed for separation of drugs and their metabolites on silica gel. Lipids can be excellently separated by a mobile phase of strong elution strength (first elution) and by a less polar mixture in the second run. Amino acids are separated on silica gel with very polar mixtures such as butanol–acetone–acetic acid–water (3.5:3.5:1:2, v/v) or pyridine–acetone–ammonia–water (80:60:10:35, v/v).

Mixing the additive in the eluent used as a mobile phase can also modify the chromatographic system (dynamic modification), but the use of modified adsorbents has led to an improvement of resolution. Example works include that by Armstrong and Zhou [11], who used a macrocyclic antibiotic as the chiral selector for enantiomeric separations of acids, racemic drugs, and dansyl amino acid on biphenyl-bonded silica.

9.3.2 Removal and Recovery of Substances from the Layer

The most obvious way of recovery of substances from the layer is to extract the adsorbed substances from the adsorbent with a suitable solvent (indirect method). The solvent is then evaporated, and the remaining sample is analyzed or

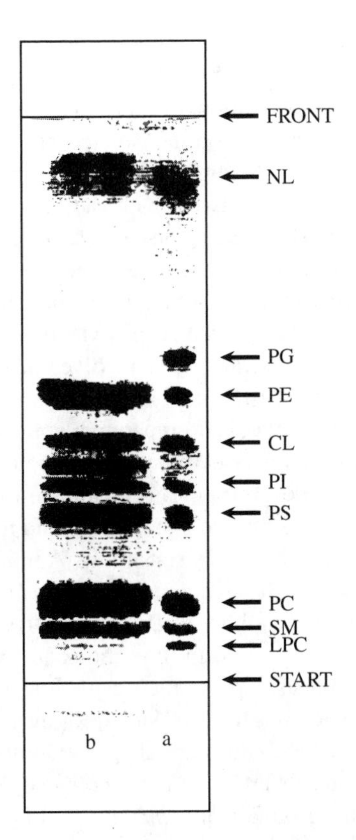

FIGURE 9.2 (a) Thin-layer chromatogram of eight phospholipid standards, (b) whole lipids of platelet from a noninsulin-dependent diabetes mellitus patients. Abbreviations: NL = neutral lipids; PG = L-α-phosphatidyl-DL-glycerol; PE = L-α-phosphatidylethanolamine; CL = cardiolipin; PI = L-α-phosphatidylinositol; PS = L-α-phosphatidyl-L-serine; PC = L-α-phosphatidylcholine; SM = sphingomyelin: LPC = L-α-lysophosphatidylcholine. Staining: iodine vapor. (From Miwa, H. et al., *J. Chromatogr. B*, 677, 217–223, 1996. With permission.)

rechromatographed. In certain hyphenated techniques, commonly referred to as direct methods, the analytes are removed from a chromatographic plate using other methods, namely, by scraping only a spot (without dissolving the analyte), blotting, or thermal, or laser desorption. Naturally, the first step of such hyphenated techniques is a semipreparative separation of the components of analyzed mixtures.

An excellent example of PLC applications in the indirect coupling version is provided by the works of Miwa et al. [12]. These researchers separated eight phospholipid standards and platelet phospholipids from the other lipids on a silica gel plate. The mobile phase was composed of methylacetate–propanol–chloroform–methanol-0.2% (w/v) potassium chloride (25:30:20:10:10, v/v). After detection with iodine vapor (Figure 9.2), each phospholipid class was scraped off and extracted with 5 ml of methanol. The solvent was removed under a stream of nitrogen, and the fatty acids of each phospholipid class were analyzed (as their hydrazides) by HPLC. The aim of this study was to establish a standardized procedure for the analysis of fatty acid compositions of different phospholipid subclasses in biological materials. The authors have shown that the TLC/HPLC coupling method provides complete separation of individual phospholipids in sufficient amounts to allow fatty acid analysis on the isolated phospholipid moieties.

The coupling of TLC with MS combines the possibilities of a particularly exact recovery and purification of the analytes with the sensitivity and specificity of a mass spectrometry detector. These properties are especially important when analyzing substances of medical interest. There are three approaches for coupling TLC with MS detection. The first involves scraping the analyte spot from the plate and extracting it into a solvent. The extracted analyte is then analyzed by MS in a conventional manner. The second approach involves analysis of a small piece of TLC cut from the plate. MS is performed in the presence of the stationary phase. These two methods pertain to classical preparative applications of TLC. The third approach involves placing the intact chromatoplate into the mass spectrometer and freeing the analytes by laser desorption. This technique (MALDI) requires a liquid matrix to be deposited directly onto the plate. The liquid matrix extracts the analyte from the stationary phase and improves sensitivity. Among the works dedicated to this topic, the achievements of Busch and coworkers [13] are especially important, as they were the first to employ a phase transition matrix that is held in the liquid state for analyte extraction but is held in the solid state for MS analysis.

MALDI is a powerful mass spectrometric method for biochemical analysis. However, this ionization method is not well suited for the analysis of low-molecular-weight compounds such as xenobiotics and their metabolites or small proteins. UV irradiation of the MALDI matrix produces a wide range of ions, and the mass spectra usually have a "noisy" background below $m/z \cong 300$. In order to eliminate the said defects, Sunner and coworkers [14] proposed a new method of laser desorption/ionization mass spectra of organic compounds, commonly referred to as SALDI (acronym derived from "surface-assisted laser desorption/ionization"). In this method, a suspension of a fine graphite powder in a solution of the analyte in glycerol was used as the matrix. The suspension was irradiated with a pulsed 337 nm UV laser and intense analyte ion signals were obtained. On developing the method further, Han and Sunner[15] showed that porous activated carbon yielded analyte mass spectra everywhere on the sample surface and from the first laser shot. The mass resolution and sensitivity were essentially the same as with graphite (Figure 9.3). Particularly good results were obtained in experiments in which substances separated on a chromatographic plate were deposited onto an active carbon strip with the help of a high-eluting-strength solvent. Crecelius and coworkers [16] used for the analysis of tetracycline antibiotics TLC/MALDI/time-of-flight mass spectrometry (TOF-MS). Particles of different materials and sizes have been investigated by applying particle suspensions to eluted plates.

At the start of the 1990s, Taki et al. [17] developed a new technique of desorption of substances separated on chromatoplates. It consists in transferring the analytes from a plate to a polyvinylidene difluoride (PDVF) membrane. The method, termed *TLC-blotting*, has found one of its applications in the separation and purification of glycosphingolipids [18]. Analytes (GlcCer prepared from the spleen of a patient with Gaucher's disease, Gb_4Cer from porcine erythrocytes, nLc_4Cer prepared from

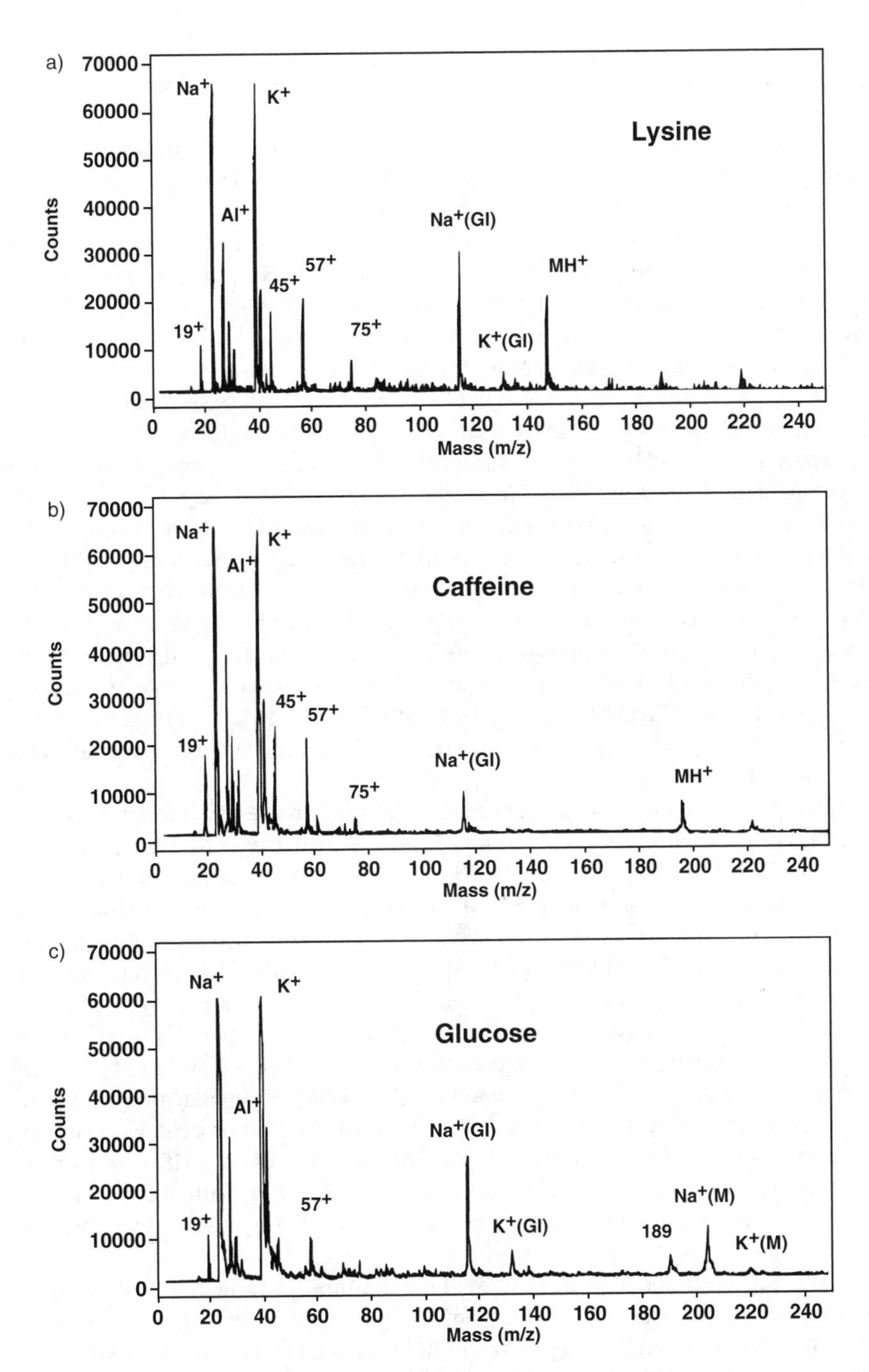

FIGURE 9.3 Active carbon substrate SALDI mass spectra of (a) DL-lysine (1.8 nmol), (b) caffeine (0.5 nmol), and (c) glucose (3.19 nmol). Solutions of these compounds were pipetted directly onto an active carbon substrate patch. (From Han, M. and Sunner J., *J. Am. Soc. Mass Spectrom.*, 11, 644–649, 2000. With permission.)

IV^{3}NeuGcαnLc$_4$Cer after mild acid hydrolysis, and gangliosides G_{M3} obtained from human placenta, and G_{M1a} obtained from bovine brain) were separated on silica gel. The solvent mixture used for developing was chloroform–methanol–0.2% aqueous $CaCl_2$ (60:35:8, v/v). After elution, the plate was cut in half. One half was sprayed with orcinol-H_2SO_4 reagent to verify the locations of the glycosphingolipids (Figure 9.4A), and the other was subjected to TLC-blotting/SIMS. To this end, each of the marked glycosphingolipid bands was excised and trimmed to form a circle 2 mm in diameter that fit the probe tip of the MS equipment. Next, a few microliters of triethanolamine (matrix) was placed on the PVDF membrane, and negative SIMS spectra were obtained (Figure 9.4B). This method proved very useful in analyzing the structures of glycosphingolipids; the signals of the protonated molecule and fragment ions are the same as those obtained by ordinary mass spectrometry. TLC-blotting/SIMS has many advantages, the most important of which are the following: (1) analysis can be done without purification by column chromatography, (2) the glycosphingolipids on the PVDF membrane can be dissolved rapidly in a matrix, (3) the method is simple, no special devices being required, (4) the sample to be analyzed can be selected easily because all the separated glycosphingolipid bands are visible, and (5) sample loss is negligible and chemical degradation of analytes does not occur. In addition, it was shown that the TLC-blotting/SIMS method should also be useful for studying phospholipids, cholesterol, and other hydrophobic products. In further works [19], this technique was applied to (1) microscale purification of complex lipids (Figure 9.5), (2) direct mass spectrometric analysis of complex lipids, (3) binding assay of microorganisms on a lipid immobilized membrane, and (4) detection of enzymes.

In a discussion on the techniques of desorption of analytes of interest in clinical chemistry from the chromatoplate, one cannot fail to mention trial coupling of TLC/NMR realized by Wilson et al. [20]. The studies were carried out using high-resolution magic-angle-spinning (HR MAS) solid-state NMR. Chromatography was performed on 10 × 20 cm silica gel with chemically bonded octadecyl plates and methanol–water (6:2, v/v) as the mobile phase. For HR MAS, the appropriate zone of the TLC plate was removed by scraping; the stationary phase was placed in the HR MAS rotor and moistened with D_2O to provide a lock solvent for spectrometer reference and stabilization. The studies showed that it is possible to obtain NMR spectra (Figure 9.6), which can be interpreted, of analytes separated on a conventional RPTLC plate, without recourse to exhaustive isolation procedures. The non-destructive nature of the technique obviously means that when NMR spectroscopy is complete, the sample is available for investigation by other methods, such as MS, etc., thus offering the potential for the analyst to deploy a suite of spectroscopic techniques on a single spot.

Another technique used in PLC is a type of affinity chromatography technology based on the use of chemically immobilized antibodies against the compounds to be extracted from a biological fluid. This technique, named *immunoaffinity chromatography* (IAC), has been used to selectively extract various substances from biological fluids. A combined strategy of PLC overlay assay and mass spectrometry for the structural characterization of immunostained glycosphingolipids in silica gel extracts was put forward by Meisen and coworkers [21].

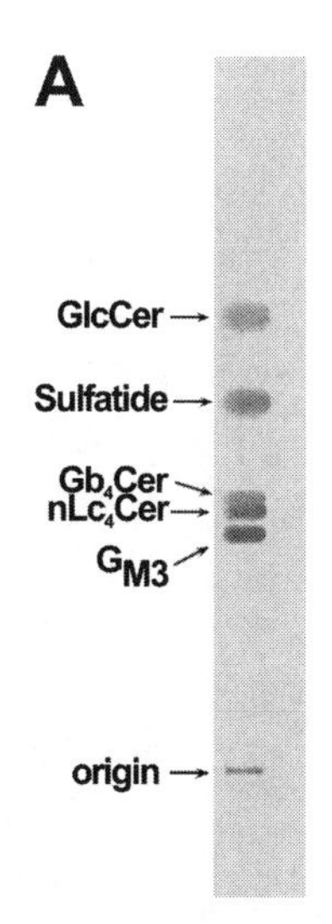

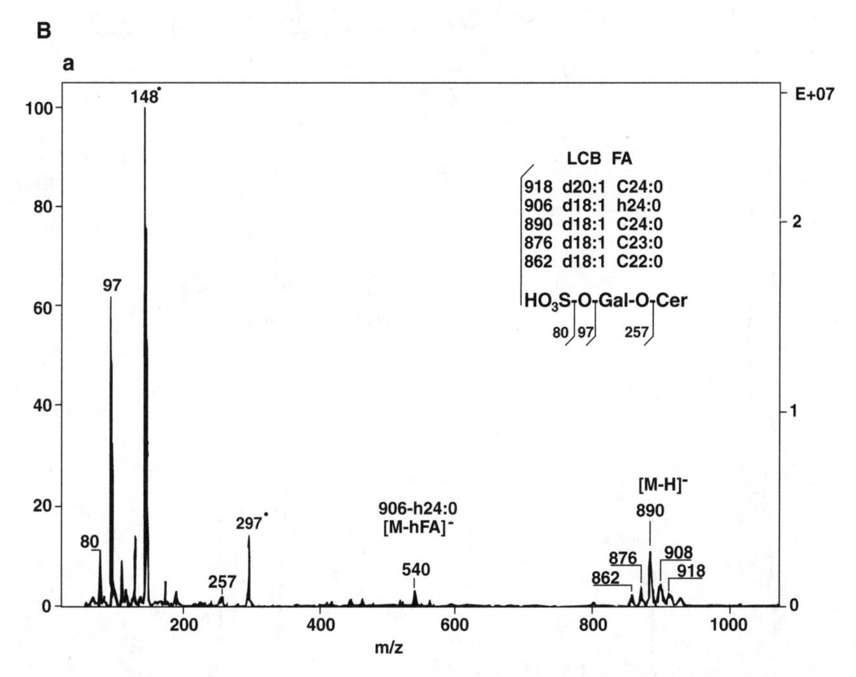

FIGURE 9.4 TLC-blotting/SIMS of various glycosphingolipids. (A) TLC profile of the separated glycosphingolipids, (B) MS profiles obtained by TLC-blotting/SIMS: (a) spectrum of sulfatide; peaks at m/z 862, 876, 890, 906, and 918 are molecular ions with different fatty acid species indicated in the figure (asterisks are peaks of triethanolamine as matrix); (b) spectrum of Gb_4Cer; peaks at m/z 1309 and 1337 are molecular ions with different fatty acid species as indicated; (c) spectrum of G_{M3}; peaks at m/z 1151, 1235, and 1263 are molecular ions with different fatty acid species as shown. Abbreviations: LCB — long chain base, FA — fatty acid, hFA — hydroxyfatty acid, Hex — hexose, HexNAc — *N*-acetylhexosamine. (From Taki, T. et al., *Anal. Biochem.*, 225, 24–27, 1995. With permission.) *Continued.*

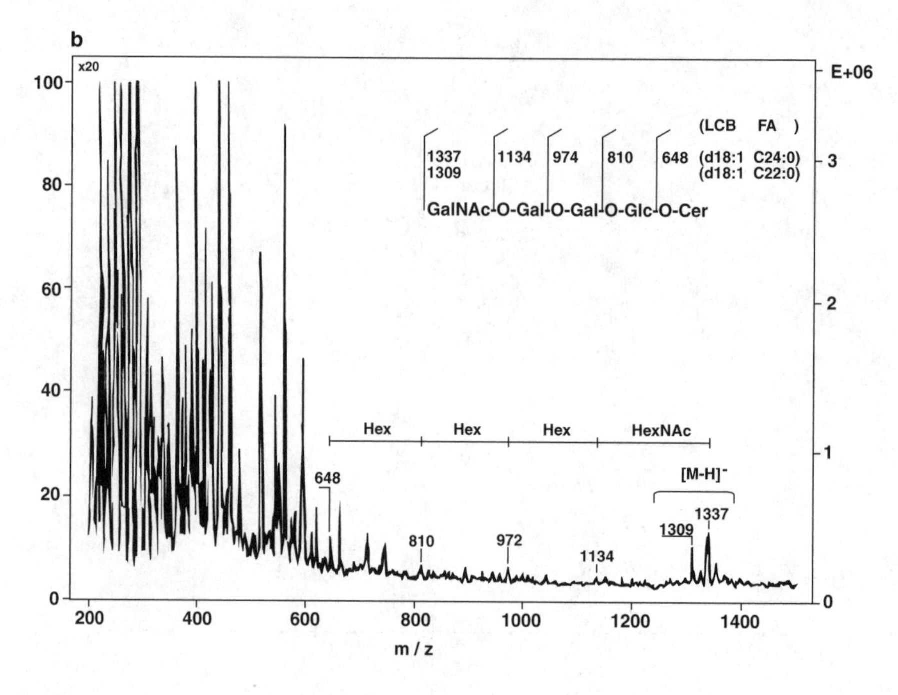

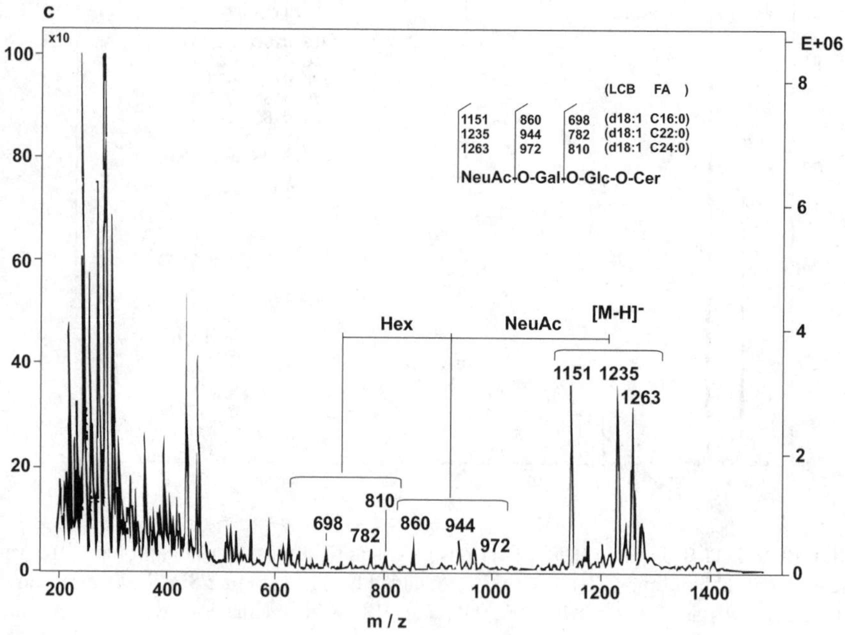

FIGURE 9.4 *Continued.*

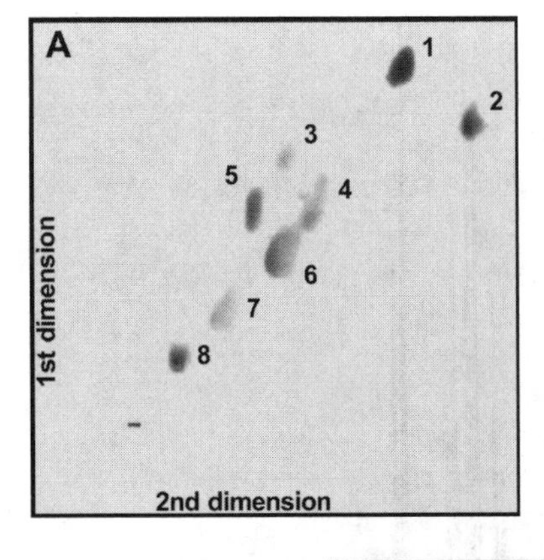

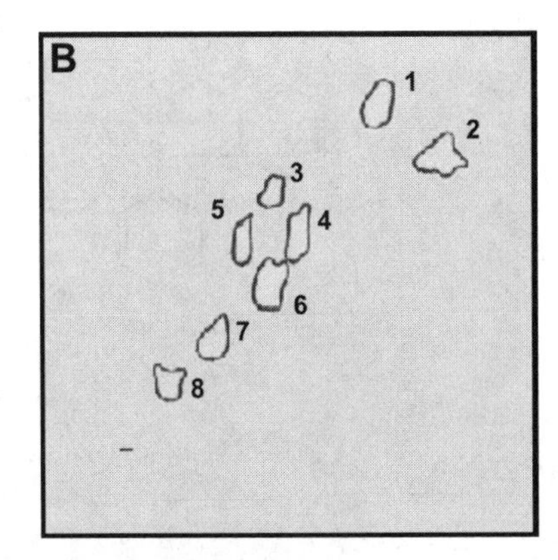

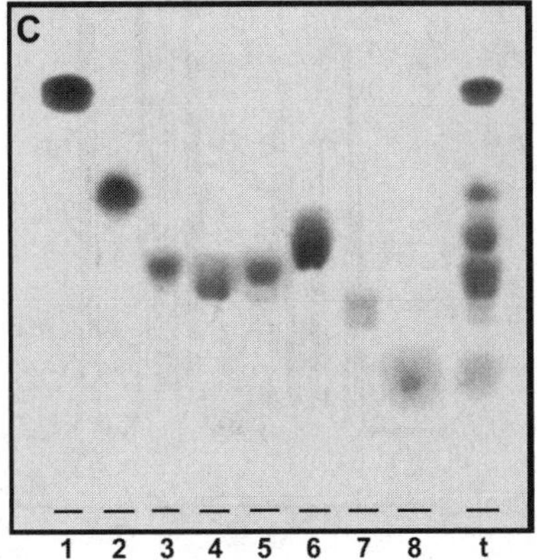

FIGURE 9.5 Phospholipid purification by TLC-blotting. (A) Profile of phospholipids separated by two-dimensional TLC. For phospholipid purification, a solvent mixture of chloroform–methanol–0.2% $CaCl_2$ (60:35:5, v/v) was used for the first direction and a mixture of chloroform–acetone–methanol–aceticacid-water (5:2:1:1:0.5, v/v) for the second direction. (B) Profile of phospholipids blotted from an HPTLC plate to a polyvinylidene difluoride membrane. Phospholipid locations are marked in drawing pencil. (C) TLC profile of phospholipids isolated by TLC-blotting. Numbers indicate bands made visible with primuline reagent; t — total phospholipid, 1 — phosphatidic acid, 2 — phosphatidylethanolamine, 3 — phosphatidylinositol, 4 — phosphatidylserine, 5 — lysophosphatidylethanolamine, 6 — phosphatidylcholine, 7 — sphingomyelin, and 8 — lysophosphatidylethanolamine. (From Taki, T. and Ishikawa D., *Anal. Biochem.,* 251, 135–143, 1997. With permission.)

Crude chloroform–methanol–water (30:60:8, v/v) extracts of immunostained TLC bands were analyzed without further purification by nanoelectrospray low-energy mass spectrometry. The authors showed that this effective PLC/MS-joined procedure offers a wide range of applications for any carbohydrate-binding agents such as bacterial toxins, plant lectins, and others. Phenyl-boronic acid (PBA) immobilized on stationary support phases can be put to similar applications. This technology, named *boronate affinity chromatography* (BAC), consists of a chemical reaction of 1,2- and 1,3-diols with the bonded-phase PBA to form a stable

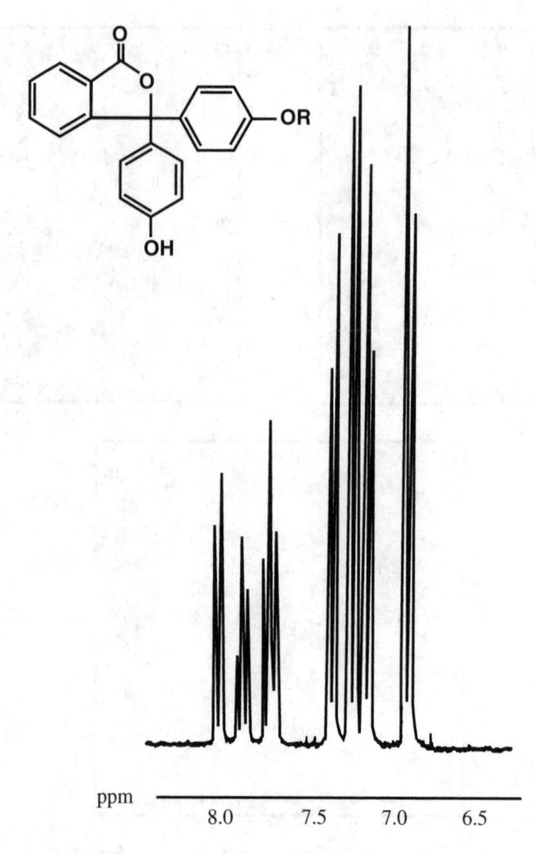

FIGURE 9.6 Partial ^{1}H NMR spectrum of phenolphthalein glucuronide in D_2O. (From Wilson, I.D., Spraul, M., and Humpfer, E., *J. Planar Chromatogr. Mod. TLC,* 10, 217–219, 1997. With permission.)

complex from which the compounds can be recovered specifically. BAC has been widely used for the selective separation of numerous dihydroxylated low-molecular-mass and high-molecular-mass compounds, including hemoglobin.

9.4 APPLICATIONS

Scientific literature offers numerous works dedicated to the applications of PLC in medical studies, which are either of research or diagnostic nature. The former concern DNA and RNA research, studies of amino acid sequence, metabolic pathways, etc. The results of these studies help to solve many problems in the field of biochemistry, histology, haematology, immunology, or molecular biology. Clinical studies consist mainly in finding the location of pathological processes in individual organs (organ diagnostics) or cells (cell diagnostics). This class of PLC applications also includes works consisting of the isolation of substances of clinical interest in order to obtain markers of endogenous substances. It was estimated [22] as early as

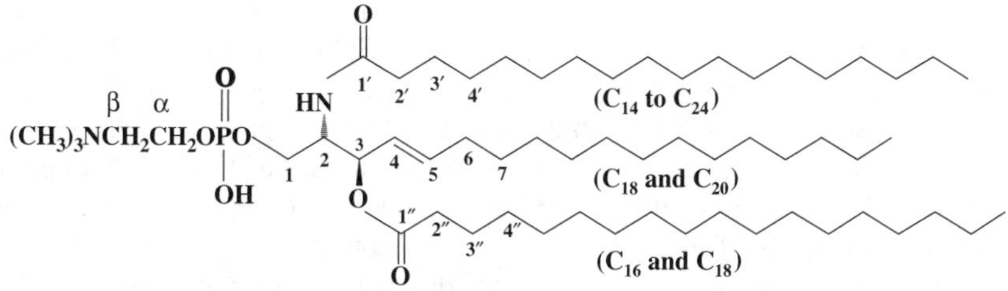

FIGURE 9.7 Structure of 3-*O*-acyl-D-*erythro*-sphingomyelin. (From Kramer, J. et al., *Biochim. Biophys. Acta,* 1303, 47–55, 1996. With permission.)

the 1980s that laboratory data formed the basis of more than 50% of diagnoses, and the annual increase in the number of analyses was 15 to 20%.

The work by Kramer and coworkers [23] constitutes a representative example of benefits issuing from the use of PLC in biomedical studies. In experiments using newborn pig and infant plasma, the researchers succeeded in isolating a new sphingolipid — a 3-*O*-acyl-D-*erythro*-sphingomyelin (Figure 9.7).

The new lipid occurred only in the plasma lipids of newborns and was not present in membrane lipids of red cell membranes or platelets. Total lipids were extracted from plasma and from red blood cell membranes and platelets. A total lipid profile was obtained by a three-directional PLC using silica gel plates and was developed consecutively in the following solvent mixtures: (1) chloroform–methanol–concentrated ammonium hydroxide (65:25:5, v/v), (2) chloroform–acetone–methanol–acetic acid–water (50:20:10:15:5, v/v), and (3) hexane–diethyl ether–acetic acid (80:20:1, v/v). Each spot was scraped off the plate; a known amount of methyl heptadecanoate was added, followed by methylation and analysis by GC/MS. The accurate characterization of the new lipid was realized using NMR technique.

9.4.1 Amino Acids and Peptides

A review on TLC and PLC of amino acids, peptides, and proteins is presented in the works by Bhushan [24,25]. Chromatographic behavior of 24 amino acids on silica gel layers impregnated tiraryl phosphate and tri-*n*-butylamine in a two-component mobile phase (propanol:water) of varying ratios has been studied by Sharma and coworkers [26]. The effect of impregnation, mobile phase composition, and the effect of solubility on hR_f of amino acids were discussed. The mechanism of migration was explained in terms of adsorption on impregnated silica gel G and the polarity of the mobile phase used.

A simple and rapid method of separating optical isomers of amino acids on a reversed-phase plate, without using impregnated plates or a chiral mobile phase, was described by Nagata et al. [27]. Amino acids were derivatized with *l*-fluoro-2,4-dinitrophenyl-5-L-alanine amide (FDAA or Marfey's reagent). Each FDAA amino acid can be separated from the others by two-dimensional elution. Separation of L- and D-serine was achieved with 30% of acetonitrile solvent. The enantiomers of threonine, proline, and alanine were separated with 35% of acetonitrile solvent and those of methionine, valine, phenylalanine, and leucine with 40% of acetonitrile solvent. The spots were scraped off the plate after the chromatography and extracted with methanol–water (1:1, v/v), and subsequently their enantiomeric configuration was determined. The absorbance of the extracts was measured at 340 nm with a spectrophotometer.

Homocysteine (HCys) easily conjugates with plasma proteins, which makes detection difficult; as a result, the diagnosis of homocystinuria in the newborn period is impeded. Ohtake and coworkers [28] have made an attempt to determine this compound in dried blood using TLC. A new compound was isolated in the study (using the PLC technique), namely, S-(2-hydroxyethylthio) homocysteine. Total HCys was extracted from dried blood spots and the extract was applied to a cellulose plate. After the development of the chromatogram (butanol–acetic acid–water; 4:1:1, v/v), the unknown fraction was harvested by scraping and analyzed using LC/MS. It was observed that *S*-(2-hydroxyethylthio) homocysteine was formed equally at room temperature and 37°C.

Successful semipreparative separations of amino acids and peptides are usually achieved with the use of planar gel electrophoresis; examples of PLC applications are presented in Table 9.1.

9.4.2 Carbohydrates

Glycoconjugates have been known for a long time to be potent antigens. Comparative studies of carbohydrate side chains of glycoconjugates have provided important information on the antigenic determinants. Following this theme, Ohene-Gyan et al. [38] conducted research on the polar glycolipid fractions of several mycobacterial strains of the closely related species *Mycobacterium avium* and *M. paratuberculosis.* The mycobacterial glycolipids have been purified by PLC using the following chromatographic system: mobile phase — chloroform–methanol–water (50:40:1, v/v) and stationary phase — silica gel. Glycolipids were eluted from plate scrapings by sonication in isopropanol–hexane–water (55:25:20 v/v). The purified substances were analyzed using GC or NMR. For NMR analyses, fractions eluted from the plates were evaporated and reevaporated several times from D_2O. Differences were shown in the oligosaccharide composition of *M. paratuberculosis* strains isolated from chronic enteritis of ruminants. The results indicate that the *M. paratuberculosis* strain 316F is more closely related to *M. avium* (from an AIDS patient) than it is to the classical *M. paratuberculosis* strain J10 and a Crohn's isolate.

A structural study on lipid A and the *O*-specific polysaccharide of the lipopolysaccharide from a clinical isolate of *Bacteroides vulgatus* from a patient with Crohn's disease was conducted by Hashimoto and coworkers [39]. They separated two potent virulence factors, capsular polysaccharide (CPS) and lipopolysaccharide (LPS), from a clinical isolate of *B. vulgatus* and characterized the structure of CPS. Next, they elucidated the structures of *O*-antigen polysaccharide (OPS) and lipid A in the LPS. LPS was subjected to weak acid hydrolysis to produce the lipid A fraction and polysaccharide fraction. Lipid A was isolated by PLC, and its structure was determined by MS and NMR.

The higher sugars have been identified as subunits in various natural products, and it is now well recognized that they play prominent roles in numerous biological processes. For example, some seven- and eight-carbon-atom sugars are important

TABLE 9.1
Amino Acids and Peptides — Separation and Purification

Chromatographic System	Desorption and Quantitation	Goal of Investigation	Reference
RP-18 Methanol–water (3:2, v/v)	Scraping and elution: MS	*N*-acylated derivatives of L-homoserine lactone (acyl-HSLs) separation	29
Cellulose Butanol–water–acetone (4:1:1, v/v)	Scraping and elution: NMR	Antitumor evaluation of cyclic peptides and macrolides from *Lissoclinum patella*; separation of cystine, threonine, and thiazole amino acids; research of the ulithiacyclamide B structure	30
Chiral plate Methanol–water–acetonitrile (5:5:20 and 5:5:3, v/v)	Scraping and elution: Rechromatography	Separation and purification of enantiomeric dipeptides, regioselective debenzylation of *C*-glycosyl compounds by boron trichloride	31
Silica gel Diethyl ether–cyclohexane (1:2, v/v)	Scraping and elution: Rechromatography	Purification and concentration of *C*-glycosyl compounds	32
Silica gel Methanol-chloroform (5:100, v/v)	Scraping and elution: NMR	Molecular studies of mono- and dicarboxylic acids in biologically active molecules	32
Silica gel Methanol–chloroform (1:10, v/v)	Scraping and elution: NMR	Preparation and studies of biological activities of dehydroxymethylepoxyquinomicin, a novel NF-B inhibitor	33
Silica gel Chloroform–methanol–2 M NH_4OH (40:10:1, v/v)	Scraping and elution: Fluorography	Lipid remodeling leads to the introduction and exchange of defined ceramides on GPI proteins in the ER and Golgi of *Saccharomyces cerevisiae*, purification of [^{3}H]dihydrosphingosine	34
Silica gel Chloroform–acetone (9:1, v/v)	Scraping and elution: UV/VIS, MS, NMR	Identification of the reaction end products — anthraquinones norsolorinic acid (NA) and averantin (AVN)	35
Silica gel Chloroform–methanol (9:1, v/v) or toluene–ethyl acetate (4:1, v/v) or ethyl acetate-petroleum ether (1:9, v/v)	Scraping and elution: NMR	Highly lipophilic benzoxazoles with potential antibacterial activity, separation of complex mixture of amino acids (Gly, Ala, Phe, Pgl, Val, Leu, Met, Tyr, Try)	36
Silica gel impregnated ethoxycarbonyl praline solution of L-ethoxycarbonyl-proline	Scraping and elution: X-ray analysis	*trans*-Benzoxanthene receptors for enantioselective recognition of amino acid derivatives	37

TABLE 9.2
Separations of Carbohydrates

Chromatographic System	Desorption and Quantitation	Objective of the Investigation	Reference
Silica gel Chloroform–methanol–water (130:50:9, v/v)	Scraping and elution: LSIMS	Core-branching pattern and sequence analysis of mannitol-terminating oligosaccharides by neoglycolipid technology; research of the dihexadecyl phosphatidylethanolamine derivatives of the periodate oxidation products of disaccharides and of oligosaccharide alditols	41
Amino–bonded silica gel Organic solvents	Immunostaining	Separation of oligosaccharides released from gangliosides by endoglycoceramidase	42
Whatman paper Butanol–ethanol–water (5:3:2, v/v)	Elution: Rechromatography	Enzymatic preparation of radiolabeled linear maltodextrins and cyclodextrins of high specific activity from [^{14}C] maltose using amylomaltase, cyclodextrin, glucosyltransferase, and cyclodextrinase	43
Silica gel Ethyl acetate–hexane (90:10, v/v)	Scraping and elution: GC/MS/MS	Specific and rapid quantification of 8-iso-prostaglandin $F_{2\alpha}$ in urine of healthy humans and patients with Zellweger syndrome	44
Silica gel Ethyl acetate–methanol (98:2, v/v)	Scraping and elution: GC/MS/MS	Quantification of 8-iso-prostaglandin $F_{2\alpha}$ and its metabolite 2,3-dinor-5,6-dihydro-8-iso-prostaglandin $F_{2\alpha}$ in human urine	45
Silica gel Ethyl acetate–2-propanol–water (3:1:1, v/v)	Elution: Rechromatography	The transglycosylation reaction of acarbose by *Bacillus stearothermophilus* maltogenic amylase in the presence of cellobiose and lactose; the cellobiose or lactose acceptor products were collected and purified	46
Silica gel Mixtures of organic solvents	Elution: NMR	Oligosaccharide structures determinations. Separation of free neutral oligosaccharides extracted from milk of a bearded seal, *Erignathus barbatus*	47
Silica gel Acetonitrile–water (85:15, v/v)	Elution: NMR, MS	Enzymatic synthesis of alkyl α-2-deoxyglucosides by alkyl alcohol resistant α-glucosidase from *Aspergillus niger*; purification of the reaction mixtures	48
RP-18 Methanol–water containing 0.1% H_3PO_4 (1:1, v/v)	Elution: NMR	Mapping the hydrolytic and synthetic selectivity of a type C feruloyl esterase (StFaeC) from *Sporotrichum thermophile* using alkyl ferulates; separation and purification of the alkyl ferulates	49
Silica gel Mixtures of organic solvents	Elution: NMR	Research of the structures of alternansucrase acceptor products arising from D-tagatose and from L-glucose; isolation of the oligosaccharide products from reaction mixtures	50

components in bacterial lipopolysaccharides; the nine-carbon sialic acids, the core class of higher sugars, occur as terminal residues of glycoconjugates involved in recognition phenomena; some C_{10} to C_{12} sugars, the rarest higher aldoses, have been identified as components of antibiotics. Baptistella and Cerchiaro reported [40] studies on the synthesis of a higher sugar from quinic acid. Preparative TLC (silica gel; mixture of organic solvent) was used for the separation and purification of the synthesis products. Other examples of PLC applications in carbohydrate analyses are presented in Table 9.2.

9.4.3 Lipids

Complete analysis of the lipid extract is usually performed for the purpose of characterizing a biological system, whereas abnormalities in known systems can be recognized from partial lipid analyses, for example, determination of fatty acids, total cholesterol, and triacylglycerols. A lipid extract usually contains water and other nonlipid contaminants, which must be removed from the sample prior to separation. In many instances, the removal of nonlipids can be combined with a preliminary segregation of the lipid classes. For many purposes, it is useful to effect a separation of the neutral and phospholipid fractions by adsorption PLC technique on silica gel. A development with a neutral solvent system retains the glycerophospholipids and sphingomyelins at the origin, whereas the neutral lipids are carried up the plate according to their R_f values. A parallel development with a polar solvent system carries the neutral lipids together to the solvent front, whereas the phospholipids are resolved according to their polarity. $AgNO_3$ PLC, and reversed-phase and chiral phase chromatographies have also served to purify lipids.

Endogenous, human-plasma-derived lipids that inhibit the platelet-stimulating activity of the platelet-activating factor (PAF) have been identified by Woodard and coworkers [51]. This is an important problem from the medical point of view, because the platelet-activating factor is a unique class of acetylated phospholipid autacoids with the most potent mediators of allergy and inflammation. A vital part of the studies consisted of the recovery and separation of lipids by PLC. Acetone supernatants containing neutral lipids were fractionated on silica gel plates using a solvent system of hexane–diethyl ether–acetic acid (50:50:1, v/v). On each plate, a mixture of neutral lipid standards (cholesterol, fatty acid, cholesteryl ester, and triglyceride) was included in peripheral lanes, which were visualized (after developing the chromatograms) by exposure to iodine vapor. For population studies, acetone supernatant containing neutral lipids from plasma was chromatographed. The sample lanes were then sectioned into 1 cm bands along the vertical axis. Lipids were extracted from the silica gel into chloroform–methanol–water (1:2:0.8, v/v). Similar separations were performed with regard to acetone precipitates containing phospholipids (lysophosphatidylcholinc, sphingomyelin, phosphatidylcholine, and phosphatidylethanolamine). Chloroform extracts of PLC fractions were stored at –20°C, and PAF inhibitory activity was measured by a bioassay method. The studies have documented the presence of multiple PAF inhibitors in human plasma.

Another set of studies on allergy and inflammation consists of a direct determination of PAF-acether after HPLC purification of human blood extracts. PAF-acether has been identified in various biological fluids as a preformed lipoprotein-bound compound (lipo-PAF). Free PAF was absent or detected in low amounts. Slightly different results were obtained by Tzzo and Benveniste [52]. Aliquot components obtained after the purification of blood (drawn from healthy volunteers) were separated on silica gel with chloroform–methanol–water (65:35:6, v/v). The areas of the bands were located by iodine staining, scraped off, extracted with 60% ethanol, and then assayed by HPLC. It was shown that sample purification using PLC allows for a detection of comparable amounts of free PAF and lipo-PAF. Probably, the free PAF could result from lipo-PAF dissociation during PLC purification.

Reactions between radical species and membrane lipids lead to the formation of lipid radicals, which interact readily with molecular oxygen and form lipid hydroperoxides. Lipid oxidation is implicated in a wide range of pathophysiological disorders, which leads to reactive compounds such as aldehydes. Among them, 4-hydroxynonenal reacts strongly with the NH_2 groups of amino acids and forms mainly Michael adducts and minor Schiff-base adducts. Guichardant et al. [53] conducted studies that showed that 4-hydroxynonenal could also react with amino phospholipids such as phosphatidylethanolamine and phosphatidylserine and characterized the resulting compounds. Total lipids were fractionated by PLC using silica gel and the mixture hexane–diethylether–acetic acid (60:40:1, v/v). Phospholipids were scraped off, extracted, and determined by HPLC/MS.

Sphingolipid metabolism during human platelet activation was the subject of the studies by Simon and Gear [54]. The analysis was focused on changes in the levels of ceramide and sphingomyelin in resting and thrombin-activated platelets. Lipids were separated on silica gel. For ceramide analysis, lipids were separated in chloroform–methanol–25% ammonia (90:10:1, v/v) or chloroform–methanol–formic acid (13:5:2, v/v). For sphingomyelin analysis, lipids were separated in chloroform–methanol–formic acid (13:5:2, v/v) or chloroform–methanol–water (60:35:8, v/v). Radioactive bands of ceramide and sphingomyelin were quantitated by scraping the bands from the plates and counting them in a scintillation counter.

Megli and Sabatini [55] studied the phospholipid bilayers after lipoperoxidation. Phospholipids were oxidized, and the oxidized phospholipid species were separated by PLC and estimated by EPR. It was shown that the early stages of lipoperoxidation brought about disordering of the phospholipid bilayer interior rather than fluidity alterations and that prolonged oxidation may result in a loss of structural and chemical properties of the bilayer until the structure no longer holds.

A very useful method for separating methyl esters of monounsaturated fatty acids by argentation PLC was put forward by Wilson and Sargent [56]. Monounsaturated fatty acid methyl esters (FAMEs) were separated from polyunsaturated and saturated fatty acid methyl esters, and the monounsaturated fatty methyl esters were resolved according to chain length; *cis* isomers are well resolved from the corresponding *trans* isomers (Figure 9.8). Silica gel 60 plates were sprayed uniformly with 20 ml of acetonitrile containing 2 g of silver nitrate until the plates were saturated. The plates were air-dried in subdued light and heated at 110°C for 30 min to achieve activation. Radiolabeled FAMEs were applied to plates as a narrow band. The plates were developed with toluene–hexane (40:60, v/v). After chromatography, areas of silica corresponding to radioactive FAMEs were located and scraped from

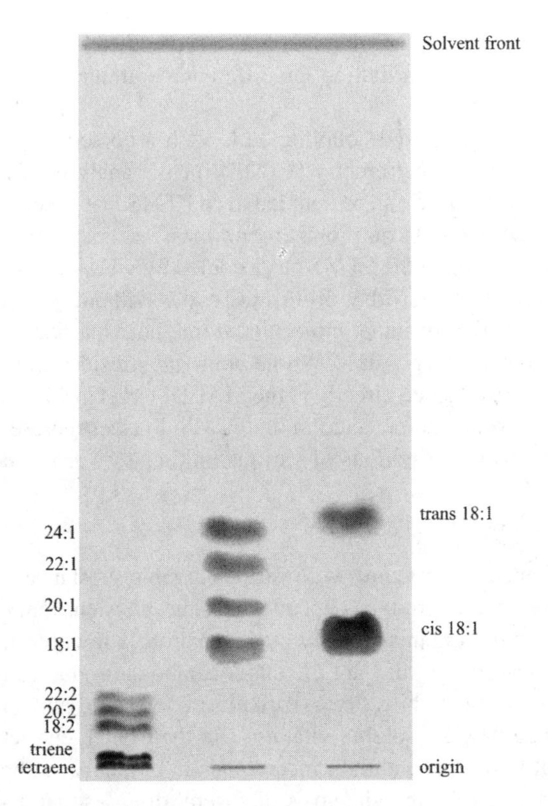

FIGURE 9.8 Chain separation of monounsaturated fatty acid by argentation thin-layer chromatography. Fatty acid methyl ester were analyzed by argentation chromatography. Mixtures of monounsaturated methyl esters are resolved according to chain length and all are clearly resolved from each other. Monounsaturated methyl esters were separated from polyunsaturated and saturated methyl esters; under these conditions saturated methyl esters did not char. The corresponding *cis*- and *trans*-ω9 monounsaturated isomers are well resolved. (From Wilson, R. and Sargent, J.R., *J. Chromatogr. A*, 905, 251–257, 2001. With permission.).

the plates into vials. Radioactivity was determined using a liquid scintillation analyzer. In other experiments, FAMEs were located, scraped from the plates, and eluted using 5 ml of ice-cold chloroform–methanol (2:1, v/v) containing 0.01% butylated hydroxytoluene; 1.25 ml of 0.88% KC1 was added and the solutions mixed. After separation, the chloroform layer was removed into a clean tube and the solvent was removed under nitrogen. FAMEs were redissolved in hexane containing 0.01% butylated hydroxytoluene and were washed with 20% NaCl to precipitate any remaining silver. The hexane layer was removed, an aliquot was counted for radioactivity, and the remainder was analyzed by radio gas chromatography.

Raith and Neubert [57] have developed a method for the profiling of human stratum corneum ceramides. The method enables the investigation of the role of ceramides in maintaining the barrier function of stratum corneum. TLC using automated multiple development was modified for semipreparative purposes. The fractionation of complex lipid extracts using this method ensured specific, sensitive, and reliable analysis by electrospray tandem mass spectrometry. Characteristic patterns of the molecular mass distribution in different ceramide fractions were easily obtained.

Ivleva et al. [58] proposed coupling TLC with a cooled by vibration MALDI Fourier transform mass spectrometry (FTMS) for the analysis of ganglioside mixtures. It was shown that TLC can be coupled to an FTMS instrument with an external-ion-source MALDI without compromising on mass accuracy and resolution of the spectra. Furthermore, when the FTMS has a cooled by vibration MALDI ion source, fragile glycolipids can be desorbed from TLC plates without fragmentation, even to the point that desorption of intact molecules from "hot" matrixes such as α-cyano-4-hydroxycinnamic acid is possible. Whole brain gangliosides were separated using TLC; the plates were attached directly to the MALDI target in which the gangliosides were desorbed, ionized, and detected in the FTMS. Further representative examples of PLC applications in separations of various lipid classes are listed in Table 9.3.

9.4.4 Bile Acids

Bile acids form micellar structures with various organic substances, the size, charge, and shape of which are strongly dependent on the physicochemical properties of bile steroids. In living organisms, they play a main role in cholesterol balance and fat digestion or absorption (they are the major catabolic products of cholesterol and facilitate the excretion of bile lipids including cholesterol and the absorption of dietary lipids including fat-soluble vitamins via their detergent action). Therefore, determination of bile acids and their metabolites in biological samples is becoming increasingly important for the diagnosis of several diseases and disorders.

Structural similarities between particular bile acid groups make their separation difficult, which justifies the employment of PLC for the purpose. It is especially difficult to perform separations of the so-called critical pairs (e.g., glycochenodeoxycholic from glycodeoxycholic or taurochenodeoxycholic from taurodeoxycholic). Scientific literature contains numerous articles on this topic, most valuable of which are the early works of Ganshirt et al. [69]. At the close of the last century, many different chromatographic systems and techniques of separation have been developed for the extraction of bile acids and related compounds from biological matrices. It has been proved that silica gel and acidic eluents are best suited for the separation of bile acids; good results are also obtained using reversed-phase systems (Table 9.4).

Detailed studies concerning the influence of several important experimental conditions on the separation of ten standard bile acids on silica gel and reversed phases (RP-8 and RP-18) were carried out by Rivas-Nass and Müllner [76]. The influence on migration distance, band shape and intensity, and resolving power exerted by conditions such as temperature, pH, support, and the addition of modifiers has been evaluated. The authors have shown that reducing the running temperature leads to an improvement not only of both band shape and intensity but also of resolution. The addition of modifiers such as EDTA, lithium chloride, boric acid, tetraethylammoniurn phosphate, tetrabutylam-monium phosphate, or ethylene glycol to the mobile phase does not improve resolution. Variation of the apparent pH and of the overall ionic strength should be carefully considered, especially if reversed-phase supports

TABLE 9.3
Lipid Separations on Silica Gel

Mobile Phase (v/v)	Desorption and Quantitation	Objective of the Investigation	Reference
Chloroform–methanol–ammonia (65:35:5)	Scraping: Scintillation counting	The *in vitro* and *in vivo* study of the binding of cyanate to erythrocyte membrane aminophospholipids; investigation of carbamylated aminophospholipids detection in the plasma membrane of native erythrocytes	59
Toluene–ethyl acetate (9:1)	Scraping and elution: Microsomal assays	Analysis of lipid metabolism in adipocytes (insulin stimulation of lipogenesis) using a fluorescent fatty acid derivative	60
Methanol–water and chloroform-methanol with additives	Scraping and elution: LC/MS	Profiling of human stratum corneum ceramides with a developing procedure optimized for human palmoplantar stratum corneum lipids	61
Chloroform–methanol–0.25% KCl in water (5:4:1)	Blotting: FAB MS	Analysis of 3- and 6-linked sialic acids in mixtures of gangliosides; separation of gangliosides and neutral glycosphingolipids of human leukocytes	62
Hexane–diethyl ether–acetic acid (70:30:1)	Blotting: Fluorescent measurement	Detection of cholesteryl linoleate hydroperoxides and phosphatidylcholine hydroperoxides	63
Chloroform–methanol-acetic acid–formic acid–water (5:15:6:2:1)	Scraping and elution: GC	Separation of phosphatidylcholine and phosphatidylethanolamine; the compositional analysis of phospholipids and their fatty acids in rabbit sarcoplasmic reticulum	64
(a) Petroleum ether–diethyl ether–acetic acid (90:15:1, v/v) (b) Chloroform–methanol–water (65:25:4, v/v)	Scraping and elution: Scintillation counting, NMR	Research of the neutrophils isolated from infected purulent exudates or obtained from healthy subjects and treated *in vitro* with bacterial lipopolysaccharide in the presence of heat-inactivated human AB serum; separation of polar and neutral lipids from whole cells serum and plasma membranes	65
Mixtures of organic solvents	Scraping and elution: GC/MS	Isolation of the lipids; utilization of a ts-sacB selection system for the generation of a *Mycobacterium avium* serovar-8 specific glycopeptidolipid allelic exchange mutant (*Mycobacterium avium* are ubiquitous environmental organisms and a cause of disseminated infection in patients with end-stage AIDS)	66
Diethyl ether–hexane (6:100, 8:100, and 1:10, v/v)	Scraping and elution: NMR and MALDI-TOF/MS	Characterization of individual mycolic acids in representative mycobacteria; separation in normal phase silica gel and argentation chromatography	67
Chloroform–methanol–water (60:16:2,v/v)	Scraping and elution: Immunoassay	Preparative TLC of crude *Mycobacterium leprae* lipids; determination of the molecular basis of CD1b-restricted T cell recognition of a glycolipid antigen generated from mycobacterial lipid and host carbohydrate during infection	68

TABLE 9.4
Separation of Bile Acids

Chromatographic System			
Stationary Phase	**Mobile Phase (v/v)**	**Fundamental Objective of the Experiment**	**Reference**
Silica gel	Four mobile phases (mixtures of water, methanol, diethyl ether, isooctane, isobutanol, isopropanol, and acetic acid)	Convenient and optimized method for sample pretreatment for the analysis of bile acids in biological matrices	70
Silica gel	Isooctane–ethyl acetate–acetic acid (4:1:1)	Separation of several bile acids and bile salts in dog bile; identification by liquid-state secondary-ion MS	71
RP-18	Two-dimensional elution 1st dimension — methanol–water (1:1) 2nd dimension — methanol–water–methyl β-cyclodextrin	Elaborating a method of separation of five major unconjugated bile acids, cholic acid, chenodeoxycholic acid, deoxycholic acid, ursodeoxycholic acid, lithocholic acid, and their conjugates; application of the method to the separation of glycine-conjugated bile acids in human bile	72
RP-18	Hexane–ethyl acetate (83:17) methanol–acetonitrile–water–formic acid (47.5:47.5:5:0.5)	Separation of 26 allo bile acids as their methyl esters; the combined use of NP and RP systems	73
Silica gel	Four mobile phases (mixtures of chloroform, isobutanol, isopropanol, water, and acetic acid) separately and in combination	The separation of the individual glycine and taurine conjugates by successive unidirectional developments in a twin-trough chamber	74
Silica gel	Chloroform–methanol (2:10)	Purification of bile acids extracted from human stool; quantitation by GC/FID	75

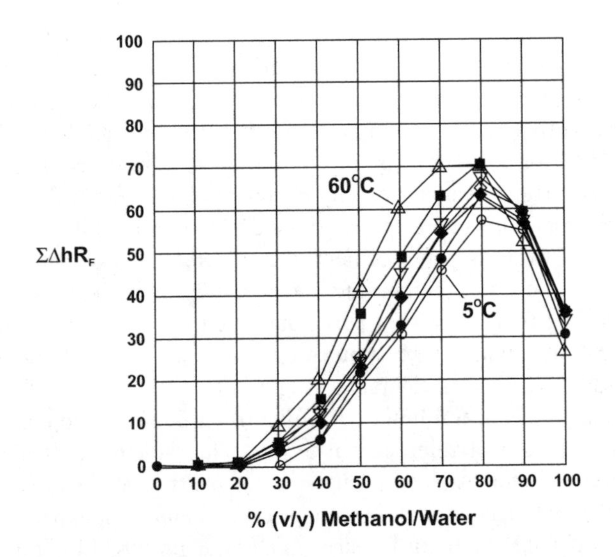

FIGURE 9.9 Relationships between ΣΔh R_f values and composition of mobile phases at different temperatures, obtained for mixtures consisting of eight steroids. ○ = 5°C, ● = 10°C, ♦ = 20°C, ◊ = 30°C, ∇ = 40°C, ■ = 50°C, Δ = 60°C. (From Zarzycki, P.K., Wierzbowska, M., and Lamparczyk, H., *J. Chromatogr. A*, 857, 255–262, 1999. With permission.)

are used. It has also been proved that substitution of glacial acetic acid with tetrahydrofuran is possible and leads to an improvement in band shape and resolution.

Zarzycki and coworkers [77] studied the influence of temperature on the separation of cholesterol and bile acids using reversed-phase stationary phases. The best chromatographic conditions for the separation of multicomponent samples of steroids were chosen. Experiments were performed on wettable plates with RP-18W and at the temperatures of 5, 10, 20, 30, 40, 50, and 60°C. The studies showed (Figure 9.9) that the degree of separation in the high-temperature region can be increased by an improvement of the efficiency of the chromatographic system. However, a relatively weak retention-temperature response for the studied steroids was observed.

9.4.5 Pharmacokinetic and Toxicological Studies of Drugs

In humans, drugs bind to plasma proteins, with the binding degree depending on their individual physicochemical properties. In general, acid and natural drugs bind primarily to albumin and basic drugs primarily to α-acid glycoprotein. Although in routine therapeutic drug monitoring total drug concentration (free and protein bound) is assayed, the use of isolation techniques such as PLC enables a separate assay of the said fractions. Information on drug–protein interactions is vital in the control of disorders; it is also significant because it is the free fraction of the drug that is thought to be the active form, because it can pass through cellular membranes to exert its action at the active site.

One can distinguish two basic aims realized during the studies of this group of compounds performed with the use of PLC: (1) screening of hard intoxication and drug abuse and (2) pharmacokinetic studies. Hard intoxication refers to the harmful effects arising in a short time after the introduction of a large dose of poison to the body. Drug abuse refers to the administration of a drug or other biologically active substance in order to produce a pharmacological effect unrelated to medical therapy. These substances form a large group consisting of various compounds, widely differing in their origin, availability, and chemical nature. From the legal point of view, they may be divided into legal and illicit drugs. The analysis of drugs of abuse is important not only in the enforcement of road traffic safety, but it also enables the differentiation between the chronic and the occasional drug user or makes it possible to identify the source of a particular batch of illegal drug. Such analysis would be not possible without the application of various forms of chromatography, among which PLC carries considerable weight.

Medical analyses concerning the screening of hard intoxication and drug abuse usually require the use of reference materials. Thus, there exists a necessity to develop methods of synthesis and purification of pharmacologically active metabolites. The work by Pelander et al. [78] may serve as an example here; in this work a method for *N*-demethylation of tertiary amines was proposed in order to produce reference material for forensic toxicological applications (the metabolism of many aliphatic tertiary amine drugs involves *N*-demethylation as a major pathway). The demethylation products were purified using PLC. To this end, the reactive mixture was transferred onto a preparative layer chromatographic plate as a 160-mm band, using an automatic sampler with the spray mode. The plate was developed over a distance of 10 cm with toluene–acetone–ethanol–concentrated ammonia (45:45:7:3, v/v). After drying, the nor-metabolite fraction was scraped off and extracted four times with dichloromethane–isopropylalcohol–concentrated ammonia (95:5:1, v/v). The extracts were combined and evaporated to dryness. The residue was weighed and analyzed using TLC and GC/MS. Similar works were also carried out by Bluhm and Li [79]. They put forward an inexpensive procedure for the preparative separation of quaternary ammonium and pyridinium compounds by flash column and PLC using normal phase system. For PLC, silica gel plates were immersed in a methanolic NaBr solution (6%, w/v). The development in a methanol–dichloromethane solvent system results in good sample movement with minimal tailing and reasonable R_f values. For more polar compounds, an aqueous mobile phase (methanol–water, 1:1, v/v) saturated with NaBr was necessary to separate the products.

The term *pharmacokinetics* refers to the movements of drugs within biological systems as affected by the processes of drug absorption, distribution, metabolism, and elimination. These processes determine how rapidly, in what concentration, and for how long the drug will appear at the target organ. Another significant pharmacokinetic parameter is bioavailability, defined as the rate and extent to which the active substance or active moiety is absorbed from a pharmaceutical form and becomes available at the site of action. Pharmacokinetic data are essential to support the safety and efficacy of new compounds.

Studies to elucidate the mode of bacteriostatic property of xanthostatin (XS), a novel depsipeptide antibiotic with an *N*-acetylglycine side chain and selective antimicrobial activity against *Xanthomonas* spp., were carried out by Kim and coworkers [80]. Two biotransformed XSs were isolated by the treatment of XS with the cell lysate of *Xanthomonas campestris* pv. *citri*, a solvent partition, PLC, and HPLC. Analysis results suggest that the *N*-acetylglycine side chain plays a critical role in the antimicrobial activity of XS.

The investigation of the levels of [^{3}H]dopamine in blood, the cerebrospinal fluid, and brain tissue samples in rats in order to determine whether the drug is transferred along the olfactory pathway to the central nervous system following nasal administration was the subject of the studies by Dahlin and coworkers [81]. The constituents of preliminarily purified samples of cerebrospinal fluid and brain tissue were separated and purified using PLC; silica gel served as the stationary phase, and a three-component mixture of butanol–acetic acid–water (4:1:1, v/v) was used as the mobile phase. After air-drying, the plates were cut into strips corresponding to the various bands and placed in glass scintillation vials. It was shown that unchanged dopamine is transferred into the olfactory bulb via the olfactory pathway and metabolism of dopamine had taken place in all samples tested.

The metabolic and pharmacokinetic profile of sucralose (this is a novel intense sweetener with a potency about 600 times that of sucrose) in human volunteers was studied by Roberts and coworkers [82]. Part of this study was realized using PLC in the following chromatographic system in which the stationary phase was silica gel and the mobile phase was ethyl acetate–methanol–water–concentrated ammonia (60:20:10:2, v/v). Separated substances were scraped off separately, suspended in methanol, and analyzed by filtration, scintillation counting, or enzymatic assay. It was shown that the characteristics of sucralose include poor absorption, rapid elimination, limited conjugative metabolism of the fraction absorbed, and lack of bioaccumulative potential.

Min and coworkers [83] studied the cytotoxicity of shikonin metabolites with biotransformation of human intestinal bacteria. Six shikonin metabolites were obtained from human intestinal bacteria, *Bacteriodes fragilis* subsp. *Thetaotus*. The metabolites were isolated by repeated silica gel column chromatography, Sephadex LH-20, and PLC. *In vitro* cytotoxicities were tested against human tumor cell lines; PC-3 (prostate), ACHN (renal), A549 (lung), SW620 (colon), K562 (leukemia), and Du 145 (prostate). The four Shikonin metabolites showed weaker cytotoxicity than the parent shikonin; one shikonin monomeric metabolite and one dimeric metabolite exhibited stronger activities compared with adriamycin, which was used as the positive control. Other applications of PLC for pharmacokinetic studies are presented in Table 9.5.

While discussing the applications of PLC in drug analyses, one cannot omit the works devoted to the identification of biologically active substances present in herbal medications. For example, the extract from an edible plant, *Actinidia arguta Planchon*, appeared to possess antitumor activity against human leukemia Jurkat T and U937 cells by inducing apoptosis. Detailed studies of the induction of apoptotic cell death by a chlorophyll derivative isolated from *Actinidia arguta Planchon* were carried out by Park et al. [96]. The constituents of the plant extract were purified by silica gel PLC and column chromatography and then analyzed by UV spectrometry and ^{1}H- and ^{13}C-NMR. It was shown that the substance belongs to the chlorophyll-derivative-like group. Some works devoted to this topic are listed in Table 9.6.

TABLE 9.5
Pharmacokinetic Studies

PLC Application	General Objective of Study	Reference
Isolation of cryptotanshinone, an active quinoid diterpene formed in callus of *Salvia miltiorrhiza* BUNGE	Effect of N(6)-benzyladenine (BA) on tanshinone formation in callus cultures of *Salvia miltiorrhiza* was examined in an attempt to increase the productivity of the medicinal compound; cryptotanshinone was characterized by NMR and MS	84
Isolation and purification of amitriptyline, nortriptyline, and 10-hydroxy (10-OH) derivatives of the drugs from rat bile	Research of glucuronides of hydroxylated metabolites of amitriptyline and nortriptyline metabolites were characterized by NMR and MS and by enzymatic or acid deconjugation with subsequent identification of aglycones and glucuronic acid	85
Separation of 2′,3′-didehydro-3′-deoxythymidine 5′-(methoxyalaninyl phosphate) triethylammonium salt	Research of the pharmacokinetics and metabolism of the experimental nucleoside reverse transcriptase inhibitor compound stampidine in mice, dogs, and cats	86
Isolation and purification of tetrahydrocurcumin (THC) and curcumin from plasma	Investigation of the pharmacokinetic properties of curcumin in mice (curcumin, the yellow pigment in turmeric and curry, has antioxidative and anticarcinogenic activities); the chemical structures of these metabolites were determined by MS	87
Purification of [^{3}H]cycloquinonee	Investigation of the influence of pharmacokinetics on pharmacodynamics with reference to three types of Mannich antimalarial compounds; effect of disposition of Mannich antimalarial agents on their pharmacology and toxicology was characterized	88
Purification of 2-(4′-[^{123}I]iodo-biphenyl-4-sulfonylamino)-3-(1*H*-indol-3-yl)-propionic acid and 2-(4′-[^{123}I]iodo-biphenyl-4-sulfonylamino)-3-(1*H*-indol-3-yl)-propionamide	Synthesis and purification of radioiodinated carboxylic and hydroxamic matrix metalloproteinase inhibitors (tumor imaging agents); research of inhibition capacities and accumulation in organs	89

Isolation of metabolites from urinary extracts	The metabolism of toborinone, (±)-6-[3-(3,4-dimethoxybenzylamino)-2-hydroxypropoxy]-2(1*H*)-quinolinone, a novel inotropic agent, was studied in rats and dogs; the structural characterization of metabolites and the metabolic pathways of toborinone ([^{14}C]toborinone) after intravenous administration were investigated	90
Purification of the anticancer drug	Research of the pharmacokinetic parameters of lipid emulsion of paclitaxel	91
Purification of the anticancer drug	Investigation of the pharmacokinetic properties of *N*-oxide analogs of WAY-100635: new high-affinity 5-HT_{1A} receptor antagonists	92
Separation and purification of metabolites	Investigation of the metabolism of ezlopitant alkene (CJ-12,458), an active metabolite of ezlopitant, in human liver microsomes	93
Separation and purification of biliary metabolites; isolation of methyl(17α-acetoxy-6-chloro-2-oxa-4,6-pregnadiene-3,20-dione-21-yl-2′,3′,4′-*O*-triacetyl-α-D-glucopyranosid)uronate	The pharmacokinetic characteristics of osaterone acetate (17α-acetoxy -6-chloro-2-oxa-4,6-pregnadiene-3,20-dione) in rats; measurements of the plasma and urine concentration profiles of metabolites	94
Separation and purification of *N*-substituted-(indol-3-yl)-carboxamides and alkanamides	Pharmacokinetic studies of *N*-pyridinyl-indole-3-(alkyl)carboxamides and derivatives as potential systemic and topical inflammation inhibitors	95

TABLE 9.6
Drugs Separation and Purification

PLC Application	General Objective of Study	Reference
Isolation cimetidine, famotidine, nizatidine, and ranitidine hydrochlorides	Studies of the histamine H_2-receptor antagonists; substances were extracted from pharmaceutical preparations with methanol or concentrated acetic acid, purified by PLC and identified on the basis of their infrared spectra	97
Separation of the components of the *Culcasia scandens* extract into seven fractions	The screening of the antiinflammatory potentials of a popular traditional antirheumatic herb, *Culcasia scandens*; the fractions were subjected to phytochemical analysis to identify the biologically active constituents	98
Separation of flavonoid and alkaloid fractions of *Synclisia scabrida*	Investigation of the biochemical evidence for the antiulcerogenic activity of *Synclisia scabrida*; the study seems to implicate alkaline phosphatase as a biochemical evidence of the antiulceroragenic activity of *S. scabrida*	99
Isolation of two quassinoids (quassin and simalikalactone D) from root bark of *Quassia africana Baill*	Studies of the activity of quassinoids against herpes simplex, Semliki Forest, Coxsackie and vesicular stomatitis viruses; structural analyses of the extracted compounds using NMR and MS	100
Separation and purification of active fraction of mulberry shoots	Studies on the antibacterial activities of mulberry phytoalexin *in vitro*; it was discovered that some constituents of phytoalexin isolated from their cortices showed a potent antibiotic activity against gram-positive bacteria	101
Separation of the components of the previously extracted fractions (indicating the presence of active compounds) of *Chaptalia nutans*	*In vitro* analysis of the antibacterial activity of a 7-O-β-D-glucopyranosylnutanocoumarin; extracts from the roots of *Chaptalia nutans*, traditionally used in Brazilian folk medicine, were screened against *Staphylococcus aureus*, *Escherichia coli*, and *Pseudomonas aeruginosa*	102
Repeated separation of antibiotics	Studies on new cell cycle inhibitors and apoptosis inducers, produced by *Streptomyces pseudoverticillus*	103
Separation and purification of icariside I, icariin, epimedroside, icariresinol-4′-β-D-glucopyranoside, and epimedin from the root of *Epimedium leptorrhizum*	Study on chemical components of the root of *Epimedium leptorrhizum* in order to obtain a more comprehensive understanding on its effective components	104

This group of PLC applications also includes works devoted to the separation and purification of the components of mixtures obtained during analyses of the influence of specific plant extracts on the growth or inhibition of cancer cells. Examples include the works of Yang and coworkers, [105] who examined the prostaglandin endoperoxide synthase activity in human oral carcinoma cell lines in the presence of the betel nut extract. After PLC, the radioactivity of the spots was determined by direct counting of the scraped spots using liquid scintillation spectroscopy. Activity of the separated substances was determined by the bioassay method. It was shown that the betel nut extract can significantly inhibit the growth of two oral cancer cell lines and one fibroblast cell line and that this inhibition is a concentration- and time-dependent phenomenon.

Recently, particular attention has been devoted to articles pertaining to classical Chinese medicine (Table 9.7). They employ off-line coupling of PLC with such analytical techniques (MS and NMR) as may provide information on the structures of individual components of the medications analyzed. The subject has been dealt with extensively in Chapter 11.

The testing of impurities in active pharmaceutical ingredients has become an important initiative on the part of both federal and private organizations. Franolic and coworkers [113] describe the utilization of PLC (stationary phase — silica gel and mobile phase — dichloromethane–acetonitrile–acetone (4:1:1, v/v)) for the isolation and characterization of impurities in hydrochlorothiazide (diuretic drug). This drug is utilized individually or in combination with other drugs for the treatment of hypertension. The unknown impurity band was scraped off the plate and extracted in acetonitrile. The solution was filtered and used for LC/MS and NMR analysis. The proposed procedure enabled the identification of a new, previously unknown impurity. It was characterized as a 2:1 hydrochlorothiazide–formaldehyde adduct of the parent drug substance.

9.4.6 Research of Enzyme Activities

D-Xylulose 5-phosphate (*d*-threo-2-pentulose 5-phosphate, XP) stands as an important metabolite of the pentose phosphate pathway, which plays a key function in the cell and provides intermediates for biosynthetic pathways. The starting compound of the pathway is glucose 6-phosphate, but XP can also be formed by direct phosphorylation of D-xylulose with *d*-xylulokinase. Tritsch et al. [114] developed a radiometric test system for the measurement of D-xylulose kinase (XK) activity in crude cell extracts. Aliquots were spotted onto silica plates and developed in *n*-propyl alcohol–ethyl acetate–water (6:1:3 (v/v) to separate *o*-xylose/*o*-xylulose from XP. Silica was scraped off and determined by liquid scintillation. The conversion rate of [^{14}C]*o*-xylose into [^{14}C]*o*-xylulose 5-phosphate was calculated. Some of the works devoted to the separation of components necessary while analyzing enzyme activity are presented in Table 9.8.

TABLE 9.7
Chemical Studies of Plant Extract Components (Separations on Silica Gel)

Matrix	Mobile Phase (v/v)	Fundamental Goal Of Experiment	Reference
Manchurian walnut, Juglans mandshurica	Different solvent systems	Study of the chemical components of matrices leaves; identification of five new compounds by MS, IR, and NMR	106
Kanguzengsheng pills	Ethyl acetate–acetone–methanol–water (10:2:2:1)	Identification of icarriine by comparison with the standard; quantitation by UV spectrophotometry after elution	107
Sanguisorba alpina	Petrol ether–acetone (different ratio)	Identification of 11 constituents by measuring their physical–chemical parameters and by IR and NMR	108
Lilac daphne, Daphne genkwa	Ethyl acetate–benzene (6:9)	Identification of daphnodorin B by IR and NMR	109
Cephalotaxus hainanesis	Ether–acetone (2:1)	Separation of cephalotaxine and its identification by IR and MS	110
Yuzhi Pingan pill	Hexane–ethyl acetate–benzene (14:3:3)	Identification of eugenol and its quantitation by GC	111
Glycyrrhiza pallidiflora	Petrol ether–ethyl acetate (8:2)	Elucidation of 42 compounds	112

TABLE 9.8
Separations Related to Enzyme Activity Analyses

Chromatographic System	Elution and Quantitation	Objective of the Investigation	Reference
Rod of silica gel 2-Propanol–ethanol–water–25% ammonia (50:25:25:1, v/v)	FI desorption: Latroscan TLC/FID	The application of the latroscan TLC/FID system to analyze chitooligosaccharide standard solutions and heat-inactivated samples from the enzymatic hydrolysis of such oligosaccharides	115
Silica gel 2D elution Mixture of organic solvents	Scraping: Scintillation counting	Characterization of the transacylase activity of rat liver 60-kDa lysophospholipase-transacylase, separation of acylation product of glycerol and 2-palmitoylglycerol	116
Silica gel Butanol–acetic acid–water (20:1:3, v/v)	Scraping: Scintillation counting	The investigation of the biopterin biosynthesis in man. Determination of GTP cyclohydrolase and D-erythro-7,8-dihydroneopterin triphosphate synthetase	117
Silica gel Butanol–formic acid–water (7:1:5, v/v)	Scraping: HPLC	Detection of peptidase activity with 6-aminoquinolyl carbamate-angiotensin I	118
Silica gel Acetone–hexane (45:55, v/v)	Scraping: Scintillation counting	Determination of cytochrome P450 2E1 activity in microsomes using [2-^{14}C]chlorzoxazone	119
Silica gel Benzene–ethyl acetate (80:20, v/v)	Scraping: GC MS	Research of the structure-function of human cytochrome P450 3A4 using 7-alkoxycoumarins	120

9.4.7 Other Applications

Menshonkova and coworkers [121] examined the effects of high-intensity ultraviolet irradiation on nucleic acids and their components. Nucleoside solutions were separated on silica gel plates in the following solvent systems: (1) ethyl acetate–isopropanol–water (75:16:9, v/v), (2) chloroform–methanol–water (4:2:1, v/v), (3) water, (4) butanol–water (86:14, v/v), (5) butanol–water–acetic acid (5:3:2, v/v), and (6) isopropanol–concentrated ammonia–water (7:2:l, v/v). Separated substances were identified by UV. Chromatograms were cut into 0.5 × 1.0 cm rectangles and the radioactivity of each rectangle was measured with the scintillation counter. The identity of UV-absorbing products was additionally confirmed spectrophotometrically after the elution of the respective spots with water. The radiochromatographic purity of [^{14}C]thymidine, [^{14}C]adenosine, and 2-deoxy-[^{14}C]adenosine amounted to 98%.

The kinetics of cyclic ADP-ribose formation in heart muscle was investigated by Meszaros et al. [122]. After purification, the reaction products were identified with the use of PLC. Chromatograms were developed on Baker-flex PEI cellulose with 70% water–30% ethanol–0.2 *M* NaCl. The R_f values for NAD glycohydrolase, cyclic ADP-ribose, and ADP-ribose were 0.51, 0.36, and 0.13, respectively. After blow-drying the plates, 0.4 × 1.0 cm squares between the origin and the solvent front were scraped and liquid scintillation counted. It was shown that the heart muscle possesses enzymatic activities that convert NAD into both ADPR and cADPR. The authors suggest that the two activities could represent alternative functions of the same enzyme, which might be controlled by endogenous ADP-ribosylation reactions.

The purine nucleoside succinyladenosine (S-Ado) is the metabolic product of dephosphorylation of intracellular adenylosuccinic acid. The presence of high concentrations of S-Ado in the human cerebrospinal fluid is a diagnostic marker for inherited enzyme deficiency of adenylosuccinate lyase. Krijt et al. [123] proposed a method for the identification and determination of succinyladenosine in human cerebrospinal fluid. Cerebrospinal fluid, acidified to pH 2.5, was purified using SPE, and S-Ado was recovered from the obtained eluate with the help of PLC. The eluate was placed on a silica gel plate and then the chromatogram was developed using a three-component mixture of ethyl acetate–acetic acid–water (13:4:4, v/v). The plate with S-Ado standard was visualized with naphtoresorcinol/sulphuric acid reagent. The R_f of the blue reacting band of S-Ado standard was determined, and the band with corresponding R_F was scraped off from the plate with the cerebrospinal fluid sample. The S-Ado was eluted from the silica with water, and the fraction was concentrated by evaporation under a stream of nitrogen at 50°C. The assay was performed using the HPLC/MS technique.

A modified nucleotide found in RNA sequencing could either be a new nucleotide of unknown chemical structure or it could correspond to an already known modified nucleotide (up to now about 90 different modified nucleotides have been identified in RNA). Keith [124] proposed preparative purifications of major and modified ribonucleotides on cellulose plates, allowing for their further analysis by UV or mass spectrometry. Separation was realized by two-dimensional elution using the following mobile phases: (1) isobutyric acid–25% ammonia–water (50:1.1:28.9, v/v) and (2) HCl–isopropanol–water (15:70:15, v/v) or (1) isobutyric acid–25% ammonia–water (50:1.1: 28.9, v/v) and (2) sodium phosphate–0.1 *M* (pH 6.8) ammonium sulphate–propanol (100:60:2, v/w/v).

Human malignant cell death by apoptosis-inducing nucleosides from the deciduas-derived $CD57^{+}HLA\text{-}DR^{bright}$ natural suppressor cell line was studied by Mori et al. [125]. Six AINs were released into 57.DR-NS cell culture media and were isolated by the combination of physicochemical procedures of C18 preparative column, HPLC, and PLC. They demonstrated that AINs could induce apoptosis in the human malignant Molt4/BeWo/GCIY cell line but not human normal WI-38 fibroblasts. Apoptosis was characterized by DNA strand breaks and activation of the caspase cascade.

The research on dehydroepiandrosterone (DHEA) is limited because of the lack of radiolabeled metabolites. Robinzon et al. [126] showed that, using pig liver microsomes, the radiolabeled metabolites of DHEA can be prepared in stable, pure form for biochemical studies. They utilized pig liver microsomal (PLM) fractions to prepare [^{3}H]-labeled 7α-hydroxy-DHEA (7α-OH-DHEA), 7β-hydroxy-DHEA (7β-OH-DHEA), and 7-oxo-DHEA substrates from 50 μ*M* [1,2,6,7-^{3}H]DHEA. The metabolites were separated by silica gel PLC plates using ethyl acetate-hexane–glacial acetic acid (18:8:, v/v) as the mobile phase, extracted with ethyl acetate, and dried under a stream of nitrogen. The purity of markers was determined with the use of TLC and GC/MS.

Cryptosporidium parvum, an apicomplexan parasite of the mammalian gut epithelium, causes a diarrhoeal illness in a wide range of hosts, and is transmitted by contamination of food or water from an infected animal. Priest and coworkers [127] identified a glycosylinositol phospholipid from the sporozoite stage of the parasite that is frequently recognized by serum antibodies from human cryptosporidiosis patients. The substances were purified by butanol extraction followed by octyl-Sepharose column chromatography and PLC, and analyzed by mass spectrometry and radiolabeled neutral glycan analysis. It was shown that the structure of the dominant glycosylinositol phospholipid antigen contained a C18:0 lyso-acylglycerol, a C16:0-acylated inositol, and an unsubstituted mannose$_3$-glucosamine glycan core. Subsequent works are being carried out so as to determine the role that these glycolipids may play in the development of the disease and in the clearance of infection.

Environmental pollution produced by incineration of wastes containing plastic products has become a serious problem on a global scale. It has been reported that 2,3,7,8- tetrachlorodibenzo-*p*-dioxin and other dioxins are possibly ubiquitous in the environment and that they are formed from combustion of plastic products containing polyvinyl chlorides (PCV), chlorobenzenes, chlorophenols and polychlorinated biphenyls, etc. Yonezawa et al. [128] attempted an evaluation of the mutagenicity of substances obtained by burning PCV product at 1000°C. The acetone extract of combustion products was fractionated into acidic, neutral, and basic by liquid–liquid distribution. The constituents of each fraction were separated using PLC (silica gel; methylene chloride). Each of the bands was scratched from the plate, extracted by $CH_3COOC_2H_5$, and provided for the mutagenicity assay. Intensive mutagenicity of the constituents of each fraction was evidenced.

ABBREVIATIONS

AC	Argentation Chromatography
BAC	Boronate Affinity Chromatography
CE	Cholesteryl Ester
CPS	Capsular Polysaccharide
DNA	Desoxyribose Nucleic Acid
EPR	Electron Paramagnetic Resonance
FA	Fatty Acid
FAME	Fatty Acid Methyl Ester
FID	Flame Ionization Detector
FTIR	Fourier Transform Infrared
GC	Gas Chromatography
GC/MS	Gas Chromatography/Mass Spectrometry
GC/MS/MS	Gas Chromatography/Tandem Mass Spectrometry
HPLC	High-Performance Liquid Chromatography
IAC	Immunoaffinity Chromatography
IEC	Ion-Exchange Chromatography
LC	Liquid Chromatography
LPS	Lipopolysaccharide
MALDI	Matrix-Assisted Laser Desorption Ionization
MS	Mass Spectrometry
NMR	Nuclear Magnetic Resonance
PBA	Phenyl-Boronic Acid
PC	Phosphatidylcholines
PS	Polysaccharide
PVDF	Polyvinylidene Difluoride Polymer
RNA	Ribonucleic Acid
SALDI	Surface-Assisted Laser Desorption Ionization
SIMS	Secondary Ion Mass Spectrometry
TLC-blotting	Transfer of the analytes from a plate to a PVDF membrane
UV	Ultraviolet light

REFERENCES

1. Papadoyannis, N. and Samanidou, V.A., Sample pretreatment in clinical chemistry, in *Separation Techniques in Clinical Chemistry*, Aboul-Enein, H.Y., Ed., Marcel Dekker, New York, 2001, chap. 1.
2. Jain, R., Thin-layer chromatography in clinical chemistry, in *Practical Thin-Layer Chromatography: A Multidisciplinary Approach*, Fried, B. and Sherma, J., Eds., CRC Press, Boca Raton, FL, 1996, chap. 7.
3. Błądek, J. and Neffe, S., Application of TLC in clinical chemistry, in *Separation Techniques in Clinical Chemistry*, Aboul-Enein, H.Y., Ed., Marcel Dekker, New York, 2001, chap. 11.

4. Sowa, J.M. and Subbaiah, P.V., *J. Chromatogr. B,* 813, 159–166, 2004.
5. Ravandi A., Kuksis, A., Myher, J.J., and Marai, L., *J. Biochem. Biophys. Methods,* 30, 271–285, 1995.
6. Rezanka, T., *J. Chromatogr. A*, 727, 147–152, 1996.
7. Aboul-Enein, H.Y., El-Awady M.I., and Heard, C.M., *J. Pharm. Biomed. Anal.,* 32, 1055–1059, 2003.
8. Whitton, P.S., Nicholson, R.A., Bell, M.F., and Strang, R.H.C., *Insect Biochem. Mol. Biol.,* 25, 83–87, 1995.
9. Luo, Q., Andrad, J.D., and Caldwell, K.D., *J. Chromatogr. A*, 816, 97–105, 1998.
10. Romano, G., Ceruso, G., Musumarra, G., Pavone, D., and Cruciani, G., *J. Planar Chromatogr. Mod. TLC*, 7, 233–239, 1994.
11. Armstrong, D.W. and Zhou, Y., *J. Liq. Chromatogr.*, 17, 1695–1707, 1994.
12. Miwa, H., Yamamoto, M., Futata, T., Kan, K., and Asano, T., *J. Chromatogr. B*, 677, 217–223, 1996.
13. Busch, K.L., *J. Planar Chromatogr. Mod. TLC*, 5, 72–79, 1992.
14. Sunner, J., Dratz, E., and Chen, Y.C., *Anal. Chem.,* 67, 4335–4342, 1995.
15. Han, M. and Sunner J., *J. Am. Soc. Mass Spectrom.*, 11, 644–649, 2000.
16. Crecelius, A., Clench, M.R., Richards, D.S., and Parr, V., *J. Chromatogr. A*, 958, 249–260, 2002.
17. Taki, T., Handa, S., and Ishikawa, D., *Anal. Biochem.,* 221, 312–316, 1994.
18. Taki, T., Ishikawa, D., Handa, S., and Kasama, T., *Anal. Biochem.,* 225, 24–27, 1995.
19. Taki, T. and Ishikawa D., *Anal. Biochem.,* 251, 135–143, 1997.
20. Wilson, I.D., Spraul, M., and Humpfer, E., *J. Planar Chromatogr. Mod. TLC,* 10, 217–219, 1997.
21. Meisen, I., Pater-Katalinić, J., and Müthing, J., *Anal. Chem.,* 76, 2248–2255, 2004.
22. Greiling, M. and Gressner, A.M., *Lerbuch der Klinischen Chemie*, Schattauer, Stuttgard, 1987, chap. 1 and chap. 2.
23. Kramer, J., Blackwell, B.A., Dugan, M.E.R., and Sauer, F.D., *Biochim. Biophys. Acta,* 1303, 47–55, 1996.
24. Bhushan, R., *Chromatogr. Sci. Ser.,* 55, 353–387, 1991.
25. Bhushan, R. and Martens, J., *Chromatogr. Sci. Ser.,* 55, 389–406, 1991.
26. Sharma, S.D., Sharma, H., and Sharma, C., *J. Indian Chem. Soc.,* 80, 926–929, 2003.
27. Nagata, Y., Iida, T., and Sakai, M., *J. Mol. Catal. B: Enzym.*, 12, 105–108, 2001.
28. Ohtake, H., Hase, Y., Sakemoto, K., Oura, T., Wada, Y., and Kodama, H., *Screening,* 4, 17–26, 1995.
29. Shaw, P.D., Gao Ping, Daly, S.L., Chung Cha, Cronan, J.E., Jr., Rinehart, K.L., and Farrand, S.K., *Proc. Natl. Acad. Sci.*, 94, 6036–6041, 1997.
30. Williams, D.E. and Moore, R.E., *J. Nat. Prod.,* 52, 732–739, 1989.
31. Wang, K.T., Chen, S.T., and Lo, L.C., *Fresenius Z. Anal. Chem.*, 324, 339–340, 1986.
32. Xie, J., Ménand, M., and Valéry, J.-M., *Carbohydr. Res.*, 340, 481–487, 2005.
33. Goswami, S., Dey, S., Maity, A.C., and Jana, S., *Tetrahedron Lett.*, 46, 1315–1318, 2005.
34. Suzuki, Y., Sugiyama, C., Ohno, O., and Umezawa, K., *Tetrahedron*, 60, 7061–7066, 2004.
35. Zhou, R. and Linz, J.E., *Appl. Environ. Microbiol.*, 65, 5639–5641, 1999.
36. Vinšová, J., Highly lipophilic benzoxazoles with potential antibacterial activity presented at 8th International Electronic Conference on Synthetic Organic Chemistry ECSOC-8. November 1–30, 2004.

37. Pérez, E.M., Oliva, A.I., Hernandez, J.V., Simon, L., Moran, J.R., and Sanz, F., *Tetrahedron Lett.*, 42, 5853–5856, 2001.
38. Ohene-Gyan, K.A., Haagsma, J., Davies, M.J., and Hounsell, E.F., *Comp. Immun. Microbiol. Infect. Dis.*, 18, 161–170, 1995.
39. Hashimoto, M., Kirikae, F., Dohi, T., Adachi, S., Kusumoto, S., Suda, Y., Fujita, T., Naoki, H., and Kirikae. T., *Eur. J. Biochem.,* 269, 3715–3721, 2002.
40. Baptistella, L.H.B. and Cerchiaro, G., *Carbohydr. Res.*, 339, 665–671, 2004.
41. Chai, W., Yuen, Ch.-T., Feizi, T., and Lawson, A.M., *Anal. Biochem.*, 270, 314–422, 1999.
42. Higashi, H., Hirabayashi, Y., Ito, M., Yamagata, T., Matsumoto, M., Ueda, S., and Kato, S., *J. Biochem. (Tokyo)*, 102, 291–296, 1987.
43. Pajatsch, M., Böck, A., and Boos, W., *Carbohydr. Res.*, 307, 375–379, 1998.
44. Tsikas, D., Schwedhelm, E., Fauler, J., Gutzki, F.-M., Mayatepek, E., and Frolich, J.C., *J. Chromatogr. B*, 716, 7–17, 1998.
45. Schwedhelm, E., Tsikas, D., Durand, T., Gutzki, F.-M., Guy, A., Rossi, J.-C., and Frolich, J.C., *J. Chromatogr. B*, 744, 99–112, 2000.
46. Lee, S.-B., Park K.-H., and Robyt, J.F., *Carbohydr. Res.*, 331, 13–18, 2001.
47. Urashima, T., Nakamura, T., Nakagawa, D., Noda, M., Arai, I., and Saito, T., *Comp. Biochem. Physiol. B: Comp. Biochem.*, 138, 1–18, 2004.
48. Kim, Y.-M., Okuyama, M., Mori, H., Nakai, H., Saburi, W., Chiba, S., and Kimura, A., *Tetrahedron: Asymmetry,* 6, 403–409, 2005.
49. Vafiadi, Ch., Topakas, E., Wong, K.K.Y., Suckling, I.D., and Christakopoulos, P., *Tetrahedron: Asymmetry,* 6, 373–379, 2005.
50. Côté, G.L., Dunlap, C.A., Appell, M., and Momany, F.A., *Carbohydr. Res.*, 340, 257–262, 2005.
51. Woodard, D.S., Ostrom, K.K., and McManus, L.M., *J. Lipid Mediators Cell Signalling,* 12, 11–28, 1995.
52. Tzzo, A.A. and Benveniste, J., *J. Pharmacol. Toxicol. Methods,* 36, 219–221, 1996.
53. Guichardant, M., Taibi-Tronche, P., Fay, L.B., and Lagarde, M., *Free Radical Biol. Med.,* 25, 1049–1056, 1998.
54. Simon, C.G. Jr. and Gear, A.R.L., *Thrombosis Res.*, 94, 13–23, 1999.
55. Megli, F.M. and Sabatini, K., *Chem. Phys. Lipids,* 125, 161–172, 2003.
56. Wilson, R. and Sargent, J.R., *J. Chromatogr. A*, 905, 251–257, 2001.
57. Raith, K. and Neubert, R.H.H., *Anal. Chim. Acta*, 418, 167–173, 2000.
58. Ivleva, V.B., Elkin, Y.N., Budnik, B.A., Moyer, S.C., O'Connor, P.B., and Costello, C.E., *Anal. Chem.,* 76, 6484–6491, 2004.
59. Trepanier, D.J. and Thibert, R.J., *Clin. Biochem.*, 29, 333–345,1996.
60. Müller, C., Jordan, H., Petry, S., Wetekam, E.-M., and Schindler, P., *Biochim. Biophys. Acta,* 1347, 23–39, 1997.
61. Raith, K., Zellmer, S., Lasch, J., and Neubert, R.H.H., *Anal. Chim. Acta*, 418, 167–173, 2001.
62. Johansson, L. and Miller-Podraza, H., *Anal. Biochem.*, 265, 260–268, 1998.
63. Terao, J., Miyoshi, M., and Miyamoto, S., *J. Chromatogr. B,* 765, 199–203, 2001.
64. Wang, Y., Xia, Q., Man, H., Geng, Ch., and Cui, Zh., *Chin. J. Chromatogr.* (*Sepu*), 17, 547–549, (1999).
65. Wright, L.C., Obbink, K.L., Delikatny, E.J., Santangela, R.T., and Sorrell, T.C., *Eur. J. Biochem.,* 267, 68–78, 2000.
66. Irani, V.R., Lee, S.-H., Eckstein, T.M., Inamine, J.M., Beliste, J.T., and Maslow, J.N., *Ann. Clin. Microbiol. Antimicrobials*, 3, 1–18, 2004.

67. Watanabe, M., Aoyagi, Y., Ridell, M., and Minnikin, D.E., *Microbiology,* 147, 1825–1837, 2001.
68. Moody, D.B., Guy, M.R., Grant, E., Cheng, T.Y., Brenner, M.B., Besra, G.S., and Porcelli, S.A., *J. Exp. Med.*, 192, 965–976, 2000.
69. Ganshirt, H., Koss, F.W., and Morianz, K., *Arzneim.-Forsch.,* 10, 943–947, 1960.
70. Dax, C.I. and Müllner, S., *Chromatographia,* 48, 681–689, 1998.
71. Dunphy, J.C. and Busch, K.L., *Talanta*, 37, 471–480, 1990.
72. Momose, T., Mure, M., Iida, T., Goto, J., and Nambara, T., *J. Chromatogr. A*, 811, 171–180, 1998.
73. Lida, T., *J. Chromatogr. A*, 366, 396–402, 1986.
74. Lamperti, L. and Vega, M., *J. Planar Chromatogr. Mod. TLC,* 5, 139–140, 1992.
75. Batta, A.K., Salen, G., Rapole, K.R., Batta, M., Batta, P., Alberts, D., and Earnest, D., *J. Lipid Res.* 40, 1148–1154, 1999.
76. Rivas-Nass, A. and Müllner, S., *J. Planar Chromatogr. Mod. TLC,* 7, 278–285, 1994.
77. Zarzycki, P.K., Wierzbowska, M., and Lamparczyk, H., *J. Chromatogr. A*, 857, 255–262, 1999.
78. Pelander, A., Ojanpera, I., and Hase, T., *Forensic Sci. Int.*, 85, 193–198, 1997.
79. Bluhm, L.H. and Li, T., *Tetrahedron Lett.*, 39, 3623–3626, 1998.
80. Kim, S.-K., Ubukata, M., and Isono, K., *J. Microbiol. Biotechnol.,* 13, 998–1000, 2003.
81. Dahlin, M., Jansson, B., and Bjork, E., *Eur. J. Pharm. Sci.,* 14, 75–80, 2001.
82. Roberts, A., *Food Chem. Toxicol.,* 38, 31–41, 2000.
83. Min, B.S., Meselhy, M.R., Hattori, M., Kim, H.M., and Kim, Y.H., *J. Microbiol. Biotechnol.,* 10, 514–517, 2000.
84. Wu, C.T., Mulabagal, V., Nalawade, S.M., Chen, C.-L., Yang, T.-F., and Tsay, H.-S., *Biol. Pharm. Bull.*, 26, 845–848, 2003.
85. Baier-Weber, B., Prox, A., Wachsmuth, H., and Breyer-Pfaff, U., *Drug Metab. Dispos.*, 16, 490–496, 1988.
86. Chen, Ch.-L., Yu, G., Venkatachalam, T.K., and Uckun, F.M., *Drug Metab. Dispos.*, 30, 1523–1531, 2002.
87. Pan, M.-H., Huang, T.-M., and Lin, J.-K., *Drug Metab. Dispos.*, 27, 486–494, 1999.
88. Ruscoe, J.E., Tingle, M.D., O'Neill, P.M., Ward, S.A., and Park, B.K., *Antimicrob. Agents Chemother.*, 42, 2410–2416, 1998.
89. Oltenfreiter, R., Staelens, L., Lejeune, A., Dumont, F., Frankenne, F., Foidart, J.-M., and Slegers, G., *Nucl. Med. Biol.,* 3, 459–468, 2004.
90. Kitani, M., Miyamoto, G., Nagasawa, M., Yamata, T., Matsubara, J., Uchida, M., and Odomi, M., *Drug Metab. Dispos.,* 25, 663–674, 1997.
91. Lundberg, B.B., Risovic, V., Ramaswamy, M., and Wasan, K.M., *J. Controlled Release,* 86, 93–100, 2000.
92. Marchais-Oberwinkler, S., Nowicki, B., Pike, V.W., Halldin, C., Sandell, J., Chou, Y.-H., Gulyas, B., Brennum, L.T., Farde, L., and Wikstrom, H.V., *Bioorg. Med. Chem.*, 13, 883–893, 2005.
93. Obach, R.S., *Drug Metab. Dispos.*, 29, 1057–1067, 2001.
94. Minato, K., Koizumi, N., Honma, S., Tsukamoto, K., and Iwamura, S., *Drug Metab. Dispos.*, 30, 167–172, 2002.
95. Duflos, M., Nourrisson, M.-R., Brelet, J., Courant, J., LeBaut, G., Grimaud, N., and Petit, J.-Y., *Eur. J. Med. Chem.*, 36, 545–553, 2001.
96. Park, Y.H., Chun, E.M., Bae, M.A., Seu, Y.B., Song, K.S., and Kim, Y.H., *J. Microbiol. Biotechnol.*, 10, 27–34, 2000.
97. Gyeresi, A., *J. Planar Chromatogr. Mod. TLC.*, 13, 296–300, 2000.

98. Okoli, C.O. and Akah, P.A., *J. Alternative Complement. Med.*, 6, 423–427, 2000.
99. Obi, E., Emeh, J.K., Orisakwe, O.E., Afonne, O.J., Ilondu, N.A., and Agbasi, P.U., *Indian J. Pharmacol.,* 32, 381–383, 2000.
100. Apers, S., Cimanga, K., Vanden Berghe, D., Van Meenen, E., Otshudi Longanga, A., Foriers, A., Vlietinck, A., and Pieters, L., *Planta Medica,* 68, 20–24, 2002.
101. Hui, Y., Pei-Lin, C., Zhao-Hui, W., Hong-He, T., Ji-min, W., Lei, D., and Ya-ming, X., *Chin. J. Antibiotics*, 26, 7–9, 2001.
102. Truiti, M.-T., Sarragiotto, M.H., De Abreu Filho, B.A., Nakamura, C.V., and Dias Filho, B.P., *Memorias Inst. Oswaldo Cruz,* 98, 283–286, 2003.
103. Hao, W., Cheng-Bin, C., Bing H., Zhi-Heng, L., You-Xin, S., and Pei-Jin, Z., *Chin. J. Antibiotics*, 26, 19–24, 2001.
104. Han, B., Shen, T., Liu, D., and Yang, J.-S., *Chin. Pharm. J.*, 37, 740–742, 2002.
105. Yang, C.Y., Meng, C.L., Van der Bijl, P., and Lee, H.K., *Prostaglandins Other Lipid Mediators*, 67, 181–195, 2002.
106. Wu, N., Chen, H., and Wang, Zh., *Chin. J. Herb Med.* (*Zhongcaoyao*), 25, 10, 1994.
107. Gu, X., *J. Chin. Trad. Patent Med. (Zhongchengyao)*, 21, 208, 1999.
108. Liu, X. and Jia, Zh., *Chin. J. Herb Med.* (*Zhongcaoyao*), 24, 451, 1993.
109. Ma, T., Liu, S., and Xu, G., *Chin. J. Herb Med.* (*Zhongcaoyao*), 25, 7, 1994.
110. He, H., *J. Chin. Trad. Herb Drugs (Zhongcaoyao)*, 32, 201, 2001.
111. Wu, Y., *Chin. J. Herb Med.* (*Zhongcaoyao*), 29, 599, 1998.
112. Kan, Y., Zhao, H., and Liu, X., *Chin. J. Herb Med.* (*Zhongcaoyao*), 25, 3, 1994.
113. Franolic, J.D., Lehr, G.J., Barry, T.L., and Petzinger, G., *J. Pharm. Biomed. Anal.*, 26, 651–663, 2001.
114. Tritsch, D., Hemmerlin, A., Rohmer, M., and Bach, T.J., *J. Biochem. Biophys. Methods*, 58, 75–83, 2004.
115. Esaiassen, M., Verb, K., and Olsen, R.L., *Carbohydr. Res.*, 273, 77–81, 1995.
116. Sugimoto, H. and Yamashita, S., *Biochim. Biophys. Acta*, 1438, 264–272, 1999.
117. Häusermann, M., Ghisla, S., Niederwieser, A., and Curtius, H.-C., *FEBS Lett.,* 131, 275–278, 1981.
118. Uchikoba, T., Ichiki, N., Yonezawa, H., Arima, K., and Kaneda, M., *J. Biochem. Biophys. Methods*, 48, 303–308, 2001.
119. Zerilli, A., Lucas, D., Berthou, F., Bardou, L.G., and Menez, J.-F., *J. Chromatogr. B*, 677, 156–160, 1996.
120. Khan, K.K. and Halpert, J.R., *Arch. Biochem. Biophys.*, 373, 335–345, 2000.
121. Menshonkova, T.N., Simukova, N.A., Budowsky, E.I., and Rubin, L.B., *FEBS Lett.*, 112, 298–300, 1980.
122. Meszaros, V., Socci, R., and Meszaros, L.G., *Biochem. Biophys. Res. Commun.*, 210, 452–456, 1995.
123. Krijt, J., Kmoch, S., Hartmannova, H., Havlcek, V., and Sebesta, I., *J. Chromatogr. B*, 726, 53–58, 1999.
124. Keith, G., *Biochimie,* 77, 142–144, 1995.
125. Mori, T., Guo, M., Li, X., and Mori, E., *J. Reproductive Immunol.,* 53, 289–303, 2002.
126. Robinzon, B., Michael Miller, K.K., and Prough, R.A., *Anal. Biochem.,* 333, 128–135, 2004.
127. Priest, J.W., Mehler, A., Arrowood, M.J., Riggs, M.W., and Ferguson, M.A., *J. Biol. Chem.,* 278, 52212–52222, 2003.
128. Yonezawa, Y., Saigusa, S., Takahagi, M., and Nishioka, H., *Mutation Res.*, 442, 97–103, 1999.

10 PLC of Hydrophilic Vitamins

Fumio Watanabe and Emi Miyamoto

CONTENTS

10.1 INTRODUCTION

Thin-layer chromatography (TLC) as a powerful separation and analytic tool is used particularly in pharmaceutical preparations, foods, and natural products. The quantification of the separated vitamins can be performed by using modern densitometry. Because amounts of most hydrophilic vitamins are low in food and natural products, bioautography or derivatization is employed before densitometry. Each spot of various vitamins and related compounds separated by TLC is removed from the plates, reextracted, and assayed or further purified by HPLC. Various high-quality precoated silica gel, cellulose, or various reversed-phase plates are available for TLC. TLC has great advantages (simplicity, flexibility, speed, and relative inexpensiveness) for the separation and analysis of hydrophilic vitamins. In this chapter, we focus on the preparation of hydrophilic vitamins and related compounds from foods, mammalian tissues, pharmaceutical preparations, and biochemical reagents (including radioactive compounds) using TLC as a powerful separation tool; further details of analysis of hydrophilic vitamins have been reviewed elsewhere [1,2].

10.2 THIAMIN (VITAMIN B_1)

Although TLC analysis of vitamin B_1 has been reported in food and mammalian tissues, [3,4] we could not find any articles for preparation of vitamin B_1 and related compounds using TLC.

10.3 RIBOFLAVIN (VITAMIN B_2)

Various TLC solvent systems were used to confirm the presence of flavins in plain yogurt and raw egg white (or egg powder) [5]. The mean content of flavin compounds (Figure 10.1) have been analyzed in plain yogurts and bioyogurts [6]. Concentrated flavin extracts were passed through a column packed with resorcinol-type resin R-15

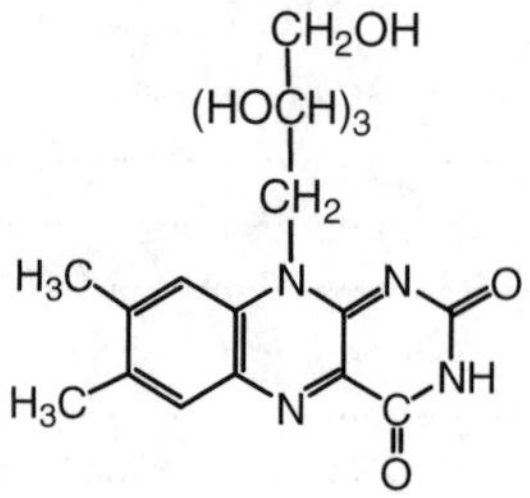

(1)

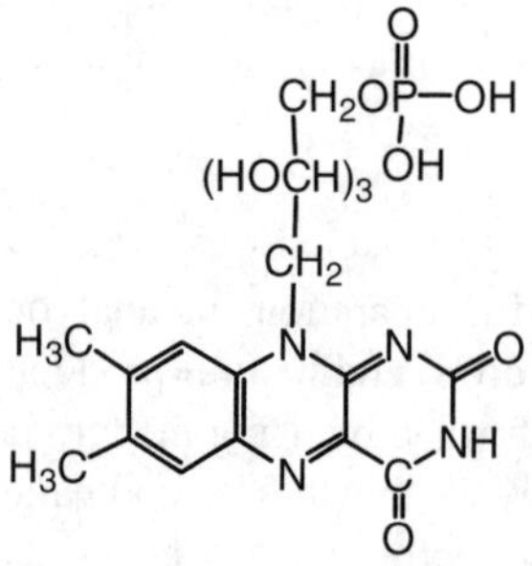

(2)

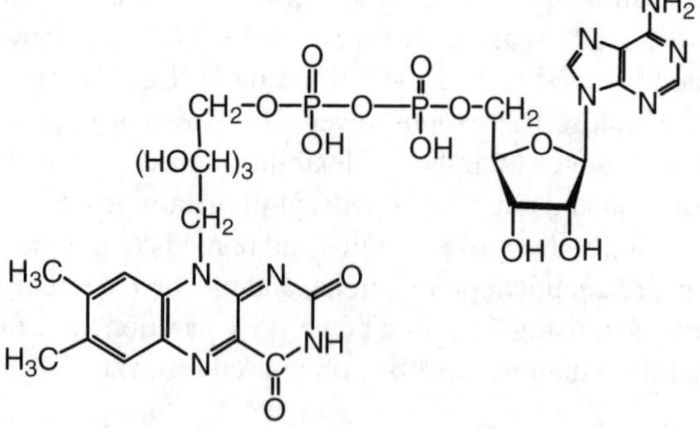

(3)

FIGURE 10.1 The structural formula of riboflavin and partial structures of riboflavin compounds. The latter show only those portions of the molecule that differ from riboflavin. 1 — Riboflavin (RF), 2 — flavin mononucleotide or 5′-riboflavin monophosphate (FMN or 5′-FMN), 3 — flavin adenine dinucleotide (FAD).

TABLE 10.1
R_f Values on TLC of Flavin Standards and the Unknown Flavin

Flavin Standards	I	IV	VII	VIII	IX
FAD	0.21	0	0	0	0
FMN	0.36	0.03	0.04	—	0.05
RFgal	0.56	0.06	0.21	0.18	0.10
RF	0.68	0.38	0.50	0.48	0.31
Unknown flavin from yogurt	0.56	0.06	0.21	0.18	0.10

Note: TLC was performed on silica gel, and the solvents were (I) *n*-butanol/glacial acetic acid/water (2:1:1, v/v), (IV) isoamyl alcohol/ethy methyl ketone/glacial acetic acid/water (40:40:7:13, v/v), (VII) *n*-butanol/2-propanol/water/glacial acetic acid (30:50:10:2, v/v), (VIII) ethyl methyl ketone/acetic acid/methanol (3:1:1, v/v), and (IX) *n*-butanol/benzyl alcohol/glacial acetic acid (8:4:3, v/v).

Source: Reprinted with permission from Gliszcznska, A. and Koziolowa, A., *J. Agric. Food Chem.*, 47, 3197–3201, 1999. Copyright (1999) American Chemical Society.

to remove all interfering nonflavin compounds. TLC on silica gel was used for preparative purposes with *n*-butanol/glacial acetic acid/water (2:1:1, v/v) as a solvent. A flavin compound, riboflavinyl galactoside (RFgal), which was prepared by incubation of riboflavin and lactose with Taka-Diastase powder, was purified using preparative TLC with *n*-butanol/glacial acetic acid/water (15:3:7, v/v) and then HPLC (Table 10.1) [6].

The various flavin phosphates and their acetyl derivatives were identified by pH titration, electrophoresis, and ^{1}H-NMR, which permit direct analysis of crude reaction products as well as rapid purity check of commercial flavin mononucleotide or riboflavin 5′-monophosphate (FMN or 5′-FMN) [7]. Riboflavin 4′-monophosphate was determined as the main by-product of commercial FMN by preparative TLC on cellulose with *n*-butanol/acetic acid/water (5:2:3, v/v) as a solvent [7].

10.4 PYRIDOXINE (VITAMIN B_6)

Vitamin B_6 and related compounds (Figure 10.2) were quantitatively separated by preparative TLC on silica gel H. After elution, the pyridoxic acid lactone method was employed for fluorimetric determination of the concentration of the vitamin forms involved [8]. Table 10.2 shows R_f values obtained for various forms of vitamin B_6, using several solvent systems. The solvent selected, ethyl acetate/pyridine/water (2:1:2, v/v), gave excellent separation of pyridoxamine, pyridoxic acid, and pyridoxine together with pyridoxal.

After the sample spots had been dried, the plates were developed with the solvent in the dark. The gel from each zone was removed from the plates dried in the dark

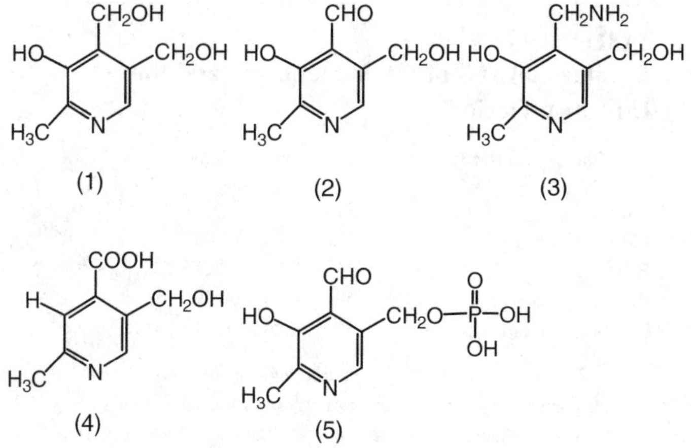

FIGURE 10.2 Structural formula of vitamin B_6 and related compounds. 1 — pyridoxine, 2 — pyridoxal, 3 — pyridoxamine, 4 — 4-pyridoxic acid; 5 — pyridoxal-5′-phosphate.

at room temperature. Elution was performed by the addition of water to the gel and filling a column with the slurry, allowing the fluid to drip through at a rate of about 0.5 ml/min. Recovery by this method was from 80 to 90% of the amount applied to the plates.

This method was applied to the determination of vitamin B_6 concentration in chicken embryo livers ranging from 8 to 14 d of development [8].

10.5 COBALAMIN (VITAMIN B_{12})

Vitamin B_{12} compounds with different upper (L) and lower (R) ligands were found in nature (Figure 10.3). Some commercially available vitamin B_{12} reagents, hydroxocobalamin and dicyanocobinamide, contain small amounts of impurities [9]. Each vitamin B_{12} reagent should be purified with silica gel 60 TLC [2-propanol/NH_4OH(28%)/water (7:1:2, v/v) as a solvent] and then used for experiments as an authentic standard material (Figure 10.4).

To determine whether the loss of vitamin B_{12} in microwave-treated foods was due to the conversion of vitamin B_{12} to some inactive vitamin B_{12} degradation products, the hydroxocobalamin that predominates in food was treated by microwave heating for 6 min and then analyzed by TLC on silica gel 60 with *n*-butanol/2-propanol/water (10:7:10, v/v) as a solvent. The treated hydroxocobalamin was separated into four red spots: intact hydroxocobalamin remained at the origin (Figure 10.5) [10]. These hydroxocobalamin degradation products were further purified to homogeneity by the use of TLC and HPLC and characterized.

To evaluate whether some foods contain true vitamin B_{12} or inactive vitamin B_{12} analogs, some vitamin B_{12} compounds were purified and characterized using silica gel 60 TLC [11–15]. An algal health food, spirulina tablets, contained considerable amounts of a vitamin B_{12} analog (pseudovitamin B_{12}) inactive for humans (Figure 10.6) [11].

TABLE 10.2
R_f Values for Vitamin B_6 Thin-Layer Chromatography, Silica Gel H

Solvent System (v/v)	Pyridoxal	Pyridoxine	Pyridoxic Acid	Pyridoxic Acid Lactone	Pyridoxamine	Pyridoxal Phosphate
n-Amyl alcohol/acetone/water (2:1:2)	0.39	0.35	0.27	0.39	0.19	—
Isoamyl alcohol/pyridine/water (2:1:2)	0.48	0.54	0.44	0.33	0.29	—
n-Butanol/1.0 mol/l acetic acid (5:1)[a]	0.57	0.50	0.20	0.22	0.08	0.23
Water/acetone/*tert*-butanol/diethyl amine (20:35:40:5)	0.71	0.57	0.80	0.48	0.66	0.44
Water/acetone/*tert*-butanol/acetic acid (20:35:40:5)	0.58	0.46	0.25	0.56	0.15	—
n-Amyl alcohol/acetone/water/diethyl amine (40:35:20:5)	0.60	0.40	0.70	0.49	0.55	0.32
Ethyl acetate/pyridine/water (2:1:2)[b]	0.51	0.49	0.37	0.37	0.12	0.07
tert-Butanol/water/89% formic acid (70:15:15)	0.79	0.46	0.45	—	0.13	—

Note: The following solvent systems were tried but gave no separation: dioxane/water (7:3, v/v); *n*-butanol saturated with water; amyl alcohol/acetone/water/benzyl amine (40:35:20:5, v/v); 2,6-lutidine/water/ethnol/diethylamine (55:25:20:1, v/v).

[a] Spreading of spots.

[b] Solvent system chosen.

Source: Reprinted with permission from Smith, M.A. and Dietrich, L.S., *Biochim. Biophys. Acta*, 230, 262–270, 1971. Copyright (1971) Elsevier.

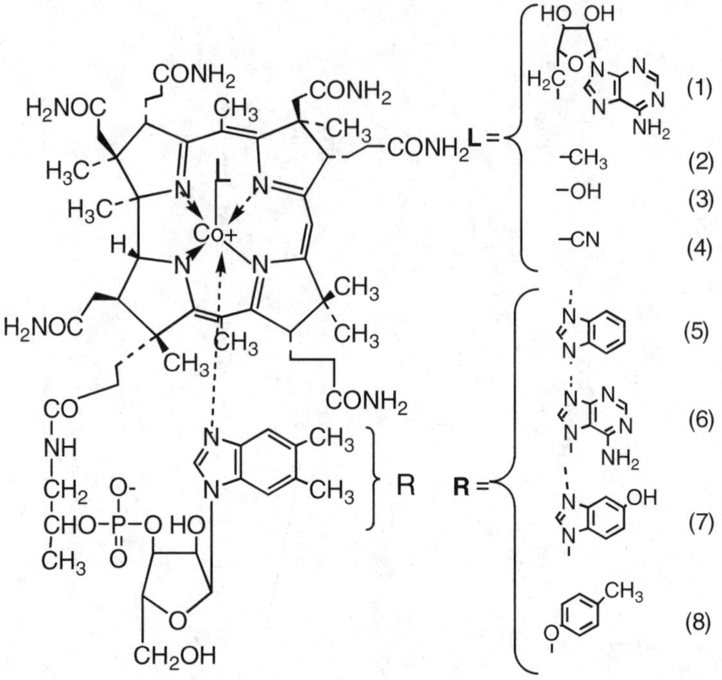

FIGURE 10.3 Structural formula of vitamin B_{12} and partial structures of vitamin B_{12} compounds. The partial structures of vitamin B_{12} compounds show only those portions of the molecule that differ from vitamin B_{12}. 1 — 5′-deoxyadenosylcobalamin, 2 — methylcobalamin, 3 — hydroxocobalamin, 4 — cyanocobalamin, 5 — benzimidazolyl cyanocobamide, 6 — pseudovitamin B_{12}, 7 — 5-hydroxybenzimidazolyl cyanocobamide, 8 — *p*-cresolyl cyanocobamide.

10.6 NICOTINIC ACID AND NICOTINAMIDE

A TLC method was developed for the estimation of nicotinic acid and nicotinamide (Figure 10.7) in pharmaceutical preparations containing other vitamins, enzymes, herbs, and drugs, etc. [16]. The percentage recoveries for nicotinic acid and nicotinamide were 100.1 ± 1.9 and 100.2 ± 1.5, respectively, with this system. Each alcohol extract of samples or standard was put on silica gel TLC plates, which were developed with distilled water. Each silica gel spot visualized under UV lamp was collected and extracted with 0.1 mol/l HCl. The optical density of each clear extract was measured at 262 nm.

10.7 PANTOTHENIC ACID

High-specific activity *D*-[3-^{3}H] panthothenic acid (Figure 10.8) was prepared from commercially available β-[3-^{3}H]alanine using *Escherichia coli* strain DV1, which converted 85 to 90% of the input β-[3-^{3}H]alanine to extracellular *D*-[3-^{3}H]panthothenate under appropriate growth conditions. The radiolabeled vitamin was purified

FIGURE 10.4 Silica gel 60 TLC pattern of commercially available B_{12} reagents, hydroxocobalamin and dicyanocobinamide. The concentrated solution (4 μl) was spotted on the silica gel TLC sheet and developed with 2-propanol/NH_4OH(28%)/water (7:1:2, v/v) at room temperature in the dark. 1 — hydroxocobalamin, 2 — dicyanocobinamide.

from the medium by TLC followed by reversed-phase HPLC. The concentrated medium was streaked across a 20-cm silica gel H TLC plate and developed with ethanol/NH_4OH(28%) (4:1, v/v), to separate panthothenate (R_f 0.61) from β-alanine (R_f 0.28) and phosphorylated panthothenate metabolites that remained at the origin (Figure 10.9) [17]. The region of the plate corresponding to *D*-[3-^{3}H]panthothenate was scraped into a glass funnel. Most (>90%) of the radioactive compound was eluted from the silica gel with 1% (v/v) acetic acid in 95% (v/v) ethanol, followed by 95% (v/v) ethanol. The overall yield of *D*-[3-^{3}H]panthothenate was 30%, and radiochemical purity was >99%.

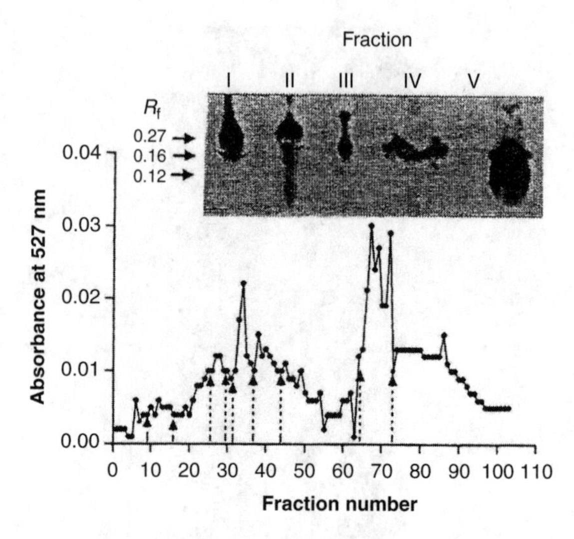

FIGURE 10.5 Elution profile on OH-B_{12} treated by microwave heating for 6 min during silica gel 60 column chromatography. Fifty milliliters of the treated OH-B_{12} solution (5 mmol/l) was evaporated to dryness and dissolved in a small amount of *n*-butanol/2-propanol/water (10:7:10, v/v) as a solvent. The concentrated solution was put on a column (1.4 × 15.0 cm) of silica gel 60 equilibrated with the same solvent and eluted with the same solvent in the dark. The eluate was collected at 4.0 ml with a fraction collector. Fractions I to V were pooled, evaporated to dryness, dissolved with a small amount of distilled water, and analyzed with silica gel TLC. Inset represents the mobile pattern of the OH-B_{12} degradation products of fractions I to V on the TLC plate. Data are typical, taken from one of five experiments. (Reprinted with permission from Watanabe, F. et al., *J. Agric. Food Chem.*, 46, 5177–5180, 1998. Copyright (1998) American Chemical Society.)

10.8 BIOTIN

Chemiluminescence energy transfer between aminobutylethylisoluminol (ABEI)-biotin and fluorescein avidin was investigated to establish a homogeneous assay for serum biotin in the physiological range [18]. ABEI-biotin was synthesized by a mixed anhydride reaction and purified by TLC and HPLC. ABEI-biotin was separated from the residual reaction products using cellulose F254 TLC with *n*-butanol/acetic acid/water (12:3:5, v/v) as a solvent. There was clear separation of ABEI (R_f 0.63) and ABEI-biotin (R_f 0.86); there was no detectable ABEI in the ABEI-biotin prepared by TLC. It was further purified by reversed-phase HPLC to remove possible contaminating biotin because the R_f for biotin (R_f 0.89) was very close to that of ABEI-biotin (R_f 0.86).

10.9 FOLIC ACID

The stability of [3′,5′,7,9-^{3}H]folic acid (Figure 10.10) was analyzed to determine whether the varied and conflicting results regarding the characteristics of folate transport in L1210 cells could be due to impurities in the labeled substrate [19]. The susceptibility of [3′,5′,7,9-^{3}H]folic acid to decomposition during storage at −20°C

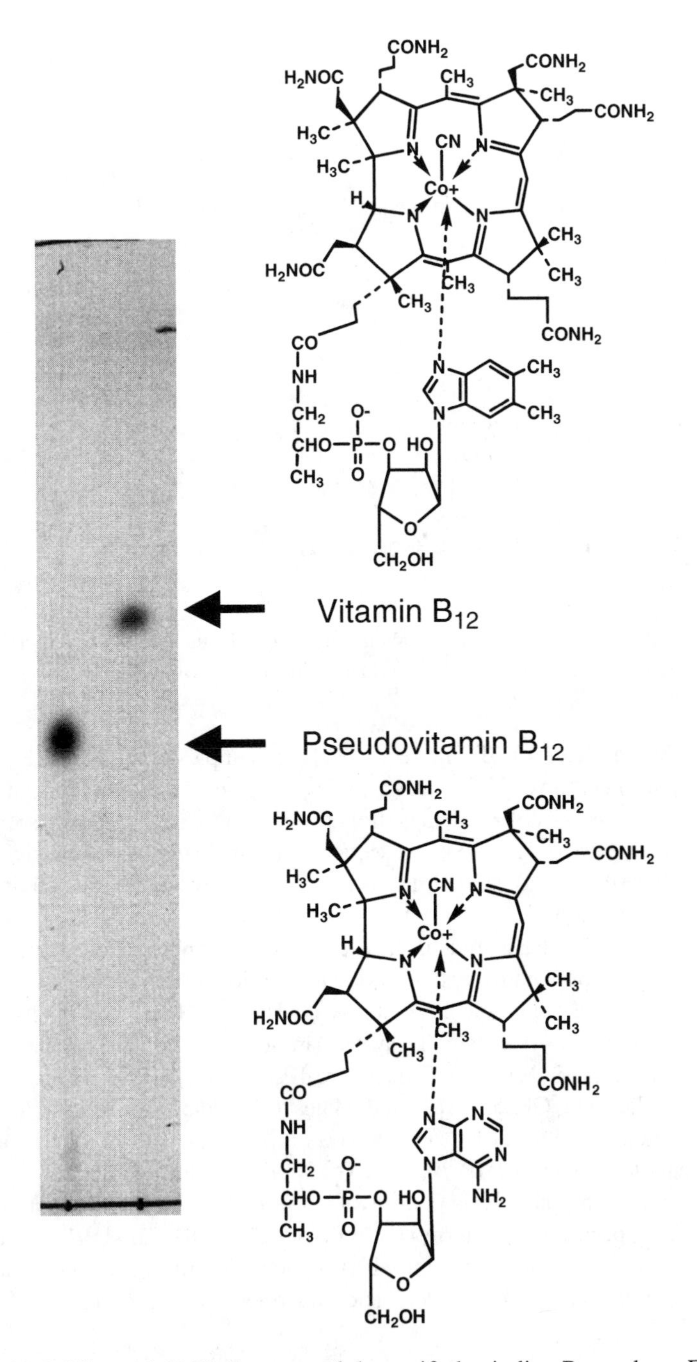

FIGURE 10.6 Silica gel 60 TLC pattern of the purified spirulina B_{12} analogs. Each final preparation (2 μl) was spotted on the silica gel 60 TLC sheet and developed with 2-propanol/NH_4OH (28%)/water (7: 1: 2, v/v) at room temperature in the dark.

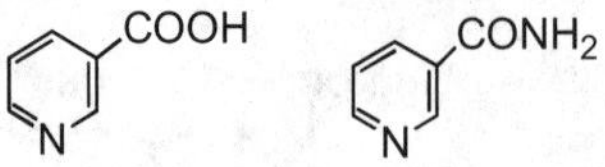

COOH CONH$_2$

N N

(1) (2)

FIGURE 10.7 Structural formula of nicotinic acid and nicotinamide. 1 — nicotinic acid, 2 — nicoinamide.

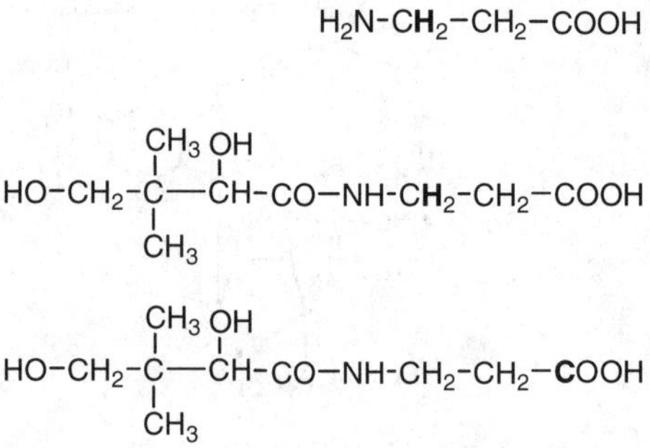

$$H_2N\text{-}\mathbf{CH_2}\text{-}CH_2\text{-}COOH \quad (1)$$

$$HO\text{-}CH_2\text{-}C(CH_3)_2\text{-}CH(OH)\text{-}CO\text{-}NH\text{-}\mathbf{CH_2}\text{-}CH_2\text{-}COOH \quad (2)$$

$$HO\text{-}CH_2\text{-}C(CH_3)_2\text{-}CH(OH)\text{-}CO\text{-}NH\text{-}CH_2\text{-}CH_2\text{-}\mathbf{C}OOH \quad (3)$$

FIGURE 10.8 Structural formula of β-alanine and pantothenic acid. 1 — β-[3-^{3}H]alanine, 2 — *D*-[3-^{3}H]pantothenic acid, 3 — *D*-[1-^{14}C]panthothenic acid. Boldfaced letter H or C denotes radioactivity.

for 6 months was determined (Figure 10.11) [19]. Samples of [3′,5′,7,9-^{3}H]folic acid were applied to cellulose TLC plates and developed in the dark with 50 mmol/l *N*-2-hydeoxyethypiperazine-*N*′-ethanesulfonate-KOH buffer (pH 7.5) as a solvent. The TLC plates were dried, cut into 1-cm sections, and analyzed for radioactivity. Purity of [3′,5′, 7,9-^{3}H]folic acid was defined as the percentage of total radioactivity present in the band at R_f 0.65.

Akhtar et al. [20] have studied the identification of photoproducts of folic acid and their degradation pathways in aqueous solution using preparative TLC. An aqueous solution of folic acid irradiated with UV at pH 2.4 to 10.0 for 6 h was subjected to TLC analysis, which gave separation of folic acid (R_f 0.67), *p*-aminobenzolyl-*L*-glutamic acid (R_f 0.78) (Figure 10.12). The photolyzed solutions were applied to silica gel GF254 precoated plates using the solvent system A, ethanol/ammonia (13.5 mol/l)/1-propanol (60:20:20, v/v) and B, acetic acid/acetone/methanol/benzene (5:5:20:70, v/v). The spots were located under the UV light. The products by the photolysis of folic acid were obtained with preparative TLC, and their UV spectra were determined in 0.1 mol/l NaOH. The UV-irradiated folic acid in aqueous solutions at pH 2 to 10 was degraded to give pterin-6-carboxylic acid and *p*-aminobenzoyl-*L*-glutamic acid under aerobic conditions.

10.10 ASCORBIC ACID (VITAMIN C)

Chromatographic evidence supporting the similarity of the yellow chromophores isolated from aged human brunescent cataract lenses and calf lens proteins modified

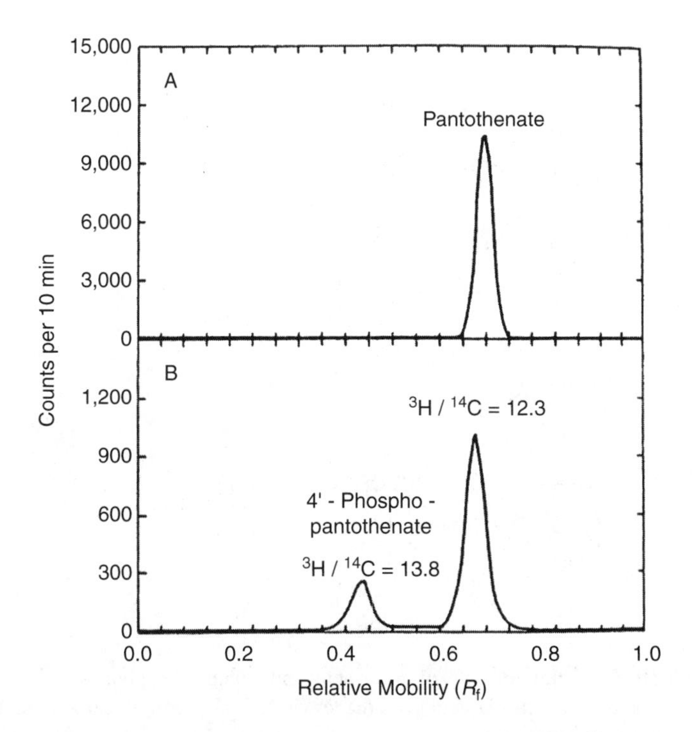

FIGURE 10.9 Radiochemical purity and biological activity of *D*-[3-^{3}H]pantothenic acid. HPLC-purified *D*-[3-^{3}H]pantothenic acid (A) and an aliquot (10 ml) from a panthothenate kinase incubation containing both *D*-[3-^{3}H]pantothenate and commercial *D*-[1-^{14}C]pantothenate (B) were chromatographed on silica gel H thin-layer developed with ethanol/NH_4OH(28%) (4:1, v/v). Radioactivity was measured using the Bioscan System 200 with a 10-min accumulation time. The ^{3}H:^{14}C ration in the peaks corresponding to panthothenate and 4′-phosphopanthothenate in (B) was determined by scraping 0.5-cm sections from the plate into vials, deactivating the silicic acid with 0.1-ml water, and scintillation counting. (Reprinted with permission from Vallari, D.S. and Rock, C.O., *Anal. Biochem.*, 154, 671–675, 1986. Copyright (1986) Elsevier.)

with ascorbic acid *in vitro* was studied. To confirm whether the fluorophores formed were because of ascorbic acid modification, each peak (2 to 7) of ascorbic-acid-modified calf lens proteins separated by Bio-Gel P-2 column chromatography was concentrated and spotted on a preparative silica gel TLC plate with ethanol/ammonia (7:3, v/v) as a solvent (Figure 10.13) [21]. The flourophores were detected by irradiation with long-wavelength (360 nm) radiation. The TLC plate was then scanned to locate the radioactive bands. The fractions isolated from the Bio-Gel P-2 column were further purified by HPLC and characterized. This study provided new evidence to support the hypothesis that the yellow chromophores in brunescent lenses represent advanced glycation end products, probably owing to ascorbic acid glycation *in vivo*.

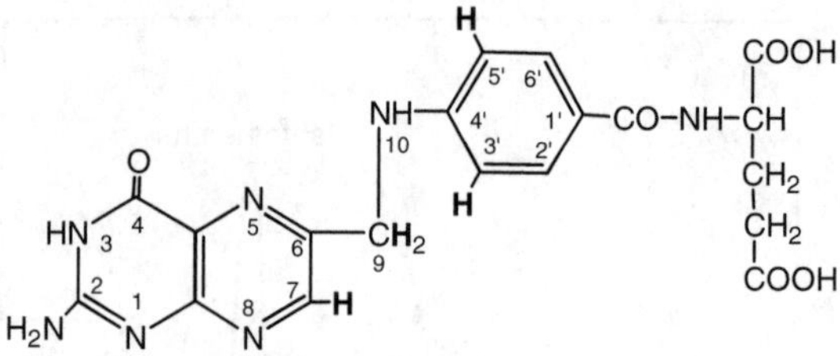

(1)

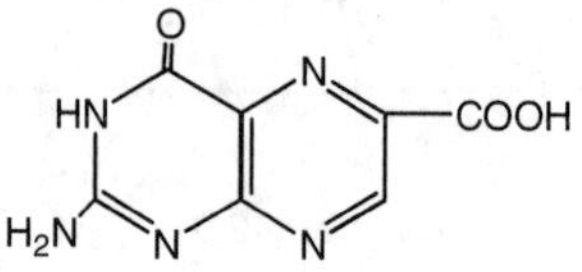

(2)

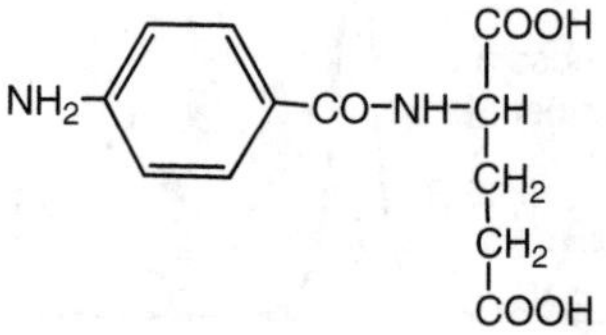

(3)

FIGURE 10.10 Structural formula of folic acid and related compounds. 1 — [3′,5′,7,9-^{3}H]folic acid (boldfaced letter H denotes radioactivity), 2 — pterine-6-carboxylic acid, 3 — *p*-aminobenzoyl-*L*-glutamic acid.

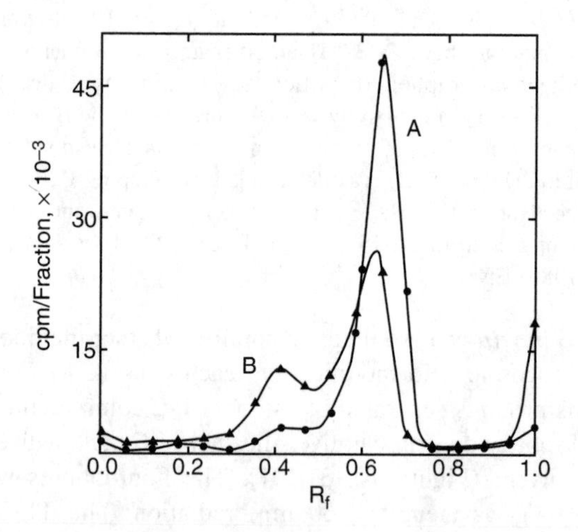

FIGURE 10.11 TLC analysis of nonpurified [3′,5′,7,9-^{3}H]folic acid. The radioactive folic acid (0.5 Ci/mmol) was analyzed when received from the supplier (A) and after storage at –20°C for 6 months in phosphate-buffered saline (B). Samples of the radioactive folic acid were applied to cellulose TLC plates and developed in the dark with 50 mmol/l *N*-2-hydeoxyethypiperazine-*N*′-ethanesulfonate-KOH buffer (pH 7.5) as a solvent. Segments (1.0 cm) of each chromatogram were removed from the plastic support and analyzed for radioactivity. (Reprinted with permission from Henderson, G.B. et al., *Cancer Res.*, 46, 1639–1643, 1986. Copyright (1986) American Association for Cancer Research.)

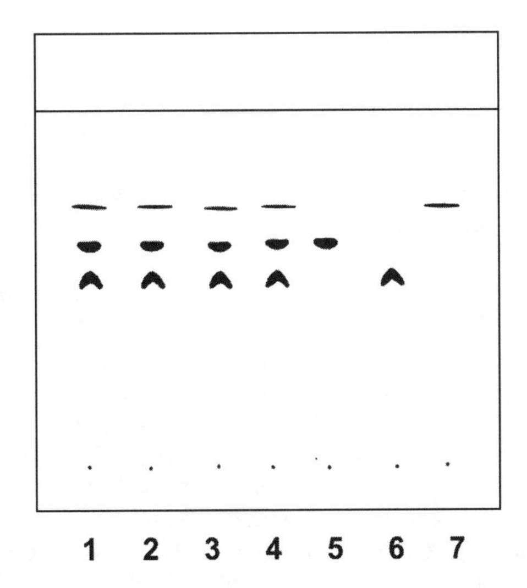

FIGURE 10.12 TLC of photolyzed solutions of folic acid along with the reference standards. The photolyzed solutions of folic acid at pH 5.0 (1), 6.0 (2), 7.0 (3) and (4); folic acid (5); pterin-6-carboxylic acid (6); and *p*-aminobenzoyl-*L*-glutamic acid (7) were applied to silica gel GF254 precoated plates using ethanol/ammnonia (13.5 mol/l)/1-propanol (60:20:20, v/v) as a solvent. (Reprinted with permission from Akhtar, M.J., Khan, M.A., and Ahmad, I.J., *Pharm. Biomed. Anal.*, 31, 579–588, 2003. Copyright (2003) Elsevier.)

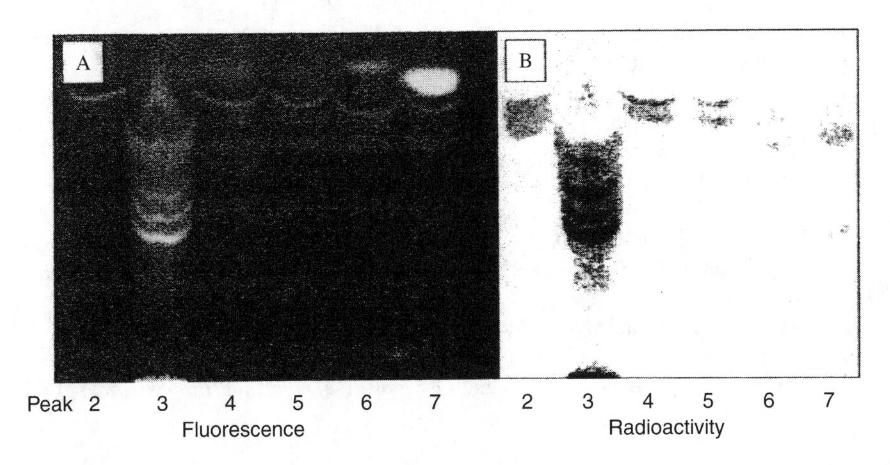

FIGURE 10.13 The TLC profiles of labeled peaks isolated from [U-^{14}C]ascorbic-acid-modified calf lens protein obtained from Bio-Gel P-2 chromatography. Peaks 2 to 7 were spotted on a preparative silica gel TLC plate and developed with ethanol/ammonia (7:3, v/v). The fluorescence in each lane was detected by irradiation with a Wood's lamp at 360 nm, and the pattern of radioactivity was determined by scanning the plate with AMBIS imaging system. (Reprinted with permission from Cheng, R. et al., *Biochim. Biophys. Acta*, 1537, 14–26, 2001. Copyright (2001) Elsevier.)

REFERENCES

1. Watanabe, F. and Miyamoto, E., Hydrophilic vitamins, in *Handbook of Thin-Layer Chromatography*, Sherma, J. and Fried, B., Eds., Marcel Dekker, New York, 2003, chap. 20.
2. Watanabe, F. and Miyamoto, E., Hydrophilic vitamins, analysis by TLC, in *Encyclopedia Chromatography*, Cazes, J., Ed., Marcel Dekker, New York, 2004 (online version).
3. Bauer-Petrovska, B. and Petrushevska-Tazi, L., *Int. J. Food Sci. Technol.*, 35, 201–205, 2000.
4. Ziporin, Z.Z. and Waring, P.P., *Methods Enzymol.*, 18A, 86–87, 1970.
5. Gliszczynska-Swiglo, A. and Koziolowa, A., *J. Chromatogr.*, A881, 285–297, 2000.
6. Gliszcznska, A. and Koziolowa, A., *J. Agric. Food Chem.*, 47, 3197–3201, 1999.
7. Scola-Nagelschneider, G. and Hemmerich, P., *Eur. J. Biochem.*, 66, 567–577, 1976.
8. Smith, M.A. and Dietrich, L.S., *Biochim. Biophys. Acta*, 230, 262–270, 1971.
9. Watanabe, F. and Miyamoto, E., Vitamin B_{12} and related compounds in food, analysis by TLC, in *Encyclopedia Chromatography*, Cazes, J., Ed., Marcel Dekker, New York, 2004 (online version).
10. Watanabe, F., Abe, K., Katsura, H., Takenaka, S., Zakir Hussain Mazumder, S.A.M., Yamaji, R., Ebara, S., Fujita, T., Tanimori, S., Kirihata, M., and Nakano, Y., *J. Agric. Food Chem.*, 46, 5177–5180, 1998.
11. Watanabe, F., Katsura, H., Takenaka, S., Fujita, T., Abe, K., Tamura, Y., Nakatsuka, T., and Nakano, Y., *J. Agric. Food Chem.*, 47, 4736–4741, 1999.
12. Watanabe, F., Takenaka, S., Katsura, H., Miyamoto, E., Abe, K., Tamura, Y., Nakatsuka, T., and Nakano, Y., *Biosci. Biotechnol. Biochem.*, 64, 2712–2715, 2000.
13. Watanabe, F., Katsura, H., Takenaka, S., Enomoto, T., Miyamoto, E., Nakatsuka, T., and Nakano, Y., *Int. J. Food Sci. Nutr.*, 52, 263–268, 2001.
14. Kittaka-Katsura, H., Fujita, T., Watanabe, F., and Nakano, Y., *J. Agric. Food Chem.*, 50, 4994–4997, 2002.
15. Watanabe, F., Michihata, T., Takenaka, S., Kittaka-Katsura, H., Enomoto, T., Miyamoto, E., and Adachi, S., *J. Liq. Chromatogr. Relat. Technol.*, 27, 2113–2119, 2004.
16. Sarangi, B., Chatterjee, S.K., Dutta, K., and Das, S.K., *J. Assoc. Off. Anal. Chem.*, 68, 547–548, 1985.
17. Vallari, D.S. and Rock, C.O., *Anal. Biochem.*, 154, 671–675, 1986.
18. Williams, E.J. and Campbell, A.K., *Anal. Biochem.*, 155, 249–255, 1986.
19. Henderson, G.B., Suresh, M.R., Vitols, K.S., and Huennekens, F.M., *Cancer Res.*, 46, 1639–1643, 1986.
20. Akhtar, M.J., Khan, M.A., and Ahmad, I., *J. Pharm. Biomed. Anal.*, 31, 579–588, 2003.
21. Cheng, R., Kin, B., Lee, K.W., and Ortwerth, B.J., *Biochim. Biophys. Acta*, 1537, 14–26, 2001.

11 Preparative Layer Chromatography of Natural Mixtures

Monika Waksmundzka-Hajnos, Teresa Wawrzynowicz, Michał Ł. Hajnos, and Grzegorz Jóźwiak

CONTENTS

11.1 INTRODUCTION

Therapeutic properties of herbs have been known for many years. They were once used in folk medicine, but recently herb therapy has become popular again. It is known that widely used herbal infusions or tinctures contain a wide spectrum of compounds ranging from very important pharmacologically active ones to those with small or sometimes undesirable activity.

Simultaneous evolution of chromatography, as a method of analysis and separation, enables the confirmation and development of chemotaxonomic investigations of new plant species, as well as the accomplishment of quality and quantitative determinations. Thin-layer chromatography (TLC) especially proved to be very useful for analysis and isolation of small amounts of some compounds. The most significant and advantageous points of the TLC technique are its speed, cheapness, and capacity to carry out the analysis of several solutes simultaneously; its continuous development under equilibrated conditions; its gradient and multiple development; and its ability to scale up the separation process.

In the chromatography of plant extracts on an enlarged scale, there are a few main problems: "general elution" because of the differentiated polarity of complex mixture components being separated; the structural and chemical analogy of compounds; and resolution decrease due to band broadening.

The optimization of preparative and even micropreparative chromatography depends on the choice of an appropriate chromatographic system (adsorbent and eluent), sample application and development mode to ensure high purity, and yield of desirable compounds isolated from the layer. For the so-called difficult separations, it is necessary to perform rechromatography by using a system with a different selectivity. But it should be taken into account that achievement of satisfactory results frequently depends on a compromise between yield and the purity of the mixture component that is being isolated.

11.2 GOALS OF PREPARATIVE LAYER CHROMATOGRAPHY OF PLANT EXTRACTS

Preparative layer chromatography (PLC) — considered as the most effective, cheapest, and simplest method for separation and isolation of small quantities of plant extracts — is widely used for different purposes. Most frequently, PLC is applied for the isolation of natural mixture components before their identification by various physicochemical methods such as hydrogen nuclear magnetic resonance (^{1}H-NMR), carbon nuclear magnetic resonance (^{13}C-NMR), IR (FTIR), and mass spectrometry. There are also numerous examples of PLC used as a method of sample preparation. Purification and fractionation of mixtures can therefore be used before gas chromatography (GC) or high-performance liquid chromatography (HPLC) analysis, with reference to environmental samples as well as crude plant extracts or biological samples. PLC is also used to obtain small quantities of standards from natural mixtures.

11.2.1 Online Purification of Plant Extracts

As plant extracts mainly comprise large amounts of ballast substances (e.g., lipids and chlorophylls), their purification is often a priority in the analysis. Such purification can be expensive in terms of both time and solvent consumed and can lead to losses of sample components. Online purification and separation of extracts contaminated with plant oil, can be readily performed by TLC in equilibrium chambers [1] that enable the use of continuous elution.

The procedure can be summarized as follows:

- TLC plates coated with the layer of polar adsorbent should be prewetted with a nonpolar solvent, such as benzene or *n*-heptane (*n*-hexane), to prevent deactivation of the adsorbent surface and to avoid glue up as a result of the penetration of the pores by lipid molecules and other impurities (i.e., wax).
- A relatively large volume of sample can be applied to the wet layer from the edge of the layer from the eluent distributor, forming a partly separated starting band by the frontal chromatography stage.
- Preliminary purification of a starting band contaminated with plant oil should be performed by predevelopment with a nonpolar solvent such as benzene or *n*-heptane, delivered from the eluent container. Weakly retained ballast substances (e.g., lipids) move with the solvent to the edge of the adsorbent layer, covering the glass plate where the volatile solvent evaporates. The contaminants can then be removed (scraped out with the adsorbent) from the layer or adsorbed on the strip of blotting paper placed on the upper edge of the layer.
- The chosen eluent can then be then used for the separation of the necessary components in the usual manner in the elution chromatography stage. The solvent flow can be monitored by spotting a dilute solution of nonadsorbed dye (e.g., azulene in case of NP systems) to the adsorbent layer.
- Localized zones can be scraped from the plate, and the fractions isolated with the strong polar solvent (acetone or methanol) can be analyzed by TLC or HPLC.

Figure 11.1a [1] shows a schematic representation of a micropreparative thin-layer chromatogram obtained on a 0.5-mm Florisil (magnesium silicate) layer prewetted with benzene of a crude extract, i.e., containing coextracted plant oil obtained from *Heracleum moelendorfi* fruit. The initial band of extract was washed with benzene and then separated by continuous development with ethyl acetate in benzene [1]. As seen from the fraction analysis presented in Figure 11.1b, small quantities of pure bergapten and xanthotoxin can be isolated in this manner.

11.2.2 PLC as a Method of Sample Preparation

Crude extracts of different plants are often rich in lipophilic substances, such as plant oils, chlorophylls, and waxes and also highly polar components such as tannines or sugars. Because the complicated liquid–liquid extraction (LLE) procedures are

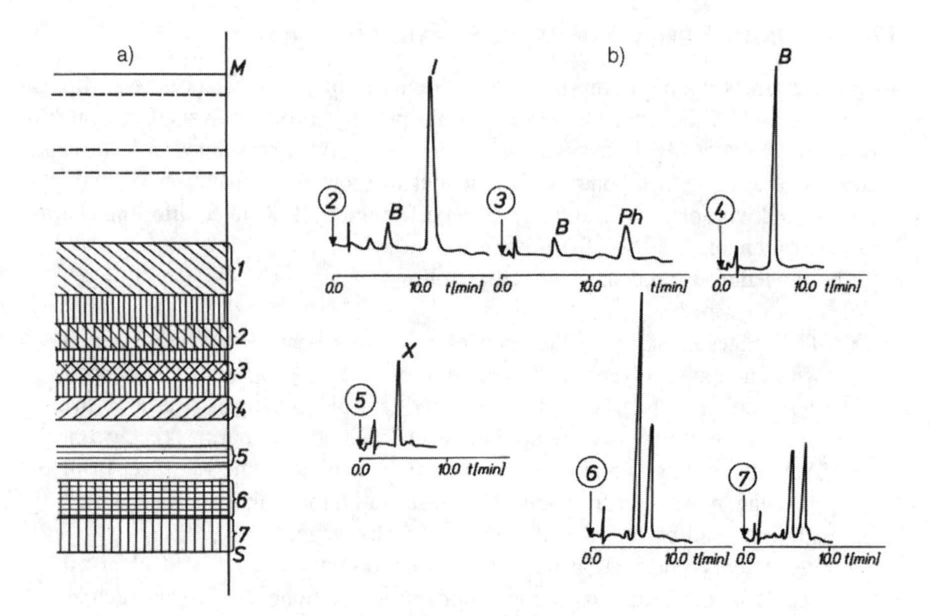

FIGURE 11.1 (a) Schematic representation of PLC of *Heracleum moellendorfi* fruit, crude extract (500-μl 2% solution), system: Florisil/AcOEt + B; plate preeluted with benzene; (b) analytical HPLC of isolated fractions, system: C18/MeOH + H_2O (6:4). Abbreviations: B — bergaptene, I — imperatorin, Ph — phelopterin, X — xanthotoxin. (For details, see Waksmundzka-Hajnos, M. and Wawrzynowicz, T., *J. Planar Chromatogr.*, 5, 169–174, 1992.)

connected with some practical problems such as emulsion formation, solid phase extraction (SPE) methods are widely applied. Unfortunately, a sample preparation in one step is often impossible. In such cases, PLC is used as a method of sample preparation [2,3].

A good example of this problem may be crude extracts of different yew species containing nonpolar (chlorophylls and waxes) as well as polar (tannines and polyphenols) ballast substances, which are difficult to purify. The difficulties are also caused by the fact that bioactive taxoids present in the extract occur in trace amounts. Because of this, sample preparation before HPLC analysis is usually a multistep procedure. In our experiment [3], the SPE procedure on C8 adsorbent does not enable separation of highly polar ballast substances, which are strongly retained by silica when nonaqueous eluents are used. Zonal TLC under these conditions results in separation of highly polar compounds and partial separation of the taxoid fraction. Combined SPE fractions of a purified extract were evaporated to 2 ml and applied to the edge of the silica layer by means of the glass distributor of the horizontal glass chamber. Starting bands were developed twice across the whole length of the plate with chloroform–benzene–acetone–methanol (85 + 20 + 15 + 7.5 v/v) as the mobile phase [4]. Bands of taxoids located under UV 254 nm, and compared with paclitaxel and 10-deacetylbaccatin standards, were scraped from the plates eluted dynamically with methanol and evaporated to 2 ml before HPLC analysis. Figure 11.2a shows a photocopy and densitogram of preparative chromatogram of the extract from needles of *Taxus brevifolia* Nutt. The analysis of fractions shown in

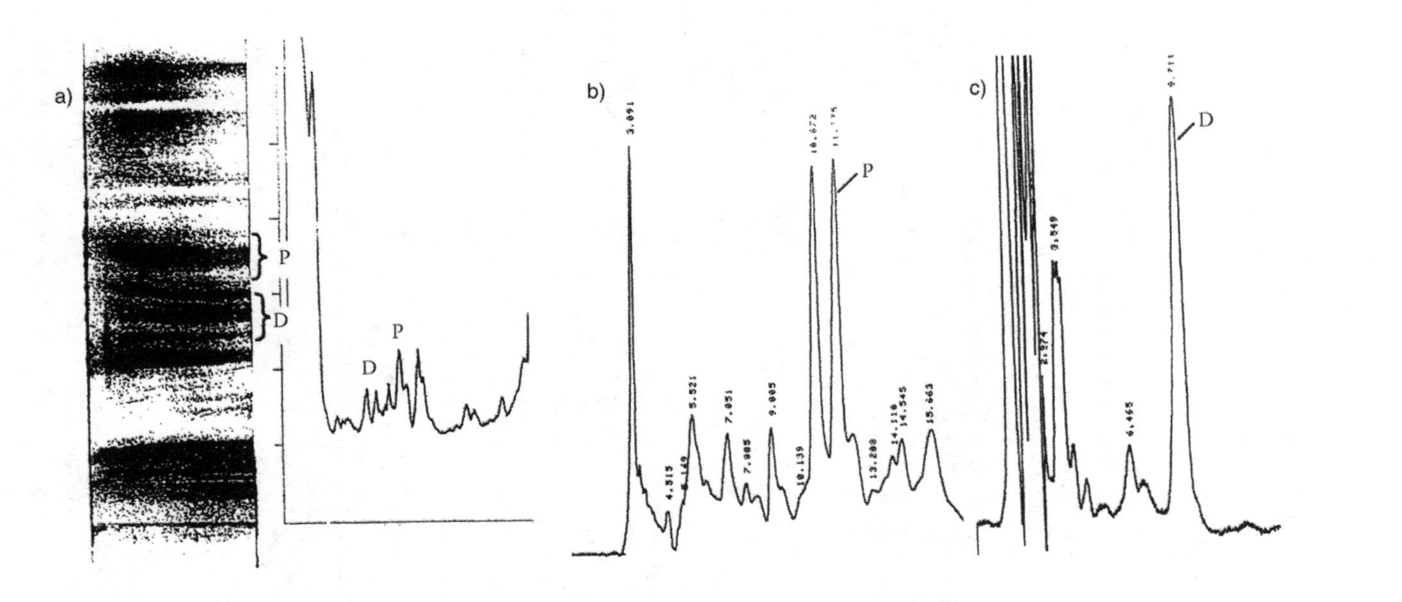

FIGURE 11.2 (a) Photocopy and densitogram of PLC of and extract of *Taxus brevifolia* (needles), system: SiO_2/$CHCl_3$ + B + Me_2CO + MeOH (85:20:15:7.5); (b) HPLC of isolated fraction containing paclitaxel (P), system: C18/MeCN + H_2O (6:4); (c) HPLC of isolated fraction containing 10-DAB III (D), system: C18/MeCN + H_2O (3:7). (For details, see Głowniak, K., Wawrzynowicz, T., Hajnos, M., and Mroczek, T., *J. Planar Chromatogr.*, 12, 328–335, 1999.)

Figure 11.2b and Figure 11.2c enables quantification of paclitaxel (T) and 10-deacetylbaccatin III (D) in isolated fractions [3].

11.2.3 PLC as a Pilot Technique for Preparative Column Chromatography

In optimization of systems for preparative column chromatography, besides the choice of optimum conditions enabling satisfactory resolution in a short time, loadability determination also seems to play an important role. TLC and PLC are usually used for the search of system selectivity for individual purposes.

For example, an alumina layer with a nonaqueous mobile phase was optimized for the separation of the taxoid fraction from ballast substances [5]. Figure 11.3 shows the densitogram obtained for *Taxus baccata* crude extract chromatographed on the alumina layer developed with nonaqueous eluents. The use of ethyl acetate and dichloromethane enables elution of nonpolar fractions (chlorophylls and waxes) and purification of the starting zone (Figure 11.3a). In this system, all taxoids are strongly retained on the alumina layer. The use of a more polar mobile phase

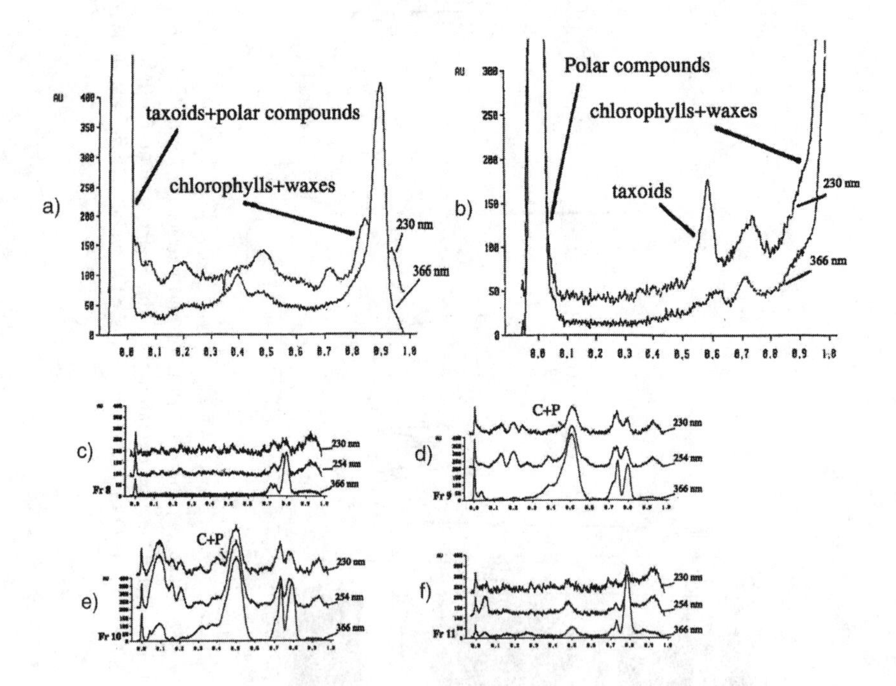

FIGURE 11.3 Densitograms obtained from chromatograms of *Taxus baccata* crude extract on an alumina layer developed with either (a) AcOEt + CH_2Cl_2 (8:2) or (b) MeOH + AcOEt + CH_2Cl_2 (0.5:2:7.5); (c-f) monitoring of the effluent fractions from the alumina column by TLC, system: silica/CH_2Cl_2 + DX + Me_2CO + MeOH. (84:10:5:1). Fractions 9 (d) and 10 (e) contain paclitaxel (P) and cephalomannine (C). (For details, see Hajnos, M., Głowniak, K., Waksmundzka-Hajnos, M., and Kogut, P., *J. Planar Chromatogr.*, 14, 119–125, 2001.)

FIGURE 11.4 Video scans of *Fumaria officinalis* chromatogram developed on alumina layer in eluent systems: (a) 10% PrOH + CH_2Cl_2; (b) first development with 10% PrOH + CH_2Cl_2 on a distance of 18 cm and second development with MeOH on a distance of 10 cm. Plates scanned after derivatization with Dragendorff's reagent.

containing a small amount of methanol causes the elution of taxoid fraction. However, polar compounds (polyphenols, tannines) are still retained by the alumina layer (Figure 11.3b). The experiments enabled selection of a procedure for purification and isolation of the taxoid fraction from crude yew extract on the alumina column and stepwise gradient elution with ethyl acetate–dichloromethane–methanol in different proportions (with increasing eluent strength). Figure11.3c to Figure 11.3f show the composition of fractions eluted from the alumina column. It is seen that the fraction of taxoids was eluted and isolated from the ballasts using the system optimized in thin-layer experiments.

In Figure 11.4 video scans for the PLC separations of *Fumaria officinalis* extract are shown. The use of the middle polar eluent — 10% propanol in dichloromethane — causes separation of the extract into two bands on the alumina layer. One zone is eluted near the eluent front, whereas the second band is strongly retained on the layer near the starting band (see Figure 11.4a). When we use strongly polar eluent (methanol) after the development with the first one, the zone of strongly polar alkaloids moved from the start (Figure 11.4b). This experiment enabled the use of the alumina column for the purification and fractionation of *Fumaria officinalis* extract into two fractions. The system can be also applied for the SPE of the extract before HPTLC or HPLC analysis.

The possibility of the application of the sample from the edge of the layer in equilibrium conditions and the application of continuous elution in sandwich chambers enable the observation of the zones' behavior from the start to the end of elution for overloaded systems. It offers the possibility to determine the loadability of the system and optimize preparative column chromatography. This method was developed for the separation of furanocoumarins contained in the *Archangelica officinalis* fruit extract during the search for the method of isolation of some biologically active compounds on a preparative scale [6]. Figure 11.5a and Figure 11.5b show the comparison of continuous TLC and column chromatography of the furanocoumarins fraction in the same system and similar overload conditions. The complete analogy of the separation can be easily observed.

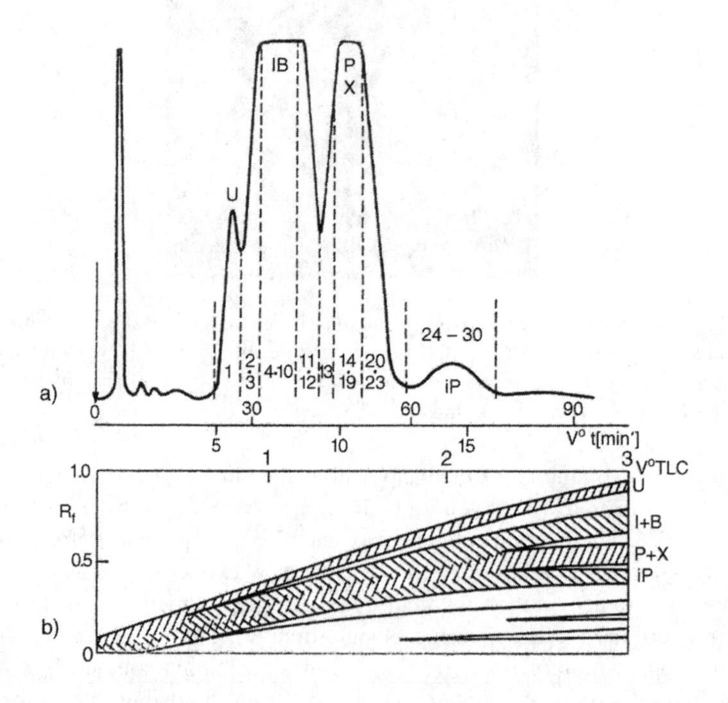

FIGURE 11.5 Comparison of (a) column chromatography and (b) continuous PLC of *Archangelica officinalis* fruit extract. System: silica gel/1.5% of MeCN in CH_2Cl_2 + H (7:3). Abbreviations: B — bergaptene, I — imperatorin, P — pimpinellin, X — xanthotoxin, U — umbelliprenin, iP — isopimpinellin. (For details, see Głowniak, K., Soczewiński, E., and Wawrzynowicz, T., *Chem. Anal. (Warsaw)*, 32, 797–811, 1987.)

11.2.4 Review of the Application of PLC for the Isolation of Natural Mixture Components

As is shown in Table 11.1, PLC is mainly used before the identification of isolated compounds with spectral methods such as NMR or IR. Figure 11.6 shows a densitogram of PLC of *Fumaria officinalis* herb extract with marked fractions being isolated (Figure 11.6a) and analytical chromatogram of isolated fractions (Figure 11.6b).

There are also numerous of papers in which the performance of biological activity of isolated compounds from natural mixtures have been described. Examples of such applications of PLC are presented in Table 11.2.

11.3 GENERAL PRINCIPLES OF THE CHOICE OF CHROMATOGRAPHIC SYSTEMS IN PLC OF PLANT EXTRACTS

For several years PLC has been used as an irreplaceable method of separation, isolation, and collection of small amounts of some components of natural mixtures, e.g., plant extracts. For most of the difficult preparative separations on a laboratory scale, these amounts are within a limited range of 5 to 100 mg and depend on the complexity of the material investigated. Apart from a number of isomeric substances, closely related compounds in plant extracts, and ballast substances influencing the separation are present. Although the principles of optimization of micropreparative or preparative chromatography separation of plant material are similar to general assumptions known either for analytical or preparative chromatography, a special strategy is needed. Obviously, it should be elaborated for each particular plant species.

Considering the principal factors affecting preparative planar chromatography, described by Nyiredy [101,102], it seems that the most important factors are the stationary and mobile phases with respect to the natural mixtures.

Apart from the choice of an appropriate stationary and mobile phase, the essential problem for PLC is to attain equilibrium in a three-phase system — between the stationary, mobile, and gas phases. In a nonequilibrated system, the velocity of the mobile phase in a thicker layer (i.e., the effect of solvent evaporation) is less in a lower part of an adsorbent. Such a situation leads to the diffusion of bands and deterioration of the adjacent bands' separation. This can be minimized or avoided by prerunning the plate with the mobile phase before spotting of the sample and the saturated chromatographic chambers.

The choice of the chromatographic system depends on the chemical character of the extracts being separated. The mobile phase should accomplish all requirements for PLC determined by volatility and low viscosity, because nonvolatile components (e.g., ion association reagents and most buffers) should be avoided. It means that, for PLC of plant extracts, normal phase chromatography is much more preferable than reversed-phase systems. In the latter situation, mixtures such as methanol–acetonitrile–water are mostly used. If buffers and acids have to be added to either the

TABLE 11.1
Application of Preparative Layer Chromatography for the Separation of Secondary Metabolites from Plant Tissues

Compounds	Sample	Plate	Eluent	Detection	Remarks	Reference
			Alkaloids			
Quaternary alkaloids	*Chelidonium maius*	SiO_2	Aqueous buffered MeOH	UV 355 nm	—	7
Tobacco alkaloid myosmine	*Arachus hypogea* *Corylus avellana*	SiO_2	$CHCl_3$ – MeOH	UV	HPLC, GC/MS	8
Lycopodium alkaloids	*Lycopodiaceae* sp.	SiO_2	$CHCl_3$ – MeOH – NH_3	Dragendorff's reagent	Identification	9
Protoberberine alkaloids	*Fissistigma balansae*	SiO_2	$CHCl_3$ – MeOH – NH_3	Dragendorff's reagent	Identification	10
Anthranilate alkaloids	*Ticorea longiflora*	SiO_2	Hx – AcOEt	—	—	11
Diterpenoid alkaloids	*Aconitum leucostomum*	SiO_2	$CHCl_3$ – MeOH – Me_2CO	Dragendorff's reagent	IR, NMR, EI-MS	12
Benzylisoquinoline alkaloids	Leaves *Anisocycla jollyana*	SiO_2, Al_2O_3	Multicomponent eluent	—	—	13
Indole alkaloids	*Strychnos icaja* roots	SiO_2	Hx + $CHCl_3$ + MeOH + H_2O; AcOEt + iPrOH + NH_3 aqueous	Dragendorff's reagent; UV	Antimalarial, cytotoxic activity	14
Pyrolizidine alkaloids	*Heliotropium crassifolium*	SiO_2	$CHCl_3$ + MeOH + NH_3 aqueous	Dragendorff's reagent; Ehrlich's reagent	Isolation	15
Isoquinoline alkaloids	*Hernandia nymphaeifolia*	SiO_2	$CHCl_3$ + Me_2CO; Hx + AcOEt; CH_2Cl_2 + MeOH	UV	Identification	16
Tazeline-type alkaloids	*Galanthus* sp.	SiO_2	B + $CHCl_3$ + MeOH; B + $CHCl_3$ + MeOH	—	Identification	17
Tetrahydroproto-berberine alkaloids	*Corydalis tashira*	SiO_2	$CHCl_3$ + EtOH; B + MeOH; $CHCl_3$ + AcOEt	UV 254 nm	Identification	18
Alkaloids	*Glaucium leiocarpum*	SiO_2	Cx + $CHCl_3$ + Et_2NH; T + Me_2CO + EtOH	—	Identification	19
β-Carboline alkaloids	*Hedyotis capitellata*	SiO_2	Cx + $CHCl_3$ + Et_2NH	—	Identification	20
Indole alkaloids	*Uncaria guianensis* bark	C18	MeOH + H_2O	—	Identification	21
Macrocyclic lactam alkaloids	*Verbascum phoeniceum*	SiO_2	AcOEt + MeOH; T + EtOH	UV 254 nm; r. Dragendorff's	Identification	22

			Phenolics			
Isoflavonoids	*Milettia griffoniana* root, bark	SiO_2	B + petroleum ether + AcOEt		Isolation, IR, NMR, MS	23
Methoxylated flavones	*Primula veris*	SiO_2	Hx + AcOEt		Isolation, HPLC	24
Flavonoid glycosides	*Astragalus* sp. roots	SiO_2 RP2	$CHCl_3$ + MeOH + H_2O; MeOH + H_2O	UV	HPLC – MS	25
Coumarins	*Ticorea longiflora*	SiO_2	Hx–CH_2Cl_2 – MeOH	UV	—	11
Coumarins, furanocoumarins	*Harbouria trachypleura*	SiO_2	Hx + AcOEt, CH_2Cl_2 + Me_2CO	—	Isolation, ^{1}H-NMR	26
Anthocyanins	Champagne vintage	—	—	—	Identification ^{1}H-NMR	27
Furanocoumarins	Plants Apiaceae	Silanized silica	MeOH + H_2O	UV	Identification by HPLC	28
Flavonoids	*Cyclopia intermedia*	SiO_2	B + Me_2CO + MeOH	Anisaldehyde, formaldehyde	Identification	29
Flavone aglycones	*Origanum vulgare*	SiO_2	T + AcOEt + formic acid	UV, $AlCl_3$	Identification	30
C-methylated flavones	*Elsholzia stauntonii*	SiO_2	$CHCl_3$ + AcOEt + MeOH	UV 254 nm	Identification	31
Flavonoids	*Chorizanthe diffusa*	SiO_2	$CHCl_3$ + MeOH	H_2SO_4 + temperature	Biological activity	32
Flavonoids	*Dorstenia manni*	SiO_2	Hx + AcOEt	UV	Identification	33
Flavonoid glycosides	*Daphniphyllum calycinum*	SiO_2	$CHCl_3$ + MeOH + H_2O	UV 254 nm	Identification	34
Isoprenylated flavones	*Morus cahayana*	SiO_2	$CHCl_3$ + Et_2O; B + MeOH; B + AcOEt	—	Isolation	35
Biflavonoids	*Calophyllum venulosum*	SiO_2	$CHCl_3$ + MeOH, $CHCl_3$ + AcOEt	UV 254 nm	Identification	36
Chlorogenic acid and related compounds	*Prunus dulcis*	SiO_2	AcOEt + AcOH + formic acid + H_2O	Diphenylboric acid ethylamino ester	Isolation	37
Chlorogenic acid derivatives	*Phyllostachus edulis*	SiO_2 C18	BuOH + MeOH + H_2O; MeOH + H_2O	UV	Identification; biological activity	38
Coumarins	*Zanthoxylum schinifolium*	SiO_2	Hx + Me_2CO; Hx + AcOEt; B + AcOEt	UV	Isolation	39
Coumarin derivatives	*Monotes engleri*	SiO_2	$CHCl_3$ + MeOH	UV 366 nm	Isolation	40

TABLE 11.1 (*Continued*)
Application of Preparative Layer Chromatography for the Separation of Secondary Metabolites from Plant Tissues

Compounds	Sample	Plate	Eluent	Detection	Remarks	Reference
Coumarin derivatives	*Cyclosorus interruptus*	SiO_2	Hx + AcOEt	Vanilin	Identification	41
Coumarins	*Peucedanum tauricum*	RP2	MeOH + H_2O	UV 366 nm	Identification	42
			Terpenoids			
Pentacyclic triterpenoids–lantadenes	*Lantana camara* var. *aculeate*	SiO_2 GF	Petroleum + AcOEt + AcOH	UV	Purification, HPLC	43
Lactone diterpenes	*Potamogeton natans*	SiO_2	B + AcOEt	—	Isolation, NMR, FABMS	44
Triterpene	*Eugenia sandwicensis*	SiO_2	$CHCl_3$ + MeOH	—	Isolation, IR NMR, MS	45
Tetraterpenoid–trianthenol	*Trianthema portulacastrum*	SiO_2	Hx + Me_2CO + AcOEt	—	Isolation, NMR, HR-EIMS	46
Euphosalicin–diterpene polyester	*Euphorbia salicifolia*	SiO_2	CHx + AcOEt + EtOH	—	Isolation, IR, NMR, MS	47
Terpenoids: danshenol-A	*Salvia gluinosa*	SiO_2	B – AcOEt	—	Quality control	48
Diterpenes	*Euphorbia segetalis*	SiO_2	CH_2Cl_2 – T – MTB	—	—	49
Furanoditerpenes	*Croton campestris*	SiO_2	Hx – AcOEt	UV 254	Quantitative analysis	50
Cytotoxic diterpenes	*Salvia miltiorrhiza*	SiO_2 RP18	Hx – AcOEt MeOH – H_2O	—	Isolation	51
Euphane triterpenes	*Schinus molle*	SiO_2	T – AcOEt – AcOH	Anisaldehyde reagent	Qualitative identification	52
Sesquiterpenes	*Chlorantus japonicus*	SiO_2	Et_2O – MeOH; $CHCl_3$ – MeOH	UV	Isolation	53
Monoterpenes, sesquiterpenes	*Lippia ducis*	SiO_2	Hx – Me_2CO	UV 254	Identification	54
Triterpene saponins	*Hedera helix*	SiO_2	$CHCl_3$ – MeOH – AcOH – H_2O	Anisaldehyde	Isolation of hederacoside	55

Triterpene diols and triols	*Chrysanthemum* flower	SiO_2	Hx + AcOEt	—	Food analysis	56
Acyclic diterpene-α-lactones	*Salix matsudan*	SiO_2	Hx + Me_2CO; Hx + $CHCl_3$ + MeOH	UV	Isolation	57
Triterpenoids	*Picea glehni* stem bark	SiO_2	Hx + AcOEt	UV	Identification	58
Sesquiterpenes	*Celastus roshornianus*	SiO_2 C18	B + Me_2CO; MeOH + H_2O	UV	Isolation	59
Diterpenes	*Ladix kaempfen*	SiO_2	$CHCl_3$ + MeOH	UV 254 nm	Identification	60
			Various Compounds			
Naphthoquinones	*Catalpa ovata*	SiO_2	Hx – AcOEt; B – AcOEt	UV 254, 366 nm	Identification	61
Acetogenin derivatives	*Annona cherimola*	SiO_2	$CHCl_3$ – MeOH; AcOEt – Me_2CO	Kedde's reagent	Identification	62
Polyacetylenes	*Bellis perennis*	SiO_2	Pentane – Et_2O	UV 254 nm	Isolation	63
Alkamides	*Echinacea purpurea* roots	C18	MeOH + H_2O	—	LC-MS	64
Pungent constituents	Ginger	SiO_2	Hx + Me_2CO	UV	Purification, HPLC-MS	65
Benzofuranones, chromanes	*Coniothyrium minitans*	SiO_2	B + AcOEt	—	Isolation, NMR, MS	66
Dibenzofurans	*Lecanora cinereocarnea*	SiO_2	T + Me_2CO	—	Isolation, IR, NMR, HR–EIMS,	67
Naphthopyranones	*Guanomyces polythrix*	SiO_2	CH_2Cl_2 + MeOH	—	Isolation, IR, NMR, MS	68
Taxoids	*Taxus baccata*	SiO_2	B – $CHCl_3$ – Me_2CO – MeOH	UV 254	Optimization of eluent	4
Taxoids	*Taxus* sp.	SiO_2	B – $CHCl_3$ – Me_2CO – MeOH	UV 254	Fractionation, HPLC	3
Taxoids	*Taxus* sp.	SiO_2		UV 254 densitometry	Fractionation, HPLC	5
Taxoids	*Taxus* sp.	SiO_2		UV 254 densitometry	Isolation	69
Guaianolides	*Anthemis carpatica*	SiO_2	Multicomponent eluents	—	Identification	70
Phthalides	*Angelica sinensis* roots	SiO_2	Hx + Me_2CO	—	Isolation, LC-MS	71
Phenylethanoides	*Buddleja cordata*	SiO_2	CH_2Cl_2 + MeOH	UV 254 nm	Identification	72
Naphthoquinone derivatives	*Bulbine capitata*	SiO_2	B+Hx; B + MeOH	—	Identification	73

TABLE 11.1 (*Continued*)
Application of Preparative Layer Chromatography for the Separation of Secondary Metabolites from Plant Tissues

Compounds	Sample	Plate	Eluent	Detection	Remarks	Reference
Coniferyl and sinapyl alcohol derivatives	*Ligularia duciformis*	SiO_2	B + Me_2CO; B + AcOEt	UV 254 nm; H_2SO_4 + temperature	Identification	74
Sesquilignans, lignans	*Tsuga heterophylla*	SiO_2	B + $CHCl_3$ + Me_2CO B + iPr_2O + MeOH	UV 254 nm, 366 nm	Isolation	75
Neolignans	Wood of *Machilus obovatifolia*	SiO_2	Hx (CH_2Cl_2) + Me_2CO; CH_2Cl_2 + AcOEt	$CeSO_4$ + temperature	Isolation	76
Neolignans	*Piper hookeri*	C18 Al_2O_3	MeOH + H_2O; Petrol + AcOEt	—	Isolation	77
Steroids	*Harrisonia abyssinica*	C2	MeOH + H_2O	UV	Identification	78
Phenylpropanes	*Acorus tatarinowii*	SiO_2	$CHCl_3$ + MeOH	UV	Isolation	79
Naphthoquinone glucosides	*Drosera gigantean*	Polyamide SiO_2	EtOH + H_2O T + HCCOH	—	Isolation	80
Secoiridoid glycosides	*Jasminum urophyllum*	SiO_2 C18	$CHCl_3$ + MeOH + H_2O; MeOH + H_2O	—	Isolation	81
Furoquinolones	*Melicope confuse*	SiO_2	$CHCl_3$ + MeOH	UV	Identification, biological activity	82
Taxoids	*Taxus walichiana* cell culture	SiO_2	B + Me_2CO	UV 254 nm	Isolation	83

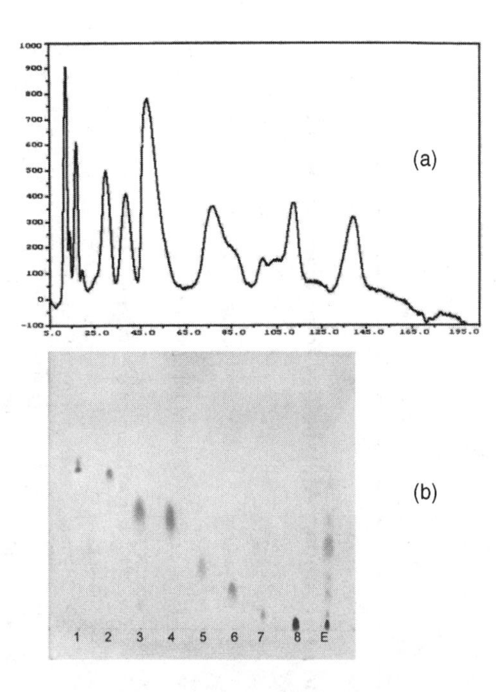

FIGURE 11.6 (a) Densitogram of PLC of *Fumaria officinalis* herb extract and (b) photocopy of isolated fractions (after spraying with Dragendorff's reagent). Numbers indicate isolated fractions. System: silica/CH_2Cl_2 + PrOH + AcOH (5:4:1). Plates double developed.

aqueous or nonaqueous mobile phase, only volatile buffers — acetate, formate, ammonium carbonate, and acids such as acetic acid and formic acid — are preferred.

Males et al. [103] used aqueous mobile phase with formic acid for the separation of flavonoids and phenolic acids in the extract of *Sambuci* flos. In a cited paper, authors listed ten mobile phases with addition of acids used by other investigators for chromatography of polyphenolic material. For micropreparative separation and isolation of antraquinone derivatives (aloine and aloeemodine) from the hardened sap of aloë (Liliaceae family), Wawrzynowicz et al. used 0.5-mm silica precoated plates and isopropanol–methanol–acetic acid as the mobile phase [104]. The addition of small amounts of acid to the mobile phase suppressed the dissociation of acidic groups (phenolic, carboxylic) and thus prevented band diffusions.

For the same reason, the alkalization of the mobile phase for the separation of basic compounds, e.g., alkaloids with addition of small amounts of ammonia, is performed [105]. Usually, eluents giving R_F values in the lower range (0.1 to 0.6) are chosen because the application of large sample volumes causes widening of the starting zone and the increase of R_F values. To form a compressed starting band, the sample should be dissolved in a solvent weaker than the eluent. Very often for normal phase chromatography, ethyl acetate is recommended as a modifier; for alkaloids, it is dichloromethane rather than chloroform. Such modifiers are good solvents for a wide group of solutes and ensure easy evaporation of the fractions.

TABLE 11.2
Application of Preparative Layer Chromatography for the Separation of Secondary Metabolites from Plant Tissues for Their Biological Activity Determination

Compounds	Sample	Plate	Eluent	Detection	Biological Activity	Reference
Phenolics	*Ocimum basikicum*	SiO_2	$CHCl_3$ + MeOH + H_2O	$FeCl_3$	Antioxidant activity	84
Phenolics	*Piper methysticum*	SiO_2	Hx + Me_2CO; T + AcOEt	UV 254 and 366 nm	Effect on cyclooxygenaze enzyme	85
Coumarins and carbazoles	*Clausena harmandiana*	SiO_2	Hx + AcOEt	UV 254 nm	Antiplasmoidal activity	86
Flavonoids	*Apocynum venetum*	Silanizsilica	MeOH + H_2O	UV	Hepatoprotective effect	87
Nortrachelogenin, daphnoretin	*Wikstroemia indica*	SiO_2	$CHCl_3$ + MeOH	$FeCl_3$ and UV	Antifungal, antimitotic, anti-HIV-1 activity	88
Coumarins	*Decatropis bicolor*	SiO_2	Hx + AcOEt	UV 254 nm	Antiinflammatory activity	89
Isoflavonoids	*Pueraria mirifica*	SiO_2 C18	$CHCl_3$ + MeOH MeOH + H_2O	UV 254 nm	Estrogenic activity	90
Lignans	*Schisandra machilus*	SiO_2	Hx + AcOEt	—	Cholesterol acyltransferaze inhibitory activity	91

Catechins	*Taxus cuspidata*	SiO_2	CH_2Cl_2 + MeOH	UV	Inhibition of chitin synthase II	92
Flavonoids	*Salvia off.* *Melissa off.* *Lavandula angustifolia*	SiO_2	Multicomponent eluents	UV 254 and 327 nm	Effect against enzyme-dependent and enzyme-independent lipid peroxydation	93
Flavonoids	*Cleome droserifolia*	SiO_2	B + Me_2CO	UV	Suppressing NO production in microphages *in vitro*	94
Limonoids	*Khaga sevegalensis*	SiO_2	Hx + CH_2Cl_2 + MeOH	—	Antisickling activity	95
Flavones	*Similacis gabrae* rhizome	SiO_2	$CHCl_3$ + MeOH	UV and $CeSO_4$	Preventing immunological hepatocyte damage	96
Coumarins	Fruit of *Citrus hystrix*	SiO_2	Hx + AcOEt	—	Inhibitors of NO generation in mouse macrophage cells	97
Alkaloids	*Cryptolepsis* leaves, roots	SiO_2	T + iPrOH + NH_3 $CHCl_3$ + MeOH	Dens. UV Dragendorff	Antiplasmoidal activity	98
Naphthalene glycosides	*Hernerocallis* flowers	SiO_2	Multicomponent eluents	UV 254 nm	Antioxidant activity	99
Rubiandin-1-methyl ether	*Stephania verosa* *Murraya simensis*	SiO_2	T + AcOEt	—	Antimalarial activity	100

In PLC of plant extracts, sorbents applied for analytical TLC are usually used. For preparative purposes, not only the chemical nature of the adsorbent (to avoid irreversible interactions) and its surface geometry but also its expense should be taken into consideration. Components of plant extracts are very often ionizable and possess an acidic or basic character (i.e., flavonoids, alkaloids). Therefore, silica or alumina are the most suitable for preparative chromatography. One must take into consideration some limitations in the adsorption on alumina. For alumina, the attachment to the active surface site (especially of the Lewis acid type Al^{3+}) by the two neighboring polar substituents is stronger than on a silica surface. That is why some compounds with the ortho or vicinal OH groups (polyphenols) or –COOH–OH grouping (phenolic acids) are capable of interacting simultaneously with the surface sites of alumina forming chelates. This "ortho effect" was described by Snyder [106] for some benzene derivatives and some nitrogen heterocycles called by Klemm et al. as "anchoring interaction" [107]. For this reason, the silica is chiefly used as an adsorbent for separation of plant extracts under overloaded conditions.

It should be mentioned that for some compounds having polar and nonpolar alkyl substituents in the ortho or vicinal position, their adsorption on polar adsorbents is weaker owing to the steric shielding effect of the alkyl group or other nonpolar bulky substituents. Thus, compounds being homologous to one another may be separated only on a nonpolar adsorbent. Preparative separation of imperatorin and its homological compound (phellopterin) were achieved using reversed-phase chromatography [108].

Investigators dealing with optimization of chromatography systems for preparative separation on laboratory or larger scales in the first stage using TLC selected stationary and mobile phases to obtain a resolution $R_S > 1.5$ for each pair of touching bands. Such a resolution permits the introduction of a 3 mg/g adsorbent.

In the second stage, the sampling mode, loadability, and system capacity are determined. For this purpose, systematic investigation should be carried out on a microscale as a model for further scaling up the separation process for PLC or preparative column chromatography [109,110]. The most advantageous method is systematic investigation of sampling of the increased sample size (concentration or volume) until one desirable zone touches the adjacent band. Further increase of the sample leads to overlapping bands, when it becomes necessary to carry out rechromatography of the isolated zone using the same (in analogy to recycling) or a more selective system. Sometimes, when large amounts of a purified product are required, this procedure seems to be more effective than a number of separations of the small samples. For PLC of multicomponent natural mixtures, rechromatography is necessary to achieve satisfactory yield and purity of the desirable compound. But in most cases, only one goal may be optimized at the expense of the other [111].

Other factors that can influence the separability of components of complex natural mixtures, such as adsorbent particle size and layer thickness, are similar to those used in analytical TLC. Mostly, adsorbents of wide dispersion of particle size — 5 to 40 μm and layers of 0.5 to 1 mm thickness — are used. Although the capacities of layers increase with their thickness, the separation efficiency decreases for thickness above 1.5 mm. Commercially available precoated preparative plates (e.g., silica, alumina, and RP2 plates) with fluorescence indicators and plates with preadsorbent zones are more convenient and commonly used.

11.3.1 Application of Different Selectivity Systems for the Separation of Closely Related Compounds

The selectivity of a chromatographic system is the main critical parameter in the result of separation in analytical and preparative chromatography. For a pair of substances, selectivity is characterized quantitatively by the separation coefficient ($\alpha = k_I / k_{II}$ for the compounds I and II); for a large number of substances the correlations $R_{M\,I}$ vs. $R_{M\,II}$ (log k_I vs. log k_{II}) are the characteristics of the selectivity system (see Figure 11.7) [112]. The following characteristic cases can be mentioned:

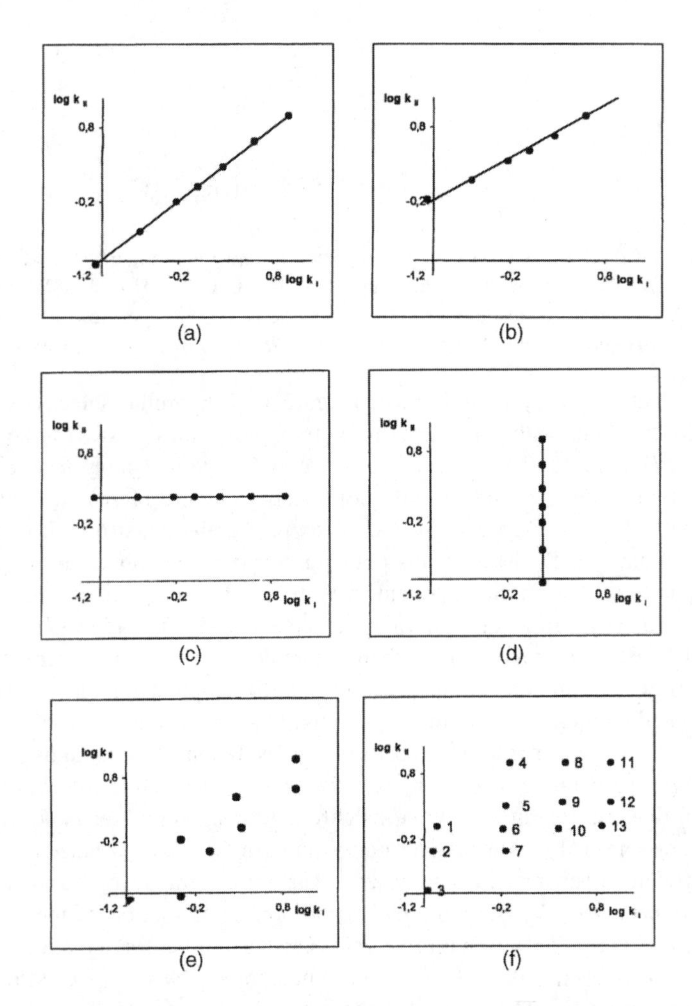

FIGURE 11.7 Characteristic cases for log k_I vs. log k_{II} (R_{MI} vs. R_{MII}) correlations: (a) identical selectivity of system I and II; (b) similar selectivity of both systems, system I more selective (higher Δ log $k_I = \Delta R_M$ values); (c) system I selective, system II nonselective; (d) system II selective, system I nonselective; (e) different selectivity of both systems; and (f) lack of correlation, strong diversity of retention. (From Soczewiński, E., Preface, in Waksmundzka-Hajnos, M., *Chromatographic Dissertations*, Medical University, Lublin, Poland, 2, 9–12, 1998.)

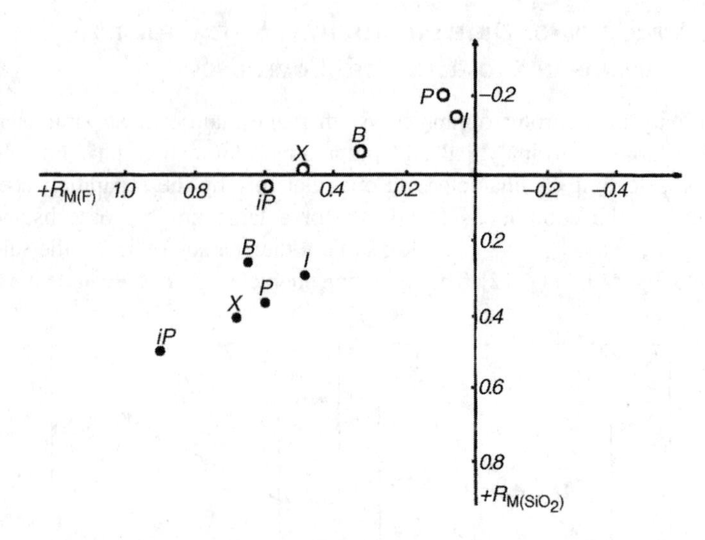

FIGURE 11.8 Correlation of R_M values of selected furanocoumarins meassured on Florisil and silica layers developed with 5% diisopropyl ether in CH_2Cl_2 + H (7:3) solid circles and with 15% AcOEt in B (open circles). For abbreviations see Figure 11.5. (For details, see Waksmundzka-Hajnos, M. and Wawrzynowicz, T., *J. Planar Chromatogr.*, 5, 169–174, 1992.)

identical selectivity of systems I and II (Figure 11.7a); similar selectivity of both systems (system I generally more selective with largest values of ΔR_M for the selected substances) (Figure 11.7b); system I selective and system II nonselective (Figure 11.7c); system II selective and system I nonselective (Figure 11.7d); different selectivity of both systems (Figure 11.7e); and lack of correlation, strong differences in retention (Figure 11.7f). Such correlation diagrams can be applied for the planning of the separation of a multicomponent mixture; in the first step, system I can be applied where the group separation can be achieved: (1+2+3), (4+5+6+7), (8+9+10), and (11+12+13). The groups can be then separated by the use of system II.

The differences in separation selectivity can be applied for the separation of closely related compounds' mixtures from natural sources. These selectivity differences are mainly important for the separation and isolation of compounds from plant extracts on a preparative scale. Figure 11.8 shows an example of $R_{M\,\mathrm{I}}$ vs. $R_{M\,\mathrm{II}}$ correlation diagram for furanocoumarins chromatographed on silica and Florisil in two eluent systems [1]. It is seen that furanocoumarins — closely related compounds — are difficult to separate in one system. Rechromatography of partly separated fractions is necessary for the complete separation and isolation of the individual compounds. Figure 11.9a to Figure 11.9c show examples of isolation of some furanocoumarins by use of PLC and rechromatography by use of classic column chromatography with low-pressure flow of eluent [28]. In Figure 11.9a, the results of PLC separation of *Angelica archangelica* fruit extract on silanized silica with aqueous methanol are presented. It enables partial separation of furanocoumarins that were rechromatographed on silica column eluted with nonaqueous eluents (see Figure 11.9b and Figure 11.9c). The isolation of pure compounds such as bergapten, isopimpinellin, imperatorin, umbeliprenin, and phellopterin was achieved (see HPLC

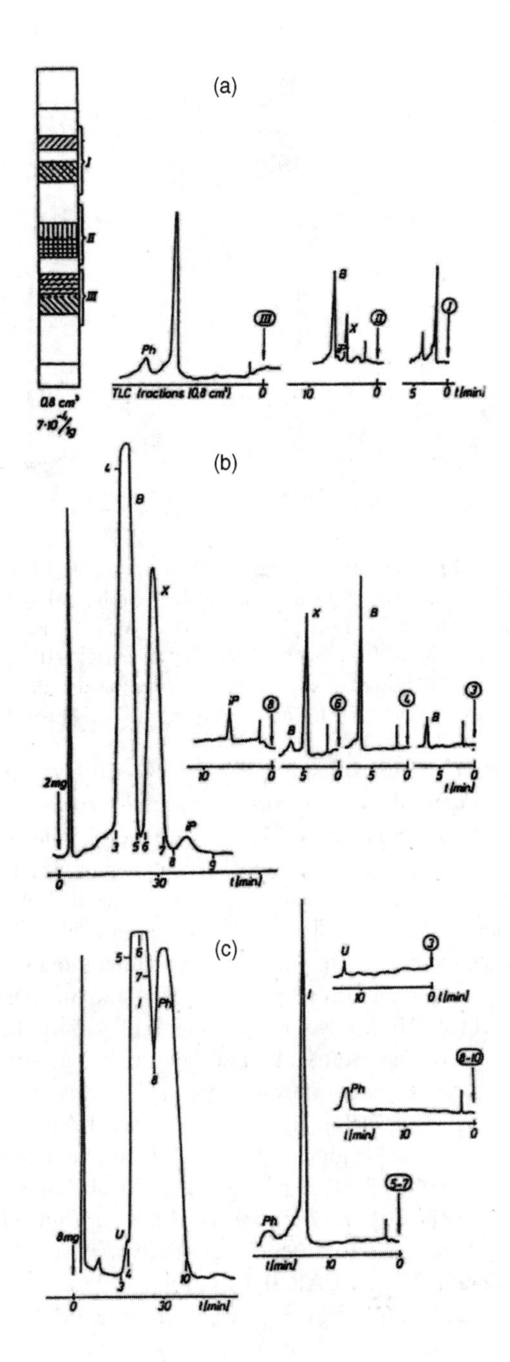

FIGURE 11.9 (a) PLC of fruit extract from *Angelica archangelica*. System: silanized silica/MeOH + H_2O (6:4); (b) rechromatography of fraction II from the plate a on Si60 Lobar A column. Mobile phase: 5% diisopropyl ether in CH_2Cl_2 + H (7:3); (c) rechromatography of fraction III from the plate a on Si60 Lobar A column. Mobile phase: 5% diisopropyl ether in CH_2Cl_2 + H (7:3); all fractions controlled by analytical HPLC in system: C18/MeOH + H_2O (6:4). For abbreviations, see Figure 11.5. (For details, see Wawrzynowicz, T. and Waksmundzka-Hajnos M., *J. Liq. Chromatogr.*, 13, 3925–3940, 1990.)

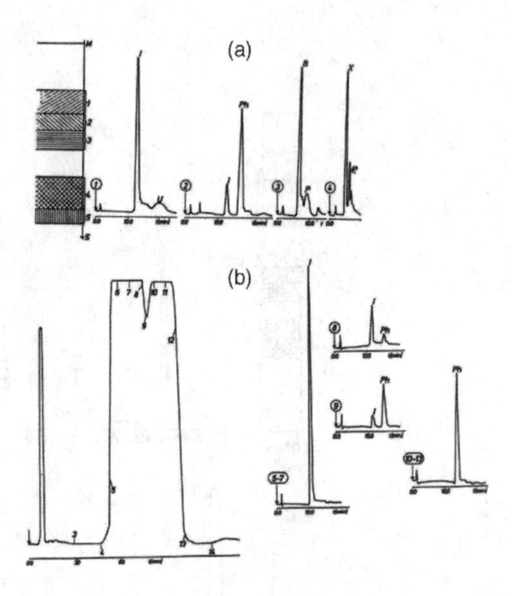

FIGURE 11.10 (a) Schematic representation of PLC of *Archangelica officinalis* fruit extract, system: Florisil/iPr$_2$O in CH$_2$Cl$_2$ + H (7:3) followed by 15% AcOEt + B; (b) rechromatography of fraction 2 from the plate a on Lobar A Si60 column eluted with 1.5% MeCN in CH$_2$Cl$_2$ + H (6:4); all fractions controlled by analytical HPLC in system: C18/MeOH + H$_2$O (6:4). For abbreviations, see Figure 11.5. (For details, see Waksmundzka-Hajnos, M. and Wawrzynowicz, T., *J. Planar Chromatogr.,* 3, 439–441, 1990.)

fractions' control). Also, PLC on Florisil layer gives partial separation of the furanocoumarin fraction from *Archangelica officinalis* (see Figure 11.10a). The use of silica column enables isolation of pure imperatorin and phellopterin (Figure 11.10b) [108]. Figure 11.11 presents the next example of preparative layer chromatography and column rechromatography. A silanized silica layer eluted with aqueous mobile phase makes partial separation of furanocoumarins possible (Figure 11.11a), which were rechromatographed using a column filled with Florisil eluted with nonaqueous mobile phase for isolation of xanthotoxin, bergapten, and imperatorin (Figure 11.11b) [108]. Also, combinations of various PLC systems can be used for this purpose. The use of a Florisil layer with various eluent systems enables separation of furanocoumarins from *Heracleum mantegazzianum* fruit and isolation of bergapten, pimpinelin, and other partly separated fractions (see Figure 11.12a, 11.12b) [1]. The chromatography on Florisil column and rechromatography on silica layers were applied for the isolation of 5,7-disubstituted simple coumarins and furanocoumarins from *Heracleum mantegazzianum* fruits [113].

Systems with different selectivity were used for the separation of 10-deacetylbaccatin III (10 DAB III) from yew extracts [69]. A silica column with stepwise gradient elution with aqueous methanolic mobile phases can be used for separation of the taxoid fraction from nonpolar materials, partial separation of the taxoid fraction into a polar one (containing 10-DAB III), and for a medium polarity taxoid fraction (containing paclitaxel and cephalomannine). Most polar material (tannins

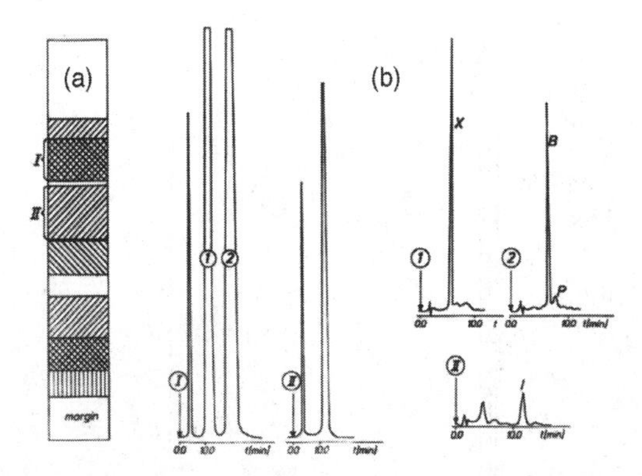

FIGURE 11.11 (a) Schematic representation of PLC of *Heracleum sosnowskyi* fruit crude extract, system: silanized silica/MeOH + H_2O (6:4); (b) rechromatography of fractions I and II from the plate a on Lobar-type column filled with Florisil eluted with 5% MeCN in CH_2Cl_2 + H (7:3); fractions controlled by analytical HPLC in system: C18/MeOH + H_2O (6:4). For abbreviations, see Figure 11.5. (For details, see Waksmundzka-Hajnos, M. and Wawrzynowicz, T., *J. Planar Chromatogr.,* 3, 439–441, 1990.)

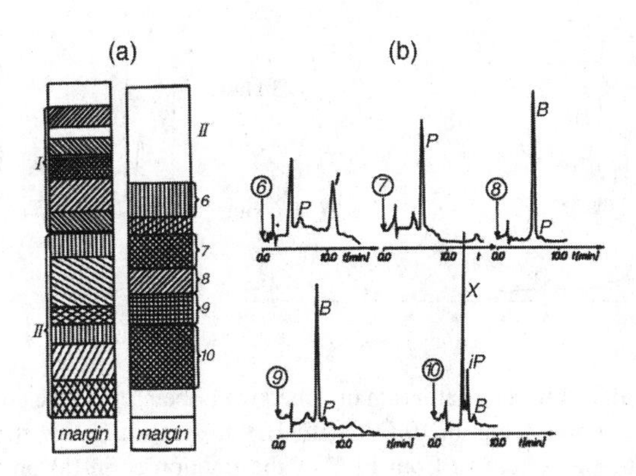

Figure 11.12 (a) Schematic representation of PLC of *Heracleum mantegazzianum* fruit extract chromatographed in system: Florisil/5% MeCN in benzene and rechromatography of fraction II isolated from the layer in system: Florisil/5% iPr_2O in CH_2Cl_2 + H (7:3); (b) fractions controlled by analytical HPLC in system: C18/MeOH + H_2O (6:4). For abbreviations, see Figure 11.5. (For details, see Waksmundzka-Hajnos, M. and Wawrzynowicz, T., *J. Planar Chromatogr.,* 5, 169–174, 1992.)

and phenolic compounds) was removed by elution with pure water, and a fraction containing 10-deacetylbaccatin III was obtained and controlled by HPLC (see Figure 11.13a). Further fractions contained other taxoids e.g., paclitaxel and cephalomannine. The partly separated fraction with 10-DAB III was rechromatographed in a normal phase system of different selectivity (silica/dichloromethane + dioxane + acetone + methanol) by the use of PLC (see densitogram in Figure 11.13b). The zone containing 10-DAB III, marked in Figure 11.13b, was removed from the plate

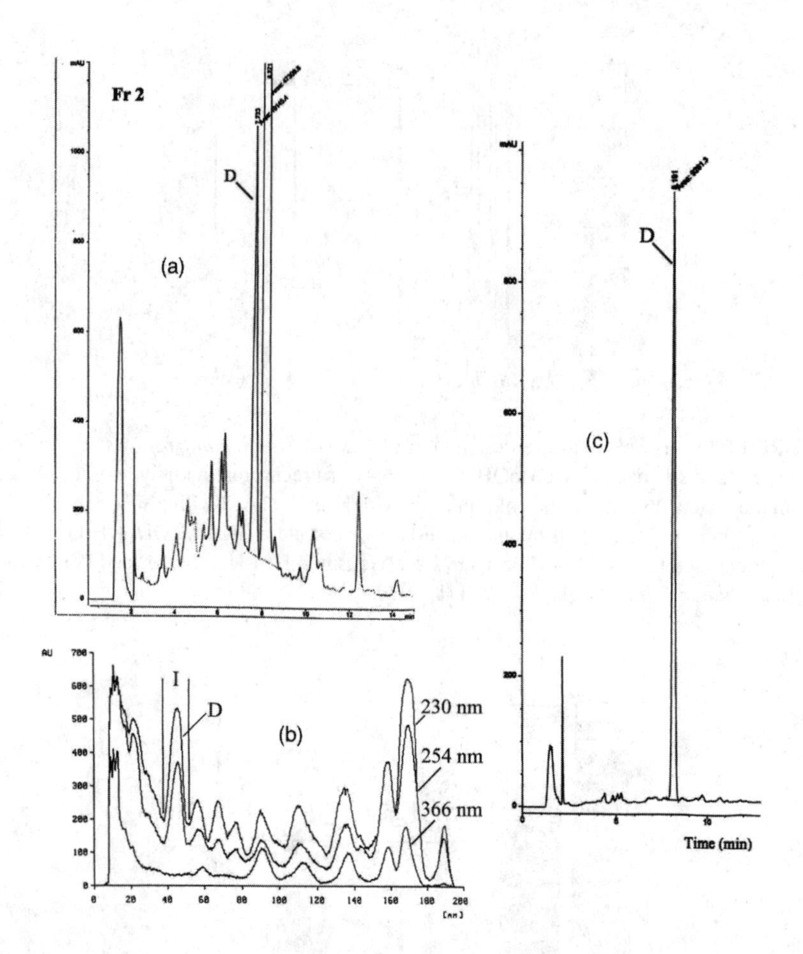

FIGURE 11.13 (a) HPLC chromatogram obtained from the preparative column (silica/aqueous methanol) fraction enriched in 10-DAB III. HPLC system: C18/MeCN + H_2O gradient elution; (b) densitogram obtained from PLC of the fraction as in (a) on silica layer eluted with CH_2Cl_2 + DX + Me_2CO + MeOH (84:10:5:1); (c) HPLC chromatogram of fraction isolated from the layer in (b) containing 10-DAB III. HPLC system: C18/MeCN + H_2O gradient elution. D – 10-DAB III. (For details, see Hajnos, M.L., Waksmundzka-Hajnos, M., Głowniak, K., and Piasecka, S., *Chromatographia*, 56, S91–S94, 2002.)

and extracted dynamically. Thus, pure 10-DAB III was isolated (see HPLC chromatogram in Figure 11.13c) [69].

Systems with different selectivity were used for the separation of cephalomannine and paclitaxel. The fractions from the alumina preparative column eluted with dichloromethane–ethyl acetate–methanol containing taxoids were applied as a band to a silica layer (0.5 mm), and preparative zonal chromatography was performed using a multicomponent mobile phase (see densitogram in Figure 11.14a) [5]. Then, the zone containing taxoids was isolated. The composition of the fraction is depicted in Figure 11.14b. The preceding procedure enables partial separation of cephalomannine and paclitaxel fractions. The isolation of paclitaxel and cephalomannine

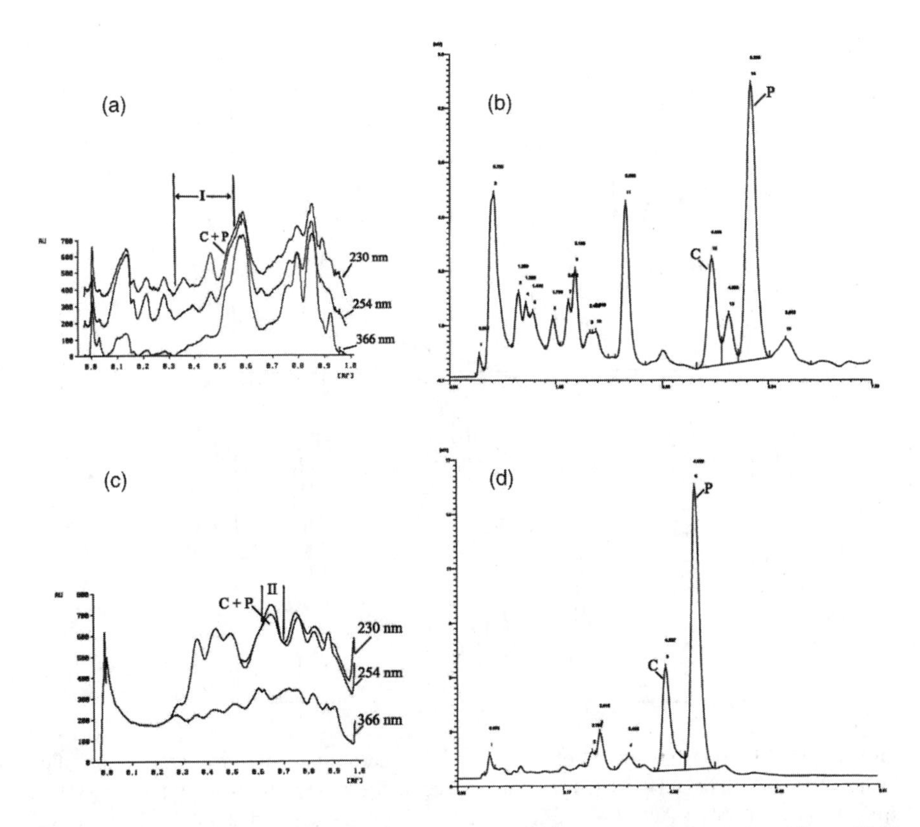

FIGURE 11.14 (a) PLC of lipophilic taxoid fraction from alumina column (see Figure 11.3) on silica layer eluted with CH_2Cl_2 + DX + Me_2CO + MeOH (84:10:5:1); (b) composition of zone I containing paclitaxel (P) and cephalomannine (C) controlled by RP18 HPLC eluted with acetonitrile + water (1:1); (c) PLC of the fraction from (a) by use of silanized silica eluted with methanol + water (7:3); (d) composition of zone II containing paclitaxel (P) and cephalomannine (C) controlled by RP18 HPLC eluted with acetonitrile + water (1:1). (For details, see Hajnos, M., Głowniak, K., Waksmundzka-Hajnos, M., and Kogut, P., *J. Planar Chromatogr.*, 14, 119–125, 2001.)

was achieved by rechromatography in a different selectivity system — PLC on silanized silica with methanol–water, 7 + 3 — as the mobile phase (see densitogram in Figure 11.14c). The composition of the isolated fraction is presented in the chromatogram in Figure 11.14d. This procedure enables isolation of a mixed fraction containing paclitaxel and cephalomannine [5].

Also, different selectivity systems of were used for the separation of the alkaloid fraction from *Corydalis solida* herb. The extract was fractionated by the use of a silica layer eluted with 10% propanol-2 in dichloromethane (see Figure 11.15a). Fraction I eluted dynamically from the adsorbent was rechromatographed by the use of silica layer and eluent of higher strength containing acetonitrille + propanol-2 + acetic acid + dichloromethane. It enables the separation of the six zones of alkaloids from fraction I (see densitogram in Figure 11.15b).

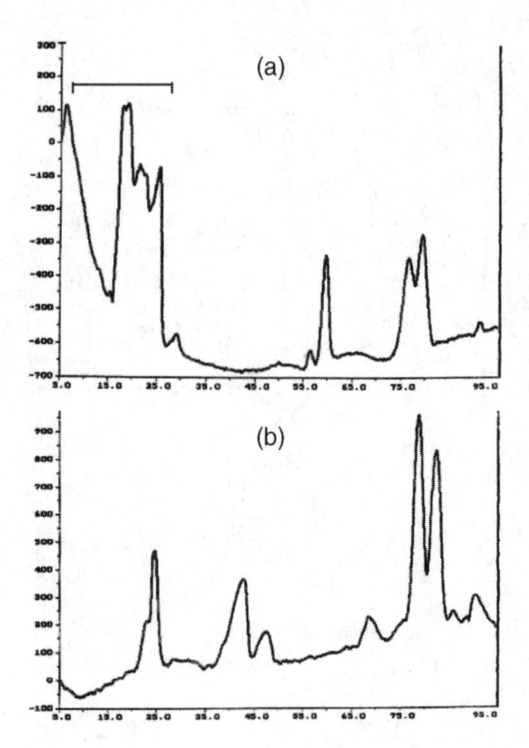

FIGURE 11.15 (a) Densitograms of *Corydalis solida* herb extract chromatographed on silica by use of 10% PrOH in CH_2Cl_2; (b) fraction marked in figure (a) rechromatographed in system silica/MeCN + PrOH + AcOH + CH_2Cl_2 (10:20:5:65); plates scanned at 400 nm after derivatization with Dragendorff's reagent.

11.4 SAMPLING OF PLANT EXTRACTS

The necessity of introducing a large sample onto a layer is a great problem, especially with reference to a type of overloading (volume or mass) and sampling mode.

11.4.1 Kind of Overloading

We deal with overloading when the introduced sample volume is so large that the eluted band (or concentrating profile) is significantly broadened, or when the concentration of an injected solute is so large that the eluted bands (or concentration profiles) become asymmetric in comparison to analytical scale sampling, owing to deviation from linearity of the adsorption isotherm. It is evident that band broadening may lead to overlapping of bands. The deviation of linearity takes place when an introduced sample is larger than 0.1 mg/g of adsorbent. Concentration overloading is more advantageous if sample solubility allows it, owing to mutual displacement effect and increasing yield. It should be taken into account that introduction of concentrated solutes may be disadvantageous owing to crystallization of substances into the adsorbent pores, which can disturb desorption–adsorption processes and lead to band or spot tailing, and to further deterioration (worsening) of separability.

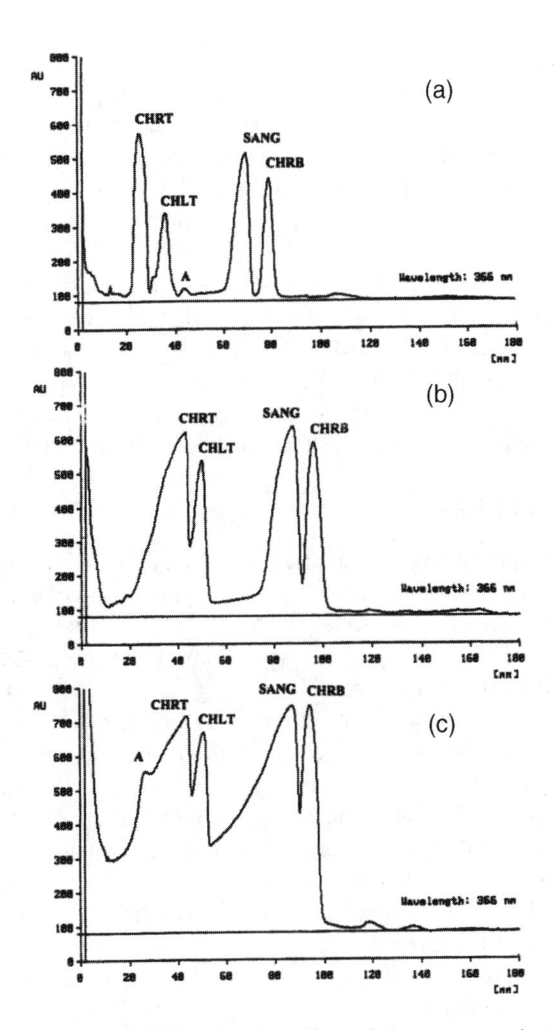

FIGURE 11.16 Densitograms illustrating the effect of the concentration of the extract of *Chelidonium majus*. System: silica/T + AcOEt + MeOH (83:15:2); (a) 0.25 mg of extract in 0.2-ml solvent; (b) 1 mg in 0.2-ml solvent; (c) 5 mg in 0.2-ml solvent. Abbreviations: CHRB — chelirubine, SANG — sanguinarine, CHLT — chelilutine, CHRT — chelerythrine, A — unidentified alkaloid. (For details, see Waksmundzka-Hajnos, M., Gadzikowska, M., and Hajnos, M., *J. Planar Chromatogr.*, 15, 289–293, 2002.)

The effect of concentration overloading on the separability of alkaloids is presented in Figure 11.16a to Figure 11.16c [114]. The same volumes (0.2 ml) of solution of increasing concentration were introduced from the edge of the silica layer. In this way, increasing amounts of the *Chelidonium majus* quaternary alkaloids from 0.25 to 5 mg were introduced and chromatographed using toluene–ethyl acetate–methanol (83 + 15 + 2). The values obtained for the resolution R_S for neighboring bands of the main components of the extract of lowest concentration are the following: for chelirubine and sanguinarine $R_{S\ 12} = 1.0$, for sanguinarine and chelilutine $R_{S\ 23} = 3.5$, and for chelilutine and chelerythrine $R_{S\ 34} = 1.7$. When 1 mg of dry extract dissolved in 0.2 ml of methanol was separated, the bands overlapped. Then, the R_S values for neighboring bands of the main components were the following: $R_{S\,12}$ = 0.56, $R_{S\,23}$ = 2.3, and $R_{S\,34}$ = 0.37. Partial separation of alkaloids is, however, still possible. When 5 mg of dry extract dissolved in 0.2 ml methanol was separated, the resolution of the main components decreased: $R_{S\,12}$ = 0.3, $R_{S\,23}$ = 1.24, and $R_{S\,34}$ = 0.2 (see Figure 11.16c), and only partial separation of two bands was possible.

Figure 11.17a to Figure 11.17d show densitograms from the separation of *Chelidonium majus* quaternary alkaloids. Various volumes of the extract solution of the same concentration, from 0.2 to 1.6 ml, were introduced to the adsorbent layer. Thus, the following portions of extract of 0.25, 0.5, 1, and 2 mg were chromatographed by the use of threefold development with the multicomponent eluent. The densitograms depict fast loss of resolution with the increase of the introduced sample [114].

11.4.2 Sampling Mode

The second factor influencing separability is the manner of delivery of larger sample volumes on the adsorbent layer. In the conventional method, the samples are applied by a syringe or pipette, using multiple spotting or streaking by a special mechanical device (e.g., programmed application), which causes the formation of a complex starting band. During the elution step, the starting band components are rearranged to a sequence according to their R_F values. This problem may be solved by the introduction of sample from the edge of the layer on a dry or equilibrated bed. Sandwich chambers permit zonal sample application from the eluent container, starting at the edge of the layer. The sample components of the starting band are already partly separated in the frontal chromatography stage. Thus, during the development (elution chromatography stage), components may be fully separated.

Figure 11.18 shows densitograms illustrating the effect of the mode of application of the sample to the adsorbent layer. *Chelidonium majus* quaternary alkaloid fractions were chromatographed in the system as the one presented in the preceding text. Plates were threefold developed. The solution of alkaloids was applied to the adsorbent layer in different ways: (1) from the edge of the layer (Figure 11.18a), (2) as a narrow band across the plate (Figure 11.18b), (3) with the set of capillaries on the bed (Figure 11.18c), and (4) by use of the plate with the preconcentrating zone (Figure 11.18d) [114]. It is apparent that application from the edge of the layer results in better separation of the neighboring zones than application by other methods, especially for strongly retained alkaloids (compare Figure 11.18a to Figure 11.18d). However, the use of the plate with the preconcentrating zone causes formation of a narrow starting band, which influences the resolution of bands. Figure 11.19 shows micropreparative TLC of aloe and its corresponding densitogram. Effective separation of aloine (Al) and aloeemodine (Ae) was possible when the plate with the preconcentrating zone was used, and pure compounds were isolated. In the other cases of application, the zones were dispersed, and the components of aloe isolated from the layer were only partly separated [104].

Very good results were obtained when the formation of the starting band was performed by use of a programmed applicator with the evaporation of the sample

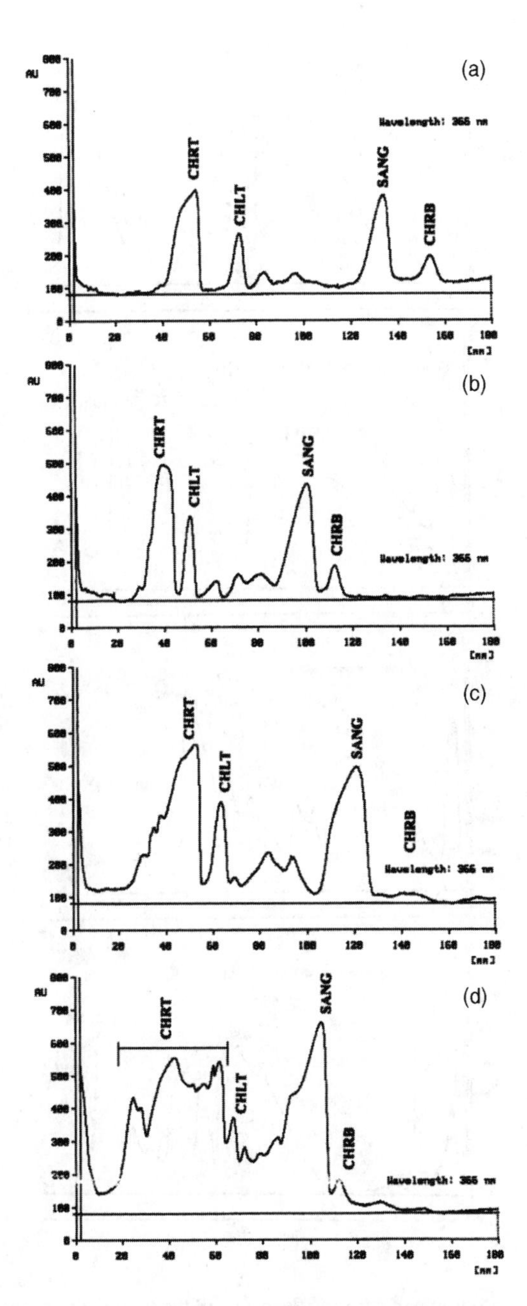

FIGURE 11.17 Densitograms illustrating the effect of the volume of the extract of the same concentration of *Chelidonium majus*. System: silica/T + AcOEt + MeOH (83:15:2). Plates triple developed; (a) 0.25 mg of extract in 0.2-ml solvent; (b) 0.5 mg in 0.4-ml solvent; (c) 1 mg in 0.8 ml solvent; (d) 2 mg in 1.6 ml solvent. For abbreviations, see Figure 11.16. (For details, see Waksmundzka-Hajnos, M., Gadzikowska, M., and Hajnos, M., *J. Planar Chromatogr.*, 15, 289–293, 2002.)

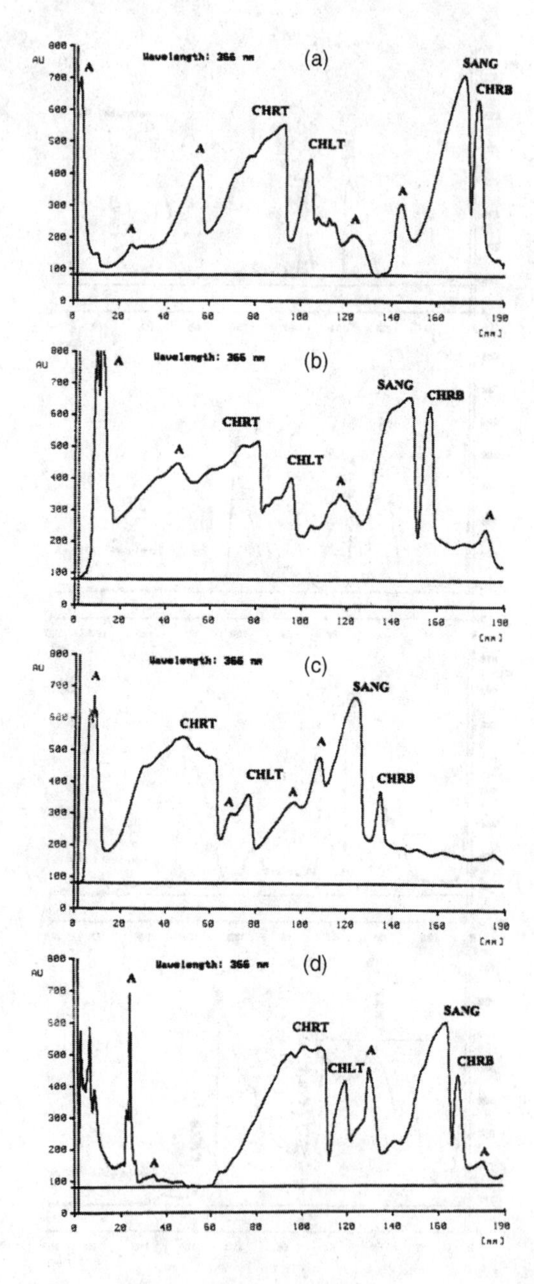

FIGURE 11.18 Densitograms illustrating the effect of the mode of introduction of the sample 2 mg of *Chelidonium majus* extract in 0.2 ml of solvent. System: silica/T + AcOEt + MeOH (83:15:2). Plates triple developed; (a) application from the edge of the layer; (b) application as a narrow band across the plate; (c) application with the set of 25 capillaries; (d) application to the preconcentrating zone. For abbreviations, see Figure 11.16. (For details, see Waksmundzka-Hajnos, M., Gadzikowska, M., and Hajnos, M., *J. Planar Chromatogr.*, 15, 289–293, 2002.)

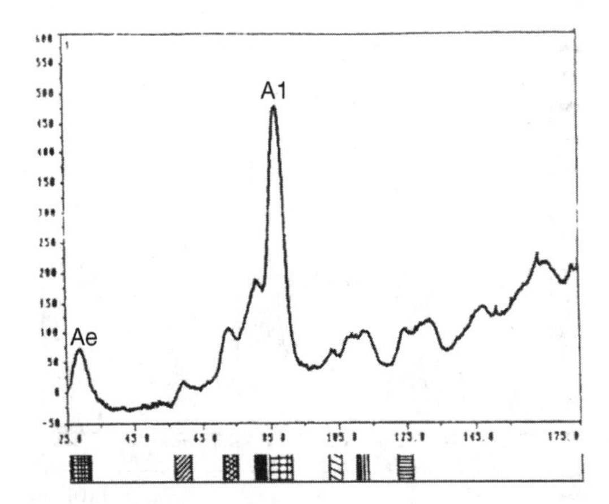

FIGURE 11.19 PLC of *Aloe* on precoated silica plate with preconcentrating zone developed with AcOEt + MeOH + water (77:13:10). Al — aloine, Ae — aloeemodine. (For details, see Wawrzynowicz, T., Waksmundzka-Hajnos, M., and Mulak-Banaszek, K., *J. Planar Chromatogr.*, 7, 315–318, 1994.)

solvent. It causes formation of an extremely narrow starting band, which gives also narrow zones of components after development. Figure 11.20 shows densitograms obtained from PLC of *Fumaria officinalis* plant extract introduced by the use of applicator with the evaporation of sample solvent (Figure 11.20a) and from the edge of the layer (Figure 11.20b) chromatographed in the same system (silica/acetic acid + 2-propanol + dichloromethane 1:4:5). It can be seen that more strongly retained alkaloids (1 – 3) are much better separated when the sample is introduced with help of an applicator.

In case large volume introduction becomes necessary (e.g., biologically active mixtures, by their nature, are diluted and thermolabile), researchers have to use multiple sampling with the evaporation of solvent from the starting band during the application, which causes preconcentration of the sample on the layer. Figure 11.21a to Figure 11.21c show the effect of multiple application various amount of taxoid fraction on the silica layer with the evaporation of solvent from the starting zone. It is seen that such a method is effective, and even large volumes introduced in such a way do not cause broadening of the zone being isolated.

11.5 INFLUENCE OF SAMPLE SOLVENTS ON FORMATION AND MIGRATION OF ZONES

Solubility of samples in analytical TLC or HPLC is important only for quantitative investigations. For other cases, it may be neglected. However, for preparative chromatography, sample solubility is very important. It should be taken into account that sample solubility may affect the bands' resolution owing to the fact that ratio of sample solvent volume to volume of eluent is greater than in analytical TLC, and may change the eluent strength significantly.

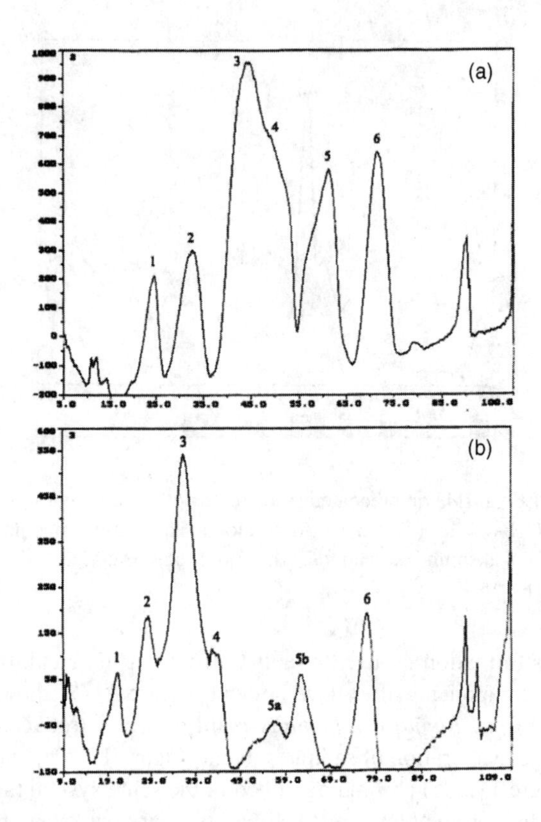

FIGURE 11.20 Densitogram of alkaloid fraction from *Fumaria officinalis* herb extract; chromatographed in system: silica/PrOH + AcOH + CH_2Cl_2 (4:1:5); (a) fraction introduced with applicator with evaporation of solvent; (b) fraction introduced from the edge of the layer with the eluent distributor.

For the sampling stage, three cases may be considered:

- The sample is dissolved in the eluent used as mobile phase.
- The sample is dissolved in diluted or more concentrated eluent.
- The sample is dissolved in a solvent of different qualitative composition.

In the first case, only widening of the starting band may be expected. Considering the second case, two possibilities should be taken into account — sample dilution (weaker solvent than mobile phase), which is advantageous because of formation of a narrow starting band, and its effect on the separability increase. Injection of a large volume of the sample, diluted with a nonpolar diluent (*n*-heptane), results in greater overloading without loss of resolution. This is the effect of the peak compression recognized in column chromatography. The use of sample solvent of higher concentration of a modifier as in the eluent causes widening of the starting band adequately to R_F values [109].

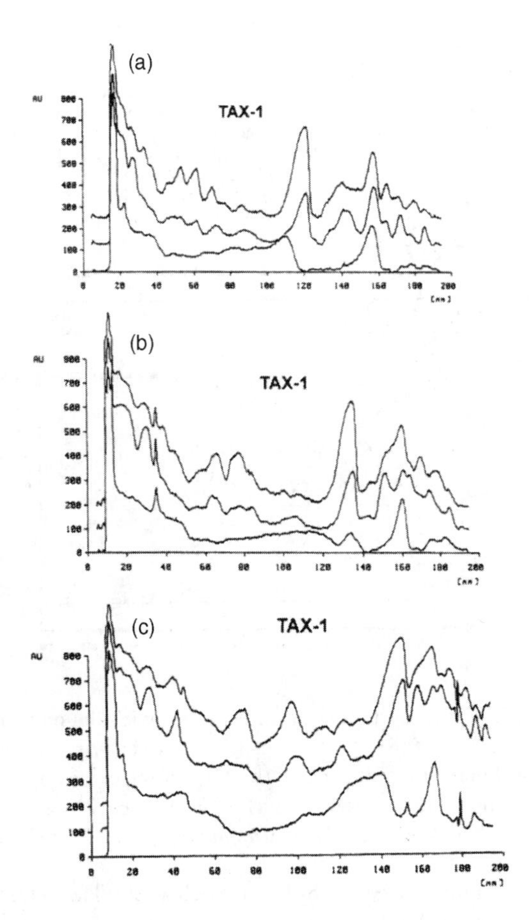

FIGURE 11.21 Densitograms of PLC of *Taxus baccata* fraction from preparative column (silica/aqueous methanol) containing unknown taxoid (Tax 1) introduced to the layer with a set of capillaries with simultaneous evaporation of solvent from the starting band. System: silica/CH_2Cl_2 + DX + Me_2CO + MeOH (84:10:5:1). Plates double developed; (a) 0.3 ml of fraction introduced; (b) 0.5 ml of fraction introduced; (c) 1 ml of fraction introduced to the layer.

The effect of the concentration of the extract solution on the separation of neighboring bands of alkaloids was also examined (Figure 11.22). Figures show the densitograms obtained from the separated bands when 2 mg of *Chelidonium majus* extract was dissolved in 0.2 ml of methanol (Figure 11.22a) and when it was diluted with methanol to 1.6 ml (Figure 11.22b). It is clearly apparent that dilution of the sample results in overlapping of the neighboring bands of alkaloids [114].

For the third case, in which the sample is dissolved in solvent different from the eluent, it may disturb the separation process owing to a change of eluent strength and the precipitation of the solutes. The substances that are stagnant in the adsorbent's pores are gradually eluted, forming elongated zones from the start line.

The second requirement referring to the mobile phase is the volatility of its components isolated from scraped adsorbent or eluted by online chromatography.

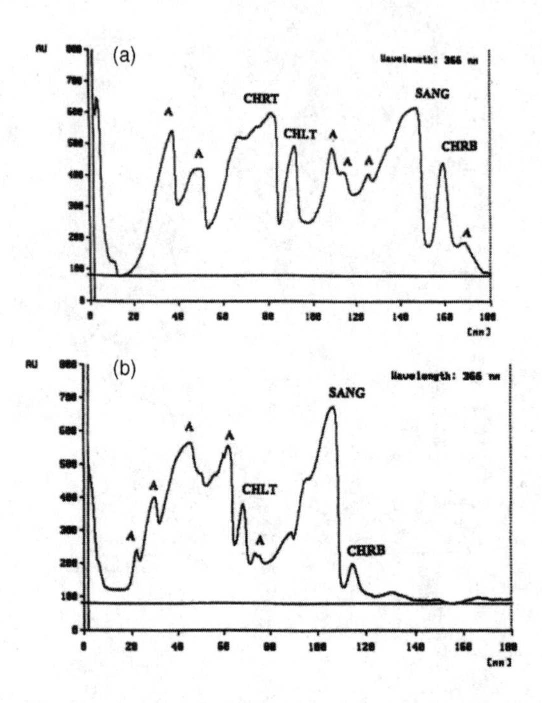

FIGURE 11.22 Densitograms illustrating the effect of sample dilution with methanol. System: silica/T + AcOEt + MeOH (83:15:2). Plates triple developed; (a) 2 mg of *Chelidonium majus* extract dissolved in 0.2 ml of methanol; (b) 2 mg of the extract dissolved in 1.6 ml of methanol. For abbreviations, see Figure 11.16. (For details, see Waksmundzka-Hajnos, M., Gadzikowska, M., and Hajnos, M.., *J. Planar Chromatogr.*, 15, 289–293, 2002.)

Eluent components should be volatile. Solvents such as ethyl acetate, isopropyl ether, diethylketone, chloroform, dichloromethane, and toluene as modifiers and *n*-hexane as diluent are recommended for normal phase chromatography. For reversed-phase systems, methanol or acetonitrile are used as modifiers. Such components as acetic acid or buffers, as well as ion association reagents, should be avoided.

11.6 DIFFERENT TYPES OF DETECTION IN PLC OF NATURAL MIXTURES

For preparative chromatography, the main goal is the recovery of pure mixture components. Therefore, the localization of separated bands is an important issue. The localization of bands directly on plates in daylight (for colored substances) or mostly in UV light is more convenient. The majority of adsorbents and commercially available precoated plates have a fluorescent indicator, e.g., silica gel 60 F254 + 366. In several cases, separated bands may be localized in iodine vapors if substances form only unstable complexes. Brown or yellow zones produced in this way should be immediately outlined.

For drastic derivatization methods in the layer after development, 1 to 2 cm wide paths are marked along two longer sides of the plate. The central part of the layer should

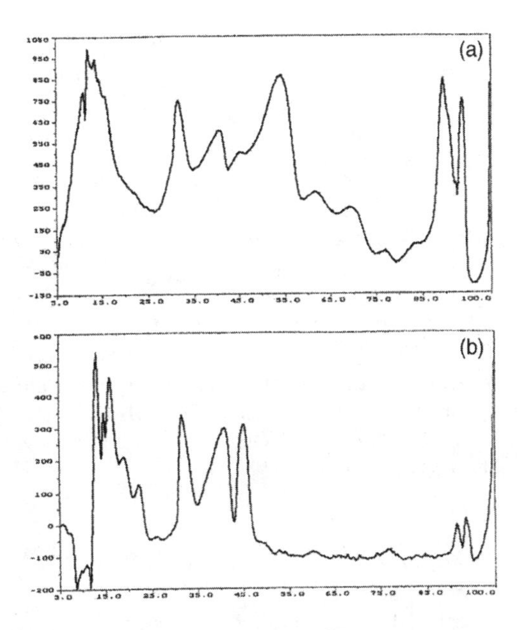

FIGURE 11.23 Densitograms of *Fumaria muralis* herb extract chromatographed in system: silica/PrOH + MeOH + AcOH + water + CH_2Cl_2 (28:2:2:2:66). Plates double developed; (a) plate scanned at 254 nm; (b) plate scanned at 400 nm after spraying with Dragendorff's reagent.

be covered and the two side segments sprayed with appropriate reagents. The two channels should be scraped and gathered into small funnels for extraction of components.

Figure 11.23a and Figure 11.23b show densitograms from the separation of *Fumaria muralis* extract scanned in UV 254 nm (Figure 11.23a) and at 400 nm after spraying with Dragendorff's reagent (Figure 11.23b). It is seen that derivatization enables determination of alkaloid zones and their isolation from the nonalkaloid extract components.

The possibility of the determination of the UV spectrum of a particular zone may lead to identification of a substance or its group. For example, the scanning of preparative chromatograms with the separated components of yew extracts at various wavelengths (360, 254, and 230 nm) enables determination of the taxoids' zones, which have a maximum of absorption at 230 nm (see Figure 11.24). Similarly, densitometry in absorbance–reflectance mode or fluorescence–reflectance mode can give information for the determination of isolated compounds. Figure 11.25 presents densitograms of alkaloids in *Chinae* extract where densitometry was performed in absorbance–reflectance mode at 327 and 290 nm, which gives the possibility of identification of some cinchona alkaloids [115].

11.7 SPECIAL MODES OF DEVELOPMENT AS A TOOL IN PREPARATIVE SEPARATION OF PLANT EXTRACTS

For the chromatography of complex mixtures (separation of isomers or closely related compounds with similar retention), the main problem is to improve the

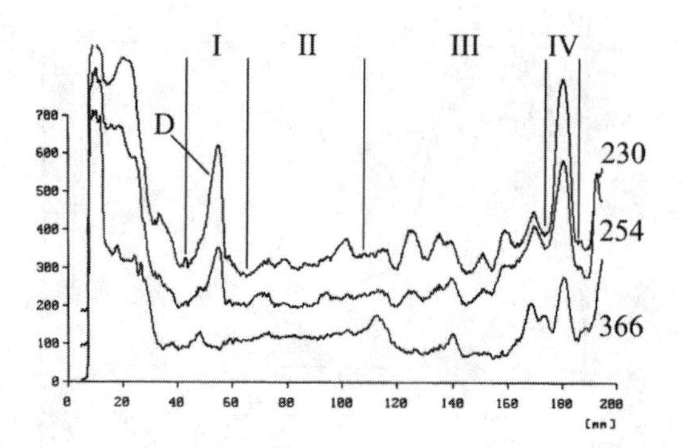

FIGURE 11.24 Densitogram of fraction obtained from *Taxus baccata* by column chromatography (silica/aqueous methanol) enriched in 10-DAB III rechromatographed by PLC in system: silica/CH_2Cl_2 + DX + Me_2CO + MeOH (84:10:5:1). Plate scanned at 366, 254, and 230 nm. Numbers indicate isolated fractions.

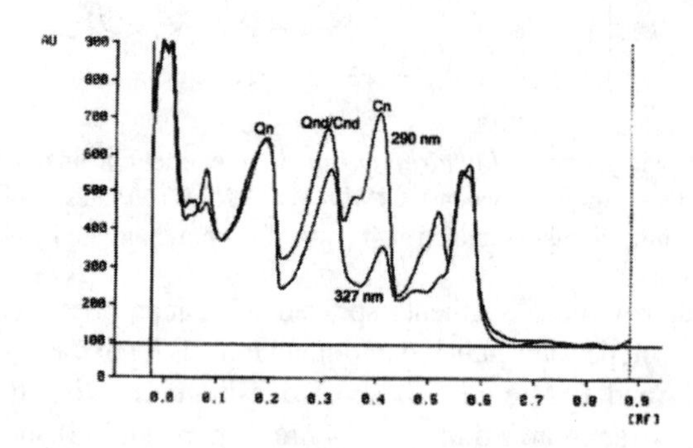

FIGURE 11.25 Assay of alkaloids in *Chinae* extract. System: silica/T + $CHCl_3$ + Et_2O + DEA (40:15:35:10). Plate scanned at 327 and 290 nm. Abbreviations: Qn — quinine, Qnd — quinidine, Cnd — cinchonidine, Cn — cinchonine. (For details see Mroczek, T. and Gowniak, K., *J. Planar Chromatogr.*, 13, 457–462, 2000.)

resolution. This might be achieved by increasing the efficiency of separation. Besides, for the analysis of such complex mixtures, it is advantageous to achieve a large peak capacity. Improvement of separation efficiency and resolution can be obtained by use of special modes of development such as gradient elution, multiple development, two-dimensional development (2-D), incremental multiple development, and gradient multiple development. All these methods influence separation efficiency and peak capacity owing to the reconcentration effect.

11.7.1 Gradient Elution

The problem of the separation of samples containing components of widely different polarities is difficult because of general elution. This can be solved by use of gradient elution. As has been observed, in TLC separation of plant extracts, gradient elution markedly improves the separation of spots owing to stronger displacement effects under conditions of numerous adsorption–desorption processes. In general, the increase of the mobile phase passing through the partly separated starting zone causes the consecutive sample components to reach optimal ranges, in order to increase R_F values. Because the lower edge of the zone is overtaken earlier than the upper edge by the mobile phase front of increasing eluent strength, the zone becomes narrower than the starting band. Figure 11.26a and Figure 11.26b show the results of separating the quaternary alkaloid fraction by stepwise gradient elution with

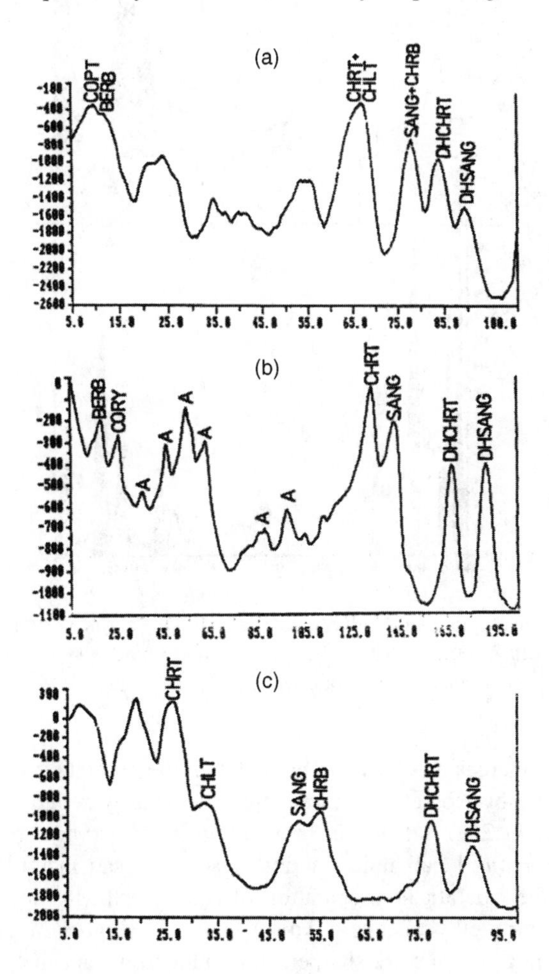

FIGURE 11.26 Densitograms obtained from PLC of quaternary alkaloids from *Chelidonium majus*; (a) silica layer, gradient elution with T + AcOEt + MeOH, 1° 70:20:10, 2° 70:15:15, 3° EtOH + $CHCl_3$ + AcOH, 57:30:3; (b) alumina layer, gradient elution with CH_2Cl_2 + MeOH, 1° 99.5:-0.5, 2° 99:1, 3° 98:2, 4° 95:5; (c) silica layer PMD with T + AcOEt + MeOH, 1° 70:15:15, 2° 70:20:10, 3° 70:25:5, 4° 70:30:0. Abbreviations: CHRB — chelirubine, SANG — sanguinarine, CHLT — chelilutine, CHRT — chelerythiyne, BERB — berberine, COPT — coptisine, DHSANG — dihydrosanguinarine, DHCHRT — dihydrochelerithrine, A — unidentified alkaloid. (For details, see Waksmundzka-Hajnos, M., Gadzikowska, M., and Gołkiewicz, W., *J. Planar Chromatogr.*, 13, 205–209, 2000.)

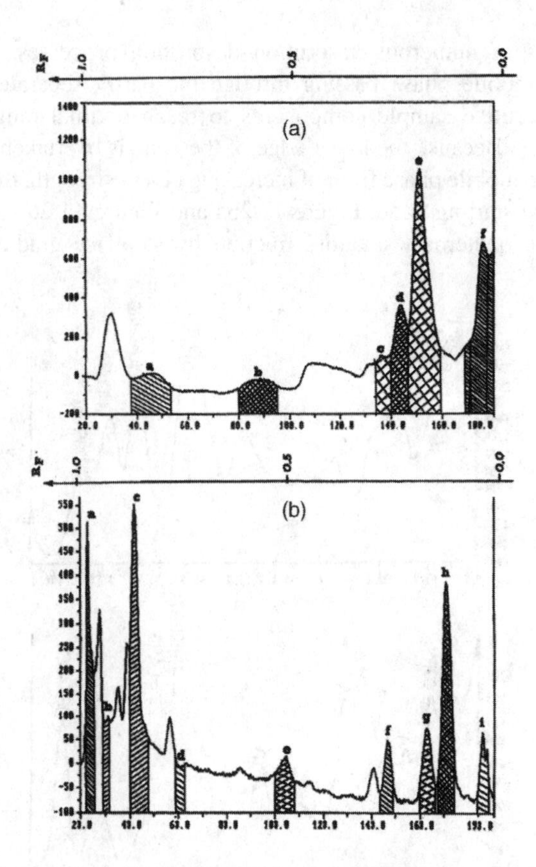

FIGURE 11.27 Densitogram of Azulan extract at 410 nm chromatographed on silica; (a) isocratic elution with AcOEt + $CHCl_3$ (1:5); (b) stepwise gradient of 10 to 40% AcOEt in $CHCl_3$. (For details, see Matysik, G., Soczewiński, E., and Polak, B., *Chromatographia*, 39, 497–504, 1994.)

mobile phases of increasing eluent strength [116]. The separation is distinctly better than that obtained by isocratic development; eight partly separated bands can be observed (Figure 11.26a). The use of basic alumina as adsorbent for the separation of *Chelidonium majus* L. alkaloid fractions also gives satisfactory results (Figure 11.26b). From the isocratically separation of Azulan extract and with a five-step gradient (chloroform–ethyl acetate mixture), it is seen (Figure 11.27) that for stepwise development the peaks are sharper, more numerous, and better spaced along the chromatogram [117]. Figure 11.26c depicts the result of reversed-gradient elution of the quaternary alkaloid mixture. The use of a reversed gradient, e.g., manual multiple development (i.e., manual operation of AMD) also gives satisfactory results (see Figure 11.26c) [116].

11.7.2 Unidimensional Multiple Development (UMD)

Unidimensional multiple development consists in repeated developments of the same plate with the solvent for the same distance and in the same direction. After each development, the plates are dried in air. Each consecutive development results in band reconcentration and thus increases the efficiency of separation. It should be emphasized that multiple development is advantageous, especially for the separation of complex mixtures and overlapped bands.

Figure 11.28 shows densitograms from the separation of *Chelidonium majus* quaternary alkaloids [114]. It is clearly apparent that multiple development results in better separation of the component bands. Threefold development results in separation or partial separation of eight alkaloid bands (Figure 11.28b), whereas after one run, only four zones — chelirubine, sanguinarine, chelilutine, and chelerithrine

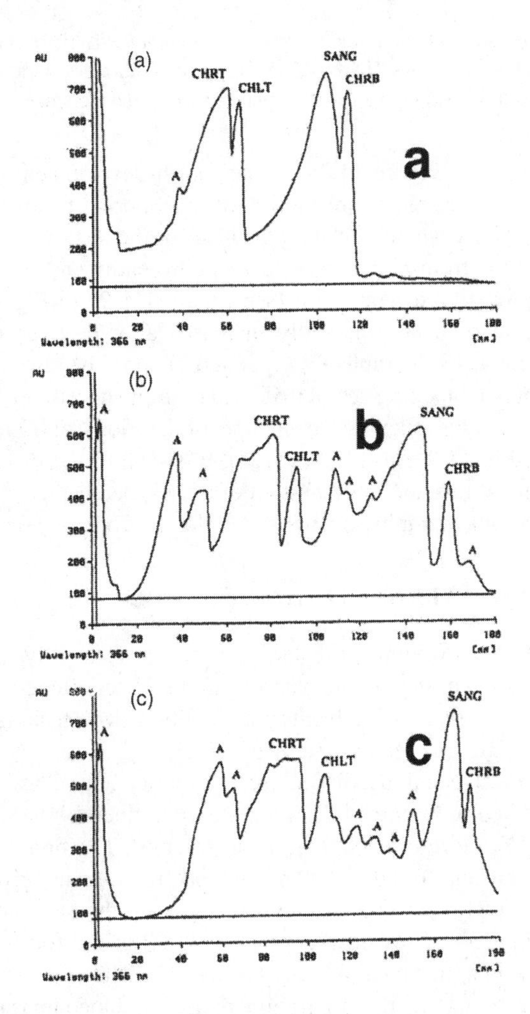

FIGURE 11.28 Densitograms illustrating the effect of the number of developments in UMD. System: silica/T + AcOEt + MeOH (83:15:2). Application of 0.2 ml solution containing 2 mg of extract of *Chelidonium majus*; (a) after single development; (b) after threefold development; (c) after fourfold development. For abbreviations, see Figure 11.16. (For details, see Waksmundzka-Hajnos, M., Gadzikowska, M., and Hajnos, M., *J. Planar Chromatogr.*, 15, 289–293, 2002.)

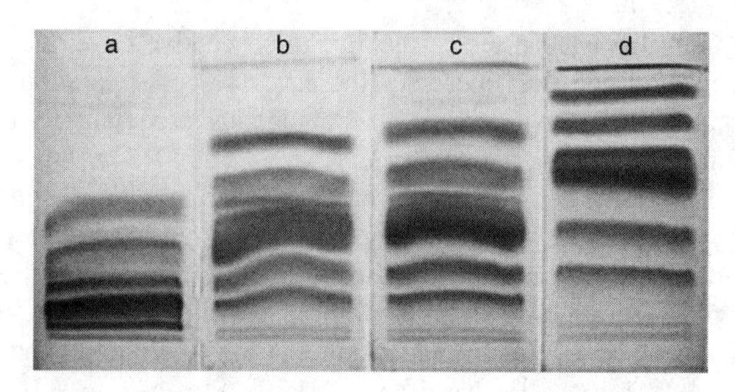

Figure 11.29 Video scan of alkaloids from *Fumaria officinalis* herb chromatographed in system: silica/PrOH + AcOH + CH_2Cl_2 (4:1:5); (a) plate single developed; (b) plate double developed; (c) plate threefold developed; (d) plate fourfold developed.

— are partly separated (Figure 11.28a). The fourth development or more did not result in a marked increase in the resolution of neighboring bands (Figure 11.28c).

Similar conclusions can be drawn on the basis of the observation of the alkaloid fraction densitograms from *Fumaria officinalis* chromatograms scanned after one-, two-, three-, and fourfold development (see Figure 11.29a to Figure 11.29d). It is seen that resolution of bands markedly increases after each development. After fourfold development, five completely separated alkaloid bands can be isolated. A ternary mobile phase containing acetonitrile and multiple development was also used for the separation of the alkaloids from one of fractions of *Fumaria officinalis* alkaloids (see Figures 11.30a to 11.30c) [105]. It is seen from the densitograms that multiple development resulted in separation of overlapped bands and could be used in a large-scale chromatography, owing to the effect of band compression.

11.7.3 Incremental Multiple Development (IMD)

In this technique, the development distance is increased linearly in 10- or 20-mm steps (with evaporation of the mobile phase from the plates after each step) by using the same solvent or a series of solvents for modified IMD technique. In gradient IMD, the eluent strength is reduced stepwise.

By use of the IMD technique, difficult separations can be achieved. For example, video scans presented in Figure 11.31 illustrate the influence of IMD technique on the separation of *Fumaria officinalis* alkaloid fraction. The first and second steps of separation on a distance of 20 and 40 mm, respectively, do not give the separation of bands (Figure 11.31a and Figure 11.31b). However, the next steps improve the separation (see Figure 11.31c and Figure 11.31d). After the fourth step, the separation of six partly separated alkaloid bands is possible. The set of next densitograms (see Figure 11.32) compare the separation of the furanocoumarin fraction from *Heracleum sphondylium* fruit by use of gradient elution (Figure 11.32a) and the IMD technique (Figure 11.32b) [118]. It is clearly seen that better results were obtained using incremental multiple development with the stepwise reduction of mobile phase strength.

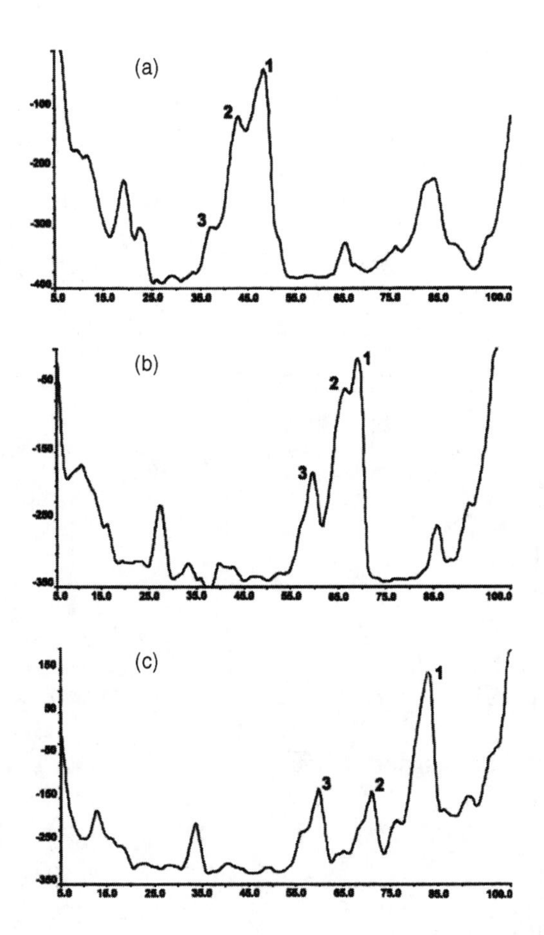

Figure 11.30 Densitograms illustrating multiple development of *Fumaria officinalis* fraction in system: silica/2% MeCN and 1% ammonia in PrOH + CH_2Cl_2 (1:9); (a) first, (b) second, (c) third stage of development. (For details, see Jóźwiak, G., Wawrzynowicz, T., and Waksmundzka-Hajnos, M., *J. Planar Chromatogr.*, 13, 447–451, 2000.)

11.7.4 Two-Dimensional TLC in Preparative Scale

Sandwich chambers can be used for 2-D separation on a large scale, rather than the traditional 2-D technique. The possibility of using solvents of different selectivities improves the separation.

Figure 11.33 shows and example of preparative 2-D separation of furocoumarins from *Archangelica officinalis* fruit extract [1]. The extract was introduced in the right corner of the Florisil layer by use of a special distributor enabling introduction from the edge of the layer, and it was developed with the eluent consisting of diisopropyl ether, dichloromethane, and *n*-heptane. Afterward, the layer was dried on air to complete evaporation of the first eluent and developed in the perpendicular direction with the eluent of different selectivity consisting of 15% ethyl acetate in benzene. It enabled the isolation of pure imperatorin, bergapten, and xanthotoxin.

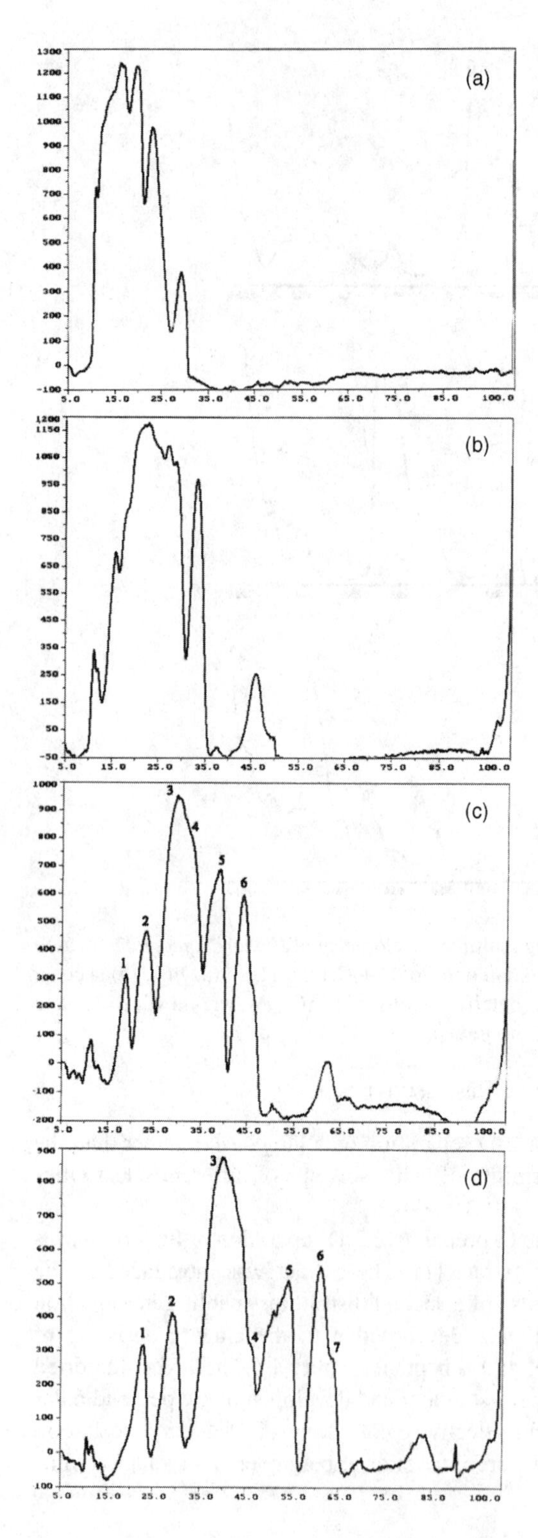

Figure 11.31 Densitograms of PL chromatograms of *Fumaria officinalis* fraction developed in system: silica/PrOH + AcOH + CH_2Cl_2 (4:1:5) by IMD method with 20-mm steps; (a) after first step, (b) after second step, (c) after third step, and (d) after fourth step.

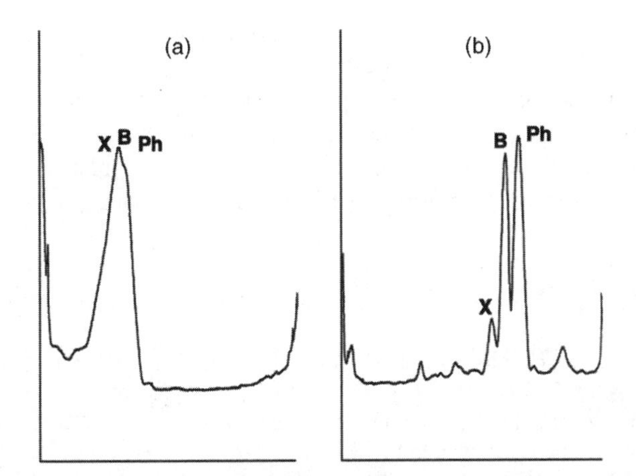

FIGURE 11.32 Densitogram illustrating various techniques of development of *Heracleum sphondylium* fruit extract on silica with AcOEt + H (1:9); (a) by isocratic method, (b) incremental multiple development with 10-mm steps. For abbreviations, see Figure 11.1. (For details, see Wawrzynowicz, T., Czapińska, K., and Markowski, W., *J. Planar Chromatogr.*, 11, 388–393, 1998.)

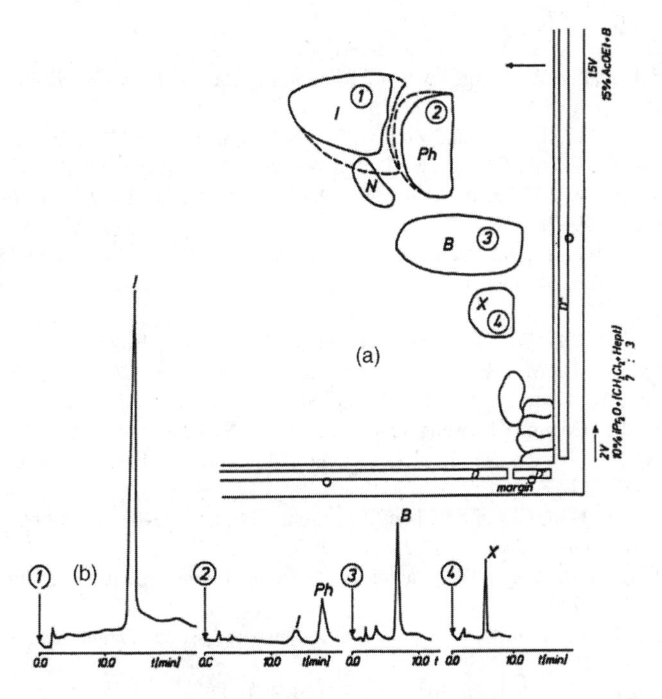

FIGURE 11.33 (a) Schematic representation of 2-D PLC separation of *Archangelica officinalis* fruit extract, system: Florisil/10% iPr_2O in CH_2Cl_2 + H (7:3) — first direction eluent, 15% AcOEt + B — second direction eluent; (b) analytical HPLC of isolated fractions. System: C18/MeOH + water (6:4). For abbreviations, see Figure 11.1. (For details see Waksmundzka-Hajnos, M. and Wawrzynowicz, T., *J. Planar Chromatogr.*, 5, 169–174, 1992.)

Preparative planar chromatography is a very important step in the complicated procedures of isolation of group of compounds or pure substances from complex matrices. The method gives additional possibilities of using various adsorbents and eluent systems to achieve complete separation of structural analogs. The method also enables combining the various methods of sample application, plate development, and derivatization to achieve satisfactory separation of isolated plant extracts' components.

REFERENCES

1. Waksmundzka-Hajnos, M. and Wawrzynowicz, T., *J. Planar Chromatogr.,* 5, 169–174, 1992.
2. Zekovic, Z., Pekic, B., Lepojevic, Z., and Petrovic, L., *Chromatographia*, 39, 587–590, 1994.
3. Głowniak, K., Wawrzynowicz, T., Hajnos, M., and Mroczek, T., *J. Planar Chromatogr.*, 12, 328–335, 1999.
4. Głowniak, K. and Mroczek, T., *J. Liq. Chromatogr.,* 22, 2483–2502, 1999.
5. Hajnos, M., Głowniak, K., Waksmundzka-Hajnos, M., and Kogut, P., *J. Planar Chromatogr.*, 14, 119–125, 2001.
6. Głowniak, K., Soczewiński, E., and Wawrzynowicz, T., *Chem. Anal. (Warsaw)*, 32, 797–811, 1987.
7. Gołkiewicz, W. and Gadzikowska, M., *Chromatographia* 50, 52–56, 1999.
8. Zwickenpflug, W., Mieger, M., and Richter, E., *J. Agric. Food Chem.*, 46, 2703–2706, 1998.
9. Ma, X.Q., Jiang, S.H., and Zhu, D.Y., *Biochem Syst. Ecol.*, 26, 723–728, 1998.
10. Chia, J.C., Chang, F.R., Li, C.M., and Wu, Y.C., *Phytochemistry*, 48, 367–369, 1998.
11. Toro, M.J.V., Müller, A.H., Arruda, M.S.P., and Arruda, A.C., *Phytochemistry*, 45, 851–853, 1997.
12. Wei, X., Wei, B., and Zhang, J., *J. Chin. Herb. Med.*, 26, 344–346, 1995.
13. Kanyinda, B., Vanhaelen-Fastre, R., and Vanhaelen, M., *J. Nat. Prod.*, 58, 1587–1589, 1995.
14. Frederlich, M., De Pauw, M.C., Llabres, G., Tits, M., Hayette, M.P., Brandt, V., Penelle, J., De Mol, P., and Angenot, L., *Planta Med.,* 66, 262–269, 2000.
15. Farsam, H., Yassa, N., Sarkhail, P., and Shafiee, A., *Planta Med.*, 66, 389–391, 2000.
16. Chen, J.J., Chang, Y.L., Teng, C.M., and Chen, I.S., *Planta Med.*, 66, 251–256, 2000.
17. Üniver, N., Noyan, S., Götzler, T., Önür, M.A., Gözler, B., and Hesse, M., *Planta Med.*, 65, 347–350, 1999.
18. Shen, J.J., Duh, C.Y., and Chen, I.S., *Planta Med.*, 65, 643–647, 1999.
19. Sari, A., *Planta Med.*, 65, 492–495, 1999.
20. Phuong, N.M., Sung, T.V., Parzel, A., Schmidt, J., and Adam, G., *Planta Med.*, 65, 761–762, 1999.
21. Lee, K.K., Zhou, B.N., Kingston, D.G.I., Vaisberg, A.J., and Hammond, G.B., *Planta Med.*, 65, 759–760, 1999.
22. Drandarov, K., *Phytochemistry*, 44, 971–973, 1997.
23. Yankep, E., Mbafor, J.T., Fomum, Z.T., Steinbeck, C., and Messanga, B.B., *Phytochemistry*, 56, 363–368, 2001.
24. Huck, C.W., Huber, C.G., and Bonn, G.K., *J. Chromatogr. A.*, 870, 453–462, 2000.
25. Lin, L., He, X., Lindenmaier, M., Nolan, G., Yang, J., Cleary, M., and Qiu, S., *J. Chromatogr. A.*, 876, 87–95, 2000.

26. Guz, N.R., Lorenz, P., and Stermitz, F.R., *Tetrahydron Lett.* 42, 6491–6494, 2001.
27. LeMeu-Olivier, L., Renault, J.H., Thepenier, P., Jacquier, M.J., Zaches-Hanrot, M., and Foucalt, A.P., *J. Liq. Chromatogr.*, 18, 1663–1670, 1995.
28. Wawrzynowicz, T. and Waksmundzka-Hajnos M., *J. Liq. Chromatogr.*, 13, 3925–3940, 1990.
29. Kamara, B.I., Brandt, E.V., Ferreira, D., and Joubert, E., *J. Agric. Food Chem.*, 51, 3874–3879, 2003.
30. Kulevanova, S., Stefova, M., Stevkov, G., and Stafilov, T., *J. Liq. Chromatogr.*, 24, 589–600, 2001.
31. Zheng, S., Kang, S., Shen, Y., and Sun, Z., *Planta Med.*, 65, 173–175, 1999.
32. Chung H.S., Chang, L.C., Lee, S.K., Shamon, L.A., van Breemen, R.B., Mehta, R.G., Fransworth, N.R., Pezzuto, J.M., and Kinghorn, A.G., *J. Agric. Food Chem.*, 47, 36–41, 1999.
33. Ngadjui, B.T., Abegaz, B.M., Dongo, E., Tambone, H., and Fogue, K., *Phytochemistry,* 48, 349–354, 1998.
34. Gamez, E.J.C., Luyengi, L., Fong, H.H.S., Pezzuto, J.M., and Kinghorn, A.D., *J. Nat. Prod.*, 61, 706–708, 1998.
35. Fukai, T., Pei, Y.H., Nomura, T., Xu, C.Q., Wu, L.J., and Chen, Y.J., *Phytochemistry,* 47, 273–280, 1998.
36. Cao, S.G., Sim, K.Y., and Goh, S.H., *J. Nat. Prod.*, 60, 1245–1250, 1997.
37. Takeoa, G.R. and Dao, L.T., *J. Agric. Food Chem.* 50, 496–501, 2002.
38. Kweon, M.H., Hwang, H.J., and Sung, H.C., *J. Agric. Food Chem.* 49, 4646–4655, 2001.
39. Tsai, I.L., Lin, W.Y., Teng, C.M., Ishikawa, T., Doong, S.L., Huang, M.W., Cen, Y.C., and Chen, I.S., *Planta Med.*, 66, 618–123, 2000.
40. Seo, E.K., Chai, H.B., Changwedera, T.E., Fransworth, N.R., Corbell, G.A., Pezzuto, J.M., and Kinghorn, A.D., *Planta Med.*, 66, 182–184, 2000.
41. Quadri-Spinelli, T., Heilmann, J., Rali, T., and Sticher, O., *Planta Med.*, 66, 728–733, 2000.
42. Głowniak, K., Bartnik, M., Mroczek, T., Zabża, A., and Wierzejska, A., *J. Planar Chromatogr.*, 15, 94–100, 2002.
43. Sharma, O.P., Sharma, S., and Dawra, R.K., *J. Chromatogr. A*, 786, 181–184, 1997.
44. Tangiano, C., Della-Greca, M., Fiorentino, A., Isadori, M., and Monaco, P., *Phytochemistry,* 56, 469–473, 2001.
45. Gu, J.Q., Park, E.J., Luygeni, L., Hawthorne, M.E., and Mehta, R.G., *Phytochemistry,* 58, 121–127, 2001.
46. Nawaz, R.H., Malik, A., and Ali, M.S., *Phytochemistry,* 56, 99–102, 2001.
47. Hohmann, J., Evanics, F., Dombi, G., Molnár, J., and Szabó, P., *Tetrahedron*, 57, 211–215, 2001.
48. Nagy, G., Yang, M.H., Günther, G., Mathi, I., Blunden, G., and Crabb, T.A., *Biochem. Syst. Ecol.*, 26, 797–799, 1998.
49. Jakupovic, J., Jeske, F., Morgenstern, T., Marco, J.A., and Berendsohn, W., *Phytochemistry,* 47, 1583–1600, 1998.
50. El Babili, F., Moulis, C., Bon, M., Respaud, M.J., and Fouraste, I., *Phytochemistry,* 48, 165–169, 1998.
51. Ryu, S.Y., Lee, C.O., and Choi, S.U., *Planta Med.*, 63, 339–342, 1997.
52. Olafsson, K., Jaroszewski, J.W., Smitt, U.W., and Nyman, U., *Planta Med.*, 63, 352–355, 1997.
53. Kawabata, J., Fukushi, E., and Mizutani, J., *Phytochemistry,* 47, 231–235, 1998.

54. Souto-Baciller, F.A., Mielendez, P.A., and Romero-Ramsey, L., *Phytochemistry*, 44, 1077–1086, 1997.
55. Trute, A., Gross, J., Mutschler, E., and Nahrstedt, A., *Planta Med.*, 63, 125–129, 1997.
56. Ukiya, M., Akihisa, T., Yasukawa, K., Yoshimasa, K., Yumiko, K., Kazuo, K., Tamotsu, N., and Michio, T., *J. Agric. Food Chem.*, 49, 3187–3197, 2001.
57. Zheng, S., Wang, J., Lu, J., Shen, T., Sun, L., and Shen, X., *Planta Med.*, 66, 487–489, 2000.
58. Tanaka, R., Kinouchi, Y., Tokuda, H., Nishino, H., and Matsunaga, S., *Planta Med.*, 66, 630–634, 2000.
59. Tu, Y.Q., *Phytochemistry*, 31, 2155–2157, 1992.
60. Ohtsu, H., Tanaka, R., Matsunga, S., Tokuda, H., and Nishino, H., *Planta Med.*, 65, 664–667, 1999.
61. Fujiwara, A., Mori, T., Iida, A., Veda, S., Hano, Y., and Nishino, H., *J. Nat. Prod.*, 61, 629–632, 1998.
62. Chen, C.J., Chang, F.R., Chin, H.F., Wu, M.J., and Wu, J.C., *Phytochemistry*, 51, 429–433, 1999.
63. Avato, P., Vitali, C., Mongelli, P., and Tava, A., *Planta Med.* 63, 503–507, 1997.
64. He, X., Lin, L., Bernart, M.W., and Lian, L., *J. Chromatogr. A*, 815, 205–211,1998.
65. He, X., Bernart, M.W., Lian, L., and Lin, L., *J. Chromatogr. A*, 796, 327–334,1998.
66. Machida, K., Trifonov, L.S., Ayer, W.A., Lu, Z.X., and Laroche, A., *Phytochemistry*, 58, 173–177, 2001.
67. Tanahashi, T., Takenaka, Y., Nagakura, N., and Hamada, N., *Phytochemistry*, 58, 1129–1134, 2001.
68. Macias, M., Gamboa, A., Ulloa, M., Toscano, R.A., and Mata, R., *Phytochemistry*, 58, 751–758, 2001.
69. Hajnos, M.L., Waksmundzka-Hajnos, M., Głowniak, K., and Piasecka, S., *Chromatographia*, 56, S91–S94, 2002.
70. Bulatovic, V., Vajs, V., Matura, S., and Milosauljevic, S., *J. Nat. Prod.*, 60, 1222–1228, 1997.
71. Lin, L.Z., He, X.G., Lian, L.Z., King, W., and Eliott, J., *J. Chromatogr. A*, 810, 71–79, 1997.
72. Acevedo, L.L., Marinez E., Castaneda, P. et al., *Planta Med.*, 66, 257–261, 2000.
73. Bezabih, M. and Abegaz, B.M., *Phytochemistry*, 49, 1071–1073, 1998.
74. Gao, K., Wang, W.S., and Jia, Z.J., *Phytochemistry*, 47, 269–272, 1998.
75. Kawamura, F., Kawai, S., and Ohashi, H., *Phytochemistry*, 44, 1351–1357, 1997.
76. Tsai, I.L., Chen, J.H., Duh, C.Y., and Chen, I.S., *Planta Med.*, 66, 403–407, 2000.
77. Pradahan, P. and Banerji, A., *Phytochem. Anal.*, 9. 71–74, 1998.
78. Balde, A.M., Apers, S., De Bruyne, T.E., Van Den Heuvel, H., Claeys, M., Vlietnick, A.J., and Pieters, L.A.C., *Planta Med.*, 86, 67–69, 2000.
79. Hu, J. and Feng, X., *Planta Med.* 66, 662–664, 2000.
80. Budzianowski, J., *Planta Med.,* 66, 667–669, 2000.
81. Shen, J.C. and Hsieh, P.W., *J. Nat. Prod.*, 60, 453–457, 1990.
82. Chen, K.S., Chang, Y.L., Teng, C.M., Chen, C.F., and Wu, Y.C., *Planta Med.*, 66, 80–91, 2000.
83. Agrawal, S., Banerjee, S., and Chattopadahay, S.X., *Planta Med.*, 66, 773–775, 2000.
84. Jaysinghe, Botch, N., Aoki, T. and Wada, S., *J. Agric. Food Chem.,* 51, 4442–4449, 2003.
85. Wu, D., Nair, M.G., and DeWitt, D.L., *J. Agric. Food Chem.,* 50, 701–705, 2002.
86. Yenjai, C., Sripontan, S., Sriprajun, P., Kittakoop, P., Jintasirikul, A., Tanticharoen, M., and Thebtaranonth, Y., *Planta Med.*, 66, 277–279, 2000.

87. Xiong, Q, Fan, W., Tezuka, Y., Adnyana, I.K., Stampoulis, P., Hattori, M., Namba, T., and Kadota, S., *Planta Med.,* 66, 127–133, 2000.
88. Hu, K., Kobayashi, H., Dong, A., Iwasaki, S., Shigeo, Y., and Yao, X., *Planta Med.*, 66, 564–567, 2000.
89. Garcia-Argaez, A.N., Apan, T.O.R., Delgado, H.P., Velazquez, G., and Martinez-Vazquez, M., *Planta Med.*, 66, 279–281, 2000.
90. Chansakaow, S., Ishikawa, T., Sekiene, K., Okada, M., Hihuchi, Y., Kudo, M., and Chaichantipyuth, A., *Planta Med.*, 66, 572–575, 2000.
91. Kwon, B.M., Jung, H.J., Lim, J.H., Kim, Y.S., Kim, M.K., Kim, Y.K, Bok, S.H., Bae, H.A., and Lee, I.R., *Planta Med.*, 65, 74–76, 1999.
92. Kim, S.U., Hwang, E.I., Nam, J.Y., Son, K.M., Bok, S.H., Kim, H.E., Kwon, B.M., *Planta Med.*, 65, 97–98, 1999.
93. Hohmann, J., Zupko, I., Redei, D., Csanyi, M., Falkay, G., Mathe, I., and Janicsak, G., *Planta Med.*, 65, 576–578, 1999.
94. Fushija, S., Kishi, Y., Hattori, K., Batkhuu, J., Takano, F., Singab, A.N.B., and Okuyama, T., *Planta Med.,* 65, 404–407, 1999.
95. Fall, A.B., Vanhaelen-Fastre, R., Vanhaelen, M., Lo, I., Toppet, M., Ferster, A., and Fondu, P., *Planta Med.*, 65, 209–212, 1999.
96. Chen, T., Li, J., Cao, J., Xu, Q., Komatsu, K., and Namba, T., *Planta Med.*, 65, 56–59, 1999.
97. Murakami, A., Gao, G., Kyung Kim, O.E., Omura, M., Yano, M., Ito, C., Furakawa, H., Jiwajinda, S., Koshimizu, K., and Ohigashi, H., *J. Agric. Food Chem.,* 47, 333–339, 1999.
98. Paulo, A., Gomes, E.T., Steele, J., Warhurst, D.C., and Houghton, P.J., *Planta Med.*, 66, 30–34, 2000.
99. Cichewicz, R.H. and Nair, M.G., *J. Agric. Food Chem.,* 50, 87–91, 2002.
100. Likhitwitayawuid, K., Dej-adisai, S., Jongbunprasert, V., and Krungkrai, J., *Planta Med.*, 65, 754–756, 1999.
101. Nyiredy, Sz., Ed., Possibility of preparative planar chromatography, in *Planar Chromatography: A Retrospective View for the Third Millenium*, Springer Scientific, Budapest, 2001, pp. 386–409.
102. Nyiredy, Sz., Preparative planar chromatography, in *Handbook of Thin Layer Chromatography*, Fried, B. and Sherma, J., Eds., Marcel Dekker, New York, 1995, pp. 307–340.
103. Maleš, Ž. and Medič-Šarič, M., *J. Planar Chromatogr.*, 12, 345–349, 1999.
104. Wawrzynowicz, T., Waksmundzka-Hajnos, M., and Mulak-Banaszek, K., *J. Planar Chromatogr.*, 7, 315–318, 1994.
105. Jóźwiak, G., Wawrzynowicz, T., and Waksmundzka-Hajnos, M., *J. Planar Chromatogr.*, 13, 447–451, 2000.
106. Snyder, L.R., *J. Chromatogr.,* 17, 73–82, 1969.
107. Klemm, L.H., Klopfenstein, C.E., and Kelly, H.P., *J. Chromatogr.*, 23, 428–445, 1966.
108. Waksmundzka-Hajnos, M. and Wawrzynowicz, T., *J. Planar Chromatogr.,* 3, 439–441, 1990.
109. Wawrzynowicz, T., Soczewiński, E., and Czapińska, K., *Chromatographia* 20, 223–277, 1985.
110. Czapińska, K., Markowski, W., and Wawrzynowicz, T., *Chem. Anal. (Warsaw)*, 83, 271–283, 1988.
111. De Stefano, J.J. and Kirkland J.J., *Anal. Chem.*, 47, 1103A–1108A, 1975.
112. Soczewiński, E., Preface, in Waksmundzka-Hajnos, M., *Chromatographic Dissertations*, Medical University, Lublin, Poland, 2, 9–12, 1998.

113. Głowniak, K., Mroczek, T., Zabza, A., and Cierpicki, T., *Pharm. Biol.*, 38, 308–312, 2000.
114. Waksmundzka-Hajnos, M., Gadzikowska, M., and Hajnos, M., *J. Planar Chromatogr.*, 15, 289–293, 2002.
115. Mroczek, T. and Głowniak, K., *J. Planar Chromatogr.*, 13, 457–462, 2000.
116. Waksmundzka-Hajnos, M., Gadzikowska, M., and Gołkiewicz, W., *J. Planar Chromatogr.*, 13, 205–209, 2000.
117. Matysik, G., Soczewiński, E., and Polak, B., *Chromatographia*, 39, 497–504, 1994.
118. Wawrzynowicz, T., Czapińska, K., and Markowski, W., *J. Planar Chromatogr.*, 11, 388–393, 1998.

12 Application of Preparative Layer Chromatography to Lipids

Weerasinghe M. Indrasena

CONTENTS

12.1 DEFINITION AND BACKGROUND INFORMATION ON THE TECHNIQUE

Preparative layer (planar) chromatography (PLC) can be defined as a liquid chromatographic technique in which the solvent–solvent composition migrates through a stationary phase either by capillary action or under the influence of forced flow, with the objective of separating and isolating compounds in interest in 10 to 1000 mg quantities for further studies such as for structure elucidation by IR, UV, MS (combined with GC, HPLC, or TLC), NMR (^{1}H, ^{13}C, ^{32}P, or any other as required), CD, etc., for any other analytical purposes, as well as for the determination of biological activity.

PLC can be further classified as forced-flow planar chromatography (FFPC) and classical PLC (CPLC), according to the solvent migration method. All methods in which the mobile phase migrates by capillary action and forced flow are categorized as FFPC. Chromatographic processes operative in CPLC resemble their respective analytical technologies fairly closely. Commonly used preparative technology in lipid analysis is preparative thin-layer chromatography (PTLC). The chromatographic process of PTLC is also quite similar to its corresponding analytical thin-layer chromatography, which includes both conventional TLC and HPTLC (high-performance thin-layer chromatography), and only very slight modifications may be required for the developing time or solvent system. It is always a common practice to experiment with analytical TLC and HPTLC methods prior to scaling up with PTLC. The main difference is that the sample size, area of adsorbents, developing chambers, and the quantities of mobile phases are proportionately larger. This chapter compiles information only on PTLC, with special emphasis on the analysis of common lipids. PTLC is widely used in lipid analysis, and a few selected applications of this remarkable, universal technology are summarized at the end of this chapter.

12.2 BASIC STRUCTURES AND NOMENCLATURE OF SIMPLE AND COMPLEX LIPIDS

Lipids are one of the most ubiquitous compounds in nature, occurring commonly in both animals and plants. They, as well as substances of lipid origin, have assumed considerable importance in the recent past with the recognition of their vital role in various biological systems of animals and plants. Lipids are part of the major components of the cell membrane structure, and some lipids in the membranes of nerve tissues aid in the transmission of electrical signals from one cell to another.

Lipids have been defined on the basis of their structure and solubility. Lipids are naturally occurring compounds consisting of fatty acids and their derivatives, bile acids, pigments, vitamins, and steroids, as well as terpenoids, which are usually soluble in organic solvents such as benzene, chloroform, ether, and alcohol, etc., with variable solubility depending on the structure of the lipid compound.

Most authors have classified lipids into two major groups such as neutral and polar lipids, whereas some define them as simple and complex lipids. Simple lipids yield a maximum of two primary products per mole after hydrolysis, whereas complex lipids yield three or more primary products.

12.2.1 Simple Lipids

12.2.1.1 Wax Esters

These are the esters of long-chain fatty acids and long-chain fatty alcohols. Usually, only the alcohols are saturated and monoenoic, whereas the fatty acids may be more highly unsaturated, as in most marine waxes. They are found in both animal and plant tissues as well as in some microorganisms. They are quite common in insects. They reserve energy in aquatic animals, aid in echolocation, and play a vital role in waterproofing.

12.2.1.2 Fatty Acids

Fatty acids of plant, animal, and microbial origin usually consist of an even number of carbon atoms in the straight chain. The number of carbon atoms of fatty acids in animals may vary from 2 to 36, whereas some microorganisms may contain 80 or more carbon atoms. Also, fatty acids of animal origin may have one to six cis double bonds, whereas those of higher plants rarely have more than three double bonds. Fatty acids also may be saturated, monounsaturated (monoenoic), or polyunsaturated (polyenoic) in nature. Some fatty acids may consist of branched chains, or they may have an oxygenated or cyclic structure.

12.2.1.3 Acylglycerols

Esters of glycerol and fatty acids are known as acylglycerols. Triacylglycerols are triesters of glycerol; all three fatty acids in the ester may be similar, or only two of them may be similar, or all three of them could be different in structure (Figure 12.1). Almost all commercially important fats and oils of both animal and plant origin mainly consist of these simple lipids. Storage fat in all animal tissues consists entirely of triacylglycerols. Acylglycerols that have only two fatty acids esterified to the (fatty acid diesters) glycerols' backbone are known as diacylglycerols, which may exist in two isomeric forms such as 1,2 (2,3) diacylglycerols and 1,3-diacylglycerols. The two fatty acids in the diesters may be different or similar.

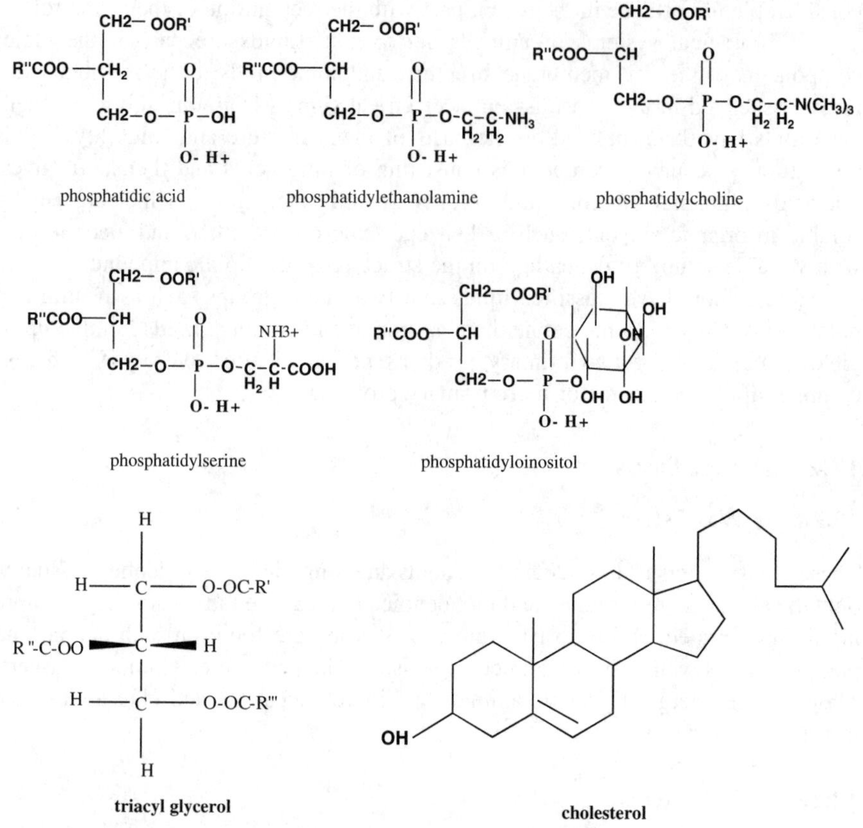

FIGURE 12.1 The structures of some principal lipids.

Monoacylglycerols are the monoesters of glycerol that consist of only one fatty acid attached either to the position 1(3) or 2 of the glycerol backbone. However, because both diacylglycerols and monoacylglycerols are hydrolytic products of triacylglycerols and phospholipids, they may be present, if any, only in negligible levels in animal and plant tissues.

12.2.1.4 Sterols and Their Esters

The most commonly found sterol in animal tissues is cholesterol, which can be found in a free state as well as in an esterified form. The structure of the cholesterol molecule is a cyclopentaphenanthrene nucleus consisting of three fused cyclohexane rings and a terminal cyclopentane ring. Cholesterol is an integral part of the cell membrane structure and plays a vital role in human nutrition, health, and diseases. Plant sterols are mainly phytosterols that include β-sitosterol, stigmasterol, ergosterol, brassicasterol, and campesterol. Some phytosterols have been recently recognized as cholesterol-lowering agents in humans, and also confer other health benefits.

12.2.2 Complex Lipids

12.2.2.1 Phospholipids

Phospholipids are also an integral part of the cell membrane structure and have a wide distribution in the nature. They consist of a phosphate moiety, either attached to a glycerol backbone to give rise to glycerophospholipids or to a spingosyl backbone to give rise to sphingophospholipids. There are different structures of glycerophospholipids based on phosphatidic acid. The most common phospholipids in nature contain nitrogenous bases or polyols and are named accordingly.

Phosphatidylcholine, commonly known as lecithin, is the most commonly occurring in nature and consists of two fatty acid moieties in each molecule. Phosphatidylethanolamine, also known as cephalin, consists of an amine group that can be methylated to form other compounds. This is also one of the abundant phospholipids of animal, plant, and microbial origin. Phosphatidylserine, which has weakly acidic properties and is found in the brain tissues of mammals, is found in small amounts in microorganisms. Recent health claims indicate that phosphatidylserine can be used as a brain food for early Alzheimer's disease patients and for patients with cognitive dysfunctions. Lysophospholipids consist of only one fatty acid moiety attached either to sn-1 or sn-2 position in each molecule, and some of them are quite soluble in water. Lysophosphatidylcholine, lysophosphatidylserine, and lysophosphatidylethanolamine are found in animal tissues in trace amounts, and they are mainly hydrolytic products of phospholipids.

Phosphonolipids (glycerophosphonolipids) found mainly in protozoans and marine invertebrates are lipids with phosphoric acid derivatives esterified to the glycerol backbone.

12.2.2.2 Glycolipids

Glycolipids consist of a sugar moiety attached to the glycerol backbone and are abundantly found in some algae, higher plant tissues, and microorganisms, whereas they may be present, if at all, only as trace constituents in animal tissues.

Monogalactosyldiacylglycerols, digalactosyldiacylglycerol, and sulfoquinovosyldiacylglycerol are glycolipids present in plant and bacterial cells. They consist of 1,2-diacyl-sn-glycerols with a glycosidic linkage at the sn-3 position to a carbohydrate. These glycolipids are soluble in acetone and can be distinguished from phospholipids, which are usually not soluble in acetone.

Glycolipids that comprise sugar attached to sphingosine are called glycosphingolipids. Sphingolipids are characterized by long-chain bases, and the bases are long-chain aliphatic amines with two or three hydroxyl groups with a unique trans double bond in position 4. These complex lipids are found in plasma membranes in animals, as well as in some plant and microbial cells. Common examples of sphingolipids are ceramides or sphingosine, sphingomyelin, glycosylceramides or cerebrosides, fucolipids or globosides, and sulfoglycosylsphingolipids or sulfatides.

Gangliosides are ceremide polyhexosides with sialic acid groups and contain glucose, galactose, and *n*-acetylgalactosamine units. Monosialogangliosides, disialogangliosides, trisialogangliosides, and tetrasialogangliosides are some examples of gangliosides and are found mainly in nervous tissues, as well as in nonneural tissues in small quantities. Gangliosides are soluble in polar organic solvents and water.

12.2.3 Ether Lipids

The chief ether lipids are commonly known as glycerol ethers. Fatty acids in the glycerol ethers are linked to the glycerol molecule by an ether bond, whereas in acylglycerols, fatty acids are linked by ester linkage. Also, as with acylglycerols, these ethers can be mono-, di-, and trialkyl ethers, according to the number of fatty acids linked to the glycerol backbone with an ether bond. Glycerol ethers are found only in trace amounts in animal tissues. However, 1-alkyl-2-acetyl-glycerophosphorylcholine, commonly known as the *platelet-activating factor*, is an important polar glycerol ether that plays a vital role in biological tissues even at trace levels [1].

12.3 PRINCIPLE AND MECHANISM OF CLASSICAL PLC TECHNOLOGY

The lipid sample is applied across the bottom of the plate as a long band, along with spots of standards at either end of the plate. After the plates are developed in an appropriate solvent system and dried, the ends of the plates where the standards were applied are sprayed with the detecting reagent so that the location of the sample band can be demarcated. Then the detected band is scrapped off for further analysis. The entire plate can be sprayed with the detecting reagent if a nondestructive reagent is used. For instance, 2′,7′-dichlorofluoroscein can be sprayed, and fluorescent bands of lipids can be marked under UV light. The fluorescent bands are then scraped for further analyses. Lipid spots can be visualized even after spraying with water. Exposing the plate to iodine vapor can also make the oil spot visible. In case of specific lipid components, destructive detective reagents can be used to spray only either end of the plate to locate the band of interest, and some of these are described in Subsection 12.4.8.

12.4 APPLICATION OF PLC METHODS

12.4.1 Solid Phase or Adsorbents and Selection of the Stationary Phase

Because a wide variety of TLC plates are available commercially, the most appropriate type of plate should be chosen considering several important factors such as the efficiency of the resolution of the lipid species of interest, cost of the plate, the size of the sample to be applied on the plate, and commercial availability of PTLC, depending on the amount of material to be separated. Silica plates are most commonly used for the analysis of lipids. If a large amount of sample material is needed to be separated, plates with a thicker layer of silica should be used. Commercial PTLC plates are available, and thicknesses of 200 to 2000 μm and 250 to 1000 μm are most commonly used in PTLC. Silica-coated glass plates can be of various sizes:

20 × 40, 20 × 20, 20 × 10, 10 × 10, 10 × 5 cm, or micro slides. Plates are prepared by applying a thin layer of sorbent consisting of a mixture of silica gel, gypsum, ammonium sulfate, and a copper compound, preferably copper acetate, to the substrate [2]. Plates can be prepared even in the lab by dipping or spreading the plates in the gel matrix slurry. Dipping is usually for micro slides. A homogeneous, air-bubble-free adsorbent slurry is usually made by mixing the adsorbent with distilled water (50 g/100 ml). Then, the slurry is spread manually or using a mechanical spreader such as a stationary or movable trough type. Plates are dried for about 10 to 20 min and 0.25-mm uniform thickness can be obtained [3].

12.4.1.1 Commercial Types of Adsorbents on Plates (Modified Sorbents and Plates)

12.4.1.1.1 Chemically Modified Sorbents

There are several types of sorbent materials in the market today, other than commonly used silica and alumina. Some of these sorbents are chemically modified to a certain extent to improve the properties of the thin layer for a wide variety of compounds, as well as for better resolution.

12.4.1.1.2 Reversed Phases

Polar surface characteristics of sorbent materials such as silica could be modified to a great extent by bonding suitable materials to the polar structure, introducing more hydrophobic interactions. Such modified sorbents are known as reversed-phase sorbents. The characteristics of the modified sorbent basically depend on the degree of modification, as well as on the characteristics of the aryl or alkyl residue that is chemically bonded to the silanol groups of the silica gel matrix. These residue groups are chemically bonded to the silanol groups of the silica gel matrix. Aliphatic functional residues most popular in reversed-phase TLC technology are RP-2 (methyl), RP-8 (octyl), RP-12 (dodecyl), and RP-18 (octadecyl), whereas phenyl residues are common as aromatic residues. These sorbents are available as precoated layers, as well as in bulk, for both analytical and PLC. Reversed-phase TLC has been used for the analysis of fatty acids [4].

12.4.1.1.3 Hydrophilic Surface

Cyano, amino, and diol residues are the most common functional groups in the hydophilically modified thin layers and not very commonly used for PTLC in lipids.

12.4.2 Mobile Phases and Selection of a Solvent System

There are several solvent systems used on different adsorbents for the separation of a vast range of lipid components in various samples, including biological tissues and oils (Table 12.1 and Table 12.2). The most suitable solvent or solvent system must be carefully chosen for specific lipid groups. Simple lipids such as wax esters, cholesteryl esters, hydrocarbons, ketones, fatty alcohols, sterols, free fatty acids, and acylglycerols can be easily separated using nonpolar solvent systems, leaving polar or complex lipids at the origin. Complex or more polar lipid fractions can be concentrated by developing the plates in 100% acetone while leaving the most polar phospholipids, glycolipids,

TABLE 12.1
Sorbents Used for PTLC in Different Commodities

Lipid	Adsorbent	Reference
Lipid hydroperoxide	Silica gel F	5
Avacado oil	Silica gel	6
Butter oil	Silica gel	7
Egg yolk	Silica argentation	8
Soybean triglycerides	Silica argentation	9
Lipids in human arteries	Silica	10
Lipids of perputial glands of mice	Silica	11
Cholesterol in rat skin	Silica	12
Complex lipids of sea squirts	Prep. Silica gel	13
Lipids in red blood cells	Silica	14
Trans fatty acids in liver of rats	Silica $AgNO_3$ argentation	15
Serum ganglioside	Silica	16
Chloropropanols and their esters in cereal products	Silica	17
Chloropropanols and their esters in baked cereal products	Silica	18
Cerebrosides of hydrocarbon-assimilating yeast	Silica	19
Microwave-heated vegetable oil tocopherol	Silica	20
FFA in spinal-cord-injury site	Silica	21
Ceramides of hydrocarbon-assimilating yeast	Silica	22
Monoglycerides in degraded butter oil	Silica gel with fluorescent indicator (Merck)	7
Phospholipids in purple membrane of *Halobacterium halobium*	Silica gel G	23

Note: H = hexane, D = diethyl ether, F = formic acid, BuOH = butanol.

as well as oxidized lipids, at the origin. However, care must be taken because the least polar phospholipids such as cardiolipin may elute with acetone. Phospholipids and glycolipids can be separated using a polar solvent system. Typical systems widely used consist of chloroform:methanol:water with different ratios.

12.4.3 Preparation of Lipid/Oil Samples for PLC: Importance of the Concentration and Dissolving Solvent

The basic technology for the preparation of sample material is similar in all TLC preparations, irrespective of the origin of the lipid and specific preparation method for a variety of biological samples [43]. The most important factor is the solubility of the sample. The lipid sample must be completely soluble in the dissolving solvent prior to the application and must be free from water. Either toluene or chloroform is commonly used as the solvent to dissolve lipid materials. The dissolving solvent should be nonpolar in nature and volatile at such a concentration that the lipid components in the sample are completely adsorbed throughout the entire thickness of the layer as quickly as possible. Although sample sizes as small as 1 to 10 μl can

TABLE 12.2
Some Selected Solvents Used for the Separation of Different Lipid Classes

Adsorbent	Solvent System	Lipid Components Separated	Reference
Silica gel	CCl_4 or H:DEE (98:2)	Wax esters	24
RP C18 plates	MeOH:AC (1:1)	Cholesteryl esters	25
Silica gel	H:HAc:H_2O (100:5:2.5)	Fatty acids	26
Silica gel	PE:DEE:HAc (between 70:40:2 and 80:20:2)	Neutral lipids	27
	PE:DEE (between 95:5 and 80:20)		
Silica gel	H:DEE:HAc(80:20:2)	Simple lipids	28
Silica gel	C:MeOH: H_2O (65:25:4)	Phospholipids	29
	C:MeOH: H_2O (25:10:1)		
	C:MeOH:HAc: H_2O		30
	AC:HAc: H_2O (100:2:1)		31
Silica gel	C:HAc: H_2O (60:34:6)	Phosphonolipids	32
Aluminum oxide	H:AC (3–4:1)	Glycols and higher alcohols	33
Silica gel	H:DEE (95:5)	Glyceryl ethers	34
Silica gel	H:EA (80:20)	Sterols	35
Aluminum oxide	IO:EA (80:20)	Sterols	35
Silica gel	CH:EA (70:3, 85:15)	Steroids	36
Silica gel	C:MeOH:2.5N NH_4OH (60:40:9)	Gangliosides	37
	C:MeOH: H_2O (65:25:4)	Glycosphingolipids	38
	C:MeOH:0.02% $CaCl_2$ (60:40:9)		39
Silica gel with 2% sodium arsenite	C:MeOH (95:5)	Ceramides	40
Silica gel	PE:MeHep(11:2)	Vitamin A isomers	41
	Chloroform or H:EA (90:20)	Vitamin D_2 and D_3	42

Note: H = hexane, C = chloroform, AC = acetone, W = water, MeOH = methanol, HAc = acetic acid, BuOH = butanol, DEE = diethyl ether, PE = petroleum ether, MeHep = methyl heptane, EA = ethyl acetate, CCl_4 = carbon tetrachloride, IO = isooctane.

be used for analytical TLC, much larger sample volumes such as 250 μl or more can be used in PTLC depending on the thickness of the plate. If the sample size is small, the sample can be streaked on a thin plate. However, it is imperative to have a large sample because further analyses are usually carried out for structure determinations, fatty acid composition, etc., after the respective lipid samples are separated by PTLC. Different methods used for the preparation of lipid samples from various tissue sources have been explained by Fried [43].

12.4.4 Sample Application: Methods of Application, Manual and Instrumental, Sample Load

Usually, the samples completely dissolved in a solvent, preferably chloroform, are applied as narrow streaks or as discrete spots about 1.5 to 2.0 cm from the bottom

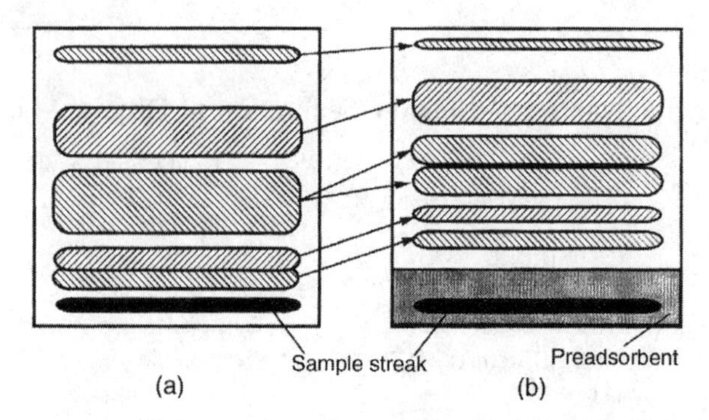

FIGURE 12.2 Diagrammatic representation of the effect of the concentration zone in preparative separations: (a) Precoated layer without concentrating zone, (b) Precoated layer with concentrating zone. (From Nyiredy, S., Preparative layer chromatography, in *Handbook of Thin Layer Chromatography,* 3rd ed., Vol. 89, Sherma, J. and Fried, B., Eds., Marcel Dekker, New York, 2003, pp. 99–133. With permission.)

of the plate either manually or mechanically using a suitable syringe. It is highly desirable to have the streak as narrow and as straight as possible for successful chromatography. It is not difficult to streak a plate correctly by hand with practice, skill, and great care. Also, the use of a Teflon tip on the end of the syringe is preferable to either pasture pipets or syringes with a metal end, because the Teflon-tipped syringes do not mechanically disturb the streaking layer. If precoated preparative layers with a concentrating zone are used, the quality of streaking is not as important because the sample is applied to a practically inert zone. Also, much better resolution can be achieved by plates with these type of layers [44] (Figure 12.2).

Application of the sample as a continuous streak is also possible using commercially available simple instruments (sample applicators), which give a sample zone for preparative separation less than 3 to 4 mm wide. It is also advisable to apply the streak across the plate starting 2 cm from both edges to avoid the edge effect, which may cause the motion of the mobile phase to be faster or slower at the edges than across the center of the plate.

The amount of lipid applied to the plate varies depending on the ease of separation of individual lipid components in the sample. Usually 25 to 50 mg of the neutral lipid sample can be applied to a 20 × 20 cm preparative plate with the silica gel G layer thickness of 0.5 mm, whereas only about 4 mg of phospholipids can be applied on these plates.

12.4.5 Development and Chamber Types

Chromatographic development chambers for analytical purposes are commercially available in several different sizes. The most commonly used ones are rectangular glass tanks with inner dimensions of 21 × 21 × 9 cm, and they can be used to develop two plates simultaneously in the preparative scale. Even bigger tanks are available for much larger plates, for preparative layer chromatography. The width of the chamber should be varied depending on the size and the number of plates to be developed.

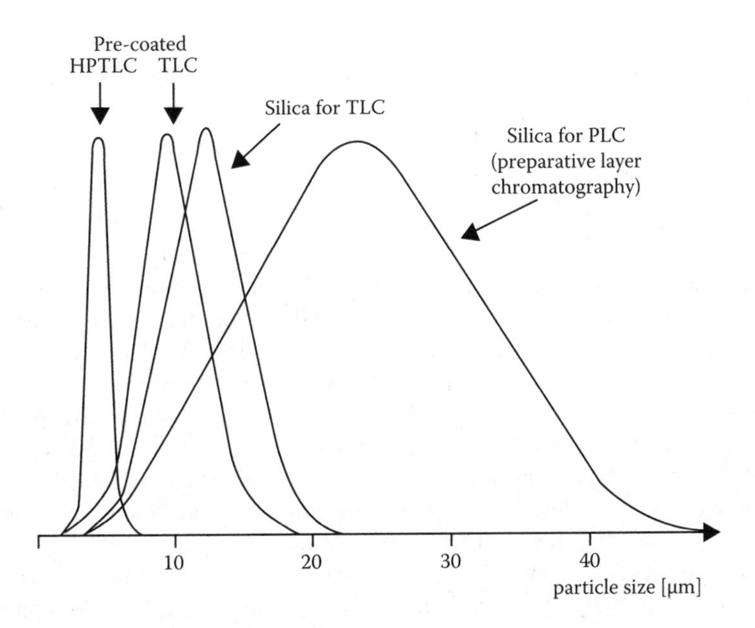

FIGURE 12.3 Particle sizes and particle size distribution of silica stationary phases for planar chromatography. (From Nyiredy, S., Preparative layer chromatography, in *Handbook of Thin Layer Chromatography,* 3rd ed., Vol. 89, Sherma, J. and Fried, B., Eds., Marcel Dekker, New York, 2003, pp. 99–133. With permission.)

12.4.6 Resolution and Factors Affecting the Resolution

There are several factors affecting the overall chromatography of both analytical and preparative TLC, and further details can be found elsewhere [44].

12.4.6.1 Stationary Phase

The quality of the plates and adsorbents, size of the particles, as well as the thickness of the layer, significantly affect the resolution of the separating samples. Thinner layers give better resolution than thicker layers. Plates with a layer thickness of 0.25 mm or less give maximum resolution. Much thicker layers such as 0.5 to 1.0 mm are used in PTLC to take heavy loads of lipids. However, the resolution becomes poor with increasing layer thickness. Also, thinner particles give better separation than thicker particles (Figure 12.3). Although the separation is not comparatively better on thinner plates, thicker plates should be used for PTLC. Precoated TLC plates with a concentrating zone give much higher resolution than conventional plates. Among the plates with a concentrating zone, taper plates (Uniplate-T taper plates) have a much higher resolution.

12.4.6.2 Solvent Systems

All solvents in the solvent system must be completely miscible, and the volatility of individual solvents must be taken into consideration during optimization because some highly volatile solvents may evaporate faster into the chamber, altering the composition of the solvent system, which would affect the elution and resolution of the sample.

12.4.6.3 Chamber Saturation

The developing chamber must be saturated with solvent vapor to get a satisfactory resolution in PTLC. Resolution of lipid classes eluting closely, such as free fatty acids and triacylglycerols as well as phospholipids such as phosphatidylserine and phosphatidylethanolamine, phosphatidic acid, and cardiolipin, may be improved by maintaining the saturated solvent vapor in the developing chamber. Chamber saturation is also quite helpful in the separation of diacylglycerol isomers (1,2- and 1,3-diacylglycerols) that have very close R_f values. This can be done by lining all four sides of the chamber with a thick filter paper thoroughly soaked with the developing solvent system by shaking prior to the development. The prepared chamber should stand at least for 1 to 2 h to enable the internal environment of the developing tank to become saturated with the vapor of the mobile phase. This chamber saturation is quite important because it increases the separation efficiency, maintains constant R_f values — especially in multiple plate development — and the development is much faster.

12.4.6.4 Effect of Developing Time, Temperature, and Humidity of the Environment

Developing time is quite important because overdevelopment may overlap some separated compounds, and some components such as hydrocarbons, wax esters, and cholesterol esters eluting near to the solvent front may be lost. Shorter development time may not elute all compounds as required, resulting in incomplete separation or poor resolution. Therefore, the correct development time must be used after careful optimization. Also, the time may be slightly different from analytical TLC. Developing time may be shorter if the chamber has been saturated for a long enough time. Temperature and humidity of the external environment (e.g., laboratory environment) directly affect the developing environment in the chamber as well. This factor influences solvent vaporization, affecting the reproducibility of R_f values of the separated zones in PTLC. The effect of slight variation in the temperature and humidity of the external environment may be negligible if the development chamber has been saturated with the solvent system.

12.4.7 Isolation of Lipid Component

12.4.7.1 One-Dimensional Development

12.4.7.1.1 Simple Lipids

Simple lipids such as CE, WE, FFA, cholesterol, alcohols, ketones, TG, DG, and MG are usually separated on silica gel plates. Depending on the complexity of the lipid material and the variety of lipid classes present in a single sample, either single- or multiple-solvent systems can be used (Figure 12.4a). Although benzene [45] or

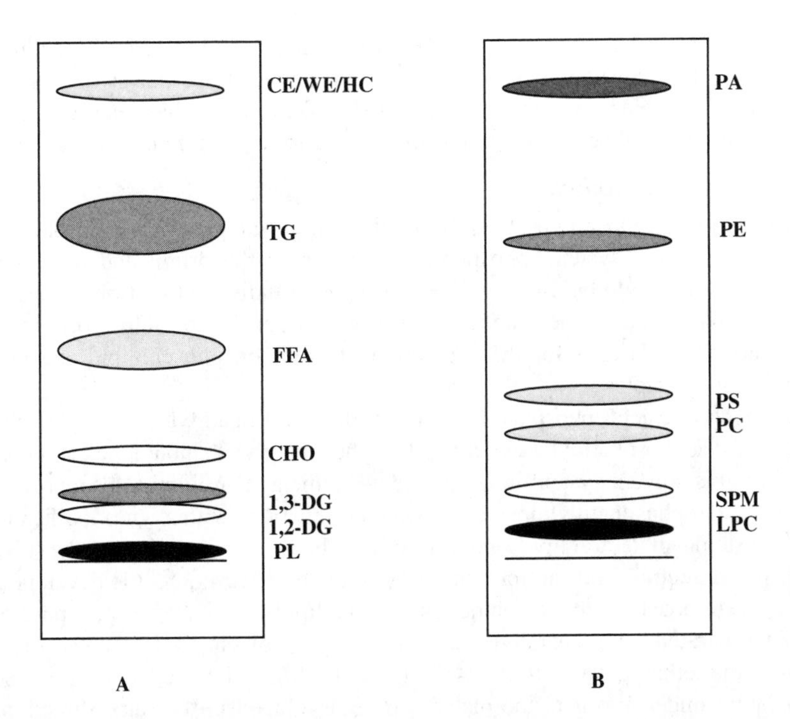

FIGURE 12.4 (A) Diagrammatic representation of the separation of major simple lipid classes on silica gel TLC — solvent system: hexane:diethylether:formic acid (80:20:2) (CE = cholesteryl esters, WE = wax esters, HC = hydrocarbon, FFA = free fatty acids, TG = triacylglycerol, CHO = cholesterol, DG = diacylglycerol, PL = phospholipids and other complex lipids). (B) Diagrammatic representation of the separation of major phospholipids on silica gel TLC — solvent sytem:chloroform:methanol:water (70:30:3) (PA = phosphatidic acid, PE = phosphatidylethanolamine, PS = phosphatidylserine, PC = phosphatidylcholine, SPM = sphingomyelin, LPC = Lysophosphatidylcholine).

chloroform [46] has been used to separate neutral lipids, it is better to use a multiple system for better resolution (Table 12.2)

The most commonly used solvent system for the separation of neutral lipids on silica gel is the Mangold [47] system, which contains petroleum ether:diethyl ether:acetic acid (80:20:2). However, wide modifications of this system such as 90:10:1, 70:30:1, and the ratios in between also have been used. Sterols in the snail *Biomphalaria glabrata* infected with *Echinostoma caproni* have been isolated by PTLC using this solvent system [48], and the isolated sterol fraction has been further analyzed by GC. However, this solvent system does not separate all commonly occurring neutral lipids in some animal tissue samples. Therefore, double development can be used for a successful separation of the entire neutral lipid profile. The Skipski system [49], which involves first development in isopropyl ether:acetic acid (65:4), drying and the second development in pet ether:diethyl ether:acetic acid (90:10:1) in the same direction for unequivocal separation of free fatty acids and glycerides from sterols, has been popular for neutral lipids. Lipid classes in *Echinostoma trivolvis* have been isolated in preparative scale TLC using this dual-solvent system, and individual lipid classes were further analyzed by other analytical methods [50]. In the analysis of marine oils and lipid samples from marine tissues using PTLC, the hexane:diethyl ether:acetic acid (80:20:2) system has been commonly applied with slight modifications to resolve entire neutral lipid classes.

12.4.7.1.2 Phospholipids

This is the most polar group of lipids in natural lipid samples. When developed in a nonpolar solvent system, phospholipids remain at the origin and more polar solvent system should be used to elute and separate individual phospholipids. The most popular system is the Wagner system, which consists of chloroform:methanol:water (65:25:4) [51] for the separation of common phospholipid species in natural tissue samples.

Phosphatidic acid, phosphatidylethanolamine, phosphatidylcholine, phosphatidylserine, lysophosphatidylcholine, and lysophosphatidylethanolamine can be separated by this solvent system in a single development. Cardiolipin, lysophosphatidylserine, and phosphatidylinosytol also can be separated with slight modifications of this system. All neutral lipids move as a single band at or very close to the solvent front. When neutral lipids are present in the sample, however, double development is more preferable for better resolution of phospholipids. For instance, phospholipids can be separated on preparative scale in the first solvent system consisting of chloroform:methanol:water (65:24:4) (Figure 12.4b). After well drying at room temperature under nitrogen, the plates can be developed in the same direction in the second solvent system consisting of hexane:diethyl ether (4:1) [52]. Separated phospholipid bands can then be scraped off from the TLC plates for further analysis by other methods. However, for lipid samples containing relatively large amounts of neutral lipids, the phospholipid fraction can be concentrated by removing all neutral lipids using solid phase extraction with silica gel on preparative scale [1]. Silica gel chromatography has been commonly used to remove neutral lipids for the enrichment of the phospholipid fraction. Then, the phospholipid fraction completely free of neutral lipids can be developed in chloroform:methanol:water (65:24:4) using PTLC, and the separated zones can be used for the determination of structures by NMR, IR, GC-MS, LC-MS-MS, etc., or fatty acid composition of individual phospholipids by GC after transesterification with BF_3-methanol or any other further analytical work.

12.4.7.2 Two-Dimensional Development

12.4.7.2.1 Simple and Phospholipids

Two-dimensional development is commonly used in analytical TLC for the separation, identification, and sometimes, quantification of complex lipid species especially when one-dimensional development does not provide satisfactory separation where some individual lipids coelute in a number of common solvent systems.

Two-dimensional systems are occasionally used to separate phospholipids (Figure 12.5). After the separation of phospholipids using the chloroform:methanol:water (65:25:4) system, the plates can be dried and turned by 90°, followed by the second development in either *n*-butanol:acetic acid:water (60:20:20) or chloroform:acetone:

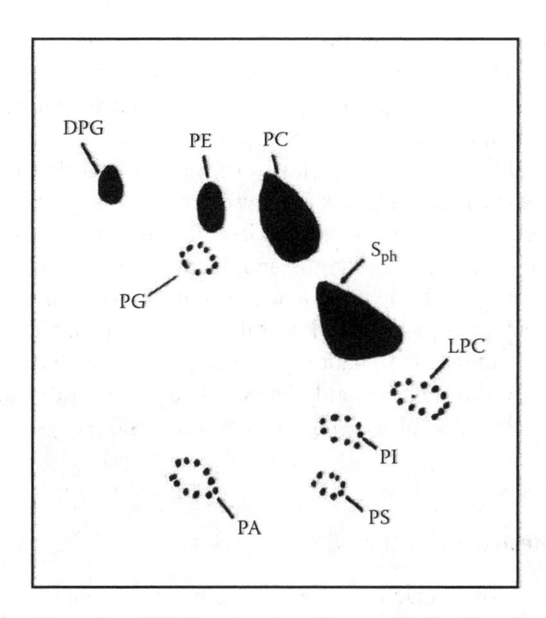

FIGURE 12.5 Two-dimensional TLC separation of serum lipids. First development in chloroform:methanol:ammonia (65:25:5), second development at 90° angle in chloroform:acetone:methanol:acetic acid:water (30:40:10:10:0.5) (DPG = diphophatidylglycerol, PE = phosphatidylethanolamine, PC = phosphatidylcholine, PG = phosphatidylglycerol, Sph = sphingomyelin, LPC = lysophosphatidylcholine, PI = phosphatidylinositol, PS = phosphatidylserine, PA = phosphatidic acid). (From Nyiredy, S., Preparative layer chromatography, in *Handbook of Thin Layer Chromatography,* 3rd ed., Vol. 89, Sherma, J. and Fried, B., Eds., Marcel Dekker, New York, 2003, pp. 99–133. With permission.)

methanol:acetic acid:water (5:2:1:1:0.5) for better resolution of phospholipids [53]. However, Fried and Shapiro [52] reported that larger spots with tailing is common, although higher resolution can be achieved when the two-dimensional systems are used.

Although not very commonly used in the separation of neutral lipids, two-dimensional systems have been used to separate hydrocarbons, steryl esters, methyl esters, and mixed glycerides that move close to each other in one-dimensional systems. Complex neutral lipids of *Biomphalaria glabrata* have been first developed in hexane:diethyl ether (80:20), dried, and the plates have been turned 90°, followed by the second development in hexane:diethyl ether:methanol (70:20:10) for complete separation of sterol and wax esters, triglycerides, free fatty acids, sterols, and monoglycerides [54].

Silica gel plates also have been used for the separation of 16 different eye pigments of *Drosophila melanogaster* using two-dimensional development in non-polar solvent systems [55]. Although not very common, two-dimensional development may be used in preparative scale on thick-layered plates for further analysis.

12.4.7.2.2 Gangliosides and Glycolipids

Gangliosides are associated with various oligosaccharides, and usually their complete separation is not satisfactory using one-dimensional developments on TLC.

HPTLC plates have been used for the separation of gangliosides by developing the plates first in chloroform:methanol:0.2% aqueous $CaCl_2$ (60:35:8), and the second development at 90° turn in *n*-propanol:water:28% aqueous ammonia (75:25:5) in the second direction [56]. Glycolipids are also difficult to separate completely using one-dimensional TLC owing to complexity of oligosaccharides attached to them. Therefore, two-dimensional TLC has been used to separate individual glycolipids. First, the silica gel plates can be developed in chloroform:methanol:7N ammonium hydroxide (65:30:4) in one direction, and then in chloroform:methanol:acetic acid:water (34:50:50:3) in the other direction. Simple lipids, monogalactosyldiacylglycerols, cardiolipin, phosphatidic acid, sterol glycosides, ceramide monohexosides, phosphatidylethanolamine, phosphatidylglycerol, digalactosyldiacylglycerols, sulfoquinovosyldiacylglycerols, phosphatidylcholine, and phosphatidylinositol have been well separated on the same plate using this two-dimensional system [43].

12.4.8 Detection

12.4.8.1 Nondestructive Method

The reagent used for the detection may be specific to a special functional group or specific lipids or may be a nonspecific reagent that makes all lipids visible. The most commonly used reagent that is nonspecific for any lipid group is 0.1% (w/v) 2′,7′-dichlorofuorescein in 95% methanol. This is mainly useful when the plates have been developed in acidic solvents. The lipid spots or bands can be visualized as yellow spots or bands under UV light. After the plates are developed in alkaline solvents, an aqueous solution of Rhodamine 6G (0.01%) can be used, and lipid spots can be seen as pink spots under UV. Because both these methods are nondestructive, they can be effectively used in PTLC so that the separated sample bands can be scraped off and used for further analysis.

12.4.8.2 Other Detection Methods

In analytical TLC, both destructive and nondestructive methods can be used, and some commonly used specific and nonspecific reagents are shown in Table 12.3 and Table 12.4. Visualization by color-forming sprays or charring are the most common. Because charring is a destructive method, it can be used only for analytical TLC and not for PTLC. The intensity of the color spot depends on the degree of unsaturation of the fatty acid attached to the phospholipids. Iodine vapor, molibdate ions, and cupric acetate/H_2SO_4 with or without dichromate give a color product with a fatty acid double bond. For general phospholipids, following reagents can be used:

1. Rosamine 6G, Rhodamine B
2. 2′,7′-Dichlorofluorescein
3. Anilino-8-naphthaline sulfonate
4. 8-Anilinonaphthaline sulfonate (8-ANS)

Monoacylglycerol in butter oil can be detected under UV light or with iodine vapor, and the detected bands extracted with hexane:isopropanol (3:2) can be used

TABLE 12.3
Nonspecific Reagents Used for the Detection of Lipids on TLC Plates or Sheets

Reagent	Procedure	Results
Iodine	Spray as 1% alcoholic solution or place a few crystals in the bottom of a closed tank	Dark brown spots on a pale yellow or tan background in a few minutes
Rhodamine B	Spray with a 0.05% solution in ethanol	Violet spots on a pink background
2′,7′-Dichlorofluorescein	Spray with a 0.2% solution in 96% ethanol	Saturated and unsaturated polar lipids give green spots on purple background
Coomassie Brilliant Blue	Spray with a 0.03% solution of the Coomassie stain in 20% methanol	Blue spots on colorless background
Phosphomolibdic acid	Spray with a 5% solution in ethanol and heat at 100°C for 5 to 10 min	Blue-black spots on yellow background
Antimonytrichloride	Spray with a 10% solution of SbCl in chloroform; heat at 110°C for 1 to 2 min	Various colors on a white background
Potassium dichromate–sulfuric acid	Spray with a 5% solution of potassium dichromate in 40% H_2SO_4; heat at 120 to 180°C for 15,040 min	Black spots on a colorless background
Sulfuric acid	Spray with 50% aqueous H_2SO_4; heat at 120 to 180°C for 15,040 min	Black spots on a colorless background
Cupric acetate–phosphoric acid	Dissolve 3 g of cupric acetate in 100 ml of an 8% aqueous phosphoric acid solution; heat at 130 to 180°C for up to 30 min.	Black spots on a colorless background
Sulfuric acid–methanol	Prepare a 50% solution of H_2SO_4 in methanol; spray and heat plates for several minutes at 100°C.	Brown-black spots on a colorless background

Source: From Fried, B., Lipids, in *Handbook of Thin Layer Chromatography,* 3rd ed., Vol. 89, Sherma, J. and Fried, B. Eds., Marcel Dekker, New York, 2003, pp. 99–133. With permission.

for structure determination using GC and GC-MS [57]. A method has been developed for the separation and detection of hydrophobic target molecules by multiple one-dimensional thin-layer chromatography. After the sequential development of the sample containing hydrophobic target molecules in hexane:diethyl ether:acetic acid (60:40:5), hexane:ether (97:3), and hexane (100%), the dried plates are sprayed with Primulin dye solution and scanned by laser-excited fluorescent detection [7]. Primulin reagent can be prepared using 0.05% solution of the dye with a 5% stock solution diluted with acetone (acetone:water, 8:2 v/v). After spraying this reagent, the plates should be kept in a tank with iodine vapor. Then the lipid spots turn to brown. However, this may not be a suitable reagent for samples with polyunsaturated fatty acids, because they react with iodine. Dehydroisoandrosterone, vitamine-E acetate and cerebral lipids in the pharmaceutical preparation (ormocephalopine)

TABLE 12.4
Specific Chemical Detection Methods Used for Various Lipids Separated by TLC

Compound Class	Reagent	Procedure	Results
Cholesterol and cholesteryl esters	Ferric chloride	Dissolve 50 mg of $FeCl_3.6H_2O$ in 90-ml water along with 5-ml acetic acid and 5-ml H_2SO_4; spray the plate and heat at 90 to 100°C for 2 to 3 min	Cholesterol and cholesteryl esters appear as red-violet spots; cholesterol spot appears before the ester spot
Free fatty acids	2′,7′-Dichlorofluorescein–aluminum chloride–ferric chloride	Spray plates with (1) 0.1% 2′,7′-dichlorofluorescein, (2) 1% $AlCl_3$ in ethanol, (3) 1% aqueous $FeCl_3$ solution; warm the plates to about 45°C briefly between sprays	Free fatty acids impart a rose color
Lipids containing phosphorus	Molybdic oxide–molybdenum "Zinzadze" reagent	Prepare solution 1: 4% solution of molybdic oxide in 70% H_2SO_4; add 0.4-g powdered molybdenum to 100 ml of solution 1; add 200-ml H_2O and filter; final spray consists of 100 ml of preceding and 200 ml H_2O and 240-ml acetic acid	Phospholipids appear as blue spots on a white background within 10 min of spraying
Choline-containing phospholipids (phosphatidylcholine and lysophosphatidylcholine)	Potassium iodide–bismuth subnitrate Dragendorff reagent	Prepare a 40% solution of KI; prepare a 1.7% solution of bismuth subnitrate in 20% acetic acid; mix 5 ml of the first solution with 20 ml of the second solution; add 75 ml of H_2O; spray the plates and warm	Choline-containing lipids appear as orange-red spots in a few minutes
Free amino groups (phosphatidylethanolamine and phosphatidylserine)	Ninhydrin	Prepare a 0.2% solution of ninhydrin in *n*-butanol and add 3-ml acetic acid; spray plates and heat for 3 to 5 min at 100 to 110°C	Lipids with free amino groups appear as red-violet spots

Glycolipids	α-Naphthol–sulfuric acid	Spray a solution of 0.5% of α-naphthol in 100-ml methanol (1:1); allow to air-dry and spray concentrated H_2SO_4 (95:5); heat at 120°C until color is maximal	Cerebrosides, sulfatides, gangliosides, and other glycolipids appear as yellow spots; cholesterol appears as a light red spot
Glycolipids	Orcinol–sulfuric acid	Dissolve 20-mg orcinol in 100 ml of 75% H_2SO_4. Spray plates lightly with the reagent, heat at 100°C for 15 min.	Glycolioids appear as blue-purple spots against a white background
Glycolipids vs. phospholipids	Iodine	Place iodine crystal in a closed tank; place the plates in the tank until color appears	Phospholipids stain distinctly, glycolipids do not
Gangliosides	Resorcinol	Prepare a 2% aqueous solution of resorcinol; add 10 ml of this solution to 80 ml of HCl containing 10.5 ml of 0.1-*M* $CuSO_4$ solution; spray the plates with the reagent and heat for few min at 110°C	Gangliosides appear as a violet-blue color; other glycolipids appear as yellow spots
Sphingolipids	Sodium hydrochlorite–benzidine reagent	Add 5 ml of sodium hypochlorite (Clorox) to 50 ml of acetic acid; prepare the benzidine solution by dissolving 0.5 g of benzidine and one crystal of KI in 50-ml ethanol–H_2O (1:1); spray the plates with the Clorox reagent and let dry; then spray with benzidine reagent	Sphingolipids (ceramids, sphingomyelin, cerebrosides, sulfatides, gangliosides) and lipids with secondary amines produce blue spots on a white backgound (CAUTION: benzidine is a carcinogen)

Source: From Fried, B., Lipids, in *Handbook of Thin Layer Chromatography,* 3rd ed., Vol. 89, Sherma, J. and Fried, B. Eds., Marcel Dekker, New York, 2003, pp. 99–133. With permission.

separated on silica gel G plates were detected by spraying the plates with ammonium molybdate solution [57]. The most commonly used method for the detection of neutral lipids separated on silica gel using Mangold systems is the spraying or dipping of plates in 5% phosphomoliblic acid in absolute ethanol or spraying the plates with 50% aqueous H_2SO_4 acid solution, followed by heating at about 110°C for about 10 min. However, analytical TLC is usually performed using suitable detection methods to arrive at a better understanding of the lipid distribution prior to the PTLC.

12.5 SPECIFIC APPLICATIONS OF PLC IN LIPIDS

12.5.1 Medicine and Pharmacological Research and Drugs

Chloroform–methanol extracts of *Borrelia burgdorferi* were used for the identification of lipids and other related components that could help in the diagnosis of Lyme disease [58]. The provitamin D fraction of skin lipids of rats was purified by PTLC and further analyzed by UV, HPLC, GLC, and GC-MS. MS results indicated that this fraction contained a small amount of cholesterol, lathosterol, and two other unknown sterols in addition to 7-dehydrocholesterol [12]. Two fluorescent lipids extracted from bovine brain white matter were isolated by two-step PTLC using silica gel G plates [59]. PTLC has been used for the separation of sterols, free fatty acids, triacylglycerols, and sterol esters in lipids extracted from the pathogenic fungus *Fusarium culmorum* [60].

In a study on the covalent interaction of chloroacetic acid and acetic acids with cholesterol in rat liver, PTLC was used for the isolation of neutral lipid species after the solid phase extraction of total lipid for the separation of neutral and polar lipids [61]. To study the formation of fatty acid conjugates of 2-chloroethanol and 2-bromoethanol, hepatic ribosomal lipids were extracted from rats after administering these compounds in mineral oil. The fatty acid esters in the lipids were isolated by PTLC for further purification and identification of the conjugates by RP-HPLC [62]. A PTLC procedure has been developed for the separation of brain lipids of experimental rats [63]. Ganglioside-containing spots were scraped and treated with a mixture of resorcinol/hydrochloric acid, followed by the measurement of color intensity to determine the brain-bound sialic acid. Nonoxidative metabolism of ethylene glycol monobutyl ether via acid conjugation in the liver of Fischer 344 male rats was studied using TLC and HPLC methods. The neutral lipid fraction was subjected to PTLC for the separation of esters and was then analyzed by HPLC to determine ethylene glycol monobutyl ether (2-butoxyethanol) [64].

After the extraction of lipids from infectious *Cryptosporidium parvum* oocytes, these lipids were fractionated into neutral lipids, glycolipids, and polar lipids using silicic acid chromatography, and they were further purified by PTLC. After methanolysis, lipid components were identified by GC-MS, NMR, and FAB-MS [65]. The serum gangliosides are of great importance. These gangliosides were isolated from serum by a step-by-step procedure involving the removal of nonpolar lipids from the total lipids using PTLC. After the removal of blood sugar, TLC and HPTLC were used again to separate the ganglioside fraction [16].

12.5.2 Food Industry

PTLC has been widely used in the food industry. Four sterol-containing lipids in finger millet (*Eleusine coracana*) phospholipids were separated by PTLC. Various amounts of fatty acids were found in these sterols [66]. PTLC was used to separate and purify the polar lipid fraction from total lipids extracted from soybean. Separated phospholipids were identified by IR spectra, and the fatty acid composition of individual phospholipids was determined by GC [67]. After the extraction of total lipids in cowpea (*Vigna unguiculata*) with chloroform, PTLC was used for the separation of free sterols, esterified sterols, and sterol glycoside in the total lipid. The fatty acid composition of each lipid class was determined, and various phytosterols, including β-sitosterol, stigmasterol, and campesterol, were found in linoleate ester [68] form.

After the extraction of total lipids from four different genotypes of flax seed (*Linum usitassimum*) differing markedly in their acyl composition, PTLC was used for the isolation of different lipid classes in the neutral lipid fraction [69]. Application of planar chromatographic methods, including PTLC, in the separation of food lipids has been reviewed with 40 references by Olsson [70]. The polar lipid fraction of niger seed (*Guizotia abyssinica* Cass.) collected from different regions of Ethiopia could be separated by PTLC on silica gel [71].

PTLC was used to enrich the polar fraction of deep-fried potato chips and vegetable oils used in industrial frying operations. After PTLC, capillary GC, GC-MS, and NMR were used to quantify sterols and sterol oxides in fried-potato products, as well as the composition of sterols in the oil used for frying [72].

Consumption of food with sterols and their oxides is a health concern. Oxidation products of phytosterol, including epimers of 7-ketositosterol and 7-hydroxycampesterol, 7-ketocampesterol, epimers of 5,6-epoxy-sitosterol, 5,6-epoxycamposterol, 24 α-ethylcholestane-3β,5,6 β-triol, and 24 α-methylcholestane-3β,5,6 β-triol, in deep-fried potato chips in palm oil, sunflower oil, and high oleic sunflower oil were quantitatively analyzed by PTLC followed by GC and GC-MS [73].

12.5.3 Biology and Biochemistry

Ethanol and choline glycerolipids were isolated from calf brain and beef heart lipids by PTLC using silica gel H plates. Pure ethanol amine and choline plasmalogens were obtained with a yield of 80% [74]. Four phospholipid components in the purple membrane (Bacteriorhodopsin) of *Halobacterium halobium* were isolated and identified by PTLC. Separated phospholipids were acid-hydrolyzed and further analyzed by GC. Silica gel G pates were used to fractionate alkylglycerol according to the number of carbon atoms in the aliphatic moiety [24]. Sterol esters, wax esters, free sterols, and polar lipids in dogskin lipids were separated by PTLC. The fatty acid composition of each group was determined by GC.

Glycerylphosphoryl inositolmannosides in total lipids of *Streptomyces griseus* were isolated by PTLC using silica gel H plates. Phosphorus-containing lipids were detected by spraying with Hanes-Ishewood spray, and carbohydrate-containing lipids were detected by spraying with α-naphthol-H_2SO_4 reagent followed by heating the plates at 100°C for 3 to 6 min. Manosides were isolated by PTLC on silica gel H plates, and glycerol, mannose, and inositol were quantified after acid hydrolysis [75]. After the extraction of total lipids from the mandibular fat body, fatty forehead, and the dorsal blubber of Pacific beaked whale (*Berardius bairdi*), lipid classes were separated by PTLC. Fatty acids in these lipid classes analyzed by GC indicated that an unusually low amount of C26-30 was present in the mandibular fat [76].

Conjugated lipids extracted with chloroform and methanol (2:1) after acetone extraction of simple lipids from the viscera of sea squirts (*Pyura michaeni*) were used to separate individual phospholipids by PTLC. Purified phospholipids and their degradation products were identified by TLC, IR, and also by quantitative analysis of phosphorus and fatty acid esters [13]. Unique hexaene hydrocarbon (all-cis-3, 6, 9, 12, 15, 18-henicosahexaene) has been isolated by PTLC on silica gel [77]. Minor galactolipids isolated from the total lipids of calf brain stem by column chromatography were separated into four major groups (monogalacatosyl diacylglycerol and its 1-O-alkyl isomer alkylgalactolipid) by PTLC. The fatty acid composition of each lipid group was determined by GC. Perbenzoylated derivatives of these lipids were further separated by RP-HPLC [78].

Using PTLC six major fractions of lipids (phospholipids, free sterols, free fatty acids, triacylglycerols, methyl esters, and sterol esters) were separated from the skin lipids of chicken to study the penetration responses of *Schistosoma cercaria* and *Austrobilharzia variglandis* [79a]. To determine the structure of nontoxic lipids in lipopolysaccharides of *Salmonella typhimurium,* monophosphoryl lipids were separated from these lipids using PTLC. The separated fractions were used in FAB-MS to determine β-hydroxymyristic acid, lauric acid, and 3-hydroxymyristic acids [79b].

PTLC was also used for the separation of lipid components in pathogenic bacteria. *Mycobacterium avium* has a requirement for fatty acids, which can be fulfilled by palmitic or oleic acid, and these fatty acids are then incorporated into triagylglycerols [80]. PTLC was used for the separation of fatty acids and triacylglycerols in the extracts of these bacterial cells to study the lipid classes in the bacterial cells cultured under different growth conditions.

After the extraction of lipid and nonlipid components from the leaves of mandarin orange *Citrus reticulata*, the lipid fraction was further separated by PTLC to determine different lipid classes that affect the chemical deterrence of *C. reticulata* to the leaf cutting ant *Acromyrmex octopinosus*. These lipids seem to be less attractive to the ants [81a]. The metabolism of palmitate in the peripheral nerves of normal and Trembler mice was studied, and the polar lipid fraction purified by PTLC was used to determine the fatty acid composition. It was found that the fatty acid composition of the polar fraction was abnormal, correlating with the decreased overall palmitate elongation and severely decreased synthesis of saturated long-chain fatty acids (in mutant nerves) [81b].

Major lipid classes in wool-colonizing microorganisms capable of utilizing wood lipids and fatty acids as the sole source of carbon and energy were isolated by PTLC [82a]. These lipids were extracted from five different fungi, *Aspergillus fumigatus, A. flavus*, *Scopulariopsis candida, Crysosporium keratinophilum*, and *Malbranchea anamorph.* Isolated lipid classes, including steryl esters, sterols, fatty acids, fatty alcohols, and monoacylglycerols, were used as the carbon source for the growth of these microorganisms. PTLC also was used for the separation of individual phospholipid species that have significant complement-inhibitory activity of the alternative complementary pathway in *A. fumigatus* [82b]. Centrifugally accelerated thin-layer chromatography was used for the separation of neutral lipids from phospholipids on a preparative scale [82c]. The isolated neutral lipids and phospholipids were then identified by ^{1}H- and ^{13}C-NMR, whereas the fatty acid composition was determined by GC and GC-MS. Heterocyst-type glycolipids in the lipids of nitrogen-fixing cyanobacterial cells extracted with chloroform:methanol (1:1) could be isolated from other complex lipids by silica gel PTLC [83]. Separated glycolipids were further purified by C_{18} solid phase cartridges and subjected to per-O-benzolization derivatization for the determination of different glycolipid species. Free and cholesteryl-ester-bound fatty acids of lipids from the wool cell membrane complex were separated from the polar fraction using PTLC. The polar fraction was then subjected to GC-MS for the identification of fatty acids in the polar lipid fraction [84].

Cripps and Tarling [85] indicated that the current PTLC procedure for the separation of lipid classes in marine zooplankton, including calanoid copepods, are subject to low recovery and are time consuming compared to their liquid chromatographic method of using silica or bonded amino-silica as the stationary phase. Studying the effect of isolation methods for the lipid composition of *Hibiscus* leaf, Chernenko and Glushenkova, however, reported that the separation of lipids by column chromatography allows one to obtain a polar lipid fraction without any need for further purification [86]. Free ceramides in *Candida lipolytica* yeast grown batchwise on two different carbon sources, glucose and *n*-hexadecane, were isolated from the spingolipid fraction and quantified by preparative TLC [87]. Yields of triterpine alcohols of nonsaponifiable matter in oil samples of the aerial parts of *Lapsana communis* L. harvested in Indre et Loire (France) at different periods were determined by PTLC after saponification [88]. Phosphonolipids contribute to the protection of cellular integrity and survival of aquatic organisms such as molluscs and cnidaria. The edible mussel *Mytilus galloprovincialis* contains about 1.27% of this material in the fresh tissues, and these compounds can be separated by 2D-TLC after separation of the polar fraction from the neutral lipids. The individual ceramide aminoethylphonate species could be isolated by PTLC for the determination of their structures by other chromatographic methods [89]. The most common lipid classes of mitochondrial and microsomal membranes are phosphotidylcholine and phosphatidylethanolamine. Free radical damage in the cultural neural cells may be due to ethanol-induced cytotoxicity because the changes in oxidative metabolism and the resulting lipid peroxidation readily modify these biological membranes, altering the cell functions. PTLC was used to separate these phospholipids in studying the effect of ethanol and acetaldehyde on microsomal and mitochondrial membrane fatty acid profiles in cultured rat astroglia [90].

Neutral gangliolipids from *Thermoplasma acidophilum* were separated by PTLC, and their tentative structures were characterized by the combination of GC, ^{1}H-NMR, and FAB-mass spectrometries [91]. The lipophilic portion of the neutral glycolipid was composed of caldarchaeol (dibiphytanyl-diglycerol tetraether), and the sugar moieties of the glycolipids were composed of glucose.

REFERENCES

1. Christie, W.W., Lipids: their structure and occurrence, in *Lipid Analysis, Isolation, Separation, Identification and Structural Analysis of Lipids,* Vol. 5, Christie, W.W., Ed., The Oily Press, England, 2003, pp. 3–33.
2. Sultanovich, Y.A., Nechaev, A.P., Koroleva, N.I., CODEN: URXXAF SU 105784 A1, CAN 100:64511, 1984.
3. Jain, R., Thin-layer chromatography in clinical chemistry, in *Practical Thin-Layer Chromatography,* Fried, B. and Sherma, J., Eds., CRC Press, New York, 1996, pp. 131–152.
4. Rebal, F.M., Sorbents and precoated layers in thin-layer chromatography, in *Handbook of Thin Layer Chromatography,* 3rd ed., Vol. 89, Sherma, J. and Fried, B., Eds., Marcel Dekker, New York, 2003, pp. 99–133.
5. Gruger, E.H. and Tappel, A.L., *J. Chromatogr.*, 40, 177, 1969.
6. Itoh, T., Tamura, T., Matsumoto, T., and Dapaigne, P., *Fruits*, 30, 687, 1975.
7. Liq, Q.T. and Kinderlerer, J.L., *J. Chromatogr. A.*, 855, 617, 1999.
8. Holub, B.J. and Kuksis, A., *Lipids*, 4, 466, 1969.
9. Roehm, J.N. and Privett, O.S., *Lipids*, 5, 353, 1970.
10. Lundburg, B. and Neovius, T., *Math. Physica*, 34, 10, 1974.
11. Mukherjea, M., *Endocrinology*, 69, 136, 1977.
12. Yasumura, M., Okano, T., Mizuno, K., and Kobayashi, T., *J. Nutr. Sci. Vitamin*, 23, 513, 1977.
13. Morita, M. and Hayashi, A., *Yukagaku*, 29, 849, 1980.
14. Hillery, C.A., Du, M.C., Montgomery, R.R., and Scott, J.P., *Blood*, 87, 4879, 1996.
15. Berdeaux, O., Sebedio, J.L., Chardigny, J.M., Blond, J.P., Mairot, T., Vatele, J.M., Poullain, D., and Noel, J.P., *Grasas y Aceites*, 47, 86, 1996.
16. Illinov, P.P., Deleva, D.D., Dimov, S.I., and Zaprianova, E.T., *J. Liq. Chromatogr. Relat. Technol.*, 20, 1149, 1997.
17. Hamlet, C.G. and Sadd, P.A., *J. Food Sci.*, 22, 259, 2004.
18. Hamlet, C.G. et al., *227th ACS National Meeting*, Anaheim, CA, U.S., March 28–April, 2004, *Am. Chem. Soc.*
19. Rupcic, J., Georgiu, K., and Maric, V., *J. Basic Microbiol.*, 44, 114, 2004.
20. Hassanein, M.M., El-Shami, S.M., and El-Mallah, M.H., *Grassa y Aceites*, 54, 343, 2003.
21. Pantovic, R., Draganio, P., Blagovic, B., Erakovic, V., Simonic, A., and Millir, C., *Period. Biolog.*, 104, 95, 2002.
22. Rupcic, J., Milin, C., and Maric, V., *Syst. Appl. Microbiol.*, 22, 486, 1999.
23. Vaver, V.A. and Shemiyakin, M.M., *Mem. Trans. Process.*, 2, 21, 1978.
24. Holloway, P.J. and Challen, S.B., *J. Chromatogr.*, 25, 336, 1966.
25. Sherma, J, O'ttea, C.M., and Fried, B., *J. Planar. Chromatogr. – Mod. TLC*, 5, 343, 1992.
26. LeTeng, J., Chen, X., and Gurrero, S., *J. Planar. Chromatogr. Mod. TLC*, 5, 64, 1992.
27. Mangold, H.K., *J. Am. Oil Chem. Soc.*, 38, 708, 1961.
28. Storry, J.E. and Tuckley, B., *Lipids*, 2, 501, 1967.
29. Pernes, J.F., Nurit, Y., and DeHeaulme, M., *J. Chromatogr.*, 181, 254, 1980.
30. Owens, K., *Biochem. J.*, 100, 354, 1966.
31. Gardner, H.W., *J. Lipid Res.*, 9, 139, 1968.
32. Kapoulas, V.M., *Biochim. Biophys. Acta*, 176, 324, 1969.
33. Kusera, J., *Coll. Czech. Chem. Commun.*, 28, 1341, 1963.
34. Schmid, H.H.O., Jones, L.L., and Mangold, H.K., *J. Lipid Res.*, 8, 692, 1967.

35. Copius, J.W. Peereboom and Beeks, H.W., *J. Chromatogr.*, 20, 316, 1962.
36. Barbier, M., Jager, H., Tobias, H., and Wyss, E., *Helv. Chem. Acta*, 42, 2440, 1959.
37. Ledeen, R.J., *Am. Chem. Soc.*, 43, 57, 1960.
38. Svennerholm, E. and Svennerholm, L., *Biochim. Biophys. Acta*, 70, 432, 1963.
39. Giorgio, S., Jasiulionis, M.G., Straus, A.H., Takahashi, H.K., and Baibieri, C.L., *Baibieri. Exp. Parasitol.*, 75, 119, 1992.
40. Karlsson, K.A. and Pascher, I., *J. Lipid Res.*, 12, 466, 1971.
41. von Planta, C., Schwieter, U., Chopard-Dit-Jean, L., Ruegg, R., Kolffer, M., and Osler, O., *Helv. Chim. Acta*, 45, 548, 1962.
42. Janecke, H. and Mas-Joebels, *Z. Anal. Chem.*, 178, 161, 1960.
43. Fried, B., Lipids, in *Handbook of Thin Layer Chromatography,* 3rd ed., Vol. 89, Sherma, J. and Fried, B., Eds., Marcel Dekker, New York, 2003, pp. 99–133.
44. Nyiredy, S., Preparative layer chromatography, in *Handbook of Thin Layer Chromatography,* 3rd ed., Vol. 89, Sherma, J. and Fried, B., Eds., Marcel Dekker, New York, 2003, pp. 99–133.
45. Sharaf, D.M., Clark, S.J., and Downing, D.T., *Lipids*, 12, 786, 1977.
46. Wertz, P.W., Stover, P.M., Abraham, W., and Downing, D.T., *J. Lipid Res.*, 27, 427, 1986.
47. Mangold, H.K., Aliphatic lipid, in *Thin-Layer Chromatography,* 2nd ed., Stahi, E., Ed., Springer-Verlag, New York, 1969, pp. 363–421.
48. Shetty, P.H., Fried, B., Sherma, J., *J. Helminthol.*, 66, 68, 1992.
49. Skipski, V.P., Smolove, A.F., Sallivan, R.C., and Barclay, M., *Biochim. Biophys. Acta.*, 106, 386, 1965.
50. Fried, B., Tancer, R., and Fleming, S.J., *J. Parasitol.*, 66, 1014, 1980.
51. Wagner, H., Hoehammer, L., and Wolff, P., *Biochem. Z.*, 334, 175, 1961.
52. Fried, B. and Shapiro, I.L., *J. Parasitol.*, 65, 243, 1979.
53. Rouser, G., Kritchvsky, G., and Yamamota, A., Column chromatographic and associated procedures for separation and determination of phosphatides and glycolipids, in *Lipid Chromatographic Analysts,* Vol. 1, Marinetti, G.V., Ed., Edward Arnold, London, 1967, pp. 99–162.
54. Thompson, S.N., *Comp. Biochem. Physiol.*, 87B, 357, 1987.
55. Fell, R.D., Thin-layer chromatography in the study of entomology, in *Practical Thin-Layer Chromatography,* Fried, B. and Sherma, J., Eds., CRC Press, New York, 1996, pp. 71–104.
56. Ohashi, M., *Lipids*, 14:52, 1979.
57. Sekules, G. and Barabino, T., *Bllet. Chim. Farmac.*, 106, 745, 1967.
58. Wheeler, C.M., Garcia, M., Juan, C., Benach, J.L., Goligtly, M.G., Habicht, G.S., and Sceere, A.C., *J. Infect. Dis.*, 167, 665, 1993.
59. Khan, A.A. and Hess, H.H., *Lipids*, 6, 670, 1971.
60. Ogierman, L., *Chemia Anal.*, 32, 363, 1987.
61. Bhat, H.K. and Ansari, G.A.S., *J. Biochem. Toxicol.*, 4, 189, 1989.
62. Kaphalia, B.S. and Ansari, G.A.S., *J. Biochem. Toxicol.*, 4, 183, 1989.
63. Deleva, D., Illinov, P.P., and Zaprinova, E.T., *Bulg. Chem. Commun.* 25, 514, 1992.
64. Kaphalia, B.S., Ghanayem, B., and Ansari, G.A S., *J. Toxicol. Envion. Health*, 49, 463, 1996.
65. White, D.C., Alugupalli, S., Schrum, D.P., Kelly, S.T., Sikka, M.K., Fayer, R., and Kaneshiro, E.S., *International Symposium on Waterborne Cryptosporidium, Proceedings*, Ed., Fricker et al., American Water Works Association, Denver, CO, 1997, pp. 53–59.
66. Mahadeevapa, V.G. and Raina, P.L., *J. Am. Oil Chem. Soc.*, 55, 647, 1978.

67. Gere, A., *Maj. Kemi. Lap.*, 34, 3, 160, 1979.
68. Mahadevappa, V.G. and Raina, P.L., *J. Agric. Food Chem.*, 29, 1225, 1981.
69. Tonnet, M.L. and Green, A.G., *Arch. Biochem. Biophys.*, 252, 646, 1987.
70. Olsson, N.U., *J. Chromatogr.*, 624, 11, 1992.
71. Chandra, H, Dutta, P.C., Helmersson, S., Kebedu, E., and Alemaw, G., *J. Am. Oil Chem. Soc.*, 71, 839, 1994.
72. Dutta, P.C. and Appelqvist, L., *Grasas y Aceites*, 47, 38, 1996.
73. Dutta, P.C. and Appelqvist, L., *J. Am.Oil Chem. Soc.*, 74, 647, 1997.
74. Paltauf, F., *Lipids*, 13, 165, 1978.
75. Khuller, G.K. and Banerjee, B., *J. Chromatogr.*, 150, 518, 1978.
76. Litchfield, C., Greenberg, A.J., Ackman, R.G., and Eaton, C.A., *Lipids*, 13, 860, 1978.
77. Lee, R.F., Nevenzel, J.C., Paffenhoefer, G.A., Benson, A.A., Patton, S., and Kavanagh, T.E., *Biochim. Biophys. Acta*, 202, 386, 1970.
78. Yahara, S. and Kishimoto, Y., *J. Neurochem.*, 36, 190, 1981.
79a. Zibulewsky, J., Fried, B., and Bacha, W.J., Jr., *J. Parasitolog.*, 62, 905, 1982.
79b. Qureshi, N., Takayama, K., and Ribi, E., *J. Biol. Chem.*, 257, 11808, 1982.
80. McCarthy, C.M., *Am. Rev. Res. Dis.*, 129, 96, 1984.
81. Jones, V., Pollard, G.V., and Seaforth, C.E., *Insect. Sci. Appl.*, 8, 99, 1987.
82a. Al Mussalm, A.A. and Radwan, S.S., *J. Applied Bacteriol.*, 69, 806, 1990.
82b. Washburn, R.G., DeHart, D.J., Agwu, D.E., Bryant-Varela, B.J., and Julian, N.C., *Infect. Immun.*, 58, 3508, 1990.
82c. Bergheim, S., Malterud, K.E., and Anthonson, T., *J. Lipid Res.*, 32, 877, 1991.
83. Davey, M.W. and Lambein, F., *Anal. Biochem.*, 206, 323, 1992.
84. Koerner, A., Hoecker, H., and Rivett, D.E., *J. Anal. Chem.*, 344, 501, 1992.
85. Cripps, G.C. and Tarling, G.A., *J. Chromatogr. A.*, 760, 2, 309, 1977.
86. Chernenko, T.V. and Glushenkova, A.I., *Khimiya. Prir. Soed.*, 4, 623, 1995.
87. Rupcic, J., Mesaric, M., and Maric. V., *Appl. Microbiol. Biotech.*, 50, 583, 1998.
88. Fontanel, D.K., Kargol, M., Gueffier, A., Viel, C., Holownia, A., Meskar, A., Menez, J.F., Ledig, M., and Brarzko, J.J., *J. Am. Oil Chem. Soc.*, 75, 1457, 1998.
89. Kariotoglou, D.M. and Mastronikolis, S.K., *Zeitsch. Nat.*, 53, 888, 1998.
90. Holownia, A. et al., *Addic. Biol.*, 3, 271, 1998.
91. Uda, I., Sugai, A., Kon, K., Ando, S., Itoh, V.H., and Itoh, T., *Biochem. Biophys. Acta*, 1439, 363, 1999.

13 The Use of PLC for Separation of Natural Pigments

George W. Francis

CONTENTS

13.1 INTRODUCTION

Natural pigments are normally small, colored, organic molecules belonging to a limited number of major structural types. The first compounds to be separated by chromatography were in fact such natural pigments: the separations were carried out on open tubular columns whose appearance after development led to the term *chromatography* from the Greek words "khromatos" and "graphos" to provide the meaning "colored writing." Further details may be found in Chapter 1 of this book. The present chapter thus follows directly the original meaning of chromatography although the compounds involved and the separations achieved are beyond the wildest dreams of the original workers.

13.1.1 Pigment Classes

Natural pigments are among the most familiar chemical compounds to the general public, who use them on a daily basis to judge the quality of their food, to color a wide variety of materials from food products and cosmetics to the clothes they wear, and to use for the internal and external decoration of their houses. There is increasing pressure today toward the wider use of natural, rather than synthetic, pigments for all purposes as consumers are becoming more informed about genuine issues of health and safety, together with increasing unspecific worries about the use of synthetic chemicals, whether identical with the natural material or not. The results of this are clearly seen in an increasing interest in natural pigments by a wide variety of scientists both in academia and the industry. A good introduction to the biology and commercial impact of plant pigments has recently appeared [1], and a variety of texts can be consulted for material on individual classes of compound [2,3].

Plants represent the major source of natural pigments, and they contain a wide variety of such pigments, mostly belonging to a few main classes, which provide almost any imaginable coloration in nature, where colors occur separately or more usually in combination. This is not to underestimate the importance of structural color as found in virtually all classes of organisms, most spectacularly in animals and insects. However, the investigation of structural coloration is largely the domain of physicists and biologists, and does not belong in the present chapter, as illustrated by recent work [4–6]. The current chapter will thus concentrate on pigments in plants. The main pigment classes to be found in plants are relatively few in number, as might be expected from the finding that metabolism operates in a way that conserves central pathways while providing a great variety of individual compounds by combining simple steps to reach the final metabolites. The pigment classes examined here include the polyene carotenoids, which provide a broad selection of colors from pale yellow to deep red by altering the extent of conjugation and functional groups present within the end groups attached to the central chain [7–9]. The carotenoids function together with the green chlorophylls in the light-harvesting reactions required for photosynthesis in plants [10]. Other important classes are the yellow-orange flavonoids and the red-blue anthocyanins, which occur most obviously in flowers and fruits [10–12]. The two classes show closely related structures, normally occur as glycosides, and are often found together, even occasionally in dimeric structures with one representative of each class. The quinones found as pigments usually belong to the class of anthraquinones of widely varying color from orange to red and blue, and these, too, are included in the current chapter [13]. Finally, the relatively unique betalain pigments are briefly discussed [14,15]. Figure 13.1 includes examples of typical structures for each class: the carotenoid zeaxanthin, chlorophyll a, the flavonoid quercetin, the anthocyanidin cyanidin, the betalain betanin, and the anthraquinone alizarin. A common feature of all of these structures is the presence of extended conjugated systems, which explains both their color and their lack of stability when isolated.

13.1.2 Distribution of Pigments

The most widespread of the pigments are the carotenoids and chlorophylls [1,16,17]. These compounds are of universal distribution in the plant world owing to their involvement with photosynthesis, although it should be remembered that they fulfill a number of other functions. The carotenoids also occur in animals [17] and microorganisms, whereas the distribution of chlorophylls is largely limited to the plant kingdom. The tetrapyrrole nucleus is also widespread in the animal world but in the form of haem and related pigments rather than as chlorophyll [1]. The flavonoids and anthocyanins are also widely distributed, being commonly found in plants, although they also occur in other organisms [11,12]. The anthraquinones, too, are widely distributed in plants, but are also found in animals, where their functions seem to be related to pigmentation as such [13,18–20]. The betalains are of very limited distribution, and although they occur to any extent in one plant order, the Caryophyllales, they are of importance as natural pigments for food applications [1]. The distribution of the individual classes will be discussed under separate headings.

13.2 TLC SYSTEMS IN GENERAL

13.2.1 Stationary Phases

The number of stationary phases available is essentially unlimited, although relatively few are, in fact, in common use. Chapter 3 discusses sorbents and precoated layers in detail. Originally, PTL was undertaken on plates made in the laboratories where they were used, and this led to the use of a wide variety of stationary phases.

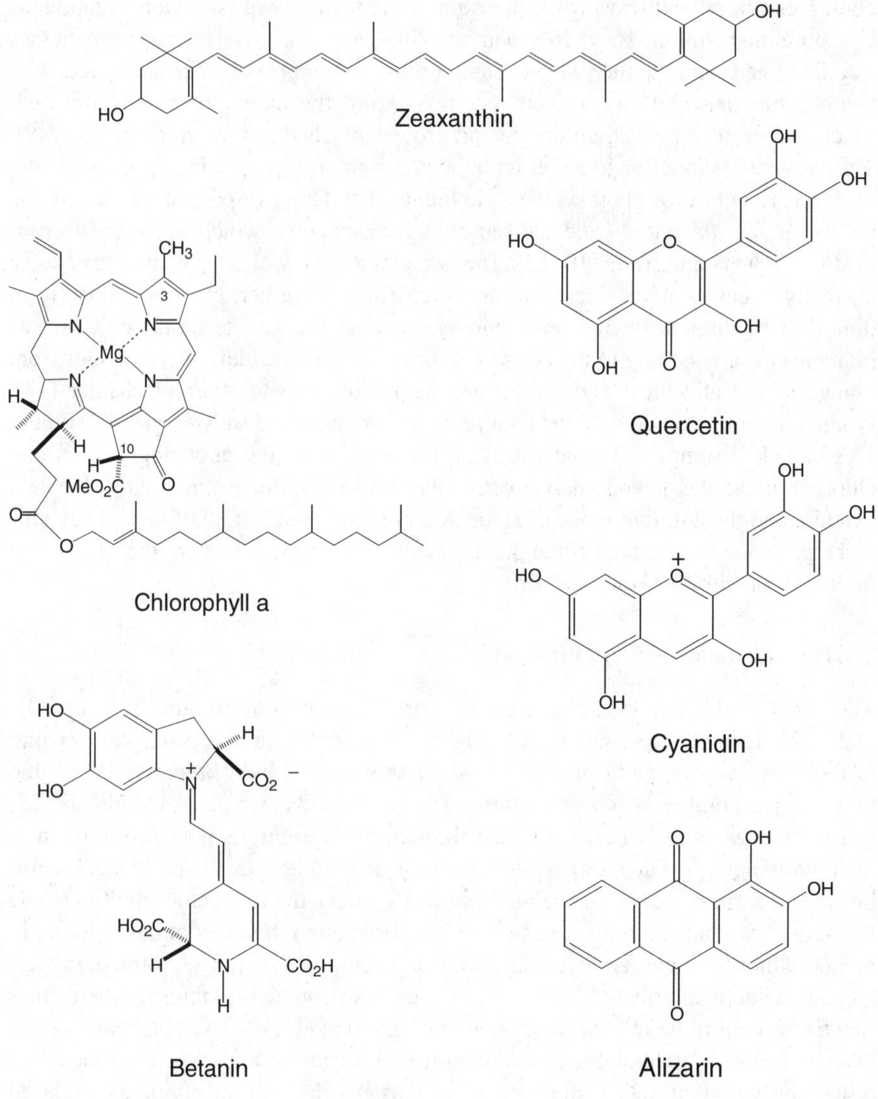

FIGURE 13.1 Typical structures for main pigment classes: zeaxanthin (carotenoid), chlorophyll a (chlorophyll), quercetin (flavonoid), cyanidin (anthocyanidin), betanin (betalain), and alizarin (anthraquinone).

Preparation of such plates requires care and practice, and whereas they were once popular, they have now been replaced by commercial plates for most, if not all, applications. The quality of the commercial plates is good, and they have the added advantage of tolerating handling, which is often a problem in the case of homemade plates in which the layers are readily damaged. The phases are attached to the surface of the carrier material, glass, plastic, or metal by means of binders that do not affect separation but stabilize the mechanical properties of the layers in commercial plates.

Commercial plates for preparative chromatography are usually made of glass, and the coating of the stationary phase varies in thickness from 0.25 to 2.0 mm, with the most usual stationary phases for normal phase chromatography being silica, alumina, cellulose, and polyamide, and RP-18 in the case of reversed-phase systems. The amount of sample that can be separated on a layer depends on the square root of the thickness of the plate; thus, a plate of four times the thickness will separate twice as much compound in a satisfactory manner. Pigments demand a large number of active sites, and loadings are strictly limited; amounts of 2 to 20 mg can usually be separated on 20 × 20 cm plate of 1-mm thickness, depending on the exact pigment mixture and the amounts of nonpigment impurities present.

Modern commercial plates are also available in tapered versions and in formats containing an application zone. These special features add greatly to their cost, and these plates are thus used only when all other methods have failed.

13.2.2 Developing Solvents

The choice of solvent for preparative layer chromatography, discussed in detail in Chapter 4, does not vary greatly from the situation for analytical systems. It is thus advisable to carry out analytical work prior to any attempt at preparative chromatography. It is also important to remember that there may be changes in retention behavior when the material of the carrier plate is changed, and if initial separation is being or has been carried out on plates made of different material, analytical separation should be repeated on glass plates, even if with thicker layers, prior to scaling up and making any adjustments required: these adjustments are not normally great but are certainly worth the work involved.

The first trials of the preparative system should be carried out using narrow strips of the preparative plate and only small amounts of pigment concentrated onto these. Narrow strips of the plate can be made by cutting larger plates using a good glass cutter. The trials may be run with fairly short developing distances, and this will suffice to reveal any problems with scaling up. Should scaling up result in poorer separation, it is generally satisfactory to reduce solvent polarity in the case of normal phase and to increase solvent polarity in the case of reversed-phase separation: such changes may also require that plates be run for somewhat longer distances, very occasionally involving intermediate drying and rerunning.

13.2.3 Recovery of Separated Components

The separated components can normally be recovered from plates by scraping the individual zones of the stationary phase off the glass plate using a scalpel or a similar tool. This sorbent material is then crushed and mixed in a flask with about five volumes of the developing solvent itself or one of somewhat greater polarity than that of the developing solvent. The flask is flushed with nitrogen, closed with a stopper, and allowed to stir in the dark for about half an hour. The mixture is then filtered and a fresh portion of solvent added in order to repeat the process. This extraction can be repeated until the stationary phase becomes largely colorless, although it should be remembered that some material may be irreversibly adsorbed, and thus the stationary phase may remain colored [21]. This last situation will be immediately apparent as the extracts will rapidly become colorless. The extracts are combined to provide the individual component and then checked to show that the separation has been satisfactory.

13.3 CAROTENOIDS

13.3.1 Structures

The carotenoids are yellow-red tetraterpenoids and thus normally contain 40 carbon atoms [7,8]. These compounds share a chromaphore provided by a polyene chain, normally consisting of at least nine double bonds. The ends of the chain are normally formed into cyclohexane rings containing one or more double bonds. The functionalities present are restricted to those containing oxygen, with alcohol, ether, epoxide, and ketone functions being usual. Occasionally, aldehydes, carboxylic acids, and esters are encountered but usually in conjunction with apo structures having reduced chain lengths. The presence of any function will, of course, change the polarity of the compound irrespective of possible changes in color. Oxygen-containing carotenoids are often called xanthophylls to distinguish them from the hydrocarbon carotenes. Structures of some common carotenoids are shown in Figure 13.2. Carotenoid structures are complicated by the fact that the double bonds present in the chain, normally all of the *trans* configuration, are readily isomerized, and during chromatography the presence of several *cis-trans* isomers may often be seen. The *cis* isomers differ from the *trans* isomers only in having a slightly shorter visible light absorption maxima. These isomers show very similar retention characteristics on chromatographic layers, some being slightly less and some slightly more polar than the all-*trans* isomers, and often result in the appearance of colored shadows on either side of the main zone. Carotenoids often contain chiral carbon atoms in their end groups, and these changes in stereochemistry vary greatly in their effects, some such stereoisomers having remained undetected for many years and others have been assumed to be completely different compounds prior to their relationships being discovered. Much more information about these subjects can be found elsewhere [7,22].

13.3.2 Sources and Distribution

Carotenoids are found in all photosynthetic tissues and are thus of nearly universal distribution in plant parts found above the ground [1]. Photosynthetic tissue, however, tends to contain only or largely those carotenoids normally associated with photosynthesis, with β-carotene, lutein, violaxanthin, and neoxanthin being the most usual and most abundant pigments, and virtually any green vegetable may be used as a source [23]. Some fruits, such as tomatoes and red and yellow paprika (bell-pepper), contain so much carotenoid and so little chlorophyll that their coloration is entirely that of their carotenoid content. Flowers, too, often contain quite large amounts of carotenoid and thus make good sources of natural carotenoids. Carotenoids often occur as esters, although these are usually subjected to procedures for the removal

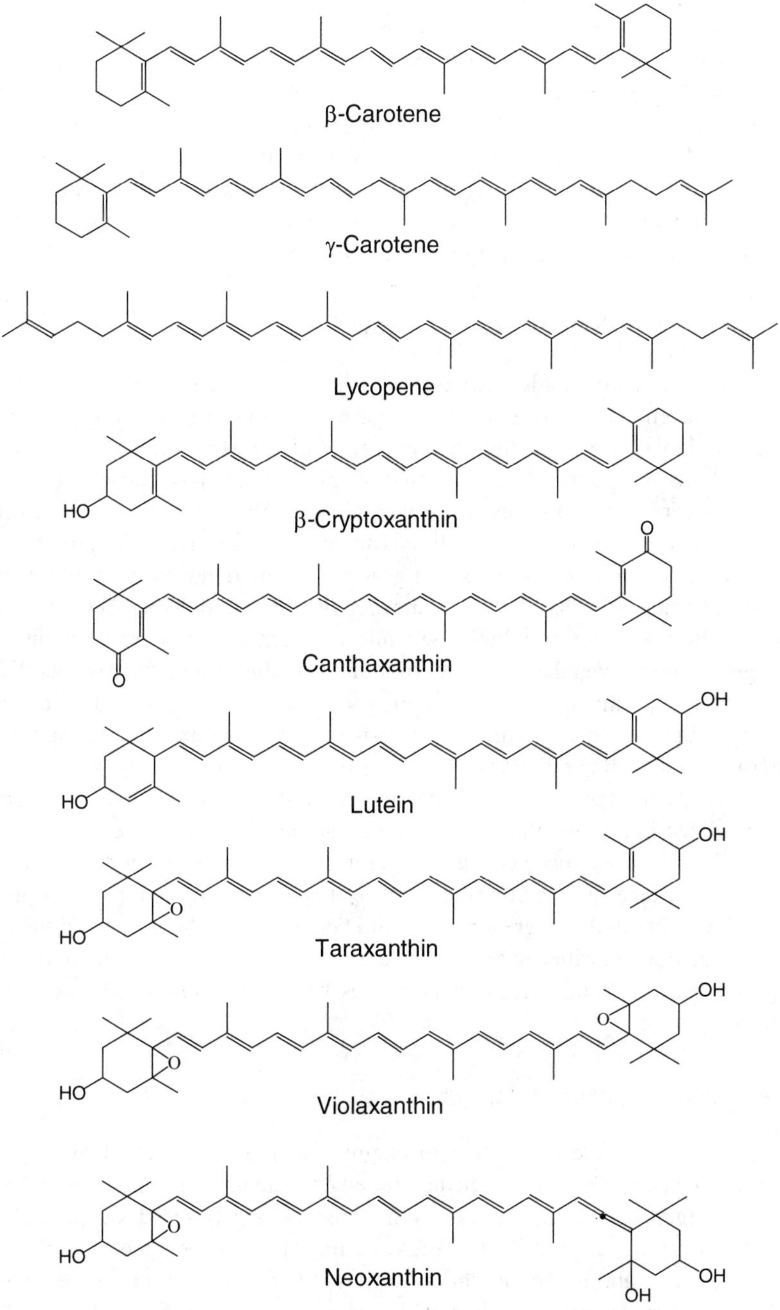

β-Carotene

γ-Carotene

Lycopene

HO β-Cryptoxanthin

O Canthaxanthin O

OH HO Lutein

OH O HO Taraxanthin

OH O O HO Violaxanthin

O HO Neoxanthin OH OH

FIGURE 13.2 Some carotenoid structures in order of increasing polarity.

of esterifying acids during isolation. Useful data on the plant sources of carotenoids should be consulted when looking for standard sources [8,16].

Carotenoids are also found in photosynthetic bacteria and in some fungi [16]. Such compounds often have quite unique structural features reflecting their biosynthetic origin. Carotenoids found in insects and animals are the result of their retention from materials making up the diet of the organism, although the compounds may have suffered minor modifications to allow of their utilization within the organism [17]. It should be noted that carotenoids in marine animals are almost always found as protein complexes, and it is not possible to separate these by thin-layer methods without prior treatment to remove or denature the protein.

13.3.3 Choice of Systems for Separation

The separation of carotenoids has been carried out on a variety of absorbents, but it is most convenient to use one of the more common ones involving a widely available stationary phase including preparative plates [13,14,24]. The choice of solvent systems then depends on the amount of work already carried out on the organism concerned because this will give a clear idea of what will work best. It is important to choose a solvent that will maximize the area of plate actually used for the separation and to avoid systems that group the compounds present into a small range of retention values. The next question is that of the thickness of layers available. Although a thicker layer will allow separation of more pigments, the advantage should be weighed against costs and possible problems associated with recovery of the pigment from the larger volumes of stationary phase involved. Experience suggests that 0.5-mm or at most 1.0-mm layers are best suited for preparative work with carotenoids.

Carotenoids are generally well separated on silica gel layers, and a plethora of data is available in the literature for such separations [24]. The developing solvent systems most commonly used consist of acetone or another polar modifier in a light hydrocarbon, hexane, petroleum ether, etc. Systems involving chlorohydrocarbons have also been reported, but great care should be taken with these to avoid the prior presence of acidic impurities in the solvent and to ensure that radicals are not formed during use, because both of these possibilities will cause rapid destruction of the carotenoids present.

13.3.4 Practical Procedures

Carotenoids are extremely sensitive to chemical degradation [25], and they will survive for extended periods in their natural environment, where they are protected from most aggressive circumstances by the surrounding tissue. Examples of this natural stabilization can be found in the time for which orange peel will retain its color after peeling and in the fact that materials stored in herbaria often retain their color for considerable periods of time. Although the destructive propensity of oxygen and acidic conditions are perhaps best understood, it is important to realize that these agents are much more active when accompanied by heat or light. All operations should thus be carried out as far as is possible under cool conditions in the dark.

This infers that the availability of nitrogen cover is a valuable adjunct to working with carotenoids.

Extraction from fresh plant tissue is generally simple. The material is chopped or cut into manageable pieces that are then subjected to primary extraction by maceration in a food mixer using a mixture of methanol–acetone as the extraction solvent. The amount of solvent should be sufficient to cover the material comfortably prior to maceration, and the proportions of the solvent are varied according to the polarity of the expected main carotenoid components. The solvent is then decanted, and the procedure is repeated until the remaining vegetable matter is almost colorless. Where very polar carotenoids, e.g., xanthophyll glycosides, are present, further extraction may be required using methanol or ethanol. Some tissues, particularly some flowers, contain large amounts of carotenes and these may require treatment with hexane–acetone mixtures to ensure complete extraction. It is thereafter almost always an advantage to combine all extracts, irrespective of the solvents involved, before further workup. The combined extracts should then be taken to dryness under reduced pressure using rotary evaporation at temperatures not in excess of 40°C. The dried residue should then be dissolved in diethyl ether and estimated by visible light adsorption spectroscopy.

Extraction is normally followed by saponification, which is intended to remove any esterifying acids but has the added advantage of resulting in the destruction of the chlorophylls and their ultimate removal during further workup. A small sample should first be subjected to saponification, as described in the following text, to check for possible changes on base treatment: generally speaking, only the esterifying fatty esters will be removed by this procedure, but it is important to realize that diosphenols will suffer oxidation under these conditions. Assuming that unwanted decomposition is not taking place, the bulk of the extract is saponified. One volume of methanol containing potassium hydroxide (10% w/v) is added to the ether pigment solution. The container is then flushed with nitrogen, closed, and allowed to stand for about 6 h in the dark. After this, the saponified mixture is washed with an approximately equal volume of water containing sodium chloride (10% w/v). The organic layer so formed is then washed with successive volumes of salt solution until neutrality: an indication of neutrality is when phase separation takes a somewhat longer time than for previous washes.

The ether solution is now taken to dryness by rotary evaporation, and the residue dissolved in fresh diethyl ether to provide a concentrated sample for further work. This sample should be flushed with nitrogen, closed, and stored in the refrigerator.

The extract should be applied to the plate in a narrow band about 1 cm from the base of the plate: because carotenoids, as with most pigments, are of limited solubility, the application will usually take place by successive passes along the length of the application zone using a pipette, syringe, or other more specialized equipment. The sample solvent is removed by blowing a current of nitrogen over the application zone. The dried plate is then placed in the development tank to which the chosen developing solvent has already been added.

After development, the plate is again dried in a stream of nitrogen, the individual zones scraped off, and the stationary phase suspended in acetone or a similar solvent for some minutes. After filtration, the process is repeated until the stationary phase

TABLE 13.1
Retention Data for Some Carotenoids

Carotenoid	System 1[a]	System 2[b]	System 3[c]	System 4[d]	DB	FG
β-Carotene	100	100	100	0.13	9	—
γ-Carotene	100	100	100	0.17	10	—
Lycopene	100	100	100	0.23	11	—
β-Cryptoxanthin	72	78	76	0.31	9	H
Canthaxanthin	71	69	66	0.38	9	KK
Lutein	44	56	53	0.55	9	HH
Taraxanthin	41	43	37	0.62	9	EHH
Violaxanthin	33	30	19	0.68	9	EEHH
Neoxanthin	18	13	8	0.72	9	EHHH

Note: R_β-values (β-carotene = 100) are used for systems 1 to 3 and R_F values for system 4. Solvent compositions by volume, p.e. = petroleum ether (40 to 60°C); DB indicates number of in chain conjugated double bonds; FG indicates functional groups: E = epoxy, H = hydroxyl, K = ketone.

[a] System 1: silica gel 60 (0.25 mm, Merck, Art. 5721) 40% acetone/p.e.
[b] System 2: silica gel 60 (0.25 mm, Merck, Art. 5721) 20% *tert*-butanol/p.e.
[c] System 3: silica gel 60 (0.25 mm, Merck, Art. 5721) 20% *tert*-pentanol/p.e.
[d] System 4: RP-18 F_{254s} layers (1.0 mm, Merck, Art 1.0543) p.e.–acetonitrile–methanol (20:40:40).

no longer yields a colored extract. The combined extract from each individual zone is then checked for chromatographic purity. If this has not been achieved, rechromatography of the zone should be undertaken using a solvent chosen on the basis of the results obtained; normally, this entails using a somewhat less polar developer.

Some retention data for carotenoids shown in Figure 13.2 are given in Table 13.1. The data are for silica layers developed using petroleum ether (p.e.) with bp 40 to 60°C and acetone, *tert*-butanol, or *tert*-pentanol. The carotenes themselves are not separated on silica systems containing more than 2% of a polar modifier, and these should be separated on reversed-phase systems: p.e.–acetonitrile–methanol in the proportions (20:40:40) will adequately separate the carotenes in the lower part of the plate [26,27]. Retention data, the polyene system, and functional groups are collected in Table 13.1 and together provide a general idea of how chromatographic retention varies with structure.

13.4 CHLOROPHYLLS

13.4.1 Structures

The chlorophylls share a common tetrapyrrole structure and in their normal occurrence contain a chelated magnesium ion [28,29]. Structures for some chlorophyll a and b are shown in Figure 13.3. The structural difference is that there is a methyl group in chlorophyll a, whereas there is an aldehyde function in chlorophyll b. The

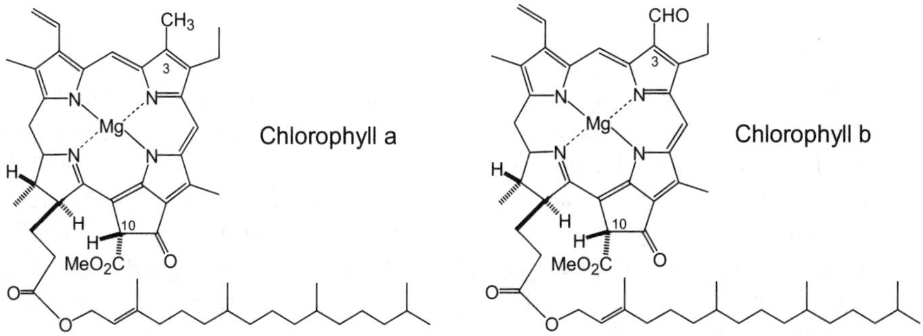

FIGURE 13.3 The structures of chlorophylls a and b. Other chlorophylls and their decomposition products can be derived from these structures as described in the text.

amount of chlorophyll a found is normally two to five times that of chlorophyll b, depending on light, with brighter conditions favoring larger amounts of chlorophyll a. The remaining chlorophylls reflect the aforementioned structural difference such that the primed chlorophylls a′ and b′ vary from the main members in having (*S*)-configuration at C-10 and are normally present in much smaller quantities than the main compounds. A variety of decomposition products are often found in chlorophyll extracts, and the debate as to the extent of their *in vivo* occurrence continues. These latter compounds have grey-brown colors rather than the blue-green hues of the major chlorophylls. Pheophytins a and b lack the magnesium atom but are otherwise unchanged from chlorophyll a and b, whereas the pheophorbides lack both magnesium atom and the phytyl chain.

13.4.2 Sources and Distribution

Chlorophylls a and b and, in smaller amounts, chlorophylls a′ and b′ can be found in any photosynthetic tissue obtained from land plants. They may thus be extracted from an almost infinite number of sources [23], although it is easier to work with material containing relatively little water as this has to be removed at later stages of the workup. Thus, woodier tissues are to be preferred as sources as compared to those containing large volumes of water, e.g., parsley rather than cucumber.

It should be noted that in many algae, chlorophyll b is replaced by chlorophyll c where the phytyl ester is replaced by a methyl ester and the side chain carrying the function is unsaturated. In a similar manner, photosynthetic bacteria contain the closely related bacteriochlorophylls rather than the normal chlorophylls [29].

13.4.3 Choice of Systems for Separation

A considerable number of systems have been used to separate chlorophylls on thin layers [30,31]. The most readily applicable layers are prepared from cellulose, silica, or sucrose and use hydrocarbon carriers with a polar modifier, usually acetone, in the developing solvent. However, silica layers cause a level of decomposition that is unacceptable for preparative work. Sucrose layers offer no particular advantages in separation and are neither commercially available nor recommended.

13.4.4 Practical Procedures

Although acetone is a suitable solvent for extracts for analytical purposes, preparative work is best accomplished by isolating the compounds as their dioxane complexes.

Plant material is macerated using two volumes of 2-propanol as the extraction solvent. The 2-propanol solvent is then decanted and filtered. Any solid material in the filter is returned to the extraction vessel, and the entire procedure is reiterated until the extract produced is colorless. A small volume of dioxane, 15% of the total extracted volume, is added to the combined extracts. The chlorophyll-containing solution is then triturated with water until it is turbid, and then it is placed in a refrigerator for 2 to 3 h.

The precipitated chlorophyll–dioxane complex is thereafter recovered by means of centrifugation. The complex is dissolved in acetone to free the native chlorophylls, and the resultant solution is then decreased to a small volume under reduced pressure to provide a concentrated sample for chromatography.

The concentrate thus prepared is applied directly to cellulose layers, and the plates are thoroughly dried in a stream of nitrogen and then developed using a solvent system consisting of p.e.–acetone (80:20).

The main compounds are well separated in this system, and recovery is accomplished by extraction of the stationary phase scraped from the plate with two successive portions of acetone. Among decomposition products often seen on the plate are the pheophytins, lacking the magnesium atom but otherwise unchanged, and the highly polar pheophorbides, lacking both the magnesium atom and the phytyl chain, but these do not interfere with the main zones. Table 13.2 shows results for some chlorophyll derivatives.

Pheophytins and pheophorbins for reference can be simply obtained by acid treatment of the intact chlorophylls: the former are obtained by treating an ether solution of chlorophyll with a half volume of 1-*M* HCl for 2 min, whereas the latter

TABLE 13.2
Retention Data for Some Chlorophylls

Chlorophyll	R_f-value	1-Methyl	1-Formyl	Mg	Phytyl
Pheophytin a	0.94	Yes	No	No	Yes
Pheophytin b	0.88	No	Yes	No	Yes
Chlorophyll a′	0.80	Yes	No	Yes	Yes
Chlorophyll a	0.76	Yes	No	Yes	Yes
Chlorophyll b′	0.60	No	Yes	Yes	Yes
Chlorophyll b	0.57	No	Yes	Yes	Yes
Pheophorbide a	0.36	Yes	No	No	No
Pheophorbide b	0.18	No	Yes	No	No

Note: System: Cellulose layer (0.1 mm, Merck Art. 5716); developing solvent p.e.–acetone (80:20); the presence or absence of structural features is indicated.

are obtained by a more intense treatment in which two volumes of 9.5-*M* HCl are used and the reaction time is extended to 5 min.

13.5 FLAVONOIDS

13.5.1 Structures

Flavonoids are made up of a number of classes of very similar groups in which two phenyl rings are connected by a three-carbon unit [10]. The open structure members are yellow in color and termed as chalcones, and simple cyclization to a furanoid structure deepens the color to the orange aurones. The most usual flavonoids are, however, the pale-yellow flavones and flavonols, or 3-hydroxyflavones, which will be treated here, and the red-blue anthocyanins, which will be treated in the next section. Figure 13.4 shows examples of these main classes and their structural relationships. The natural compounds of all classes often occur as glycosides and as methyl ethers.

13.5.2 Sources and Distribution

The flavonoids universally occur in vascular plants, in which they are often responsible for the colors of flowers and fruits, although they are also present (often less apparently) in roots, stems, and leaves [10]. The number of possible sources from which these compounds can be isolated is very large and much useful information on this can be obtained in reviews [10–12].

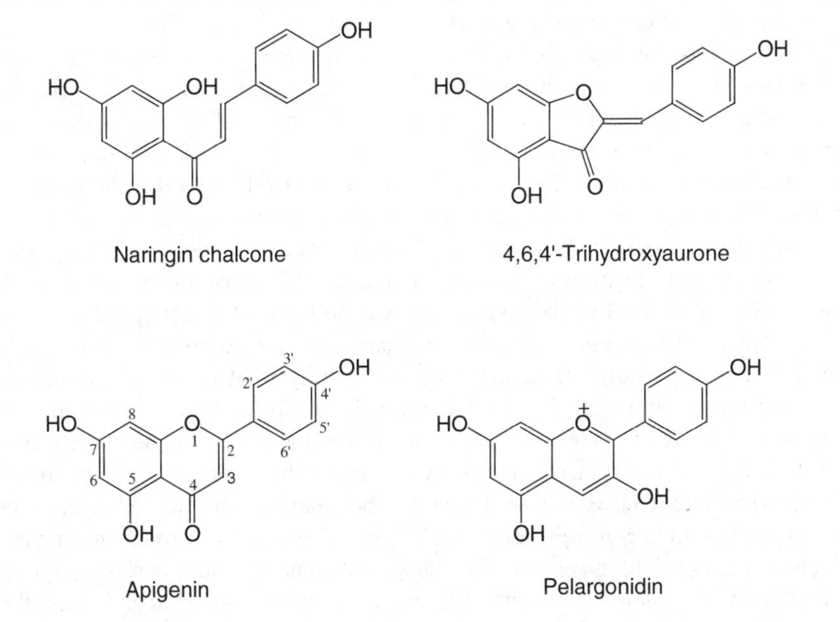

FIGURE 13.4 Typical structures for main classes of flavonoids: naringin chalcone, 4,6,4′-trihydroxyaurone, apigenin (flavone), and pelargonidin (anthocyanidin).

13.5.3 Choice of Systems for Separation

Flavonols and flavones have been analyzed on a variety of thin-layer systems [32–37]. However, for preparative purposes, silica layers are readily applicable to a wide variety of these compounds. Developing solvents normally contain weak organic acids, usually formic acid, to ensure that the flavonoids are retained in a single stable form during separation and otherwise contain esters together with water in the case of glycosides, and aromatic hydrocarbons as carriers when aglycones are being analyzed.

13.5.4 Practical Procedures

Flavonoids may be extracted from fresh or frozen plant tissues or from herbarium material, although freeze-dried material may also be utilized [34]. It is very important to ensure that the material to be extracted is finely divided, whether by cutting or crushing, to ensure proper extraction. Extraction can be carried out successively with methanol containing some 10% of water and then with a 1:1 mixture of methanol and water. Each extraction should be carried out for a period of about 2 h, shaking or stirring to facilitate the process. The extracts are then combined for chromatographic separation.

It should be pointed out that flavonoids and flavonols are only weakly adsorbing in the visible area of the spectrum, and they are most easily seen in UV light (366 nm) after spraying plates with naturstoffreagenz A followed by polyethylene glycol (PEG-4000) [34].

When carrying out preparative work, care should be taken to avoid spraying more than a minimal amount of the plate surface and that part of the plate must be discarded to avoid contamination.

Separation of the intact glycosides may be carried out using a developing solvent containing ethyl acetate–formic acid–acetic acid–water (100:11:11:27). Flavone and flavonoid glycosides are to be found interspersed over much of the plate surface. Lest this be taken as an indication that separation is not systematic, Table 13.3 shows values for a small set of flavonol glycosides, in which it is readily seen that the pattern of hydroxylation on the flavonol nucleus and the sugars present operate on retention in a logical manner: extra hydroxyl groups increase retention as does the change from rhamnose through glucose to galactose. Structures of the flavonols involved can be found in Figure 13.5. More data of this type can be found elsewhere [24].

Analysis and separation are often facilitated by first examining the free aglycones. These can readily be obtained by the hydrolysis of the glycoside mixture [34]. The initial extract is taken to dryness, redissolved in methanol, and an equal volume of 2-*M* HCl is added, and the resultant solution heated on the waterbath for 1 h. Free aglycones are then obtained by extracting the cooled hydrolysis mixture with ethyl acetate. These compounds can then be separated on silica gel layers using benzene–ethyl acetate–formic acid (40:10:5) or toluene–ethyl formate–formic acid (50:40:10) as developing solvents. The observed retention values correspond to the pattern of hydroxylation and methoxylation present in the compound and thus reflect polarity in a general way. Retention data for a variety of these compounds are available [24].

TABLE 13.3
Retention Data, R_f, for Some Flavonol 3-Glycosides

Aglycone	3-Rhamnoside	3-Glucoside	3-Galactoside
Kaempferol	0.72	0.65	0.59
Quercetin	0.69	0.53	0.51
Myricetin	0.58	0.46	0.45

Note: System: layer silica gel 60 F_{254} (0.25 mm, Merck Art. 5716); developing solvent ethyl acetate–formic acid–acetic acid–water (100:11:11:27).

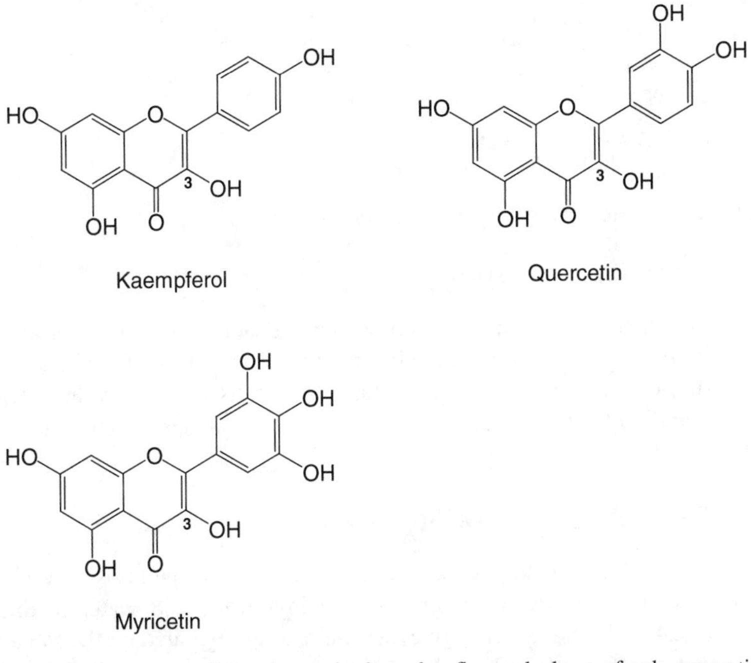

FIGURE 13.5 Structures of the increasingly polar flavonols kaempferol, quercetin, and myricetin.

13.6 ANTHOCYANINS

13.6.1 STRUCTURES

Anthocyanins are the deeply colored glycosides of the aglycone anthocyanidins. Only the six aglycones shown in Figure 13.6 are common, although others occur sporadically [11,12].

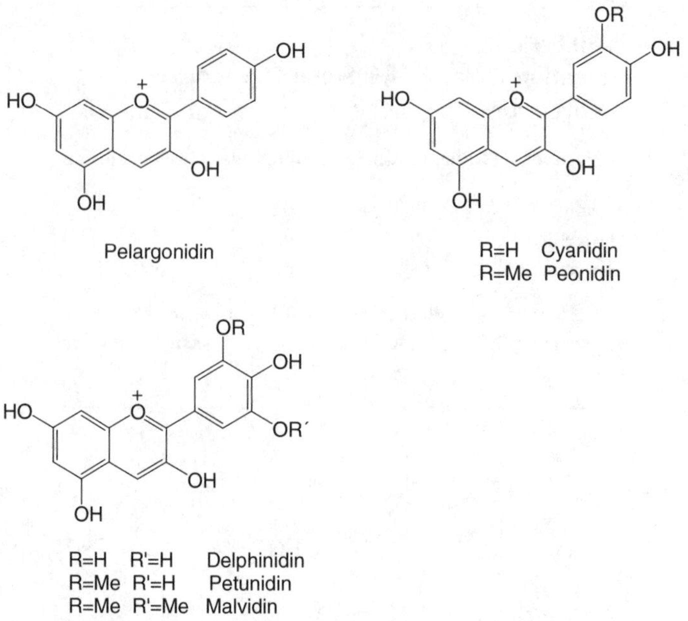

Pelargonidin

R=H Cyanidin
R=Me Peonidin

R=H R'=H Delphinidin
R=Me R'=H Petunidin
R=Me R'=Me Malvidin

FIGURE 13.6 Structures of the most common anthocyanidins.

13.6.2 Sources and Distribution

As with other flavonoid classes, the anthocyanins generally occur in higher plants [11,12]. Their presence is immediately apparent as a result of their color, and this is readily confirmed by treating a small piece of tissue with acidic methanol, and thus many red fruits and some flowers of red-blue shades are suitable sources of anthocyanins.

13.6.3 Choice of Systems for Separation

Anthocyanins and anthocyanidins are readily separated on cellulose layers using developing solvents consisting hydrochloric acid–formic acid–water in differing proportions [24]. Glycosides and aglycones are readily separated on the same plate, provided that a sufficiently polar solvent is employed.

13.6.4 Practical Procedures

Fresh plant material is mixed with two volumes of methanol containing 0.3% concentrated HCl. If particularly labile compounds are known or found to be present, methanol containing 5% acetic acid may be used to advantage because this prevents hydrolysis of the acyl groups sometimes present on the sugar moieties while maintaining the low pH required to retain the pigments in the flavinium ion form.

The extract may be used immediately for the separation of the anthocyanins or hydrolyzed to provide anthocyanidins.

TABLE 13.4
Retention Data, R_f Values, for Some Anthocyanidins and Anthocyanins

Pigment	System 1[a]	System 2[b]
Delphinidin	0.11	—
Petunidin	0.20	—
Cyanidin	0.22	—
Malvidin	0.27	—
Peonidin	0.31	—
Pelargonidin	0.35	—
Delphinidin-3-glucoside	0.38	0.08
Petunidin-3-glucoside	0.49	0.13
Cyanidin-3-glucoside	0.51	0.17
Malvidin-3-glucoside	0.64	0.22
Peonidin-3-glucoside	0.64	0.25
Pelargonidin-3-glucoside	0.65	0.32
Cyanidin-3,5-diglucoside	0.70	0.38
Peonidin-3,5-diglucoside	0.81	0.49

[a] System 1: cellulose (0.1 mm, Merck) HCl–formic acid–water (7:51:42).
[b] System 2: cellulose (0.1 mm, Merck) HCl–formic acid–water (19:19:62).

Hydrolysis is readily accomplished by mixing the extract with an equal volume of 4-*M* HCl and refluxed for 0.5 h. The hydrolyzed mixture is allowed to cool and then subjected to extraction with a small volume of 1-pentanol to obtain the aglycones, the solvent then being removed by rotary evaporation. The anthocyanidins are then dissolved in a small volume of acidic methanol (0.3% concentrated HCl).

The pigment mixtures, either aglycones or glycosides, may be separated on cellulose layers using a system consisting of concentrated HCl–formic acid–water (7:51:42). If aglycones are not present, a somewhat less polar developer may be used by changing the solvent mixture proportions to 19:19:62. The observed retentions depend on the pattern of hydroxylation and methoxylation and the sugars present in the glycosidic pigments. Some representative data can be found in Table 13.4.

13.7 BETALAINS

13.7.1 STRUCTURES

Betalains consist of two groups of pigments of similar structure, the yellow compounds classed as betaxanthins and the purple as betacyanins [14]. Both groups contain a betalamic acid residue coupled to an amine or an amino acid to form an iminium salt. Whereas the amino components of the betaxanthins vary considerably,

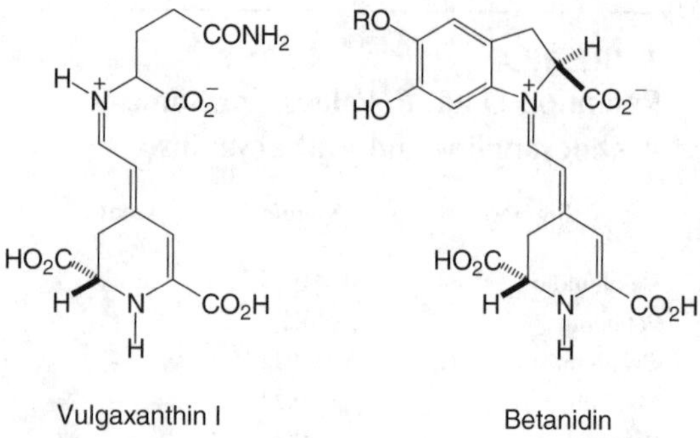

Vulgaxanthin I

Betanidin

FIGURE 13.7 Structures for the representative betalains vulgaxanthin I (betaxanthin) and betanidin (betacyanin).

the betacyanins all contain a cyclized 3,4-dihydroxyphenylalanine with the resulting extended conjugation giving their deep purple color. Representative betalain structures are shown in Figure 13.7.

13.7.2 Sources and Distribution

Betalains are largely confined to higher plants belonging to species under families of the order Caryophyllales, although they are also found in some fungi [14]. The pigments may be found in flowers, roots, and fruits of suitable species. Betacyanins do not co-occur with anthocyanins, which they resemble closely in color; it has been speculated that they may have similar functions [14].

13.7.3 Choice of Systems for Separation

Betalains are unusually sensitive, even for pigments, and are difficult to obtain in pure form. Preparative thin-layer chromatography on cellulose seems to be a relatively successful method of separating these pigments. Relatively polar developing solvents are required, examples being ethyl acetate–formic acid–water (33:7:10), which has been successfully utilized to separate betacyanins from Bougainvillea [38], 2-propanol–ethanol–water–acetic acid (6:7:6:1 or 10:4:4:1) [39], and 2-propanol–water–acetic acid (75:20:5) [40].

13.7.4 Practical Procedures

Fresh tissue is macerated in a mixture of methanol–water (1:1). This resulting mixture is stirred under nitrogen cover in the dark for 2 h. After filtration, the residue is reextracted with water for a further 2 h under the same conditions. The combined filtrates are taken to small volume under reduced pressure at not more than 40°C. The concentrate is then applied to the chromatography plate as a band and developed with one of the solvent systems described in the preceding text [38–40]. Pigments can be recovered by scraping the plates and suspending the cellulose fractions in methanol–water mixtures depending on zone polarity.

13.8 ANTHRAQUINONES

13.8.1 STRUCTURES

Anthraquinones occur as their oxygenated derivatives, with the substitution pattern reflecting their origins from the polyketide or shikimate pathway [13,18–20]. They are typically rather polar and, in many cases, their polarity is increased by their glycosidation. Although dimeric forms are known, most of the compounds identified are monomeric and have a relatively simple substitution of the central nucleus. Figure 13.8 shows some typical anthraquinones. The compounds are generally orange to red in color, but the color ultimately depends on the other substituents present.

13.8.2 SOURCES AND DISTRIBUTION

Anthraquinones are the largest class of quinones, and they occur in a wide variety of organisms, including many plants, fungi, and microorganisms. Anthraquinones are often present in woody tissues, stems, and roots. Good examples are madder, which has long been used in dyeing, and rhubarb, in which the roots are bright red as a result of the presence of anthraquinones. Sources for anthraquinones can be found in a number of books on quinones in general [13,18–20].

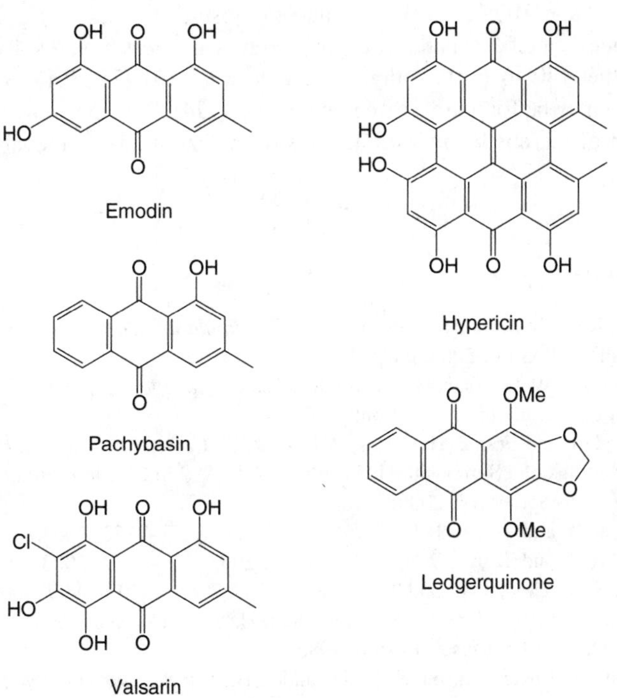

OH O OH
HO
O
Emodin

OH O OH
HO
HO
OH O OH
Hypericin

O OH
O
Pachybasin

O OMe
O
O
O OMe
Ledgerquinone

OH O OH
Cl
HO
OH O
Valsarin

FIGURE 13.8 Structures for some typical anthraquinones.

13.8.3 Choice of Systems for Separation

Anthraquinone glycosides and aglycones can be readily separated on silica layers using moderately polar developing solvents [41–43]. The best such solvents consist of ethyl acetate modified to increase polarity by the addition of alcohols or water for the glycosides or changed to decrease polarity by inclusion of hydrocarbon components.

13.8.4 Practical Procedures

The material to be extracted is crushed or chopped and then macerated with two to three volumes of methanol. The extract is removed and the procedure repeated several times.

Difficult cases may be dealt with by using warm methanol or, indeed, using Soxhlet extraction. The extracts are then collected and taken to small volume.

Glycosides can be hydrolyzed by taking the extract to dryness and treating the dry residues with 7.5% hydrochloric acid under reflux for 30 min. The freed aglycones can then be recovered by serial extraction with portions of diethyl ether. The extracted aglycones are then redissolved in a small volume of a suitable solvent, usually acetone or methanol.

Glycosidic anthraquinones may be developed using ethyl acetate–methanol–water systems (100:10:10) with suitable adjustments made for polarity. Similarly, aglycones can be separated using a somewhat less polar solvent such as petroleum ether (40 to 60°C)–ethyl acetate–formic acid (75:25:1). Some chosen retention data may be found in a recent monograph [24]. Pigments may be recovered by extraction of the absorbant with acetone or methanol after removal of the individual zones.

REFERENCES

1. Davis, K.M., in *Plant Pigments and Their Manipulation*, Davis, K.M., Ed., Blackwell Publishing, Oxford, 2004, chap. 1.
2. Goodwin, T.W., Ed., *Chemistry and Biochemistry of Plant Pigments,* Vol. 1 and Vol. 2, 2nd ed., Academic Press, London, 1976.
3. Lauro, G.J. and Francis, F.J., Eds., *Natural Food Colorants: Science and Technology* (Proceedings of a Symposium held July 23–24, 1999 in Chicago, Illinois.), IFT Basic Symposium Series, 14, 2000.
4. Prum, R.O. and Torres, R.H., *J. Exp. Biol.*, 207, 2157–2172, 2004.
5. Prum, R.O. and Torres, R.H., *J. Exp. Biol.*, 206, 2409–2429, 2003.
6. Prum, R.O., Cole, J.A., and Torres, R.H., *J. Exp. Biol.*, 207, 3999–4009, 2004.
7. Britton, G., Liaaen-Jensen, S., and Pfander, H., Eds., *Carotenoids, Vol. 1A: Isolation and Analysis*, Birkhauser, Boston, 1995.
8. Britton, G., Liaaen-Jensen, S., and Pfander, H., *Carotenoids Handbook*, Birkhauser, Boston, 2004.
9. Cuttriss, A. and Pogson, B., in *Plant Pigments and Their Manipulation*, Davis, K.M., Ed., Blackwell Publishing, Oxford, 2004, chap. 3.
10. Schwinn, K.E. and Davies, K.M., in *Plant Pigments and Their Manipulation*, Davis, K.M., Ed., Blackwell Publishing, Oxford, 2004, chap. 4.

11. Harborne, J.B. and Baxter, H., *The Handbook of Natural Flavonoids*, John Wiley & Sons, Chichester, 1999.
12. Markakis, P., Ed., *Anthocyanins as Food Colors,* Academic Press, New York, 1982.
13. Thomson, R.H., Ed., *Naturally Occurring Quinones IV: Recent Advances,* 4th ed., Blackie, London, 1997.
14. Zrÿd, J.-P. and Christinet, L., in *Plant Pigments and Their Manipulation*, Davis, K.M., Ed., Blackwell Publishing, Oxford, 2004, chap. 6.
15. Von Elbe, J.H. and Goldman, I.L., in *Natural Food Colorants: Science and Technology,* Lauro, G.J. and Francis, F.J., Eds., IFT Basic Symposium Series, 14, 2000, pp. 11–30.
16. Goodwin, T.W., *The Biochemistry of the Carotenoids, Vol. I: Plants*, Chapman and Hall, London, 1980.
17. Goodwin, T.W., *The Biochemistry of the Carotenoids, Vol. II: Animals*, Chapman and Hall, London, 1984.
18. Thomson, R.H., *Naturally Occurring Quinones*, Academic Press, New York, 1957.
19. Thomson, R.H., *Naturally Occurring Quinones, II*, Academic Press, London, 1971.
20. Thomson, R.H., *Naturally Occurring Quinones III: Recent Advances*, Chapman and Hall, New York, 1987.
21. Ferenczi-Fodor, K., Lauko, A., Wiszkidenszky, A., Vegh, Z., and Ujszaszy, K., *J. Planar Chromatogr. Mod. TLC,* 12, 1999, 30–37.
22. Liaaen-Jensen, S., *Pure Appl. Chem*., 69, 2027–2038, 1997.
23. Gross, J., *Pigments in Vegetables*, Van Nostrand Reinhold, New York, 1991.
24. Francis, G.W. and Andersen, Ø.M., in *Handbook of Thin-Layer Chromatography,* 3rd ed., Revised and Expanded, Sherma, J. and Fried, B., Eds., Marcel Dekker, New York, 2003.
25. Schoefs, B., *J. Chromatogr. A*, 1054, 217–226, 2004.
26. Francis, G.W. and Isaksen, M., *J. Food Sci.*, 53, 979–980, 1988.
27. Andersen, Ø.M. and Francis, G.W., in *Plant Pigments and Their Manipulation*, Davis, K.M., Ed., Blackwell Publishing, Oxford, 2004, chap. 10.
28. Hendry, G.A.F., in *Natural Food Colorants: Science and Technology,* Lauro, G.J. and Francis, F.J., Eds., IFT Basic Symposium Series 14, 2000, pp. 227–236.
29. Willows, R.D., in *Plant Pigments and Their Manipulation*, Davis, K.M., Ed., Blackwell Publishing, Oxford, 2004, chap. 2.
30. Holden, M., in *Chemistry and Biochemistry of Plant Pigments,* Vol. 2, 2nd ed., Goodwin, T.W., Ed., Academic Press, London, 1976, chap. 18.
31. Minguez-Mosquera, M.I., Gandul-Rojas, B., Gallardo-Guerrero, L., and Jaran-Galan, M., in *Methods of Analysis for Functional Foods and Neutraceuticals,* Hurst, W.J., CRC Press, Boca Raton, FL, 2002, pp. 101–157.
32. Hostettmann, K. and Hostettmann, M., in *The Flavonoids: Advances in Research,* Harborne, J.B. and Mabry, T.J., Eds., Chapman and Hall, London, 1982, chap. 1.
33. Markham, K.R., in *The Flavonoids*, Harborne, J.B., Mabry, T.J., and Mabry, H., Eds., Chapman and Hall, London, 1975, chap. 2.
34. Markham, K.R., *Techniques of Flavonoid Identification*, Academic Press, London, 1982.
35. Stahl, E. and Schorm, P.J., *Z. Physiol. Chimie*, 325, 263–270, 1961.
36. Hiermann, A. and Kartnig, T., *J. Chromatogr.*, 140, 322–326, 1977.
37. Hiermann, A., *J. Chromatogr.*, 174, 478–482, 1979.
38. Heuer, S., Richter, S., Metzger, J.W., Wray, V., Nimtz, M., and Strack, D., *Phytochemistry,* 37, 761–767, 1994.
39. Bilyk, A., *J. Food Sci.*, 46, 298–299, 1981.

40. Strack, D., Schmitt, D., Reznik, H., Boland, W., Grotjahn, L., and Wray, V., *Phytochemistry*, 26, 2285–2287, 1987.
41. Rauwald, H.W., *PZ Wissenschaft*, 3, 169–181, 1990.
42. Wagner, H., Bladt, S., and Zgainski, E.M., *Plant Drug Analysis*, Springer-Verlag, Berlin, 1984, p. 93.
43. Khafagy, S.A., Girgis, A.N., Khayal, S.E., and Helmi, M.A., *Planta Medica*, 21, 304–311, 1972.

14 Application of PLC to Inorganics and Organometallics

Ali Mohammad

CONTENTS

14.1 INTRODUCTION

Thin-layer chromatography (TLC), a subdivision of liquid chromatography, has grown much in status and has experienced a dramatic surge in use since its inception in 1938 by Izmailov and Schraiber, who used a thin layer of aluminum oxide (2-mm thick) on glass plate for the separation of certain medicinal plants [1]. In addition to being an off-line technique in which the various procedural steps can be carried out independently, several other interesting features such as possibility of direct observation, use of specific and colorful reactions, minimal sample cleanup, wider choice of mobile and stationary phases, reasonable sensitivity, excellent resolution power, high sample loading capacity, low solvent consumption, capability of handling a large number of samples simultaneously, applicability of two-dimensional separation, and disposable nature of TLC plates have maintained its continuing popularity as an analytical technique.

After the pioneering work of Stahl [2,3], TLC became an important tool for the study of samples not amenable to analysis by gas chromatography (GC). However, the influence of environmental conditions on the reproducibility of R_F values has been the major weakness of TLC. The rapid growth of TLC slowed down in the 1970s with the corresponding emergence of high-performance liquid chromatography (HPLC) and ion chromatography (IC). To counter this situation, several new sister techniques such as high-performance thin-layer chromatography (HPTLC), overpressured thin-layer chromatography (OPTLC), centrifugal-layer chromatography (CLC), and high-pressure planar liquid chromatography (HPPLC) were developed. Though HPTLC is faster than HPLC, it has lower sensitivity. These techniques are now considered complementary rather than competitive. Numerous publications appearing in the literature attest to the versatility of TLC and its applicability in almost all fields of chemical science including pesticides, environmental pollution, fuels, foods, pharmaceuticals, industrial chemicals, natural products, radioactive compounds, and metal complexes.

The first report describing the application of TLC in inorganic analysis appeared in 1949 when two American chemists, Meinhard and Hall [4], successfully separated Fe^{2+} from Zn^{2+} on a layer prepared from a mixture of aluminum oxide (adsorbent) and celite (binder). However, the importance of inorganic TLC did not receive recognition until the beginning of the 1960s when Seiler separated inorganic substances [5–7]. After the publications of Seiler, TLC of metal ions received a great impetus and a rapidly increasing number of publications dealing with the separation of inorganic species appeared in the literature. The work on TLC of inorganics published up to the end of 1972 has been excellently reviewed by Brinkman et al. [8], and that which appeared during 1972 to 1980 has been presented by Kuroda and Volynets [9]. The work on TLC of inorganics and organometallics covering the period 1978 to 2003 has been well documented in chapters of the *Handbook of Thin-Layer Chromatography* edited by Sherma and Fried [10–12]. The excellent coverage of literature of TLC and HPTLC has appeared in biennial reviews of planar chromatography dating back to 1970, and the latest review of this series was published in 2004 [13].

The main uses of TLC include (1) qualitative analysis (the identification of the presence or absence of a particular substance in the mixture), (2) quantitative analysis (precise and accurate determination of a particular substance in a sample mixture), and (3) preparative analysis (purification and isolation of a particular substance for subsequent use). All these analytical and preparative applications of TLC require the common procedures of sample application, chromatographic separation, and

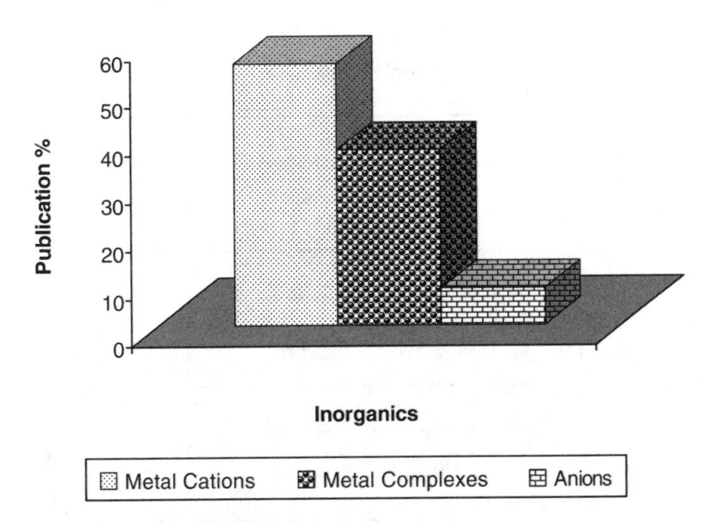

FIGURE 14.1 Percentages of TLC/HPTLC publications on metal cations, metal complexes, and anions that appeared during the period 1980 to 2004.

component visualization. However, analytical TLC differs from preparative layer chromatography (PLC) in that microgram amounts of mixtures are separated by analytical TLC, whereas larger weights or volumes of sample are applied to thicker (1.0 to 5.0 mm) layers in PLC in order to isolate milligram quantities (10 to 1000 mg) of material for further analysis. However, in micro PLC sample amounts (2 to 25 mg) are isolated. The amount of sample that can be applied to the preparative layer depends on the nature of mixture to be separated.

The survey of literature on inorganic TLC of last 25 yr indicates that most of the studies have been done on metal cations as compared to metal complexes and anions (Figure 14.1). Though a lot of work has been reported on inorganic analytical TLC applied for identification, separation, and quantification of microgram quantities of inorganic species [14–21], little work has appeared on PLC dealing with the isolation and separation of inorganics at the milligram level.

The aim of this chapter is to encapsulate the work performed on inorganic PLC during the last 25 yr.

14.2 METHODOLOGY

The methodology involved in PLC is similar to analytical TLC. The ascending technique is usually used for the development of TLC plates in closed cylindrical or rectangular chambers at room temperature. Most of the workers use layers of 0.2- to 2-mm thickness to handle sample amounts ranging from 0.15 to 5.0 mg. A definite volume from a series of sample solutions of different concentrations are spotted with the aid of a micropipette or syringe on activated laboratory-made TLC plate (glass sheet coated with an adsorbent layer) at about 2 to 3 cm above the lower edge of the plate. After drying the spot, the plate is developed, usually to a distance of 10 cm, in a glass jar or chromatographic chamber of appropriate size at room

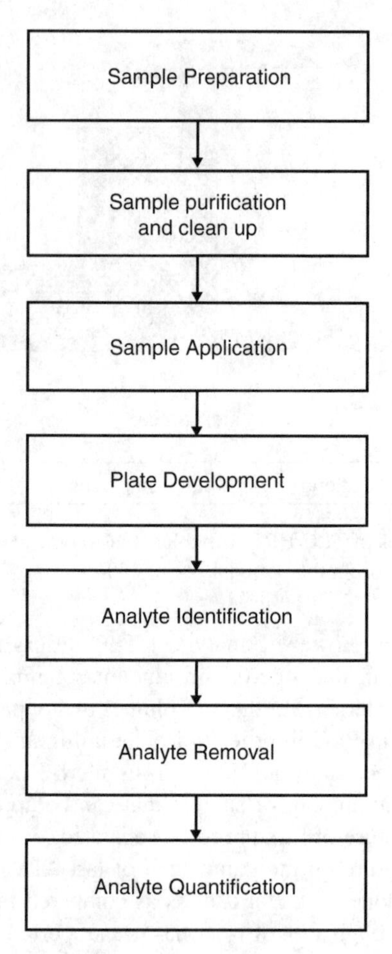

FIGURE 14.2 Flowchart of the steps involved in preparative inorganic thin-layer chromatography.

temperature using a one-dimensional ascending technique. After development, the TLC plate is dried in air. Chromogenic and fluorogenic reagents are sprayed onto the TLC plate to locate the separated species. The various steps involved in PLC have been depicted in Figure 14.2.

14.3 SAMPLE PREPARATION

Metal solutions are generally prepared by dissolving their corresponding salts (nitrates, chlorides, or sulfates) in distilled water, 0.1-*M* HNO_3, or 0.1-*M* HCl. Sample solutions of rare earths are prepared by dissolving their nitrates in 0.1-*M* HNO_3 or by fusion of their oxides followed by dissolution in 0.5- to 6.0-*M* HNO_3. Inorganic anion solutions are prepared from sodium, potassium, or ammonium salts of the corresponding acids using distilled water, dilute acid, or alkali. The solutions of metal complexes are generally prepared by dissolving appropriate amounts in ethanol, acetone, chloroform, or distilled water. In a few cases, metal complexes were produced directly in the separating zone of the TLC plate. The complexation of metals (or analyte), either in solution prior to spotting or *in situ* at the origin by overspotting of a complexing agent, has been recommended for obtaining improved chromatographic resolution and enhanced selectivity [22–24]. The prechromatographic derivatization of analyte on chelating sorbent layer has also been suggested. This approach has been utilized in the determination of heavy metals present in electroplating waste according to the *in situ* color reaction in the cellulose sorbent chelated with azopyrocatechol groups [25].

Special sample preparation methods have been used for isolation and separation of inorganics from complex matrices such as textile materials [26], biological samples [24,27,28], plants and food items [29–31], geological [32–34], pharmaceuticals [35], alloys [36], and environmental samples [37,38], etc. The methods for sample preparation were designed to provide a final spotting solution of adequate concentration and purity. Samples found to be below the sensitivity of the test are pretreated by extraction, evaporation, and solvent partitioning. The selection of a proper solvent for extraction enhances the specificity of the procedure and isolates the analyte of interest from interfering materials. Liquid waste samples are spotted directly on the plate or subjected to pretreatment (if analyte concentration is low) by solvent extraction following cleanup of the extract. The selective precipitation of the analytes of interest has been another effective technique of sample purification [39,40]. Solid samples are digested with acid and adjusted to a suitable volume [34,36,41,42].

14.4 STATIONARY PHASE

Similar to analytical-level TLC, PLC has been performed mostly on silica layers. Cellulose, alumina, and kieselguhr, alone or in combination with silica gel, have also been used. Most of the workers have used laboratory-made TLC plates (layers' thickness > 2.0 mm) obtained by applying the slurry of silica, alumina, kieselguhr, and cellulose or their combinations in different proportions on glass plates (20 × 20, 10 × 20 or 5 × 10 cm) with the aid of layer-spreading equipment. Precoated HPTLC plates have also been occasionally used. The impregnation of the adsorbent (LK cellulose) layer with hydrochloric acid (5 to 10 *M*) has improved the separation of molybdenum from tantalum by reducing their spot sizes [43]. Chelating cellulose sorbent with azopyrocatechol groups [25], irregularly shaped silufol chromatographic plate [44], tributyl-phosphate-impregnated silica plates [45], and thorium (Th^{4+})-modified silica gel layers [46] have been used for quantitative and preparative TLC. Thorium was introduced into the silica surface by utilizing the cation-exchange properties of the surface group. The silanol groups, being weakly acidic, undergo cation-exchange reaction when silica gel is immersed in an aqueous solution of thorium nitrate, resulting in the formation of thorium-exchanged silica gel:

$$Th^{4+} + 4\ (\text{–SiOH}) = Th^{4+}\ (OSi^-)_4 + 4H^+$$

A new layer material (mosambi skin powder + silica gel + alumina, 10:45:45)for the isolation of nickel from industrial wastes has been reported recently [47].

14.5 MOBILE PHASE

The selection of proper mobile phase in TLC exerts a decisive influence on the separation of inorganic ions. With a particular stationary phase, the possibility of separation of a complex mixture is greatly improved by the selection of an appropriate mobile phase system. In general, the mixed aqueous-organic solvent systems containing an acid, a base, or a buffer have been the most favored mobile phases for the separation of ionic species. The mobile phases used as developers in inorganic PLC include:

1. Inorganic solvents (acids, bases, salt solutions, mixtures of acids bases, and their salts).
2. Organic solvents (acids, bases, ketones, alcohols, chloroform, amines, esters and their mixtures in various proportions). Among these, acetone, which favors the formation of nondissociated metal complexes, has been the most preferred component of mixed organic solvents in inorganic PLC.
3. Mixed aqueous-organic solvents (organic solvents mixed with water, mineral, acids, inorganic bases, or salt solutions). The introduction of an organic solvent into aqueous mobile phase improves the flow hydrodynamics and facilitates the formation of more compact spots, especially for the metal cations, which tend to undergo hydrolysis.
4. Surfactant-mediated solvents (micellar and nonmicellar solutions, and the microemulsion systems of cationic, anionic, and nonionic surfactants). Micellar mobile phases with added organic or inorganic modifiers have been found more useful from the separation point of view. The use of acidified sodium dodecyl sulfate (an anionic surfactant) as part of a surfactant-containing mobile phase system for the separation of coexisting nickel, iron, and copper cations in synthetic ores on a silica gel layer has been reported recently [40].

14.6 DETECTION AND IDENTIFICATION

To locate the exact position of the analyte after development with the mobile phase, chromogenic and fluorogenic reagents capable of forming colored products with the separated species were sprayed onto the dried TLC plate. Compared to the use of universal detection reagents, selective chromogenic reagents are more frequently used. In some cases, detection is completed by inspecting the TLC plate under UV light, after spraying with a suitable reagent or exposing the plate to ammonia vapors.

In addition to using conventional chromogenic reagents, some new detection reagents such as arsenazo M, phenylarsonic acid, metanilic acid, phenolazotriaminorhodanine, sulfochlorophenol azorhodanine, 4-(2-pyridylazo) resorcinol, m-nitrochlorophosphonazo, and tribromochlorophosphonazo have been utilized for selective detection of metal ions. Selenium in food samples [30] has been detected as the 2,3-diaminonaphthalene–Se complex, which produces pink fluorescence under UV radiation (366 nm). For the detection of anions, saturated silver nitrate solution in methanol, diphenylamine (0.2 to 5.0%) in 4-*M* H_2SO_4, 1% bromocresol purple containing dilute NH_4OH, ferric chloride (10%) in 2-*M* HCl, alcoholic pyrogallol (0.5 %), and aqueous potassium ferrocyanide (1%) have been used. Inorganic anions have also been detected by 0.5% $FeCl_3$ solution as highly colored iron (III) diantipyrilmethane complexes or by viewing under UV light as fluorescent zones in the case of terbium (III) diantipyrilmethane complexes [48].

For the detection of rare earth elements (REEs) (La, Ce, Pr, Nd, Sm, Eu, Gd, Tb, Dy, Ho, Er, Tm, Yb, Lu, and Y), dilute solutions (0.02 to 1%) of tribromoarsenazo, arsenazo (III), and saturated ethanolic solution of alizarin have been used. REEs have also been detected by (1) spraying the TLC plate first with 0.1% arsenazo (III) solution and then with aqueous ammonia, followed by gentle heating, (2) exposing the plate to ammonia after spraying with tribromochlorophosphonazo or xylenol orange solution, and (3) heating the plate for 10 min at 70°C after spraying with 0.02% chlorophosphonazo solution.

The identification of separated compounds is primarily based on their mobility in a suitable solvent, which is described by the R_F value of each compound. Kowalska et al. have nicely discussed in greater detail the theory of planar chromatography and separation efficiency parameters in Chapter 2 of the third edition of the *Handbook of Thin-Layer Chromatography*, published in 2003.

14.7 QUANTIFICATION

There are three main methods for quantitative estimation of an analyte after isolation on thin-layer chromatoplates:

1. Measurement of spot area and visual estimation
2. Elution of analyte and spectrophotometry
3. Measurement of optical density directly on the layer (i.e., *in situ* densitometry)

The spot-area method is the simplest, most rapid, and most inexpensive, though it is less accurate. Usually, a linear relationship between the amount of analyte and the size (or area) of the spot exists. This relationship has been utilized by several workers for semiquantitative determination of inorganic species [22,25,40,44,45,47–51].

This procedure has been utilized to determine metal cations and anions in water sample [48,50,51], titanium in high-speed steel at a concentration level of 25 ± 3 mg/g [22], heavy metals (20 to 400 mg/l) in electroplating waste waters [25], copper and nickel (5 mg/l) in metal electroplating baths on wedge-shaped plates [44], copper, lead, cadmium, or mercury in vegetable juices [29], and nickel (1 to 3.8 mg/l) in electroplating waste water of lock industries [42,47].

The zone elution method has been used for quantitative estimation or recovery of heavy metals in plants and vegetable juices [29], mercury (II) in river and waste waters [52], zinc in different environmental samples [46], nickel and copper in alloys [53], zirconium in Mg–Al alloys [22], cobalt, zinc, nickel, and copper in natural water and alloy samples [54], thiocyanate in spiked photogenic waste water [55], and aluminum in bauxite ores [42].

In situ densitometry has been the most preferred method for quantitative analysis of substances. The important applications of densitometry in inorganic PLC include the determination of boron in water and soil samples [38], NO_3^- and $Fe(CN)_6^{3-}$ in molasses [56], Se in food and biological samples [28,30], rare earths in lanthanum, glass, and monazite sand [22], Mg in aluminum alloys [57], metallic complexes in ground water and electroplating waste water [58], and the bromate ion in bread [59]. TLC in combination with *in situ* fluorometry has been used for the isolation and determination of zirconium in bauxite and aluminum alloys [34]. The chromatographic system was silica gel as the stationary phase and butanol + methanol + HCl + water + HF (30:15:30:10:7) as the mobile phase.

In addition to the aforementioned methods, TLC in combination with other instrumental techniques have also been used for quantification of inorganic species. For example, two-dimensional TLC coupled with HPLC has been utilized for the separation and quantification of REEs in nuclear fuel fission products using silanized silica gel as layer material [60]. In another interesting method, REEs in geological samples have been determined by ICP-AAS after their preconcentration by TLC on Fixion plates [32]. TLC in combination with neutron activation has been used to determine REE in rock samples on Fixion 50 × 8 layers with the sensitivity limit of 0.5 to 10 μg/g for 10- to 30-mg samples [41]. A combination of TLC and AAS has been utilized for the isolation and determination of zinc in forensic samples [27].

14.8 SPECIFIC APPLICATIONS

14.8.1 Analysis of Synthetic Mixtures

In order to utilize TLC on preparative scale, microgram amounts of an analyte are separated from milligram quantities of another analyte from their mixtures. The results are summarized in Table 14.1 and Table 14.2. At higher loading of an analyte, separation possibilities are reduced owing to the formation of diffused spots.

14.8.2 Analysis of Food Samples

A novel TLC spectrofluorometric method for identification and determination of selenium in different food samples of animal and vegetable origin has been proposed [30]. The procedure involves the digestion of food sample (1 to 5 g) in the presence of conc. HNO_3 (5 ml), 70% $HClO_4$ (10 ml), and H_2O (10 ml) in a 250-ml Kjeldahl flask; reduction of Se(VI) into Se(IV); complexation of the isolated selenium with 2,3-diaminonaphthene (DAN); extraction of the resultant Se–DAN complex with cyclohexane; and spectrofluorometric determination followed by confirmation of the presence of Se in the sample by TLC using thin layers of MN-300 cellulose powder.

TABLE 14.1
Microgram-to-Milligram Separation of Inorganics from Synthetic Mixtures

Stationary Phase	Mobile Phase	Quantitative Separation	Reference
Silica gel G	Formic acid (22 *M*) + propanol-2-ol (1:9)	Separation of IO_3^- (500 µg) from Cl^- (20.0 mg), Br^- (50 mg) or I^- (100.0 µg) and the separation of Cl^- (500 µg) from IO_3^- (9.0 mg) or Br^- (100 µg) from IO_3^- (1.0 mg) or I^- (100 µg) from IO_3^- (10.0 mg)	61
Silica gel G	Formic acid + propanol + acetone (25:15:60)	Separation of nickel chlorosulfate (5 µg, R_F = 0.38) from cobalt chlorosulfate (5.0 mg, R_F = 0.88)	62
Silica gel G impregnated with 0.3-*M* aqueous sodium molybdate	1.0-*M* formic acid in butanol	Separation of microgram amounts of Zn^{2+}, Cd^{2+}, Al^{3+}, Fe^{2+}, Bi^{3+}, Pb^{2+}, or Ag^+ from Tl^{3+} (0.4 to 1.0 mg, R_F 0.8 to 0.85)	63
Silica gel G	1.0-*M* aqueous sodium formate + 1.0-*M* aqueous KI (9:1)	Separation of Cd^{2+} (50 to 100 µg) from Zn^{2+} (0.5 to 4.0 mg) and of Cu^{2+} (50 to 100 µg) from Ni^{2+} and Co^{2+} (0.5 to 4.0 mg)	64
Silica gel G	5% Aqueous sodium dodecyl sulfate + 1% aqueous CH_3COOH (1:1)	Separation of Cu^{2+} (10 µg, R_F = 0.46) from Ni^{2+} (0.8 mg, R_F = 0.74)	40
Silica gel G impregnated with 1% ammonium thiocyanate	1.0-*M* aqueous formic acid + 1.0-*M* aqueous sodium formate (3:7)	Separation of Ni^{2+} or Co^{2+} (5 to 100 mg) from 50 µg of Zn^{2+}, Fe^{2+}, $UO^{2+}{}_2$, and Ti^{4+}	65

After completing the spectrofluorometric determination, the cyclohexane phase containing the Se–DAN complex was concentrated nearly to dryness and the residue was dissolved in 0.5 ml of cyclohexane. This solution (0.5 ml) was spotted on the TLC plate and chromatographed along with a standard selenium sample using ethanol + 25% ammonia solution (7:3) as the mobile phase. The Se–DAN complex produces a pink fluorescence upon the exposure of the chromatogram to UV radiation (360 nm). It was always possible to confirm the presence of selenium in food samples by TLC in all cases in which the fluorometric response was positive. The results are shown in Figure 14.3 to Figure 14.5.

Another interesting TLC method for the isolation and determination of bromate ion in flour dough and breads has been developed [59]. It involves extraction of BrO_3^- from foodstuff, purification on alumina column, TLC separation on silica gel layer developed with water + *n*-butanol + *n*-propanol (1:1:3), and quantification by densitometry. Bromate ion down to 0.1 µg in bread (1.0 g) was detected with tolidin–HCl reagent.

TABLE 14.2
Quantitative Separation of Ni from Zn, Cd,Cu, Fe^{3+}, and Pb

Separation	Volume Loaded	Amount Loaded	R_L and R_T Values (Average of Triplicate Results)
Ni–Zn	20 µl (1 ml)	20 µg (10 mg)	1.00–0.95 (0.30–0.00)
	0.05 ml (0.05 ml)	5 × 102 µg (5 mg)	1.00–0.80 (0.40–0.00)
	0.05 ml (0.3 ml)	5 × 102 µg (30 mg)	1.00–0.80 (0.70–0.00)
	0.1 ml (0.5 ml)	1 mg (5 × 102 µg)	1.00–0.80 (0.45–0.00)
	0.5 ml (5 µl)	5 mg (50 µg)	1.00–0.80 (0.25–0.00)
	0.2 ml (5 µl)	20 mg (50 µg)	1.00–0.70 (0.20–0.00)
	0.3 ml (10 µl)	30 mg (100 µg)	1.00–0.75 (0.50–0.00)
	0.3 ml (30 µl)	30 mg (300 µg)	1.00–0.75 (0.48–0.00)
	0.2 ml (5 µl)	20 mg (50 µg)	1.00–0.70 (0.60–0.00)
Ni–Cd	0.3 ml (5 µl)	30 mg (50 µg)	1.00–0.70 (0.65–0.00)
	0.3 ml (10 µl)	30 mg (100 µg)	1.00–0.70 (0.50–0.30)
Ni–Cu	0.3 ml (0.3 ml)	30 mg (30 mg)	1.00–0.70 (0.50–0.30)
Ni–Fe^{3+}	0.3 ml (0.3 ml)	30 mg (30 mg)	1.00–0.70 (0.50–0.00)
Ni–Pb	0.3 ml (0.3 ml)	30 mg (30 mg)	1.00–0.70 (0.35–0.00)

Note: Stationary phase: Plain silica gel G; Mobile phase: 1.0-*M* aqueous formic acid + 1.0-*M* aqueous sodium molybdate (1:9); Quantities in parentheses denote those of the second element of the mixture, i.e., 20 µl (1 ml) in the first row indicates that the volume of Zn was 1 ml.

Source: From Ajmal, M., Mohammad, A., Fatima, N., and Ahmad, J., *J.Planar Chromatogr. Mod. TLC*, 1, 329–335, 1988. Reproduced with permission of Editor-in-Chief, JPC, Hungary.

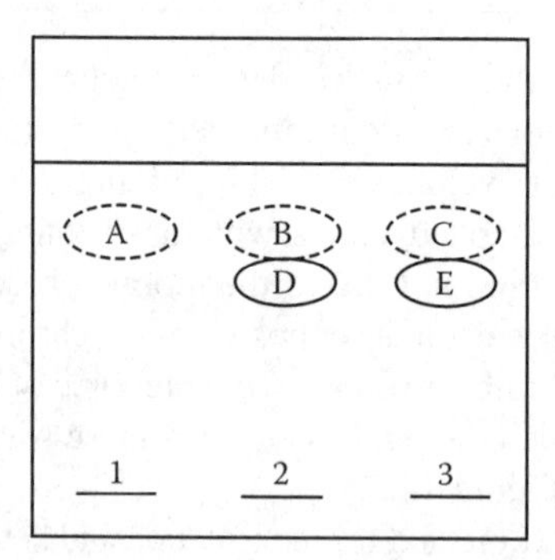

FIGURE 14.3 Thin-layer chromatogram. 1, blank; 2, sample; and 3, standard selenium–DAN complex. Broken codes represent areas of green fluorescence and solid codes, areas of pink fluorescence. The support was cellulose powder MN-300 and the developing solvent ethanol-25% ammonia solution (70 + 30). On exposure to UV light at 360 nm, the selenium–DAN complex gives a pink fluorescence. (Reproduced from Moreno-Dominguez, T., Garcia-Moreno, C., and Marine-Font, A., *Analyst*, 108, 505–509, 1983. With the permission of The Royal Society of Chemistry, U.K.)

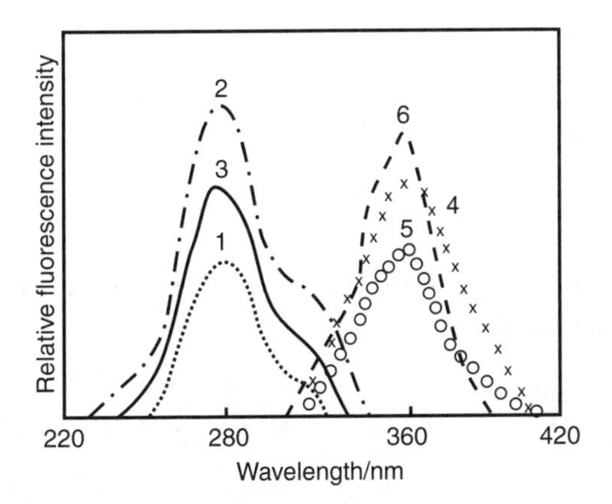

FIGURE 14.4 Fluorescence excitation spectra for the removal from cyclohexane of substances shown in the chromatogram in Figure 14.3 and the standard selenium–DAN complex not chromatographed previously: 1, substance A (in the chromatogram gives a green fluorescence); 2, substance B (in the chromatogram gives a green fluorescence); 3, substance C (in the chromatogram gives a green fluorescence); 4, substance D (in the chromatogram gives a pink fluorescence); 5, substance E (in the chromatogram gives a pink fluorescence); and 6, standard selenium–DAN complex not chromatographed previously. (Reproduced from Moreno-Dominguez, T., Garcia-Moreno, C., and Marine-Font, A., *Analyst.*, 108, 505–509, 1983. With the permission of the Royal Society of Chemistry, U.K.)

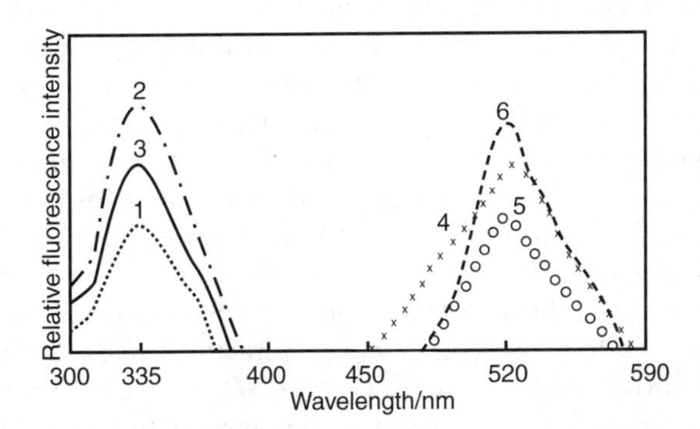

FIGURE 14.5 Fluorescence emission spectra for the removal from cyclohexane of substances shown in the chromatogram in Figure 14.3 and the standard selenium–DAN complex not chromatographed previously: 1, substance A (in the chromatogram gives a green fluorescence); 2, substance B (in the chromatogram gives a green fluorescence); 3, substance C (in the chromatogram gives a green fluorescence); 4, substance D (in the chromatogram gives a pink fluorescence); 5, substance E (in the chromatogram gives a pink fluorescence); and 6, standard selenium–DAN complex not chromatographed previously. (Reproduced from Moreno-Dominguez, T., Garcia-Moreno, C., and Marine-Font, A., *Analyst.*, 108, 505–509, 1983. With the permission of the Royal Society of Chemistry, U.K.)

14.8.3 Analysis of Geological Samples

TLC has very important applications in the isolation and determination of REEs in rocks, ores, and monazite sand.

TABLE 14.3
Determination of Noble Metals in Ore Samples by GFAAS–HPTLC

Metal Concentration (μg/g)	Ore Sample			
	SARM 7	SARM 8	CHR-Pt +	CHR-Bkg
Pt	3.68	2.50	13.0	0.04
Pd	1.51	0.08	64.0	0.06
Rh	0.20	—	—	—

14.8.3.1 Analysis of Rocks and Minerals

To analyze silicate rocks and minerals, a rock sample (≈ 2 g) was taken in a platinum dish along with few drops of water, conc. $HClO_4$ (2 to 5 ml), and conc. HF (20 to 50 ml). The contents were boiled to dryness, the residue dissolved in dil. HNO_3 or HCl (0.1 *M*), and the resultant sample solution was used for chromatography. For rock samples containing large quantities of phosphorous and lead, only HF was used for acid treatment, and lead fluoride was removed by filtration. Pertinent examples of rock analysis by TLC include preconcentration of REEs from rock samples (10 to 30 mg) by circular TLC on Fixion 50 × 8 and their determination by neutron activation analysis [41]; separation and determination of rare earths present in ores and rock samples using diethylether + bis (2-ethyl hexyl) phosphate + nitric acid (100:1:35) as mobile phase [33]; and the identification of Fe^{3+} in rock sample by reversed-phase TLC system comprising 0.2 *M* of tributyl-phosphate-impregnated stannic arsenate–silica gel mixed layer as stationary phase and 1.0-*M* KSCN + 5.0-*M* HCl + 1.0-*M* NaCl (8:1:1) as mobile phase [66].

Mineral samples with high content of alumina are decomposed by fusion in the presence of excess NaOH or Na_2O_2 in a corundum crucible, using a mixture of Na_2CO_3 and $Na_2B_4O_7$ as fluxing agent. A mixture of conc. HNO_3 and HCl is used to decompose carbonates. Tatanates and niobotantanates are decomposed by fusion with potassium pyrosulfate or by heating with a mixture of conc. H_2SO_4 and $(NH_4)_2SO_4$. The rare earths and yttrium in rocks, ores, and minerals were determined by atomic emission spectroscopy coupled with inductively coupled plasma [32] after preconcentration on Fixion TLC plates (circular variant) using aqueous solution of oxalic acid (0.1 *M*), ammonium chloride (2.0 *M*), ammonium citrate (0.5 *M*), or HCl (5.0 *M*) as eluent.

The strong selectivity of *N,N*-dialkyl *N*-benzoylthiourea toward platinum metals has been favorably exploited to determine noble metals (Rh, Pd, Pt, and Au) in samples of ore and rocks by graphite furnace atomic absorption spectroscopy (GFAAS) and UV detection after liquid chromatographic separation on silica HPTLC plates [23]. The results are presented in Table 14.3.

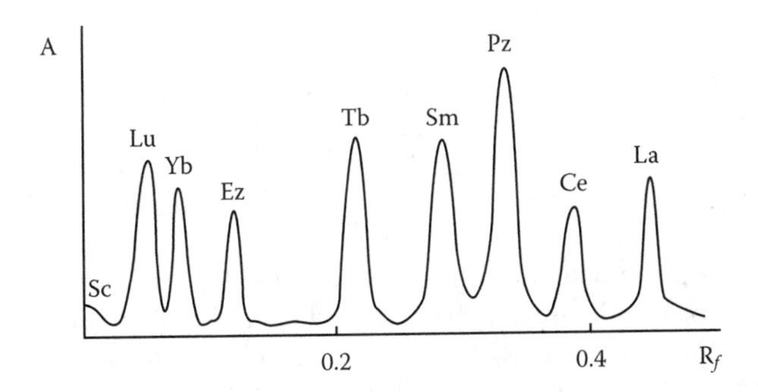

FIGURE 14.6 Densitogram of the mixture of rare earth element complexes. Spraying reagent 0.1% arsenazo III; wavelength: 590 nm. (Reproduced from Timerbaev, A., Shadrin, O., and Zhivopistsev, V., *Chromatographia*, 30, 436–441, 1990. With the permission of Vieweg Publishing GWV-Fachverlage GmbH, Wiesbaden, Germany.)

14.8.3.2 Analysis of Monazite Sand

TLC spectrophotometry and TLC densitometry have been very useful for determination of REEs in monazite sand. A monazite sample (1 g) is heated with a mixture of conc. H_2SO_4 and HF (1:1), and the contents boiled until SO_3 vapors appear. The REEs are separated from the matrix by extraction with 0.2-*M* diantipyrilmethane (DAM) solution in chloroform at pH 5 in the presence of 5.0-*M* NH_4NO_3. The extraction process is repeated to ensure the complete extraction of all REEs, and the combined extract is evaporated to dryness. The residue is dissolved in propanol saturated with NaCl and 0.5-μl aliquot was used for chromatography on silufol plate using propanol + 0.07-*M* HCl (4:1) as developer. The separated REEs were detected by spraying plates with arsenazo(III) solution (0.1%), spots are scanned at 590 nm, and the concentrations of REEs are determined from a calibration curve obtained by plotting peak area vs. the amount of REEs (linear range, 0.6 to 4.9 μg and detection limit 0.2 μg). With this procedure, the amounts of La and Tb in monazite sand were found to be 4.0 ± 2 and 1.1 ± 0.2 mg/g, respectively [22]. Similarly, La content in lanthanum glass was found to be 179 ± 5 mg/g. The representative densitogram is shown in Figure 14.6. Other interesting TLC methods developed for the analysis of monazite sand include (1) simultaneous densitometric determination of light rare earths on thin-layer chromatograms using diisopropyl ether + diethyl ether + bis (2-ethyl hexyl) phosphate + HNO_3 (8:8:0.4:0.07) as eluent [67]. The R_F values of the isolated rare earths such as lanthanum, cerium, praseodymium, neodymium, samarium, and yttrium on thin layers of silica gel mixed with 1% carboxymethylcellulose (binder) containing 4% ammonium nitrate were 0.13, 0.39, 0.55, 0.69, 0.90, and 0.98, respectively. The densitometric calibration graphs were linear in the range 0.015 to 0.60 μg of individual rare earth and (2) TLC separation, identification, and determination of lanthanum, cerium, praseodymium, neodymium, and samarium on silica layers using diisopropyl ether + tetrahydrofuran + tetrabutyl phosphate + HNO_3 (10:6:1:1) as eluent [68].

14.8.3.3 Analysis of Bauxite Ore

Preparative TLC has been used to determine zirconium and aluminum in bauxite ore samples. Zirconium was first isolated from bauxite on silica gel layers using butanol + HCl + methanol + water + HF (30:30:15:10:7) as the mobile phase and then determined by *in situ* fluorometry [34]. An interesting micellar TLC method for the separation of Al^{3+}, Ti^{4+}, and Fe^{3+} from bauxite ore with consequent spectrophotometric determination of aluminum has been developed [42]. Bauxite ore solution (1%) was obtained by treating bauxite sample (1 g) with conc. HCl (10.0 ml), followed by adding 5 ml conc. HNO_3 and 15 ml of dil. H_2SO_4 to it and then heating the contents at 100°C for 1 h. The SiO_2 present in the sample was removed by filtration. The filtrate was completely dried, the residue dissolved in 1% HCl, and the total volume increased to 100 ml with 1% HCl. An aliquot (0.01 ml) of ore solution was spotted on silica gel layers, and the plates were developed with 10^{-3}-*M* aqueous sodium bis (2-ethyl hexyl) sulfosuccinate + 1.0-*M* aqueous formic acid (1:1, v/v). The resolved spots of Al^{3+}, Ti^{4+}, and Fe^{3+} were identified by spraying with appropriate chromogenic reagents.

Aluminum was quantitatively determined by spectrophotometry (λmax 540 nm) after extracting from silica layer with double-distilled water and developing the color with 0.1% aqueous aluminon.

14.8.4 Analysis of Alloys

TLC spectrophotometry is used to determine zirconium in Mg–Al alloy. For this purpose, the alloy sample (2 g) is dissolved in HNO_3 (20 ml, 6 *M*), and zirconium is extracted in 6 ml of 0.02-*M* diantipyrilmethane (DAM) solution in chloroform. The extract was concentrated to 0.4 ml and an aliquot (10 μl) was chromatographed on silica gel LS plate using 4-*M* HCl + dimethylformamide (1:2) as the mobile phase. After development, the portion of the sorbent layer containing the zirconium–DAM complex was removed, and the metal was extracted with 6-*M* HCl. The zirconium present in this solution was determined in the form of a xylenol orange complex (λmax, 540 nm) by spectrophotometry [22].

TLC atomic emission spectrometric (AES) method was developed to determine tantalum in molybdenum alloys containing ≥ 0.5% tantalum. The procedure involves the separation of tantalum from molybdenum alloy on a silica gel layer (0.3 mm thick) using 10.0-*M* HCl + acetone (1:1) as developer and the subsequent determination of tantalum by AES [43].

A circular TLC spectrophotometric method for the determination of lanthanum and yttrium at concentration level of 0.01 to 1.0% in molybdenum-based alloys has also been developed. It involves the separation of lanthanum and yttrium on cellulose layers impregnated with 0.2-*M* trioctylamine using aqueous HCl as developer, extraction from sorbent layer, and determination by spectrophotometry [69].

TLC is used to determine copper in aluminum alloys. The process involves the sampling of the investigated material by anodic dissolution, development of TLC plate with acetone + HCl + H_2O (70:15:15), and the identification of analyte by 1-(2-pyridylazo)-2-naphthol [70]. A TLC system comprising silica gel as stationary phase and a mixture of aqueous solutions of sodium dodecyl sulfate (5%) and acetic acid (1%) in 1:1 ratio has been utilized to isolate copper from brass, bronze, and german silver alloy samples [40].

14.8.5 Analysis of Aquatic Plants

TLC has been used for the identification of heavy metals in aquatic plants. For this purpose, plants were mineralized with conc. H_2SO_4, HNO_3, and H_2O_2, extracted with water, derivatized with dithizone, and chromatographed. The identified metals were zinc, copper, mercury, and lead [31].

14.8.6 Analysis of Cotton Materials

Metal-free cotton material samples were impregnated with standard solutions of iron, copper, and manganese ions containing 10 to 100 μg of metal per gram of cotton material. It was completely dried and 1 g of cotton sample was combusted with a mixture of conc. H_2SO_4 and H_2O_2. The excess H_2SO_4 was volatilized, and the residue was dissolved in 1 ml of HCl (2.0 *M*). An aliquot (10 μl) of each solution was chromatographed on precoated microcrystalline cellulose plates using acetone + HCl + H_2O (8:1:2) as the mobile phase. Spots were visualized by spraying the TLC plate with rubeanic acid (0.5% ethanolic solution) followed by exposure to ammonia vapors. In cotton material, the metals were detected at a lower limit of 20 μg/g of material [26]. The results are presented in Figure 14.7.

14.8.7 Analysis of Pharmaceuticals

Iron(II) has been isolated from pharmaceutical formulations on microcrystalline cellulose layers using chloroform + propanol + acetic acid + HNO_3 + 4-*M* HCl (10:30:10:5:5) as the mobile phase. After isolation, the amount of Fe^{2+} present in the pharmaceuticals was determined by spectrophotometry using 1,10-phenanthroline as chromogenic reagent [71]. A TLC method for identification of inorganic anionic and cationic impurities in drugs of the Austrian pharmacopoeia has been developed [35]. Aqueous drug solutions (10%) were spotted on cellulose microplates and developed with MeOH + HCl (8:1) for cations or with MeOH + water + *n*-butanol (2:1:1) for anions. Inorganic impurities were detected as colored spots by spraying the developed plates with 16 spray reagents.

14.8.8 Analysis of Biological Samples

14.8.8.1 Human Placenta Sample

Human placenta (20 g) was completely dried at 105°C, crumbled, and a portion (5 g) was mineralized by treating with nitric acid (12 *M*, 15 ml) at 110°C in a Teflon bomb. After mineralization, the contents were evaporated to dryness and the residue was dissolved in 1.0 ml of distilled water (termed sample A). An aliquot (10 μl) was chromatographed on RP-18 using MeOH + H_2O + CH_3COOH (25:15:2) as the mobile phase. The separated spots of the metals were visualized by spraying the

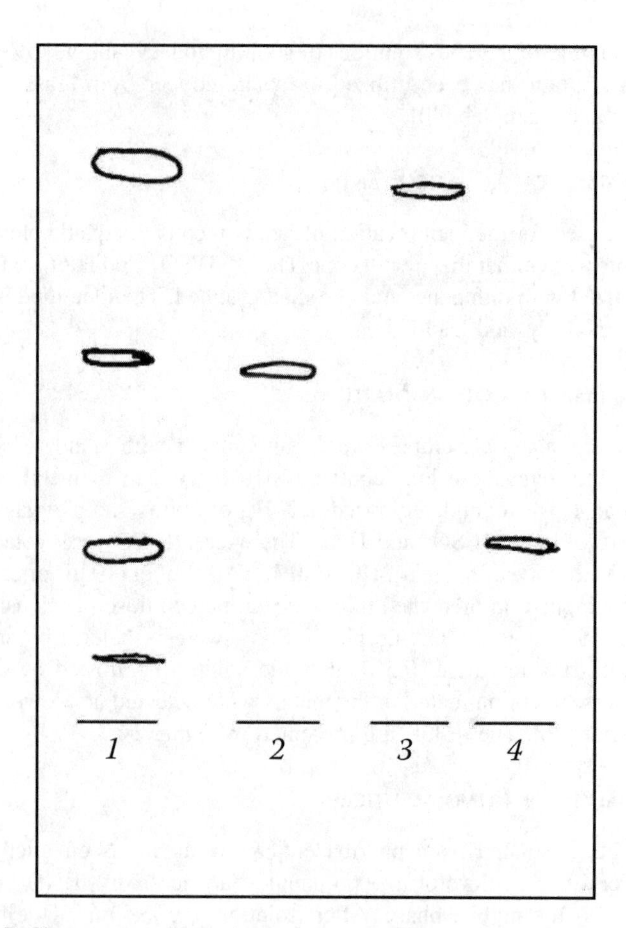

FIGURE 14.7 Chromatogram from impregnated samples of cotton material with detected spots: 2, copper; 3, iron; and 4, manganese. The amount of applied copper, iron, and manganese ions is 0.2 μg/1.5 cm. Unknown samples (1) of cotton material shows sharp-edged spots of copper ($R_F = 0.65$), iron ($R_F = 0.95$), manganese ($R_F = 0.343$), and a black spot at $R_F = 0.11$ (probably nickel). (Reproduced from Kastelan-Macan, M., Bokic, Lj., Cerjain-Stefanovic, S., and Moskaliuk, K., *Chromatographia*, 22, 19–20, 1986. With the permission of Vieweg Publishing GWV-Fachverlage GmbH, Wiesbaden, Germany.)

plates with 0.25% alcoholic solution of 1-(2-pyridylazo)2-naphthol followed by exposure to ammonia vapors.

Alternatively, an aqueous solution of sodium diethyldithiocarbamate (3.5%, 2 ml) or freshly prepared solution of dithizone in chloroform (0.1%, 10 ml) was added to sample A. The metal diethyldithiocarbamates (termed sample B) or metal dithizonates (termed sample C) thus formed were extracted in chloroform. The volume of chloroform extract was reduced to 1.0 ml. Aliquots (10 μl) each of sample B and sample C were chromatographed on plates coated with 0.25-mm layer of silica gel G using benzene + methyl isopropylketone (50:1) and toluene + chloroform (50:1), respectively, as mobile phases. Metal dithizonates were self-detected. The natural colored metal diethyldithiocarbamates were converted into brown spots by spraying with copper sulfate solution. The presence of various metals (R_F values are given in parentheses for metal ions, metal dithizonate, and metal diethyldithiocarbamate, respectively) such as zinc (0.58, 0.48, 0.72), lead (0.1, 0.25, 0.0), copper (0.28, 0.43, 0.68), and manganese (0.70, not detected, 0.46) in most of the placenta samples and that of nickel (0.88, 0.38, 0.6) in a few samples was established. However, the existence of cadmium and cobalt could not be confirmed in any sample [24].

14.8.8.2 Human Viscera Sample

Human viscera containing zinc was subjected to wet digestion. Tissue (2 to 5 g) was kept overnight in contact with a mixture of conc. H_2SO_4 (4 ml), conc. HNO_3 (20 to 25 ml), and 3 ml of $HClO_4$ (60 to 70%). The contents were digested on a hot plate, the solution filtered, and the filtrate boiled to remove SO_3, HNO_3, and $HClO_4$. After cooling, 5 ml of H_2O was added to the extract and heated again to dryness. The residue was extracted in 0.05-*M* HCl and analyzed by silica TLC using 0.01-*M* aqueous solution of sodium thioglycolate as the mobile phase. The identified zone of Zn^{2+} on the TLC plate was removed from the silica layer, and the amount of zinc was determined by AAS [27].

Some of the TLC methods that have not been covered in the preceding text and have been used for isolation, identification, and quantification of inorganics in water, milk, soft drinks, and edible oils are listed in Table 14.4 and Table 14.5.

TABLE 14.4
Results of Isolation, Identification, and Quantification of Inorganics from Water Samples

		TLC System			
Species	Sample	Stationary Phase	Mobile Phase	Remark	Reference
$CuSO_4$, $HgCl_2$ $CdSO_4$, $AgNO_3$	Fresh water	Silica gel G	0.1–1% aqueous NaCl	Identification of ionized free metal compounds	72
$HgCl_2$, $CuSO_4$	Fresh and sea water	Silica gel G	Distilled water	Detection of $HgCl_2$ (100 ng) and $CuSO_4$ (0.5–5 μg) by micro TLC enzymatic detection method	73
Hg(II)	Fertified river and industrial waste waters	Oxalic-acid-treated silica gel G	Ethyl acetate + acetone + H_2O (8:7:4:1)	Recovery of mercury (5–15 μg/10 ml) from river and industrial waste water by TLC spectrophotometry	52

(Continued)

TABLE 14.4 (*Continued*)
Results of Isolation, Identification, and Quantification of Inorganics from Water Samples

Species	Sample	TLC System: Stationary Phase	TLC System: Mobile Phase	Remark	Reference
Cu, Cd, Co, Pb	Groundwater and electroplating waste water	Silica gel, sodium carboxy methyl. cellulose	0.2-M acetic acid + 0.2-M sodium acetate (1:1)	Recovery of metal ions (96–104%) by TLC densitometry	58
Cu, Fe, Zn	Waste water	Silica gel G	0.1-M sodium malate solution (pH = 4)	Waste water samples received from decarbonization plant contains Fe^{3+} (0.2–0.56 mg/l, R_F = 0.77), Cu^{2+} (0.1–0.57 mg/l, R_F = 0.93), and Zn^{2+} (0.03–1.6 mg/l, R_F = 0.69)	37
Zn	Electroplating waste water	Th^{4+}-impregnated silica gel	0.1-M aqueous sodium formate	Zinc (0.68 mg/l in industrial waste water released from electroplating units was determined after separation from nickel by TLC spectrophotometry	46
Fe, Mn, Zn	Tube well water	Silica gel G	0.1-M aqueous sodium acetate (pH = 5)	The metal contents (mg/l) in different tube well water samples were iron (15.2–16.7), manganese (0.6–3.2), and zinc (0.6–3.6)	75
Cu	Spiked water and industrial waste water	Silica gel G	1.0-M aqueous sodium formate + 1.0-M aqueous KI (1:9)	TLC coupled with atomic absorption spectrometry and titrimetry was used for the recovery of copper	76

TABLE 14.5
Use of TLC in Isolation and Identification of Inorganics in Milk, Edible Oil, and Soft Drinks

Inorganic Species	Sample	Procedure	Remark	Reference
Human milk	Fe, Mn, and Co	Metals were extracted with isobutyl methyl ketone + amyl acetate (2:1) and chromatographed on cellulose layer	Separation and identification of iron, manganese, and cobalt	76
Spiked human milk	Li and Bi	Human milk was spiked with Li_2Co_3 and $Bi(NO_3)_3$, ashed, dissolved in proper solvent, and chromatographed on cellulose plate developed with methanol + 10-M HCl (3:2)	Identification of lithium and bismuth with detection limit of 0.25 and 1.5 μg, respectively.	77
Cheese and milk	Polyphosphoric acids	Acids were extracted from food samples with 25% trichloric acid and separated on polyamide plates with formic acid + *n*-butanol (1:1)	Separation, identification, and determination of polyphosphoric acids	78
Soft drinks	Ortho and polyphosphates	Two-dimensional TLC and ion-exchange chromatography	Separation and determination of polyphosphates	79
Edible oils	Silicones	Silicon is separated on silica gel layer using light petroleum + diethyl ether (98:2) as the mobile phase and detection with rhodamine B	Isolation and detection of silicones from vegetable oils (detection limit 2 μg)	80

ACKNOWLEDGMENT

I am highly grateful to the Almighty who bestowed on me this knowledge. I wish to express my sincere thanks to all the publishers and editors who granted permission to reproduce figures and tables from published articles in their journals. My special thanks are due to T. Kowalska for his valuable suggestions and constructive criticism. My loving thanks to my research group — Hina Shahab, Showkat Ahmad, and Rubi Gupta for their help during the literature search. I am indebted to my colleagues M.Z.A. Rafiquee and R.A.K. Rao for their help during the preparation of the manuscript.

REFERENCES

1. Izmailov, N.A. and Schraiber, M.S., *Farmatsiya*, 3, 1–7, 1938.
2. Stahl, E., *Pharmazie*, 11, 633–637, 1956.
3. Stahl, E., *Chem. Ztg*., 82, 323–329, 1958.
4. Meinhard, J.E. and Hall, N.F., *Anal. Chem*., 21, 185–188, 1949.
5. Seiler, H. and Seiler, M., *Helv. Chim. Acta*, 43, 1939–1941, 1960.
6. Seiler, H. and Seiler, M., *Helv. Chim., Acta*, 44, 939–941, 1960.
7. Seiler, H. and Seiler, M., *Helv. Chim., Acta*, 45, 381–385, 1962.
8. Brinkman, U.A. Th., de Vries, G., and Kuroda, R., *J. Chromatogr*., 85, 187–326, 1973.
9. Kuroda, R. and Volynets, M.P., in *CRC Handbook of Chromatography: Inorganics,* Vol. I, Qureshi, M., Ed., CRC Press, Boca Raton, FL, 1987.
10. Mohammad, A. and Varshney, K.G., in *Handbook of Thin-Layer Chromatography,* Vol. 55, Sherma, J. and Fried, B., Eds., Marcel Dekker, New York, 463–539, 1991.
11. Mohammad, A., in *Handbook of Thin-Layer Chromatography,* Vol. 71, Sherma, J. and Fried, B., Eds., 1996, pp. 507–619.
12. Mohammad, A., in *Handbook of Thin-Layer Chromatography,* Vol. 89, Sherma, J. and Fried, B., Eds., 2003, pp. 607–634.
13. Sherma, J., *Anal. Chem*., 76, 3251–3261, 2004.
14. Mohammad, A. and Khan, M.A.M., *Chem. Environ. Res*., 1, 3–31, 1992.
15. Mohammad, A. et al., *J. Chromatogr.,* 642, 445–453, 1993.
16. Mohammad, A. and Tiwari, S., *Sep. Sci. Technol*., 30, 3577–3614, 1995.
17. Mohammad, A. et al., *J. Planar Chromatogr. Mod. TLC*, 9, 318–360, 1996.
18. Sherma, J., *Anal. Chem*., 72, 9R–25R, 2000.
19. Horvat, A.J.M. et al., *J. Planar Chromatogr. Mod. TLC*, 14, 426–429, 2001.
20. Sherma, J., *Anal. Chem*. 74, 2653–2662, 2002.
21. Forgacs, E. and Cserhati, T., *J. Liq. Chromatogr. Relat. Technol*., 25, 2023–2038, 2002.
22. Timerbaev, A., Shadrin, O., and Zhivopistsev, V., *Chromatographia*, 30, 436–441, 1990.
23. Schuster, M., *Fresenius J. Anal. Chem*., 342, 791–794, 1992.
24. Baranowska, I. et al., *J. Planar Chromatogr. Mod. TLC*, 5, 469–471, 1992.
25. Myasoedova, G.V. et al., *Zh. Anal. Khim*., 41, 662–665, 1986.
26. Kastelan-Macan, M., Bokic, Lj., Cerjain-Stefanovic, S., and Moskaliuk, K., *Chromatographia*, 22, 19–20, 1986.
27. Deshmukh, L. and Kharat, R.B., *J. Liq. Chromatogr*., 14, 1483–1494, 1991.
28. Funk, W. et al., *J. High Resol. Chromatogr. Chromatogr. Commun*., 9, 224–235, 1986.
29. Volynets, M.P., Gureva, R.F., and. Duvrova, T.V., *Zh. Anal. Khim*., 46, 1595–1600, 1991.
30. Moreno-Dominguez, T., Garcia-Moreno, C., and Marine-Font, A., *Analyst*., 108, 505–509, 1983.
31. Kavetskii, V.N., Karnaukhov, A.I., and. Palienko, I.M., *Gidrobiol. Zh*., 20, 65–68, 1984.
32. Kuzmin, N.M. et al., *Zh. Anal. Khim*., 48, 898–910, 1993.
33. Jung K., Specker H., and Yenmez, I., *Chim. Acta Turc*., 8, 179–186, 1980.
34. Kastelan-Macan, M. and Cerjan-Stefanovic, S., *Chromatographia*, 14, 415–416, 1981.
35. Buchbauer, G. and Vasold, I., *Sci. Pharm*., 51, 54–58, 1983.
36. Asolkar, A. et al., *J. Liq. Chromatogr*., 15, 1689–1701, 1992.
37. Deshmukh, L. and Kharat, R.B., *J. Chromatogr. Sci.,* 28, 400–402, 1990.

38. Touchstone, J.C. et al., in *Thin Layer Chromatography: Quantitative Environmental Clinical Applications,* Touchstone, J.C. and Rogers, D., Eds., John Wiley & Sons, New York, 1980, pp. 151–157.
39. Stephens, R.D. and Chan, J.J., in *Thin Layer Chromatography: Quantitative Environmental and Clinical Applications,* Touchston, J.C. and Rogers, D., Eds., John Wiley & Sons, New York, 1980, pp. 363–369.
40. Mohammad, A., Agrawal, V., and Hena S., *Adsorption Sci. Technol.,* 22, 89–105, 2004.
41. Ryabukhin, V.A., Volynets, M.P., and Myasoedov B.F., *Zh. Anal. Khim.,* 45, 279–288, 1990.
42. Mohammad, A. and Hena, S., *Sep. Sci. Technol.,* 39, 2731–2750, 2004.
43. Volynets, M.P. et al., *Zh. Anal. Khim.,* 43, 842–845, 1988.
44. Volynets M.P., Kitaeva, L.P., and Timerbaev, A.P., *Zh. Anal. Khim.,* 41, 1989–1994, 1986.
45. Mohammad, A., Sirwal, Y.H., and Hena, S., *Chromatography,* 24, 135–145, 2003.
46. Mohammad, A. and Khan, M.A.M., *J. Chromatogr. Sci.,* 33, 531–535, 1995.
47. Mohammad, A. and Agrawal, V., *Acta Universitatis Cibiniensis Seria F. Chemia,* 5, 5–17, 2002.
48. Shadrin, O., Zhivopistsev, V., and Timerbaev, A., *Chromatographia,* 35, 667–670, 1993.
49. Mohammad, A. and Fatima, N., *Chromatographia,* 22, 109–116, 1986.
50. Mohammad, A. and Tiwari, S., *Microchemical J.,* 44, 39–48, 1991.
51. Fatima, N. and Mohammad, A., *Sep. Sci. Technol.,* 19, 429–443, 1984.
52. Ajmal, M. et al., *Microchemical J.,* 39, 361–371, 1989.
53. Shukla, R. and Kumar, A., *Zh. Anal. Khim.,* 46, 1550–1556, 1991.
54. Morosanova, E.I., Maksimova, I.M., and Zolotov Yu. A., *Zh. Anal. Khim.,* 47, 1854–1863, 1992.
55. Mohammad, A. and Khan, M.A.M., *Indian J. Environ. Health,* 38, 100–104, 1996.
56. Franc, J. and Kosikova, E., *J. Chromatogr.,* 187, 462–465 1980.
57. Petrovic, M. et al., *J. Liq. Chromatogr.,* 16, 2673–2684, 1993.
58. Gao, L. et al., *Shendong Daxue Xuebao, Ziran Kexueban,* 24, 69–73, 1989.
59. Nagayama, T. et al., *Shokuhin, Eiseigaku Zasshi,* 23, 253–258, 1982.
60. Specker, H., *Chem. Labor. Betr.,* 32, 519–520, 1981.
61. Ajmal, M. et al., *J. Planar Chromatogr. Mod. TLC,* 3, 396–400, 1990.
62. Mohammad, A., *Indian J. Chem. Technol.,* 2, 233–235, 1995.
63. Ajmal, M., Mohammad, A., and Fatima, N., *Indian J. Chem.,* 28A, 91–92, 1989.
64. Ajmal, M., Mohammad, A., and Fatima, N., *Microchemical J.,* 37, 314–321, 1988.
65. Ajmal, M. et al., *J. Planar Chromatogr. Mod. TLC,* 3, 181–185, 1990.
66. Mohammad, A., Iraqi, E., and Sirwal, Y.H., *Sep. Sci. Technol.,* 38, 2255–2278, 2003.
67. Hsu, Z.F., Jia, X.P., and Hu, C.S., *Talanta,* 33, 455–457, 1986.
68. Ding, W., Qian, B., and Yang, G., *Zhongguo Xitu Xuebao,* 3, 79–80, 1985.
69. Kitaeva, L.P. et al., *Zh. Anal. Khim.,* 36, 102–107, 1981.
70. Djeli, A., Turina, S., and Kovacicek, F., *Kim Ind.,* 33, 1–12 1984.
71. Hanai, L.W., Longo, A., and Zuanon, N.J., *Rev. Cienc. Farm.,* 3, 11–16, 1981.
72. Prameela Devi, Y. and Nand Kumar, N.V., *J. Assoc. Off. Anal. Chem.,* 64, 1301–1304, 1981.
73. Nand Kumar, N.V. and Prameela Devi, Y., *J. Assoc. Off. Anal. Chem.,* 64, 729–732, 1981.
74. Desmukh, L. and Kharat, R.B., *Int. J. Environ. Anal. Chem.,* 16, 1–6, 1989.
75. Mohammad, A., *J. Planar Chromatogr. Mod. TLC,* 8, 463–466, 1995.

76. Borkowska, Z., Klos, J., and Tyfeznska, J., *Farm. Pol.*, 35, 551–552, 1979.
77. Borkowska, Z. et al., *Farm. Pol.*, 26, 599–601, 1980.
78. Krist, E., *Nahrung.*, 29, 391–396, 1985.
79. Tonogai, Y. and Iwaida, M., *J. Food Prot.*, 94, 275–281, 1981.
80. Kundu, M.K., *Fettle Seifen Anstrichm.*, 83, 155–156, 1981.

15 PLC in a Cleanup and Group Fractionation of Geochemical Samples: A Review of Commonly Applied Techniques

Monika J. Fabiańska

CONTENTS

15.1 INTRODUCTION

The analysis of geochemical liquids such as crude oils, hydrothermal bitumens, extracts of coals, and host rocks containing dispersed organic matter or pyrolysates of sedimentary organic matter of various origins involves several fields. There are investigations of trends in organic matter evolution in sediments including biological origin of organic matter, its thermal maturity (rank) reflecting the degree of its alteration, routes of bitumen migration, and secondary alterations of organic matter, together with problems of geological prospecting of fossil fuel deposits, assessment of fuel quality, and environmental protection. The composition of such mixtures is very complex, with numerous groups of compounds of different polarity, various functional groups, mass weights, and concentrations. Moreover, many compounds of significance in geochemical analysis — among them biomarkers (called also molecular fossils or biological markers) — occur in very low amounts, being efficiently concealed by signals of compounds present in higher concentrations, for example, *n*-alkanes. Biomarkers, compounds derived from their biological precursors that are synthesized by living organisms in the biosphere, are widely applied in geochemistry for the assessment of source biological organic matter type, its depositional environment, and thermal maturity of organic matter. Their detection, identification, and quantification are commonly performed by means of gas chromatography (GC) or gas chromatography–mass spectrometry (GC–MS). To eliminate or at least to reduce such problems as poor quality of mass spectra caused by low levels of biomarker concentrations and contamination of analytical instruments by nonvolatile and highly polar components of organic matter, it is necessary to obtain concentrates of biomarkers. Moreover, the identification of molecular structures of novel biomarkers preformed with such spectral techniques as nuclear magnetic resonance requires the prior isolation of single compounds from a sample, which usually contains compounds of similar chemical properties. Thus, the cleanup of geochemical samples and fractionation into groups of substances showing similar chemical properties and more narrow bands of molecular masses are the necessary first steps in geochemical investigation before any other spectral, pyrolytic, or chromatographic methods can be applied [1–4].

Preparative thin-layer chromatography (PTLC) has several advantages over the other proposed methods of soluble organic matter fractionation. The technique requires only small amounts of solvents and provides a method not only for the isolation and recovery of commonly separated compound groups, but also of the heaviest (immobile) fractions. Moreover, it is much less expensive than medium- or high-pressure liquid chromatographic variants of fractionation requiring advanced instrumentation. It is also much more flexible in the research of less conventional samples, such as heavily biodegraded tars or hydrothermal crude oils and bitumens. PTLC also provides the possibility to separate one selected fraction of an organic matter sample. The repeatability of the method is high. In the most commonly described version of the technique, the separation is effected according to polarity of the solvent, but for complex geochemical mixtures, the mobility of a sample on the plate appears to depend on the overlapping effects of the polarity of both the solvent and solute and of the molecular mass [4,5].

There are two main aims in applications of PTLC in organic geochemistry: (1) assessment of the bulk group composition of soluble organic matter by its fractionation and (2) separation of a particular selected group of compounds with geochemical meaning. The important factor in technique selection should be the repeatability of the technique, enabling separation of well-defined compound fractions of comparable features, particularly polarity of the fractions and their general chemical composition.

15.2 BULK FRACTIONATION INTO GROUPS OF COMPOUNDS

15.2.1 SOLVENT FRACTIONATION

Sophisticated preparative methods are required to fractionate geochemical mixtures because of their complexity. Numerous proposals have been described in the literature since the problem was first mentioned, but at present two or three main techniques are generally used such as solvent fractionation, nonpressured column liquid chromatography, medium-pressure liquid chromatography (MPLC), and PTLC. Solvent fractionation into solubility classes (so-called oils, maltenes, asphaltenes, resins, and preasphaltenes) has been the first proposed technique, in which after removal of the low-boiling components at a certain temperature and under reduced pressure, asphaltenes are precipitated by the addition of a nonpolar solvent (for example, *n*-hexane). The soluble (maltene or oil) portion is separated into aliphatic hydrocarbons, aromatic hydrocarbons, and a fraction containing the polar, heteroatomic compounds (nitrogen-, sulfur-, and oxygen-containing compounds usually called "NSO compounds" or resins) [3]. This method was the one most often applied in the past as the simplest and least expensive, but it is also time- and solvent-consuming, and lacks repeatability [4]. Though more advanced chromatographic methods have been developed, solvent fractionation has been utilized even in the most recent investigations [6].

As further subfractionation facilitates subsequent studies at a molecular level, further separation into compound groups is applied. For example, the saturated hydrocarbon fraction can be treated with 5 Å molecular sieves or urea for the removal of *n*-alkanes, leaving behind a fraction of branched and cyclic alkanes [7,8]. The procedure is described in the following text in detail.

15.2.2 COLUMN CHROMATOGRAPHY IN SOLUBLE ORGANIC MATTER FRACTIONATION

The more advanced stage of fractionation technique development applies chromatographic methods for one or all steps of group fractionation. Nonpressured open-column liquid chromatography as the least complicated chromatographic method is the most often performed method of this type. In this technique, glass columns (for example, 100 × 5 mm or 250 × 8 mm) or Pasteur pipettes packed with prewashed and activated silica gel are utilized, binary packed with 2/3 silica gel (the upper part) and 1/3 alumina (the lower part), or only alumina (the recent applications are, for instance, described in Reference 9 to Reference 20). The activation temperature is in the range 140 to 180°C (2 to 12 h) for silica and 250 to 350°C (up to 12 h) for alumina; in some cases, adsorbents are deactivated with 5% water [18,21]. A sample, usually adsorbed on neutral alumina, is placed on the top of a dry packed column or a column filled an adsorbent slurry in an apolar solvent such as *n*-pentane, *n*-hexane, *n*-heptane, or petroleum ether (boiling point about 40 to 60°C). To obtain three fractions of aliphatic, aromatic, and polar compounds, the following solvent sets containing eluents of increasing polarity may be utilized:

- *n*-Hexane (aliphatics), CH_2Cl_2 (aromatics), methanol (polar compounds) [12,14]
- *n*-Hexane (aliphatics), CH_2Cl_2 (aromatics), methanol:chloroform:water (9:15:1; v:v:v) [6]
- *n*-Pentane (aliphatics), *n*-pentane:dichloromethane (4:1, v:v) (aromatics), dichloromethane: methanol (1:1, v:v) (polar) [22,23]
- *n*-Hexane (aliphatics), toluene (aromatics), chloroform:methanol (98:2; v:v) (polar) [15]

These examples do not cover all possible variations of the solvents or column types described in the geochemical literature. The combinations of solvent separations, ion-exchange chromatography, and adsorption chromatography have also been proposed.

The open-column technique is commonly applied in the case of crude oils (being the least complex geochemical organic mixtures). MPLC, high-pressure liquid chromatography (HPLC), and PTLC are more often applied to more complex samples, especially those dominated by more polar compounds, such as hydrothermal bitumens or samples showing terrestrial organic matter input, such as extracts or pyrolysates of coals of various ranks.

In order to reduce the time-consuming open-column chromatographic processes, conventional methods of hydrocarbon-group-type separation have been replaced by MPLC and HPLC. Flash column chromatography is a technique less commonly applied than open-column version, but several applications have been described [2,24–27]. The common technique version is to use a silica-gel-filled column for example, 230 to 400 mesh; 20 × 1 cm column size with a back pressure of 1.5×10^5 Pa of an ambient gas such as nitrogen. Solvents are similar to the ones applied in the case of open-column chromatography fractionations.

Radke et al. [28] described an automated medium-pressure liquid chromatograph, now commonly called the Kohnen–Willsch instrument. At present, the method is widely used to isolate different fractions of soluble organic matter (for instance, as described in Reference 29 to Reference 31). A combination of normal phase and reversed-phase liquid chromatography has been used by Garrigues et al. [32] to discriminate between different aromatic ring systems and degrees of methylamine in order to characterize thermal maturity of organic matter.

Another variation of the preceding method is to apply HPLC to fractionate the cleaned-up aliphatic–aromatic fraction from flash column separation of soluble organic matter as it is performed in the Chevron laboratory, for example, as described in Reference 2. A Waters HPLC system equipped with a preparative Whatman Partisil 10 silica column (9.4 × 500 mm), a HPLC pump, and two detectors for separation monitoring (a UV and refractive index detector) are used, giving three fractions of aliphatic hydrocarbons, mono-, di-, and triaromatics and polar compounds. The first two fractions are eluted with hexane, whereas polar compounds are eluted with methylene chloride. The cut point between saturates and aromatics is determined based on retention times of cholestane and monoaromatic steroid standards. The cut point between triaromatics and polars is made after elution of dimethylphenanthrene [2,33] or at valleys between the major peaks in the absorbance trace [34]. The other possibility is to fractionate aromatic hydrocarbons into ring classes by semipreparative HPLC [35].

15.2.3 PTLC in Group Composition Investigation

The most common approach in geochemical PTLC is to fractionate a sample into three main fractions of aliphatic, aromatic, and polar heterogenic compounds (NSO fraction), yielding information about the bulk group composition of the soluble organic matter [the examples in Reference 36 to Reference 52 have been selected to show the wide range of sample types fractionated using this method). The results are shown on a ternary diagram. Compounds group fractions analogical to those from solvent fractionation enable rough comparison of results described in the geochemical literature, despite the differences in methodology. The same information may be obtained using the instrumental thin-layer chromatography/flame ionization detection (TLC-FID) technique called Iatroscan widely described in the geochemical literature which, however, is not covered in this chapter, which is limited to noninstrumental preparative techniques [53–55].

A few preparatory steps should be performed prior to preparative layer chromatography (PLC) fractionation. Suitable precautions should be taken to prevent contamination of samples during sampling and transport from the sampling site to a laboratory. Wrapping of samples in plastic bags or newspapers, one of the commonest faults, should not be done, because it may result in serious contamination of organic matter. Despite their weight, glass containers or aluminum foil are the best. Prior to the analyses, all solvents applied in fractionation should be distilled to remove plasticizers (mainly phthalates) from bottle caps, and their grade should be confirmed with GC analysis.

Free sulfur, a commonly occurring constituent of soluble organic matter, should be removed by heating a dissolved sample with elemental activated copper powder, turnings, or sheets. In the case of host rock or coal extracts obtained by Soxhlet extraction, it is sufficient to add copper to a round bottom flask of the apparatus. Because elemental sulfur dissolves well in common solvents, it is usually found in aliphatic hydrocarbon fractions and may be identified on GC or GC–MS chromatograms as a wide hump. A high background of sulfur mass spectrum can make it impossible to identify many important biomarkers. Moreover, free sulfur decreases the life of a gas chromatographic column. Prior to the fractionation, asphaltenes may be precipitated in one of the nonpolar solvents; however, this stage is sometimes omitted because they form a residue irreversibly adsorbed on TLC plates.

To remove any adsorbed organic substances, the TLC plates should be predeveloped with a polar solvent such as acetone, ethyl acetate, pyridine, or dichloromethane–methanol mixture (4:1; v:v), and activated in a temperature range of 105 to 120°C for 30 to 60 min in a drying oven [4,41,56,57], more seldom in higher temperatures, for example, 140°C for 12 h as described by Bastow et al. [45,58].

Preparative TLC glass plates precoated with silica gel are most often applied as a stationary phase with *n*-hexane, *n*-pentane, *n*-heptane, or petroleum ether (boiling point about 40 to 60°C) as a developer (for example, as described in Reference 47, Reference 49, Reference 56, and Reference 57). The most popular types of preparative TLC plates are coated with Kieselgel 60G or Kieselgel 60 PF_{254} (Merck). Aluminum oxide is a much less commonly applied adsorbent [39,45,59]; however, it is indispensable for ring class fractionation of aromatic compounds (see Section 15.3.1). The information concerning thickness of an adsorbent bed is usually not provided in papers; however, some exceptions may be found in Reference 61 (0.25 mm), Reference 45 (0.6 mm), and Reference 42, Reference 47, Reference 56, Reference 57, Reference 61, and Reference 62 (0.5 mm in all these papers). A sample of soluble organic matter dissolved in a volatile solvent, such as dichloromethane or a dichloromethane–methanol mixture (4:1; v:v, for instance) is applied onto the plate as a narrow band or a series of tightly overlapping spots [57]. The concentration of a sample should not exceed 5% (wt.:wt.). As a general guideline, about 50 to 70 mg of soluble organic matter can be separated on a 0.5 mm thick 20 × 20 cm silica gel plate. This amount of total sample enables fractionation in weights sufficient for GC or GC–MS analyses. Special care should be taken to remove solvents, before the development in the case of samples containing lightweight components, which can easily evaporate. Drying for several minutes at room temperature in a horizontal position seems to be the best procedure in such cases. However, it should be mentioned that sometimes it is difficult to remove solvents completely; thus, a gentle stream of cold air or nitrogen may be used.

In geochemical applications, the PLC plates are developed in a TLC chamber in saturated vapor conditions. There are various visualization techniques. UV light visualization is the most common because many components show strong fluorescence. This technique is also recommended by the author because it does not introduce any possible reagent into a sample [47,52,56,63]. The spraying reagents sometimes used to enhance UV fluorescence include berberine sulfate in methanol [9], 1% berberine hydrochloric in ethanol solution [64], 1% rhodamine 6G in methanol [43,47,65], fluorescein-bromine reaction for olefins [66], or iodine vapors [67–70]. It is possible to spray only a narrow band along one of the plate edges.

Reference compounds developed on the same plate together with UV light are used to locate the bands of the particular compound groups [38,39,57,70–73]. The applications with reference compounds developed on analytical TLC plates prior to the preparative separations to find R_f values are far less numerous [74,75].

Depending on activation and development conditions, the R_f values of three main separated fractions may change but their ranges are approximately as follows: 0.40 to 1.00 (aliphatic hydrocarbons), 0.05 to 0.40 (aromatic compounds), and 0.00 to 0.05 (polar compounds fractions) [49,72,76].

After the identification of the suitable compound bands, silica gel is scraped off the plates, placed in short glass columns, Pasteur pipettes, or sintered filter funnels, and fractions are recovered with such volatile solvents as ethyl acetate or dichloromethane.

The general scheme of bulk group fractionation is shown in Figure 15.1.

Aliphatic hydrocarbon fractions are sometimes fractionated further into the fractions of normal and branched/cyclic alkanes by urea clathration or molecular sieves

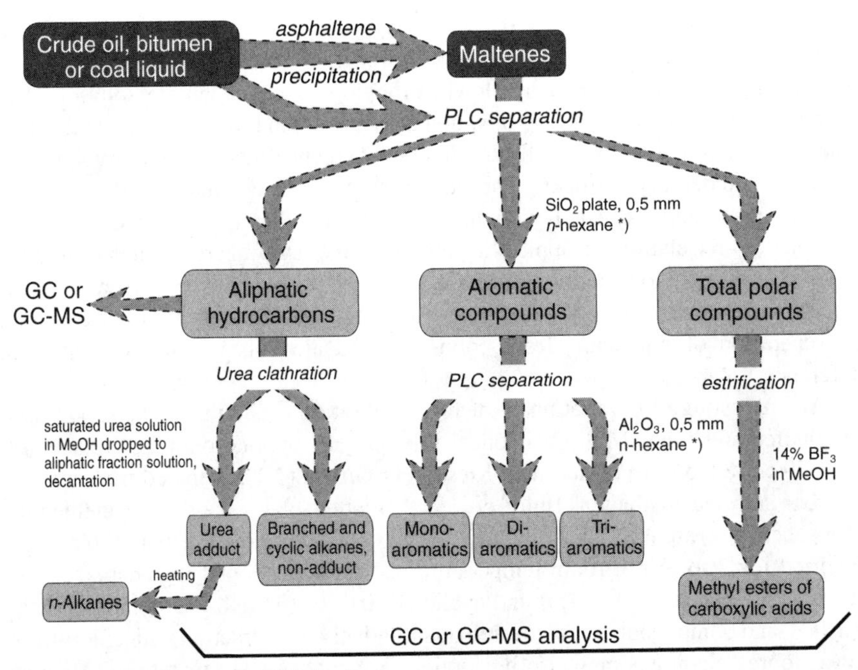

FIGURE 15.1 Scheme showing PLC group fractionation of soluble organic matter into fractions of aliphatic hydrocarbons, aromatic compounds with application of urea clathration, and methylation of carboxylic acids in polar fractions; based on experimental data given in Reference 36 to Reference 52, Reference 77 to Reference 81, and Reference 88 to Reference 89.

(Figure 15.1). In the case of highly waxy samples, *n*-alkanes present in very high concentrations can conceal signals of cyclic biomarkers occurring in low concentrations. Their removal enables better geochemical interpretation of GC–MS data.

Crystalline urea clathrates are formed from urea as host and unbranched alkanes or their derivatives as guests. From a saturated solution, urea crystallizes in tetragonal form. If *n*-alkanes are present in the solution, hexagonal crystals of the adduct precipitate much more easily, forming needles. In the hexagonal urea crystal, there is a channel 5.3 Å across, fitting the molecular dimensions of straight-chain saturated hydrocarbons, halides, alcohols, aldehydes, ketones, or carboxylic acids. During crystallization the molecules of urea precipitate around each linear molecule in the form of a tight helix. The presence of one double or triple bond still permits clathration with urea while branching, or presence of the ring structure prevents urea clathrate formation. This phenomenon is used for the separation of straight-chain substances from other organic compounds. The formation of clathrates is spontaneous after agitating the sample in the milieu. The molar ratio of urea to the substrate is determined by the length of the hydrocarbon chain, not by any functional group present in the guest molecule. The urea clathrates are usually thermolabile; they are irreversibly decomposed in boiling water [77,78].

To perform urea clathration, the aliphatic hydrocarbon fraction is dissolved in a methanol–toluene mixture (1:4; v:v) in a centrifuge tube. The saturated solution of urea in methanol is added by slowly dropping it into a sample solution. The mixture is stirred for 10 min. The oily, nonadduct fraction is separated from the urea clathrate containing *n*-alkanes by decantation or centrifugation, and solvents are evaporated under dry nitrogen. The adduct fraction (precipitate) is dissolved in distilled water and extracted with light petroleum ether or *n*-hexane [77]. The examples of urea clathration applications for *n*-alkane removal are given in Reference 8 and Reference 79 to Reference 80 and many other publications; however, the exact procedure description is not usually given in papers. Applications of urea clathration for separation of *n*-alkanols from polyisoprenoid alcohols are also described in Reference 81.

An interesting PLC variation of the urea clathration technique has been proposed by Chaffee and Johns [82]. Component mixtures are applied onto TLC plates (20 cm × 20 cm × 0.5 mm) coated with Kieselguhr G/urea (2:1), prepared from a slurry in urea-saturated methanol (1ml/g powder). Spotted plates are left in methanolic atmosphere overnight to allow clathrate formation. To remove methanol, plates are air dried for 2 to 3 h at room temperature and then developed in *n*-heptane. Two bands of acyclic (R_f 0.9 to 1.0) and cyclic (R_f 0.0 to 0.1) hydrocarbons are distinguished, and components are recovered quantitatively by extraction with chloroform.

Thiourea clathrates may also be applied for geochemical separations. They are similar to the molecular complexes of urea but owing to the larger dimensions of the thiourea crystal channels, only branched alkanes, cycloalkanes, and their derivatives participate in clathrate formation. In contrast, the unbranched molecules are too small and hence cannot be held rigidly [77].

Another way to fractionate aliphatic hydrocarbons into more narrow fractions is to apply molecular sieves. These adsorbents are characterized by a narrow distribution of small pores that possess some specific properties which can be utilized to advantage in the separation of complex mixtures. The most important is the molecular-sieve effect, which operates with a different intensity if the diameters of the most numerous pores agree with the molecular dimensions of the adsorbates. In this method zeolites are used, being crystalline aluminosilicates forming a regular system of cavities connected mutually with openings of uniform dimensions. The cavities in the crystal lattice are filled with water molecules after the zeolite has crystallized. It is eliminated in higher temperatures. Zeolites contain two types of cavities: smaller inside the cubo-octahedrons (not accessible for hydrocarbon molecules) and between them (accessible for hydrocarbon molecules). Separation of *n*-alkanes on the sieve 5D can be carried out both in the gas and in the liquid phases. In the liquid phase, *n*-alkanes are separated either by filtration of the fractions through a column packed with a powder sieve or by refluxing a solution of the hydrocarbon fraction in an inert solvent in the presence of an excess of the sieve. In both instances *n*-alkanes remain adsorbed on the sieve, and other hydrocarbon types pass into the filtrate. In the following examples of investigations the preceding procedure was applied to remove *n*-alkanes [37,39,63,71,83,84]. There were also proposals to apply dealuminated modernite molecular sieves of low polarity for fractionation of aromatic hydrocarbons [85].

Lastly, it is worth mentioning that there are applications of two-or-more-step preparative TLC in soluble organic matter fractionations; however, they are rarely described [4,86]. Each plate is developed successively in a series of solvents such as tetrahydrofurane, $CHCl_3$/MeOH (4:1; v:v), toluene, and pentane such that the solvent front advancers approximately 4 cm with each successive solvent; the plate dries up between the solvent, and the development tank atmosphere is allowed to equilibrate for at least 30 min after adding a new solvent, and before inserting the plates. Fractions represented immobile material in tetrahydrofurane (THF) and mobile compounds in successive solvents.

15.3 ADVANCED PLC SEPARATION OF THE SELECTED GROUPS OF COMPOUND

15.3.1 Fractionation of Aromatic Compounds

The general PLC approach described in the preceding text can be modified when the research aim is to investigate one selected compound (or a group of compounds), or in the case of some less common samples, such as highly biodegraded or hydrothermal petroleum, tars, immature sedimentary organic matter, or even products of biomass combustion [70,86,87].

Fractionation of total aromatic compound fractions into ring classes prior to GC or GC–MS analysis will generally improve the determination of aromatic compound distribution. Such subfractionation may be performed on nonactivated silica plates developed with *n*-hexane, enabling classes of mono-, di-, and triaromatic hydrocarbons to be obtained [63]. However, alumina-coated glass plates are more popular. As alumina shows much better selectivity than silica in relation to the number of aromatic rings, it is also often used with *n*-hexane as a developer (Figure 15.1). Alumina-coated plates are activated at 120°C for at least 12 h prior to use. The particular compound bands are located relative to such reference compounds as tridecylbenzene, naphthalene, and phenanthrene or anthracene. The R_f values for mono-, di-, and triaromatic hydrocarbon fractions are in the range of 1 to 0.6, 0.4 to 06, and 0.2 to 0.4, respectively [23,39,88,89].

The fractionation of aromatic hydrocarbons according to the number of rings may also be achieved using argentation thin-layer chromatography. Silver-nitrate-impregnated silica gel is a special adsorbent used for the various compound separations, both in open-column and PLC versions. Substances containing one or more double bonds with π-electrons, such as olefins, aromatic hydrocarbons, or heterocyclic sulfur compounds, react with silver ions with complexation, which enables their proper separation from saturated hydrocarbons on this adsorbent. The reaction with silver ions is reversible, and organic compounds can be eluted from the adsorbent.

Silica-gel-precoated preparative TLC plates impregnated with $AgNO_3$ are widely applied in this technique [60,90]. To impregnate a silica-precoated plate with silver nitrate, the plate is dipped for 3 min in a 1% $AgNO_3$ solution in a methanol–water mixture (4:1, v:v). The plate is then dried, first in air with exclusion of light, and next, it is activated at 80°C in a drying oven. If impregnated plates are not used immediately, they should be stored wrapped in black plastic film in a desiccator [90–92].

The TLC plates prepared in a similar manner may be used to separate monoaromatic hydrocarbon fractions with R_f values in the range 0.29 to 0.78 from polyaromatic hydrocarbon fractions (R_f = 0.06 to 0.29) [84].

15.3.2 Sulfur Compound Separation

Argentation TLC is often applied in various sulfur compound fractionation, for example, to obtain two fractions as follows: (1) aliphatic sulfides ($R_f < 0.3$) and (2) the fraction containing saturates, aromatic hydrocarbons, and heteroaromatic sulfur compounds ($R_f > 0.9$) (Figure 15.2). The latter one is further fractionated on silica gel plates to yield aliphatic hydrocarbons ($R_f > 0.9$), and two polar fractions, containing aromatic and thiophenic triterpenes ($0.9 < R_f < 0.7$ and $0.7 < R_f < 0.5$) [90,93]. In the same way, it is possible to fractionate previously obtained apolar fractions into aliphatic hydrocarbons (R_f 0.9 to 1.0), thiophenes (R_f 0.6 to 0.9), benzothiophenes (R_f 0.1 to 0.6), and sulfides (R_f 0.0 to 0.1) [60,91,94–98]. The most common developer in such fractionations is *n*-hexane.

To obtain alkanes and alkenes from aliphatic hydrocarbon fractions, argentation PLC was proposed that utilized silica-gel-60-precoated plates impregnated with 5% or 10% of $AgNO_3$ [37,80,99,100]. In some applications, TLC plates impregnated

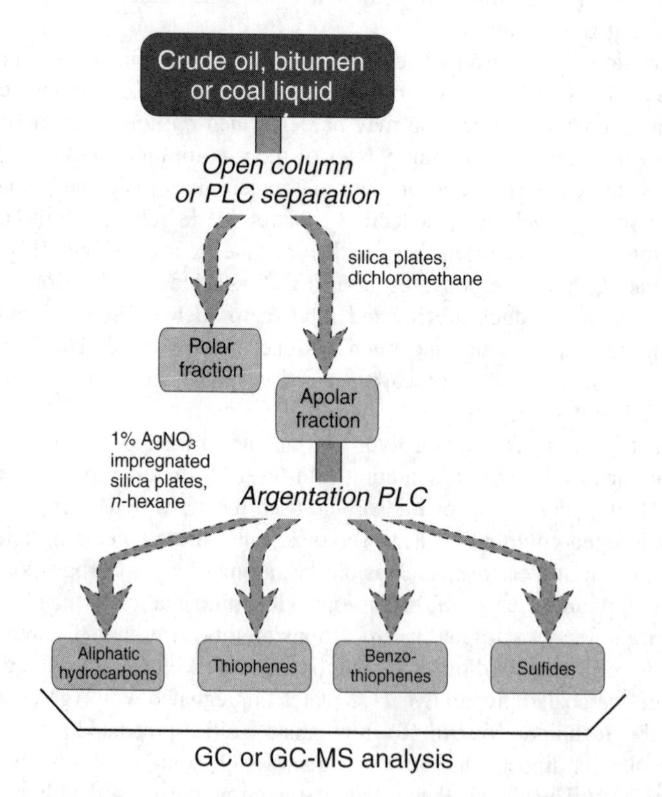

FIGURE 15.2 Group fractionation of soluble organic matter with application of argentation PLC; based on Reference 61 and Reference 89 to Reference 93.

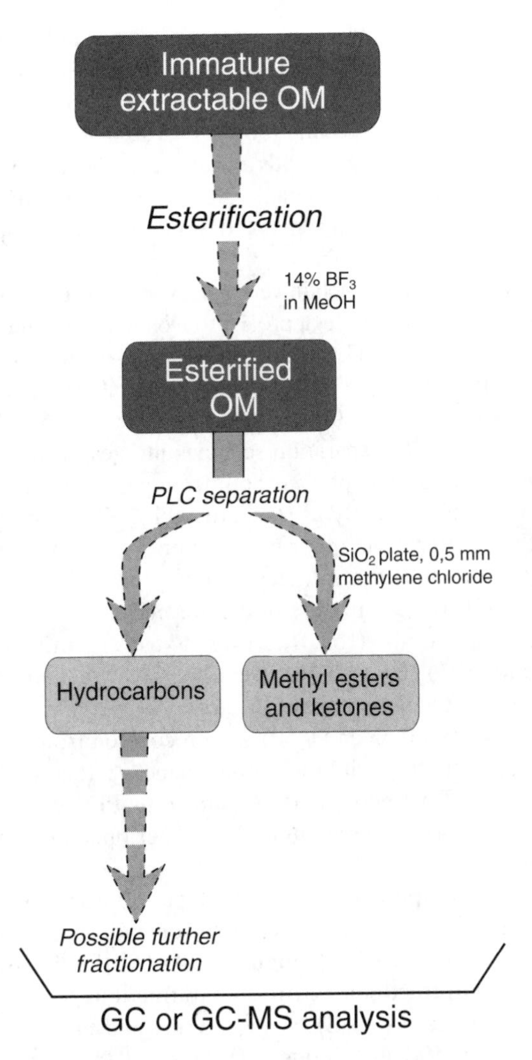

FIGURE 15.3 PLC separation of carboxylic acids from soluble organic matter with the use of carboxylic acids esterification; based on experimental data given in Reference 36, Reference 67 to Reference 69, and Reference 102 to Reference 104.

with 5% $AgNO_3$ are applied for separation of aliphatic hydrocarbons from the total extracts [101].

15.3.3 Polar Compounds Fractionation

In the case of very immature organic matter, or when the main research aim is to investigate polar fractions, a different analytical scheme may be applied (Figure 15.3). Prior to the fractionation, total concentrated extracts are treated with 14% BF_3 in methanol or diazomethane in ether to esterify free carboxylic acids, and then they are subjected to silica gel TLC using methylene chloride or a mixture of hexane–diethyl ether (9:1; v:v) as a developer. Two bands are distinguished, corresponding to hydrocarbons ($R_f \sim 0.9$) and methyl esters of carboxylic acids together with ketones ($R_f \sim 0.5$). The separated fractions are recovered from scraped-off silica gel with ether or ethyl acetate [36,67–69,102–104]. In the same way, it is also possible to esterify carboxylic acids occurring in polar compound fractions obtained in bulk fractionation of organic matter into three fractions: aliphatic hydrocarbons, aromatic hydrocarbons, and polar (NSO) compounds, as was described in Reference 105.

PTLC may be also applied to obtain very narrow organic compound groups such as petroporphyrins. The most simple approach to isolate nickel and vanadyl porphyrin fractions is to use silica gel TLC plates and hexane–dichloromethane (1:1; v:v) or toluene–chloroform (3:1; v:v) as a developer. The R_f values for nickel and vanadyl porphyrins are 0.40 to 0.85 and 0.05 to 0.40, respectively [106–109].

One of the most complex separation schemes utilizes flash liquid chromatography and PLC to obtain petropophyrins both from geochemical samples or those synthesized and used subsequently as standards [110]. Ocampo and Repeta [111] described the scheme of petroporphyrins isolation in which at the first step the sediment extract is fractionated into ten fractions on silica gel using dichloromethane (fractions 1 to 4), a mixture of dichloromethane–acetone with increasing acetone concentrations (for fractions 5 to 9), and, at last, dichloromethane:methanol (4:1) (fraction 10). Next, the fifth fraction was separated on silica PLC plates using dichloromethane–acetone (97.5:2.5; v:v:v) as a developer. Two purple bands (with R_f 0.53 and 0.50) were recovered from silica and purified further on a silica gel column with dichloromethane–acetone (97.5:2.5, v:v:v) as an eluent. The enriched fraction was then separated by PLC with the same solvent mixture, and the purple bands containing two bacteriopheophytin allomers were recovered with acetone.

Another scheme was applied to isolate both petroporphyrins and maleimides deriving from bacteriochlorophyll c (green sulfur bacteria), which were found in Kupferschiefer extracts [73,112] (Figure 15.4). At the first step, flash chromatography was used to fractionate extracts into fractions containing the following: (1) aliphatic hydrocarbons, (2) Ni porphyrins and aromatic hydrocarbons, (3) VO porphyrins, and (4) maleimides. The second fraction was fractionated into aromatic hydrocarbons and Ni porphyrins, and preparative TLC was used to purify maleimides present in the fourth fraction. To perform it, the fraction dissolved in acetone was applied onto an activated silica gel plate (150°C, 24 h), and, next, the plate was developed with hexane–acetone (4:1) with maleimide standards applied on the same plate. When GC–MS analyses suggested coelution of maleimides, the fraction was derivatized with *N*-tertiary-butyl-(methylsilyl)-*N*-methyltrifluoracetamide. The derivatized fraction was then fractionated on unactivated preparative silica plates with hexane/acetone (4:1) as a developer (Figure 15.4).

Preparative TLC may be applied to cleanup selected compound fractions separated from geochemical samples by such methods as HPLC, as Aries et al. [113] has described. To analyze phospholipids and nonphospholipids in sediments, organic matter was extracted and extracts LC-fractionated to obtain polar fractions. At the

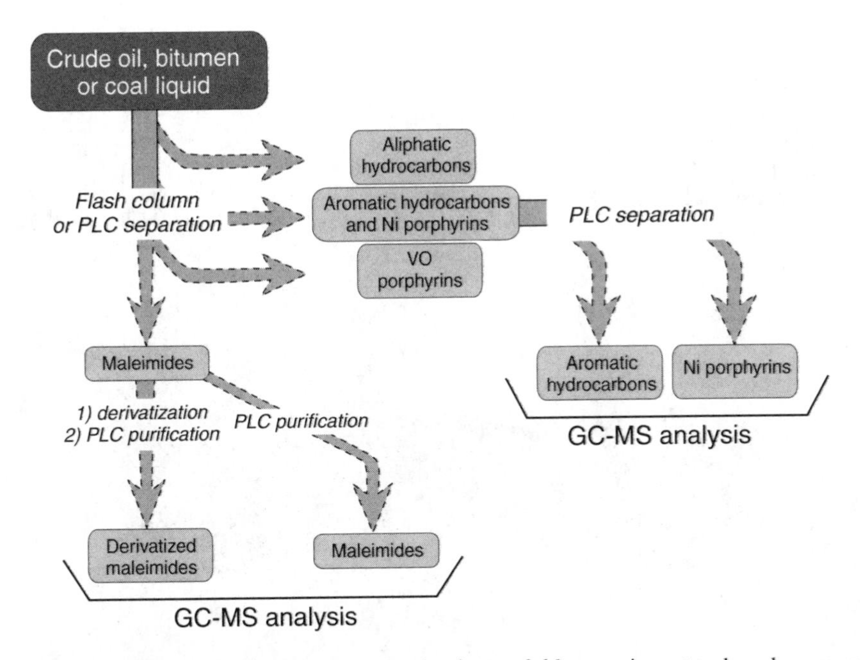

FIGURE 15.4 Scheme of porphyrins separation from soluble organic matter; based on experimental data given in Reference 73 and Reference 112.

next step of separation, polar fractions were fractionated by HPLC. Finally, preparative TLC was carried out on silica gel F plates using $CHCl_3$–MeOH–glacial acetic acid (65:25:8, v:v:v) as a developer to separate phospholipid and nonphospholipid fractions, with R_f values being in the range of 0 to 0.85 and 0.85 to 1.0, respectively. Compounds were visualized using Vaskovsky's reagent.

15.4 WHAT GEOCHEMICAL INFORMATION PLC RESULTS YIELD AND HOW TO INTERPRET THEM

It is important to remember that there are many primary and secondary factors influencing a group composition of geochemical samples. As a result, PLC yields information that can be applied only as the additional source for organic matter evaluation, combining its results with those given by other geochemical methods such as GC, GC–MS, and Fourier transformation infrared spectroscopy or Rock Eval analysis. It does not diminish the importance of the technique, because the most professional geochemical approach is to use at least two or three different analytical methods in the research. However, the PLC results should be considered carefully and compared with other methods' results to draw valid conclusions.

The detailed information concerning geochemical factors, which influence the group composition of soluble organic matter, can be found in numerous general geochemical publications [e.g., 2,114–117]. The most important factors are described briefly in the following subsections.

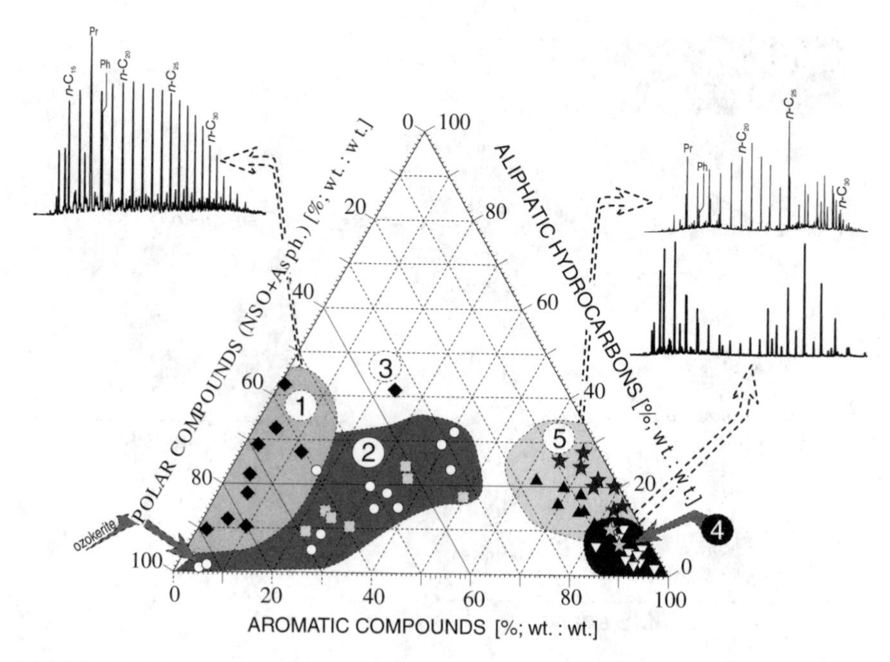

FIGURE 15.5 Group compositions of selected types of soluble organic matter: 1 — typical Polish crude oils of the Gorlice region, 2 — hydrothermal bitumens (○) found in calcite veins or druses and host rock extracts (□) (the Upper Devonian; the Holy Cross Mountains, Poland), 3 — heavily biodegraded oil (the Gorlice region, Poland), 4 — typical Polish brown coal extracts, 5 — extracts of bituminous and subbituminous coals (★) and coal pyrolysates (▲); gas chromatograms show variability of *n*-alkane distributions.

15.4.1 Composition of Primary Biogenic Matter

Generally, coals and dispersed organic matter with higher terrestrial inputs give extracts showing higher concentration of polar compounds and aromatic hydrocarbons, whereas organic matter of algal or algal/bacterial origin is enriched in aliphatic hydrocarbons such as predominating alkanes deriving from lipid components (Figure 15.5). However, lignin and cellulose do not participate in crude oil formation. Many oils of terrestrial origin have highly waxy composition with long-chain *n*-alkanes derived from cuticular waxes predominating. Marine organic matter produces oils of paraffinic-naphthenic or aromatic-intermediate type with higher concentrations of isoprenoids than oils from terrestrial organic matter. Thus, careful consideration of GC–MS data is required to assess the type of biogenic matter of oils [1,2,61,71,118].

15.4.2 Expulsion Type

The method by which crude oils are expelled from source kerogen also influences oil group composition. In the case of hydrothermal oils or hydrothermal bitumens, hot water can enhance the content of lighter aliphatic and aromatic hydrocarbons (benzene, toluene, ethylbenzene, and xylenes). Typical hydrothermal oils have a

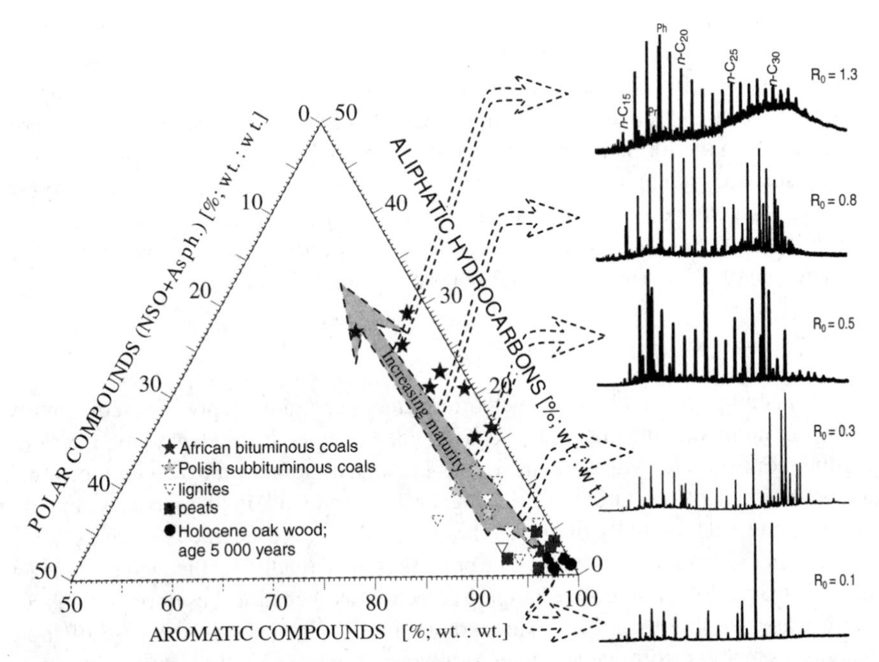

FIGURE 15.6 Changes with increasing thermal maturity in group composition of extracts of organic matter with high terrestrial matter input: Holocene oak wood and peats (the Ustka cliff, Poland, the Quaternary), lignites (Miocene, the Chapowo-Wadysawowo cliff, the Konin Brown Coal Basin, Poland,), subbituminous coals (the Upper Vistula Basin, Poland) and bituminous coals of various rank (R_o = 0.5 to 1.3%), age and locality (the Karoo Basin, South Africa, the Upper Silesia and Upper Vistula coal basins, Poland); gas chromatograms show changes in *n*-alkane distributions with increasing vitrinite/huminite reflectance (R_o).

significant content of heteroatomic compounds and high polyaromatic hydrocarbon (PAH) concentrations. Their group composition seems to be highly dependent on the source kerogen type; however, even crude oils derived from algal/bacterial kerogen types contain relatively lower concentrations of aliphatic hydrocarbons compared to conventional oils [119] (Figure 15.5).

15.4.3 Thermal Maturity

Thermal maturity of organic matter affects the composition of soluble organic matter; however, the type of primary biogenic organic matter remains important at almost all stages of thermal evolution in a deposit. The general trends are shown in Figure 15.6 for the case of terrestrial organic matter. These changes include the decrease in the content of total polar compounds together with the increase of aliphatic and aromatic compounds concentrations. Less mature oils are considered to be more resinous and nonwaxy and to have higher concentration of NSO compounds than more mature ones. Subsequently, their evolution leads to lighter oils rich in resin components, thence to naphthenic oils of moderate gravity, and, finally, to light paraffinic oils [114,115,119]. This means that at the higher stages crude oils occurring in an oil trap show a decrease in lighter aromatic hydrocarbon content because of their lower thermal stability compared to the stability of aliphatic hydrocarbons. Lighter aromatics tend to condense forming large aromatic compounds, whereas aliphatic hydrocarbon molecules crack to produce gaseous aliphatic hydrocarbons. Finally, thermal evolution of oils leads to the formation of two very stable products: methane and pyrobitumen. During secondary migration, crude oils tend to be enriched in aliphatic hydrocarbons, and lose polar components, probably as a result of geochromatography on argillaceous minerals. As a result, residual bitumen left in rocks on migration pathways tends to be heavy and rich in asphaltenes and polar compounds [114].

15.4.4 Water Washing

Water washing occurs wherever meteoric waters penetrate deeply into sedimentary beds containing organic matter. It is usually accompanied by biodegradation, removing the more soluble hydrocarbons. It aids in concentration of the heavier molecules in the residual oil or bitumen. Moving past a crude oil field or coal seam, the groundwater preferentially dissolves the most soluble hydrocarbons, such as methane, ethane, benzene, and toluene. The process is most readily recognized by changes in composition of the gasoline range hydrocarbons, because these compounds are more water soluble than C_{15+} hydrocarbons [120]. However, in the case of more heavily water-washed organic matter, *n*-alkanes in the range of C_{15} to C_{20} may also be affected [121]. In some cases, all *n*-alkanes are removed, especially in less thermally mature and weathered coals [52,122]. Degraded crude oils associated with meteoric waters have been found as deep as 1830 m (the Gulf of Mexico), 2134 m (the Niger delta), and 3048 m (Bolivar coastal fields, Venezuela) [114].

Water washing also affects the bulk group composition of soluble organic matter, tending to decrease concentration of aromatic hydrocarbons compared to aliphatic hydrocarbons and polar compounds (Figure 15.7). The latter ones are usually adsorbed on coal macromolecule or siliclastic rocks, and this prevents their removal in water solution. Generally, the extent of changes depends on permeability of the host rocks and their adsorptive properties, and thermal maturity of organic matter in some cases. The main changes are shown in Figure 15.7. It shows that a group composition may be applied to indicate altered samples, particularly when the typical composition of soluble organic matter of a region is known or further chromatographic investigation is performed [52,122]. Such indicators are the loss of lighter *n*-alkanes, loss of methylnaphathalenes, and depletion of ethylnaphthalenes compared to dimethylnaphthalenes investigated by GC or GC–MS [122,123].

15.4.5 Biodegradation

The process of biodegradation concerns all types of organic matter at all evolution stages, which are exposed to bacteria, both at the earth's surface and in deposits. Bacteria act as oxidizing agents. To degrade organic matter successfully, the presence of oxygen in circulating water is required, temperatures in the range of 20 to 50°C, and absence of H_2S [114,115]. In the case of crude oils, biodegradation produces

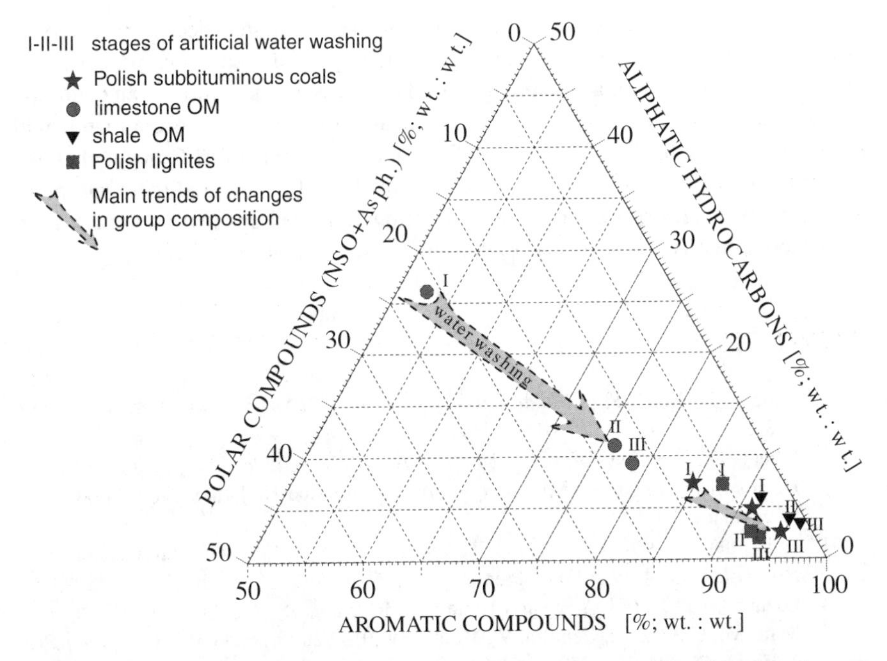

FIGURE 15.7 Secondary changes in the total group composition of soluble organic matter as a result of simulated water washing; extracts of the Miocene lignite and shale (both of the Bechatów open cast mine, Poland), the Upper Devonian shale (the Holy Cross Mountains, Poland), and the Upper Carboniferous bituminous coal (the Upper Silesia Coal Basin, Poland).

heavy, low-API-gravity oils depleted in hydrocarbons and enriched in the nonhydrocarbon NSO compounds, and asphaltenes.

Several researchers [120,124–126] have presented removal of hydrocarbon compound classes in order of increasing resistance to biodegradation and a scale of the degrees of biodegradation. Bacterial attack leads to formation of ketones, alcohols, and acids from hydrocarbons, altering group composition of organic matter significantly. Generally, smaller molecules are consumed before the larger ones. The same type of changes may be observed in bioremediated oils or tars, coals biodegraded in deposits in the natural way, or microbially desulfurized coals [127–132]. Group composition, together with the other techniques, may be used to indicate these alterations [123].

15.5 CONCLUSIONS

Over the last 20 yr, organic geochemistry has undergone great development because of the application of advanced chromatographic methods, among them, techniques aimed at precise fractionation and separation of geochemical liquids. They enabled geochemical procedures to be worked out that are used in evaluation of a wide range of organic matter types to find their thermal maturity (rank), primary sources of biogenic organic matter, conditions of depositional environments, and secondary alterations occurring in a deposit such as weathering or biodegradation. In these analytical procedures, PLC plays an important role as a step necessary to obtain well-defined, narrow fractions of compounds having similar chemical properties, or even single substances, or as the method used to assess group composition of soluble organic matter yielding additional geochemical information when compared with results obtained from other techniques. The wide application of PLC in geochemical investigation has significantly improved final analytical results and their interpretation. The advantages of PLC over other fractionation methods that have been mentioned above justify the assumption that its use will increase in future.

REFERENCES

1. Philp, R.P., *Fossil Fuel Biomarkers. Application and Spectra.* Elsevier, Amsterdam. 1985.
2. Peters, K.E. and Moldowan, J.M., *The Biomarker Guide. Interpreting Molecular Fossils in Petroleum and Ancient Sediments*, Prentice Hall Inc., Englewood Cliffs, New Jersey, 1993.
3. Rullkotter, J., 1993, in: Engel, M.H. and Macko, S.A. Eds., *Organic Geochemistry. Principles and Applications*, Plenum Press, New York. 1993, ch. 16. pp. 389–396.
4. Lazaro, M-J., Herod, A.A., and Kandiyoti, R., *Fuel,* 78, 795–801, 1999.
5. Matt, M., Galvez, E., Cebolla, V., Membrado, L., Vela, J., and Gruber, R., *Fuel Proc. Technol,.* 77–78, 245–253, 2001.
6. Stefanova, M., Oros, D.R., Otto, A., and Simoneit, B.R.T., *Org. Geochem.*, 33, 1078–1091, 2002.
7. O'Connor, J.G., Burow, F.H., and Norris, M.S., *Anal. Chem.,* 34, 82–85, 1962.
8. Richardson, J.S. and Miller, D.E., *Anal. Chem.,* 54, 765–768, 1982.
9. Stefanova, M, Simoneit, B.R.T., Sojanova, G., Nosyrev, I.E., and Goranova, M., *Fuel,* 74, 768–778, 1995.
10. Dzou, L.I. P. Noble, R.A., and Senftle, J.T., *Org. Geochem.*, 23, 681–697, 1995.
11. Pieri, N., Jaquot, F., Mille, G., Planche, J.P., and Kister, J., *Org. Geochem.*, 25, 51–68, 1996.
12. Sun, Y-Z., Wang, B-S., and Lin, M.Y., *Org. Geochem.,* 29, 583–591, 1998.
13. Stefanova, M., Marinov, S.P., and Magnier, C., *Fuel* 78, 1395–1406, 1999.
14. Papanicolaou, C., Dehmer, J., and Fowler, M., *Int. J. Coal Geol.*, 44, 267–292, 2000.
15. Li, M., Lin, R., Liao, Y., Snowdon, L.R., Wang, P., and Li, P., *Org. Geochem.,* 30, 15–37, 1999.
16. Duan, Y. and Ma, L., *Org. Geochem.*, 32, 1429–1442, 2001.
17. Fleck, S., Michels, R., Ferry, S., Malartre, F., Elion, P., and Landais P., *Org. Geochem.*, 33, 1533–1557, 2002.
18. Grimalt, J.O., Campos, P.G., Berdie, L., Lopez-Quintero, J.O., and Navarrete-Ryes, L.E., *Appl. Geochem.,* 17, 1–10, 2002.
19. Kotarba, M.J., Clayton, J.L., Rice, D.D., and Wagner, M., *Chem. Geol.* 184, 11–35, 2002.
20. Kotarba, M.J. and Clayton, J.I., *Int. J. Coal Geol.*, 73–94, 2003.
21. Grimalt, J. and Albaiges, J., *Geochim. Cosmochim. Acta,* 51, 1370–1384, 1987.
22. Zhiguang, S., Batts, B.D., and Smith, J.W., *Org. Geochem.* 29, 1469–1485, 1998.
23. Warton, B., Alexander, R., and Kagi, R.I., *Org. Geochem.,* 30, 1255–1272, 1999.
24. Cabrera, L., Hegemann, H.W., Pickel, W., and Saez, A., *Int. J. Coal Geol,.* 27, 201–226, 1995.

25. Borrego, A.G., Blanco C.G., and Puttmann, W., *Org. Geochem.*, 26, 219–228, 1997.
26. Otto, A., Walther, H, and Puttmann, W., *Org. Geochem.*, 26, 105–115, 1997.
27. Otto, A. and Simoneit, B.R.T., *Geochim. Cosmochim. Acta,* 65, 3505–3527, 2001.
28. Radke, M., Willsch, H., and Welte, D.H., *Anal. Chem,* 52, 406–411, 1980.
29. Radke, M., Welte, D., and Willisch, H., *Org. Geochem,.* 10**,** 51–63, 1986.
30. Petersen, H.I., Depositional environments of coals and associated siliclastic sediments in the Lower and Middle Jurassic of Denmark, *Geological Survey of Denmark*, series 33, Copenhagen, 1994.
31. Araujo, C.V., Loureiro, M.R.B., Barbanti, S.M.B., Goncalves, F.T.T., Miranda, A.C.M.L., and Cardoso, J.N., *Revista Latino Americana de Geoquimica Organica,* 4, 25–39, 1998.
32. Garrigues, P., De Sury, R., Angelin, M.L., Bellocq, J., Oudin, J.L., and Ewald, M., *Geochim. Cosmochim. Acta* 52, 375–384, 1988.
33. Isaksen, G.H., in: *Paleogeography, Paleoclimate, and Source Rocks,* Huc, A.-Y. Ed., *AAPG Studies in Geology,* No 40, 1995, 81–104.
34. Freeman, K.H., Boreham, ChJ., Summons, R.E., and Hayes, J.M., *Org. Geochem.*, 21, 1037–1049, 1994.
35. Radke, M., Willsch, H., and Welte, D.H., *Anal Chem.* 56, 2538–2546, 1984.
36. Grimalt, J.O., Simoneit, B.R.T., Hatcher, P.G., and Nissenbaum, A., *Org. Geochem.* 13, 677–690, 1988.
37. Mycke, B., Michaelis, W., and Degens, E.T., *Org. Geochem.*, 13, 619–625, 1988.
38. Azevedo, D.A., Aquino Neto, F.R., and Simoneit, B.R.T., *Org. Geochem.*, 22, 991–1004, 1994.
39. Alexander, R., Bastow, T.P., Fisher, S.J., and Kagi, R.I., *Geochim. Cosmochim., Acta,* 59, 4259–4266, 1995.
40. Killops, S.D., Raine, J.I., Woolhouse, A.D., and Weston, R.J., *Org. Geochem.*, 23, 429–445, 1995.
41. Elias, V.O., Cardoso, J.N., and Simoneit, B.R.T., *Org. Geochem.*, 25, 241–250, 1996.
42. Farrimond, P., Bevan, J.C., and Bishop, A.N., *Org. Geochem.*, 25, 149–164, 1996.
43. Innes, H.E., Bishop, A.N., Head, J.M., and Farrimond P., 1997, *Org. Geochem.*, 26, 565–576.
44. Laureillard, J., Pinturier, L., Fillaux, J., and Saliot, A., *Deep-Sea Research II,* 44, 1085–1108, 1997.
45. Bastow, T.P., Alexander, R., Kagi, R.I., and Sosrowidijojo, I.B., *Org. Geochem.* 29, 1297–1304, 1998.
46. Suzuki, A., Ishida, K., Shinomiya, Y., and Ishiga, H., *Palaeogeography, Palaeoclimatology, Palaeoecology,* 141, 53–65, 1998.
47. Fabiańska, M., *J. Planar Chrom.*, 13 (1), 20–24, 2000.
48. Reddy, C.M., Eglington, T.I., Palić, R., Benitez-Nelson, B.C., Stojanowić, G., Palić, I., Djordjević, S., and Eglington, G., *Org. Geochem.*, 31, 331–336, 2000.
49. Marynowski, L., Czechowski, F., and Simoneit, B.R.T, *Org. Geochem.*, 32, 69–85, 2001.
50. Czechowski, F., Stolarski, M., and Simoneit, B.R.T., *Fuel* 81, 1933–1944, 2002.
51. Fabiańska, M., and Kruszewska, K.J., *Int. J. Coal Geol.* 54, 95–114, 2003.
52. Ćmiel, S., and Fabiańska, M., *Int. J. Coal Geol.* 57, 77–97, 2004.
53. Ray, J.E., Oliver, K.M., and Wainwright, J.C., in: *Petroanalysis* 81, IP Symposium, London, Heyden and Son, London, 1982, 361–388.
54. Karlsen, D. and Larter, S., in: *Correlation in Hydrocarbon Exploration*, Norwegian Petroleum Society, Graham and Trotman, London, 1989, 77–85.
55. Bharati, S., Patience, R., Mills, N., and Hanesand, T., *Org Geochem.*, 26, 49–57, 1997.

56. Fabiańska, M., Bzowska, G., Matuszewska, A., Racka, M., and Skręt, U., *Chemie der Erde (Geochemistry),* 63, 63–91, 2003.
57. Heath, D.J., Lewis, C.A., and Rowland, S.J., *Org. Geochem.*, 26, 769–785, 1997.
58. Bastow, T.P., Alexander R., and Kagi, R.I., *Org. Geochem.*, 26, 79–89, 1997.
59. Alexander, R., Cumbers, K.M., and Kagi, R.I., *Org. Geochem.*, 10, 841–845, 1986.
60. Rospondek, M., Koster, J., and Sinninghe-Damsté, J.S., *Org. Geochem.*, 26, 295–304, 1997.
61. Yangming, Z. and Zhongyi, Z., *Chinese J. Geochemistry*, 10, 80–87, 1991.
62. Bzowska, G., Fabiańska, M., Matuszewska, A., Racka, M., Skręt, U., *Geological Quarterly*, 44, 425–437, 2000.
63. Norgate, C.M., Boreham, C.J., and Wilkins A.J., *Org. Geochem.*, 30, 985–1010, 1999.
64. Tritz, J.P., Herremann, D., Bisseret, D., Connan, J., and Rohmer, M., 1999, *Org. Geochem.*, 30, 499–514.
65. Sinninghe Damsté, J.S., Betts, S., Ling, Y., Hofmann, P.M., and de Leeuw J.W., *Org. Geochem.*, 20, 1187–1200, 1993.
66. Curiale, J.A. and Frolov, E.B., *Org. Geochem.*, 29, 379–408, 1998.
67. Simoneit, B.R.T., and Mazurek, M.A., in: Sibuet, J.-C & Ryan, W.B.F, Eds., *Init. Reports of the DSDP*, vol. XLVII , 1979, 541–545.
68. Simoneit, B.R.T., in: Curray, J.R. and Moore , D.G., Eds., *Init. Reports of the DSDP*, vol. LXIV, 1982, 877–880.
69. Simoneit, B.R.T, and Philp R.P., in: Curray, J.R. and, Moore, D.G., Eds., *Init. Reports of the DSDP*, vol. LXIV, 1982, 883–904.
70. Simoneit B.R.T., Aboul-Kassim, T.A.T and Tiercelin, J.-J., *Appl. Geochem.*, 355–368, 2000.
71. Yongsong, H., Ansong, G., Jiamo, F., Guoying, S., Biqiang, Z., Yixian, Ch., and Mafen, L., *Org. Geochem.*, 19, 29–39, 1992.
72. Marynowski, L., Pięta, M., and Janeczek, J., *Geological Quarterly*, 48, 169–180, 2004.
73. Pancost, R.D., Crawford, N., and Maxwell, J.R., *Chem. Geol.*, 188, 217–227, 2002.
74. Sperline, R.P., Song, Y., Ma, E., and Freiser, H., *Hydrometallurgy* 50, 1–21, 1998.
75. Kuroda, K.-I., and Dimmel, D.R., *J. Anal. Appl. Pyrolysis,* 62, 259–271, 2002.
76. Behrens, A., Schaeffer, P., Bernasconi, S., and Albrecht, P., *Geochim. Cosmochim. Acta* 64, 3327–3336, 2000.
77. Ma, T. S., and Horak, V., *Microscale Manipulation in Chemistry* (Series: Chemical Analysis: A Series of Monographs on Analytical Chemistry & Its Applications), John Wiley & Sons Inc, 1976, 504pp.
78. *Comprehensive Analytical Chemistry,* vol. XIII, *Analysis of Complex Hydrocarbon Mixtures,* part A, *Separation Methods,* Elsevier Sci. Pub. Comp., Amsterdam, 1981.
79. Fowler, M.G., Abolins, P., and Dougles, A.G., *Org. Geochem.*, 10, 815–823, 1986.
80. Miranda, A.C.M.L, Loureiro. M.R.B., and Cardoso J.N. *Org. Geochem.*, 1027–1038, 1998.
81. Sauer, P.E., Eglington, T.I., Hayes, J.M., Schimmelmann, A., and Sessions, A.L., *Geochim. Cosmochim. Acta* 65, 213–222, 2001.
82. Chaffee A.L., and Johns R.B., *Org. Geochem.*, 8, 349–365, 1985.
83. Stefanova, M., Magnier, C., and Velinowa, D., *Org. Geochem.*, 23, 1067–1084, 1995b.
84. Koopmans, M.P., Schouten, S., Kohnen, M.E.L., and Sinninghe-Damsté, J.S., *Geochim. Cosmochim Acta,* 60, 4873–4876, 1996.
85. Ellis, L., *Aromatic hydrocarbons in crude oil and sediments: Molecular sieve separations and biomarkers*, PhD thesis, Curtin University of Technology, 1994.

86. Lazaro, M.-J., Moliner, R., Suelves, J., Herod, A.A., and Kandiyoti, R., *Fuel,* 80, 179–194, 2001.
87. Simoneit, B.R.T., *Appl. Geochem.,*17, 129–162, 2002.
88. Ellis, L., Langworthy, T.A., and Winians, R., *Org. Geochem.*, 24, 57–69, 1996.
89. Ellis, L., Singh, R.K., Alexander, R., and Kagi, R.J., *Org. Geochem.*, 23, 197–203, 1995.
90. Poinsot, J., Adam, P., Trendel, J.M., Connan, J., and Albrecht, P., *Geochim. Cosmochim. Acta,* 59, 4653–4661, 1995.
91. Van Kaam-Peters, H.M.E. and Sinninghe-Damsté, J.S., *Org. Geochem.*, 27, 371–397, 1997.
92. Hahn-Deinstrop, E., *Applied Thin Layer Chromatography. Best Practice and Avoidance of Mistakes.* Wiley-VCH, Weinheim, 2000, pp.304.
93. Poinsot, J., Schneckenburger, P., Adam, P., Schaeffer, P., Trendel, J.M., Riva, A., and Albrecht, P., *Geochim. Cosmochim Acta,* 62, 805–814, 1998.
94. Schouten, S., Sinninghe-Damsté, J.S., and de Leeuw, J.W., *Geochim. Cosmochim. Acta,* 59, 953–958, 1995.
95. Schouten, S., Sinninghe-Damsté, J.S., and de Leeuw, J.W., *Org. Geochem.*, 23, 125–138, 1995.
96. Köster, J., van Kaam-Peters, H.M.E., Koopmans, M.P., de Leeuw, J.W., and Sinninghe-Damsté J.S., *Geochim. Cosmochim. Acta*, 61, 2431–2452, 1997.
97. Van Kaam-Peters, H.M.E., Rijpstra, W.I.C., de Leeuw, J.W., and Sinninghe-Damsté, J.S., *Org. Geochem.*, 28, 151–177, 1998.
98. Köster, J., Rospondek, M., Schouten, S., Kotarba, M., Zubrzycki, A., and Sinninghe-Damsté, J.S., *Org. Geochem.,* 29, 649–669, 1998.
99. Frolov, E.B., *Org. Geochem.*, 23, 447–450, 1995.
100. Dunlop, R.W. and Jefferies, P.R., *Org. Geochem.*, 8, 313–320, 1985.
101. Fowler, M.G. and Douglas, A.G., *Org. Geochem.*, 11, 201–213, 1988.
102. Simoneit, B.R.T., *Geochim. Cosmochim. Acta,* 41, 463–476, 1977.
103. Putschew, A., Schaeffer-Reiss, Ch., Schaeffer, P., Koopmans, M.P., de Leeuw, J.W. Lewan, M.D., Sinninghe-Damsté, J.S. and Maxwell, J.R., *Org. Geochem.*, 29, 1845–1856, 1998.
104. Simoneit, B.R.T., *Atm. Env.*, 31, 2225–2233, 1997.
105. Fabiańska, M., *Chemometrics and Intelligent Laboratory Systems,* 72, 241–244, 2004.
106. Bonnett, R. and Czechowski, F., *J. Chem. Soc., Perkin Trans.,* I, 125–131, 1984.
107. Bonnett, R., *Int. J. Coal Geol.*, 32, 137–149, 1996.
108. Premović, P.I., Tonsa, I.R., Dordević, D.M., and Premović, M.P., *Fuel* 79, 1089–1094, 2000.
109. Grosjean, E., Adam, P., Connan, J., and Albrecht. P., *Geochim. Cosmochim. Acta,* 68, 789–804, 2004.
110. Boggess, J. M., Czernuszewicz, R.S., and Lash, T.D., *Org. Geochem.*, 33, 1111–1126, 2002.
111. Ocampo, R. and Repeta, D.J., *Org. Geochem,*, 33, 849–854, 2002.
112. Grice, K., Gibbson, R., Atkinson, J.E., Schwark, L., Eckard, Ch. B., and Maxwell, J.R., *Geochim. Cosmochim. Acta.* 60, 3913–3924, 1996.
113. Aries, A., Doumenq, P., Artaud, J., Molinet, J., and Bertrand, J.C., *Org. Geochem.*, 32, 193–197, 2001.
114. Hunt, J.M., *Petroleum Geochemistry and Geology,* 2nd ed., Freeman, W.H. and Company, New York, 1996.
115. North, F.K., *Petroleum Geology,* 2nd ed. Unwin Hyman, Boston, 1990, 631pp.

116. *Organic Geochemistry. Principles and Applications*, Engel, M.H. and Macko, S.A. Eds., Plenum Press, New York. 1993, 862pp.
117. van Krevelen, D.W., Coal. *Typology, Physics, Chemistry, Constitution,* 3rd ed. Elsevier, Amsterdam, 1993.
118. Killops, S.D. and Killops, V.J., *An Introduction to Organic Geochemistry*, Longman Scientific & Technical, Harlow, UK, 1993.
119. Simoneit, B.R.T., in: *Organic Geochemistry, Principles and Applications*, Engel, M.H. and Macko, S.A., Eds, Plenum Press, New York, 1993, ch. 17, pp. 397–418.
120. Palmer, S.E., in: Engel, M.H. and Macko, S.A. Eds., *Organic Geochemistry. Principles and Applications*, Plenum Press, New York. 1993, ch. 23, 511–533.
121. Schultze, T. and Michealis, W., *Org. Geochem.*, 16, 1051–1058, 1990.
122. Baranger P. and Disnar J.R., *Org. Geochem.*, 647–653, 1988.
123. Fabiańska, M., and Nowak, I., Zeszyty Naukowe Politechniki Śląskiej, seria Górnictwo, z. 249, s.19–28, 2001.
124. Connan, J., In: Brooks, J. and Welte, D., Eds, *Advances in Petroleum Geochemistry*, 1, 300–335, 1984.
125. Volkman, J. K., *Org. Geochem.,*. 6, 619–632, 1984.
126. Dajiang, Z., Difan, H., and Jinchao, L., *Org. Geochem.*, 1, 295–302, 1988.
127. Lin, L.H., Michael, G.H., Kovachev, G., Zhu, H., Philp, R.P., and Lewis, C.A., *Org. Geochem.,* 14, 511–523, 1989.
128. Fetter, C.W., *Contaminant Hydrogeology,* 2nd ed. Prentice Hall, Upper Saddle River. 1996.
129. Williams, J.A., Bjøroy, M., Dolcater, D.L., and Winters, J.C., *Org. Geochem.* 10, 451–461, 1986.
130. Ahmed, M., Smith, J.W., and George, S.C., *Org. Geochem.*, 30, 1311–1322, 1999.
131. Fabiańska, M., Lewińska-Preis, L., and Galima-Stypa, R., *Fuel,* 82, 165–179, 2003.
132. Ewbank G., Manning D.A.C., and Abbott G.D., *Org. Geochem.,* 20, 579–598, 1993.

16 The Use of PLC for Isolation and Identification of Unknown Compounds from the Frankincense Resin (Olibanum): Strategies for Finding Marker Substances

Angelika Koch, Rita Richter, and Simla Basar

CONTENTS

16.1 OLIBANUM RESINS

Olibanum (frankincense) is one of a group known as the oleogum resins (mono-, sesqui-, di-, and triterpenes and mucous substances) that exude from incisions in the bark of the *Boswellia* trees (fam: Burseraceae), the most common species of which are *B. carterii* (Sudan, Somalia, and Ethiopia) and *B. serrata* (India), whereas *B. frereana* (Oman, Somalia) and *B. sacra* (Arabia) belong to the rare resins on the market.

Frankincense is one of the oldest fragrant and medicinal resins. The most widely traded olibanum belongs to the "Ethiopian-type" resin produced by the Ethiopians and Sudanese as reported by Helfer [1], whereas the most valuable olibanum resin comes from Oman (≈ 50 to 70$/kg). The Indian type is comparable to the Ethiopian type with respect to price (≈ 5 to 10$/kg). The quality is indicated by geographical trade names such as "Oman first choice," "Somalia peasize," or "Indian siftings" and not by the botanical classification. As mentioned in the preceding text, the great demand for the Oman type (this resin belongs to *B. frereana*) is due to its excellent fragrance.

Apart from the use and need as fragrances, the resins are marketed for medicinal use as antiarthritic and antiinflammatory pharmaceutical products. The pharmacological effects are mainly attributed to the presence of the nonvolatile pentacyclic triterpenoid boswellic acids. This class of ingredients is not present in the actual valuable *B. frereana* species; hence, *B. frereana* plays no part in the pharmaceutical area.

For pharmaceutical purposes, one of the main problems will be to define the botanical origin of the different olibanum resins. Up to now, there are no scientifically current pharmaceutical monographs on olibanum, and pharmaceutical companies that want to develop new medicinal products have an urgent need of analytical methods for the botanical identification and quality assurance of the resins. Attempts had been made by Hahn-Deinstrop et al. [2].

16.2 HOW TO DIFFERENTIATE BETWEEN *BOSWELLIA CARTERII* AND *BOSWELLIA SERRATA*

16.2.1 Looking for Striking Differences in the Gas Chromatogram

To identify the volatile components, gas chromatography–mass spectrometry (GC–MS) is still the method of choice. A comparison of the GC fingerprints of *B. carterii* and *B. serrata* reveals the different composition of the volatile fractions (Figure 16.1). Common monoterpenes, aliphatic, and aromatic compounds of olibanum are, e.g., pinene, limonene, 1,8-cineole, bornyl acetate, and methyleugenol (Figure 16.2).

A perceptible marker of the olibanum fragrances coming from Somalia, Sudan, or Ethiopia is the aliphatic octyl acetate marked by a distinct acrid smell. *Boswellia carterii* contains up to 50% of the aliphatic octyl acetate, demonstrated by the strong stinging smell of the fume, whereas *Boswellia serrata* (its common name in India is salai guggul) contains none or only small amounts of it and, consequently, does not have such a harsh smell.

Another striking difference relates to the diterpenes (Figure 16.3): *B. carterii* reveals strong peaks that have been assigned to verticilla-4 (20),7,11-triene (compound 1), incensole acetate (compound 2), and incensole (compound 3) and, on the other hand, *Boswellia serrata* shows peaks of *m*- and *p*-camphorene (compound 4 and compound 5) as well of cembrenol (= serratol) (compound 6).

As there exists such a distinct difference in the spectrum of the volatile diterpenic components of both species, it would be advantageous to transfer these results to TLC in order to display these taxonomic marker substances as a tool for quality assurance operations.

16.2.2 Comparison of the Results Obtained by GC–MS to Those Obtained by TLC

Figure 16.4 displays the separation of the resins of *B. carterii* and *B. serrata*, their hydrodistillates, and three commercially available “olibanum essential oils” on a Merck LiChrospher plate in the mobile phase heptane–diethylether–formic acid (7 + 3 + 0.3; v/v/v) without chamber saturation after derivatization with anisaldehyde reagent.

Lane 1 displays the separation of the resin of *B. carterii* and lane 2 and lane 3 its volatile fractions received by hydrodistillation. Accordingly, lane 8 displays the

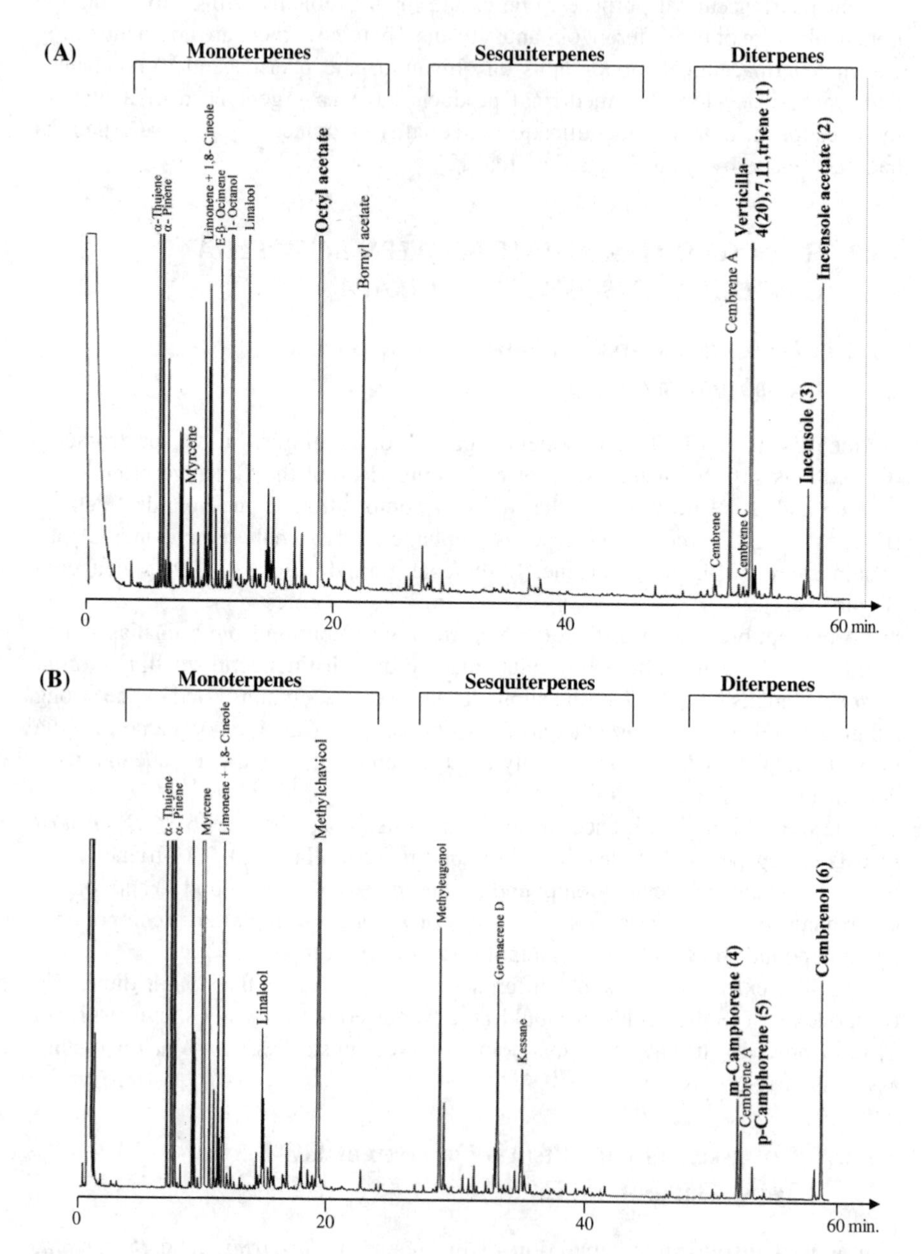

FIGURE 16.1 (A) Gas chromatographic analysis of the volatile fraction of *Boswellia carterii* (25-m capillary column with CPSil 5 CB, 50°C, 3°C/min up to 230°C, injector at 200°C, detector at 250°C, carrier gas 0.5 bar H_2). (B) Gas chromatographic analysis of the volatile fraction of *Boswellia serrata* (25-m capillary column with CPSil 5 CB, 50°C, 3°C/min up to 230°C, injector at 200°C, detector at 250°C, carrier gas 0.5 bar H_2).

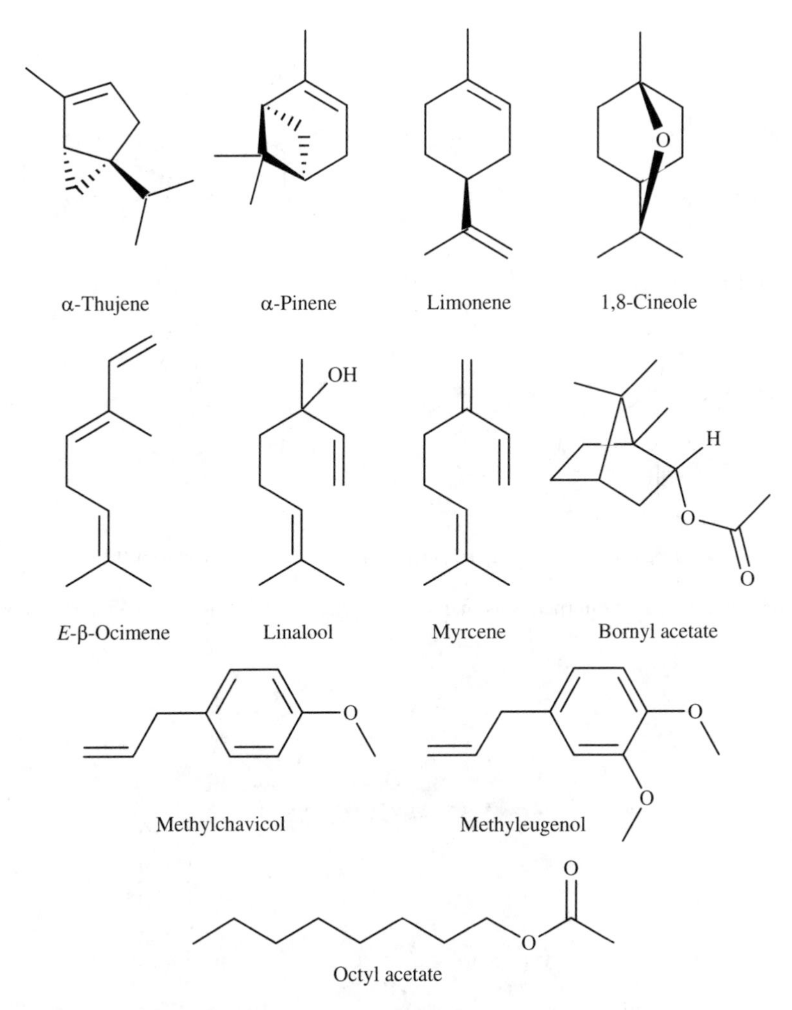

FIGURE 16.2 Common monoterpenic, aromatic, and aliphatic constituents of olibanum resins.

separation of the resin of *B. serrata* and lane 7, its volatile fraction. The lanes 4, 5, and 6 display the separation of the commercial olibanum essential oils.

The brownish colored zone (R_f 0.28) of incensole (compound 3), which occurs in both the resin and the volatile fractions of *B. carterii*, draws the line between the volatile diterpenes and the nonvolatile triterpenes. *B. carterii* reveals two further colored prominent spots, a yellowish-ochre (R_f 0.65) of incensole acetate (compound 2) and a violet-colored spot (R_f 0.98) of verticilla-4(20),7,11-triene (compound 1). Lane 2 and lane 3 reveal a light blue "area" (R_f 0.60) of 1,8-cineol that is only visible in freshly distilled oils.

B. serrata is specified — apart from a likewise violet spot in the front (R_f 0.98) — by a dark ochre-greenish spot (R_f 0.40) of cembrenol (compound 6). The commercial oils on lane 4 to lane 6 have their origin in *B. serrata* resins spiked with

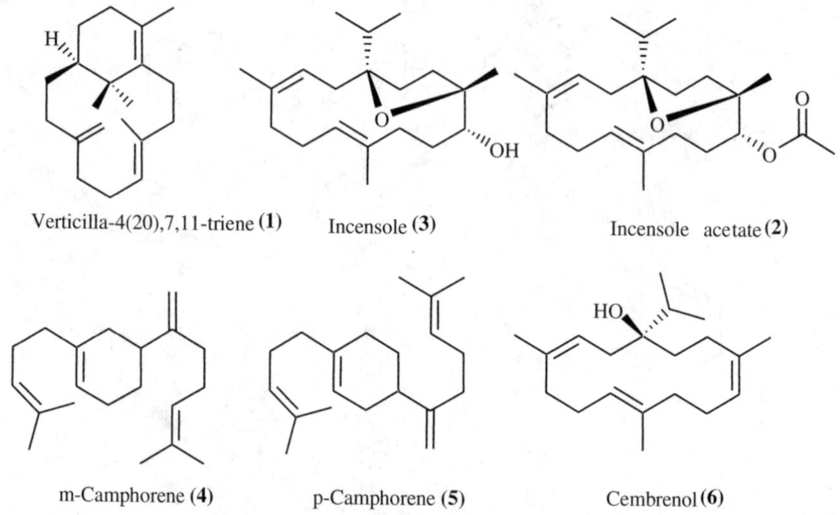

Verticilla-4(20),7,11-triene (**1**) Incensole (**3**) Incensole acetate (**2**)

m-Camphorene (**4**) p-Camphorene (**5**) Cembrenol (**6**)

FIGURE 16.3 Decisive diterpenes as marker substances of *B. carterii* and *B. serrata* resins.

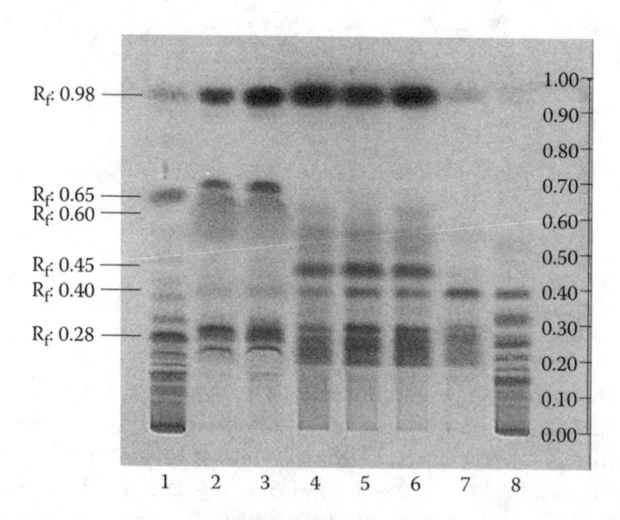

FIGURE 16.4 TLC separation of some *Boswellia* samples. Sequence of application: lane 1: *B. carterii* resin, lane 2 and 3: Hydrodistillates of *B. carterii*, lane 4, 5, and 6: merchandized oils, lane 7: hydrodistillate of *B. serrata* and lane 8: *B. serrata* resin. Allocation of the R_f values: R_f 0.98 = verticilla-4(20),7,11-triene (compound 1), R_f 0.65 = incensole acetate (compound 2), R_f 0.60 = 1,8-cineol, R_f 0.45 = caryophyllene, R_f 0.40 = cembrenol (compound 6), R_f 0.28 = incensole (compound 3).

well-scented components such as caryophyllene (R_f 0.45) and various components at R_f 0.98.

The following section describes the isolation of these substances, which are qualified as taxonomic markers, in order to compare their identity with the mentioned diterpenes of the GC.

16.3 ISOLATION OF THE DITERPENES AS TAXONOMIC MARKERS

16.3.1 Fractionation of the Resin Extracts

The solutions of the olibanum resins reveal an unpleasant property of stickiness. This must be considered with all steps of analyzing and isolation. That is why a preceding column chromatography (CC) is recommended, to enrich the diterpenes of interest. The further purification of the supposed marker substances was carried out by PLC.

A concentrated hexane extract (11.5 g/4 ml) of *B. carterii* resin and *B. serrata* resin, respectively, was applied to a column filled with 80-g silica gel (Merck LiChroprep Si 60 No. 9390) conditioned by hexane. The fractionation was achieved by a gradient of 200-ml hexane followed by 200-ml hexane–dichloromethane (1 + 1; v/v), 200-ml dichloromethane and 200-ml dichloromethane–acetone (1 + 1; v/v) as eluents to give four subfractions of 200 ml each. These fractions were collected, concentrated, and applied to a TLC plate for a screening with the mobile phase dichlormethane–diisopropylether (9 + 1; v/v).

16.3.2 *Boswellia carterii*: Verticilla-4(20),7,11-Triene (Compound 1) and the Other Cembrenes

The TLC screening of the silica gel fractions of *B. carterii* revealed in fraction 1, after derivatization with anisaldehyde reagent, a single violet zone (R_f 0.85). The purification steps of this zone are illustrated in Figure 16.5. The zones of the following fractions 2 and 3 were not separated from each other. Their separation and purification steps are shown in Figure 16.6, whereas the appropriate steps of fraction 4 are demonstrated in Figure 16.7.

16.3.2.1 Isolation of Verticilla-4(20),7,11-Triene (Compound 1)

The GC–MS data of fraction 1 revealed a strong peak of verticilla-4(20),7,11-triene (compound 1) accompanied by small amounts of cembrane A and cembrane C. To purify the violet spot and isolate compound 1, it was necessary to reduce the solvent strength. In the mobile phase dichloromethane–hexane (9 + 1; v/v), the development time decreases, which leads to minor diffusion of the zone. The zone of (compound 1) was marked by $\lambda = 254$ nm UV light. To exclude the impurities, the separation process had to be repeated several times. The zone was removed from the glass plate and eluted from the adsorbent with dichloromethane. The concentrated solution achieved was applied onto a TLC plate as well as injected onto a GC column; the

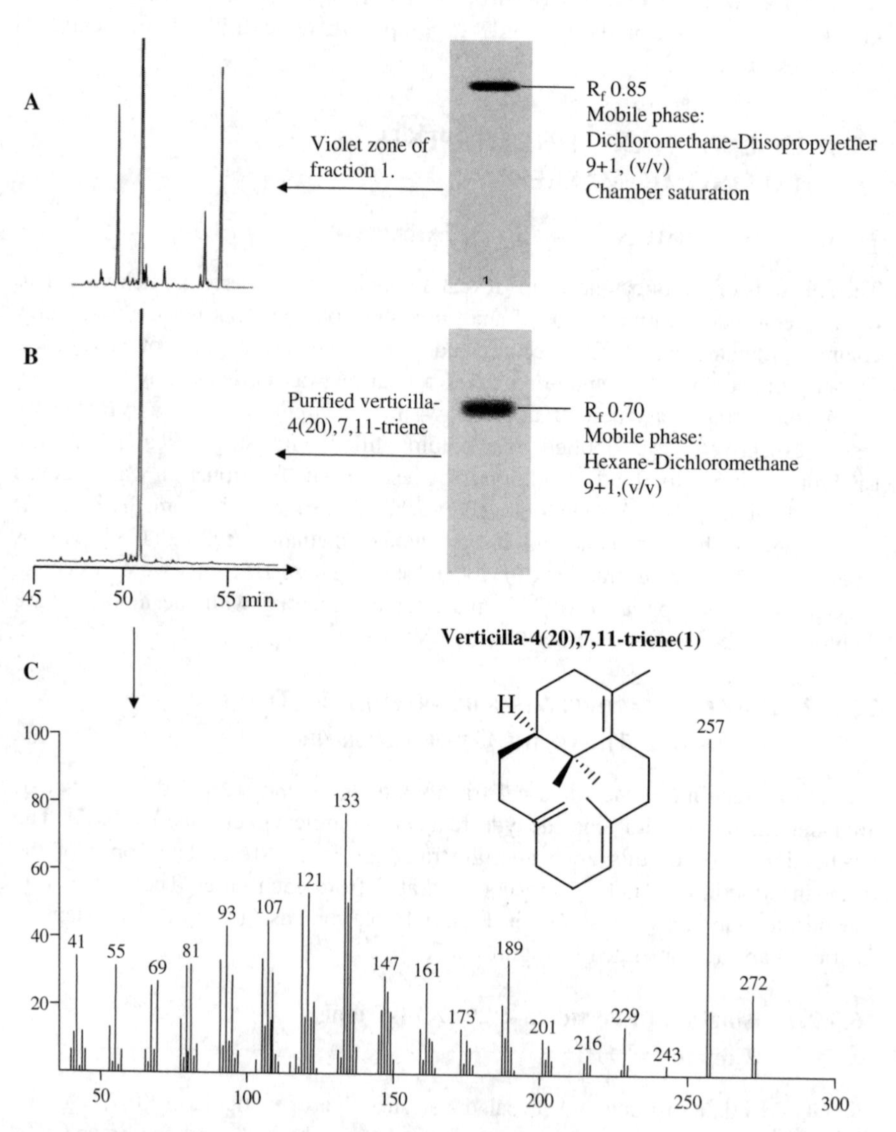

FIGURE 16.5 Purification steps of fraction 1 from *B. carterii.* (A) GC data of the roughly collected violet zone (R_f 0.85). (B) GC data of the purified verticilla-4(20),7,11-triene (compound 1). (C) Mass spectrum of verticilla-4(20),7,11-triene (compound 1).

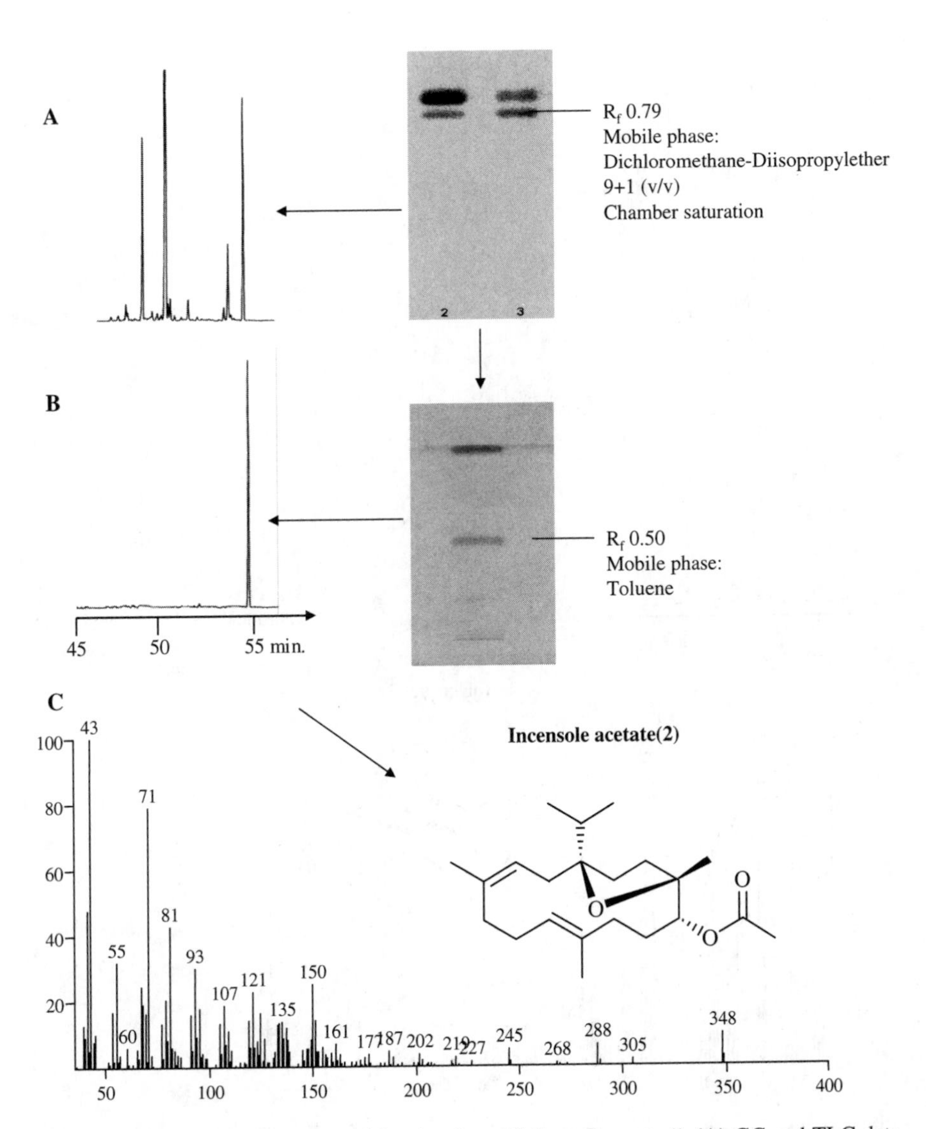

FIGURE 16.6 Purification steps of fraction 2 and 3 from *B. carterii*. (A) GC and TLC data of fractions 2 and 3. (B) GC data of incensole acetate after separation and purification of the zone R_f: 0.50. (C) Mass spectrum of incensole acetate (compound 2).

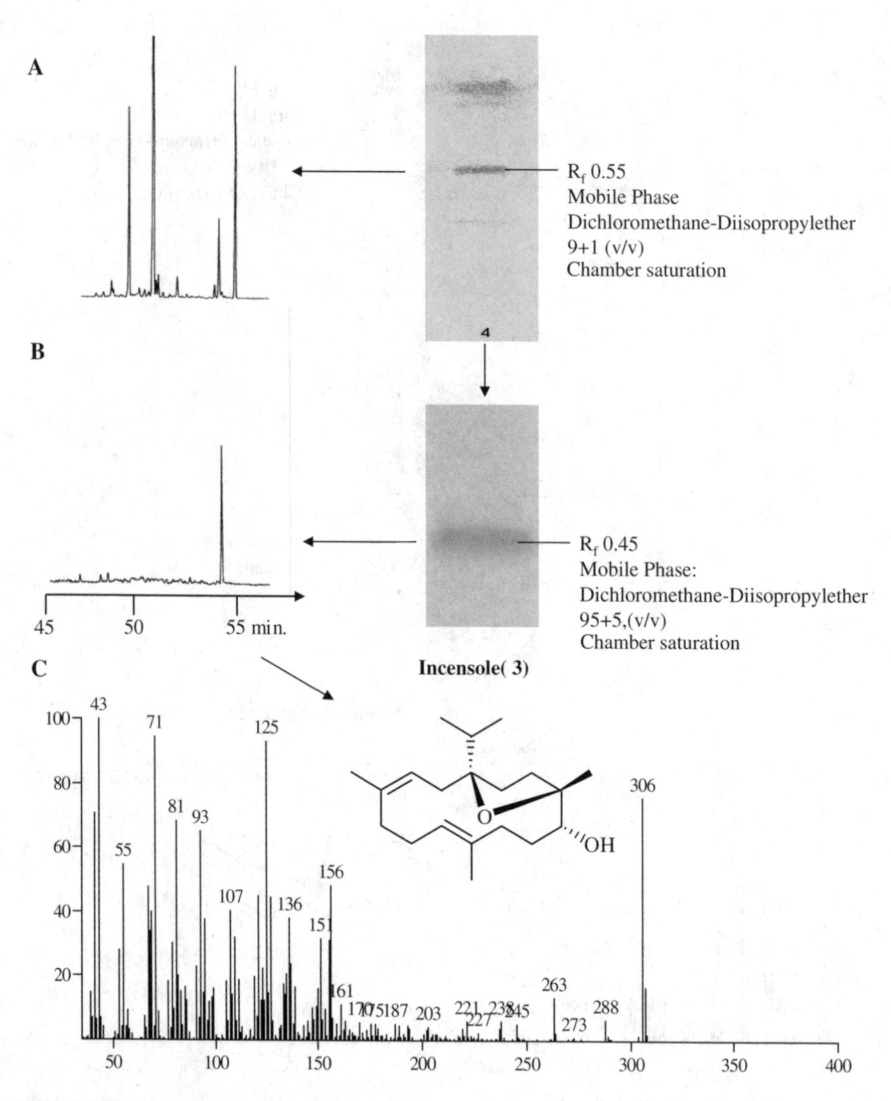

FIGURE 16.7 Purification steps of fraction 4 from *B. carterii*. (A) GC and TLC data of fraction 4. (B) GC data of incensole after separation and purification. (C) Mass spectrum of incensole (compound 3).

violet spot of the hexane fraction turned out to be compound 1 (Figure 16.5). The structure elucidation has been described extensively by Basar [3].

16.3.2.2 Isolation of Incensole Acetate (Compound 2)

The yellowish-ochre spot of incensole acetate (compound 2) in fraction 2 and fraction 3 was separated from compound 1 by using toluene as the mobile phase. The zone of incensole acetate is not visible under λ = 254 nm UV light and had to be derivatized by anisaldehyde reagent. After marking the position of the zone on the remaining part of the plate, the zones were treated as described for compound 1. Figure 16.6 illustrates purification steps that were necessary to reveal one spot to match with incensole acetate (compound 2).

16.3.2.3 Isolation of Incensole (Compound 3)

To isolate the brownish spot of incensole in fraction 4, the mobile phase, dichloromethane–diisopropylether (95 + 5; v/v) was favored. The marked zones were scraped off in a larger scale. After collection and extraction of the zones, the roughly separated substance was purified several times until the collected extracts revealed one brownish spot, which matched the incensole (compound 3) peak on GC. Figure 16.7 illustrates the purification steps.

16.3.3 *Boswellia serrata*: *m*-Camphorene (Compound 4) and *p*-Camphorene (Compound 5) and Cembrenol (Serratol) (Compound 6)

According to the descriptions given for the isolation of the diterpenes of *B. carterii*, the isolation of the corresponding substances of *B. serrata* started with a preceding enrichment by CC.

16.3.3.1 Isolation of *m*-Camphorene (Compound 4) and *p*-Camphorene (Compound 5)

The separation and collection of the violet zone (R_f 0.98) of fraction 1 led to the isolation of *m*- and *p*-camphorene (compound 4 and compound 5) and cembrane A as well. Though *m*- and *p*-camphorene are taxonomic marker substances of the species *B. serrata*, it is pointless to use them in TLC, because the corresponding violet spot on the chromatogram of *B. carterii* stands for verticilla-4(20),7,11-triene, the taxonomic marker substance of *B. carterii* and cembrene A, too. These “violet” substances should be reserved for GC investigations.

16.3.3.2 Isolation of Cembrenol (Compound 6)

Figure 16.8 illustrates the separation of cembrenol (compound 6), the prominent colored spot (greenish-brown at R_f 0.65), which was accompanied by a blue spot (R_f 0.62). The upper zone of compound 6 was separated, isolated, and purified by using a sequence of mobile phases, beginning with dichloromethane–diisopropyl

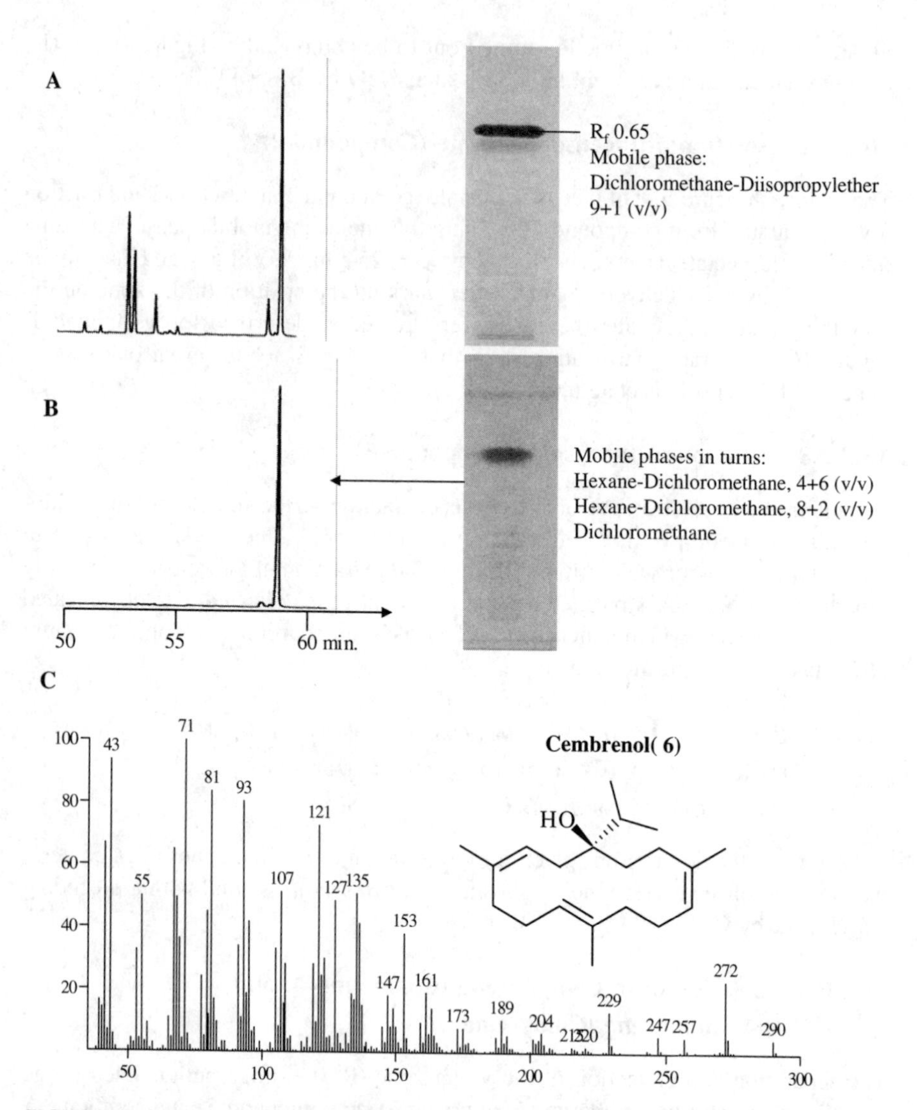

FIGURE 16.8 Purification steps of fraction 2 and 3 from *B. serrata*. (A) GC data of zone R_f 0.65. (B) GC data of the separated zone after several separation steps. (C) Mass spectrum of cembrenol (compound 6).

ether (90 + 10; v/v). The following mobile phases, for example, dichloromethane–hexane (40 + 60; v/v, 80 + 20; v/v) and dichloromethane were used in turn until the initial roughly separated substance appeared pure. The marked zones (by anisaldehyde reagent) were scraped off and after collection and extraction of the zones, the concentrated extract revealed one zone. The check on the successful separation and purification was done by TLC and GC.

To summarize, generally, the diterpenes of both investigated *Boswellia* species are suitable for marker substances. Figure 16.9 clearly demonstrates that for TLC

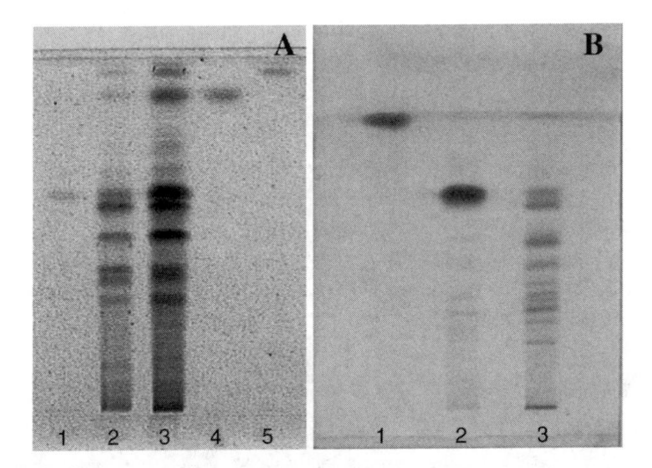

FIGURE 16.9 An overview of the marker substances of (A) *B. carterii* and (B) *B. serrata*. (A) Lane 1: incensole (compound 3), lane 2 and 3: hexane extract of *B. carterii*, lane 4: incensole acetate (compound 2), lane 5: verticilla-4(20),7,11-triene (compound 1). (B) Lane 1: *m*-camphorene (compound 4) and *p*-camphorene (compound 5), lane 2: cembrenol (compound 6), lane 3: Hexane extract of *B. serrata*.

purposes, only the oxygenated and more polar components such as incensole, incensole acetate, and cembrenol (serratol) are of great analytical value.

16.4 OLIBANUM PYROLYSATES

16.4.1 GENERATING THE FUME AND ITS ABSORPTION

The investigation of the pyrolysates of the olibanum resins was initiated by an article in a newspaper. The author of this article seemed to be apprehensive about the pharmacological and toxicological effects of the fume of the resins, which are used in religious ceremonies, on the health of people. The resins used in churches named as "Pontifical" or "Olibanum König" mainly consist of *B. carterii*, whereas those of inferior quality contain *B. serrata* resins. For this reason, the fumes of both resins were investigated by Basar [4].

The fume, which is generated in a censer at high temperatures < 600°C when the resins are shaken onto the glowing embers, consists of substances of unknown composition. For investigation purposes, the fume had to be absorbed *in statu nascendi* as it was coming out of the censer. 100 mg of adsorption material (Super Q® by Alltech Chemicals) was filled into glass cartridges and the absorbed fume was eluted with ethyl acetate, concentrated, and applied to GC–MS and TLC analyses as well to get an overview of the substances that resulted from the pyrolysis.

16.4.2 SELECTION AND SEPARATION OF THE PYROLYZED TRITERPENES

After development to a distance of 7 cm with cyclohexane–diethyl ether (80 + 20; v/v) as the mobile phase and derivatizing with anisaldehyde reagent (Figure 16.10),

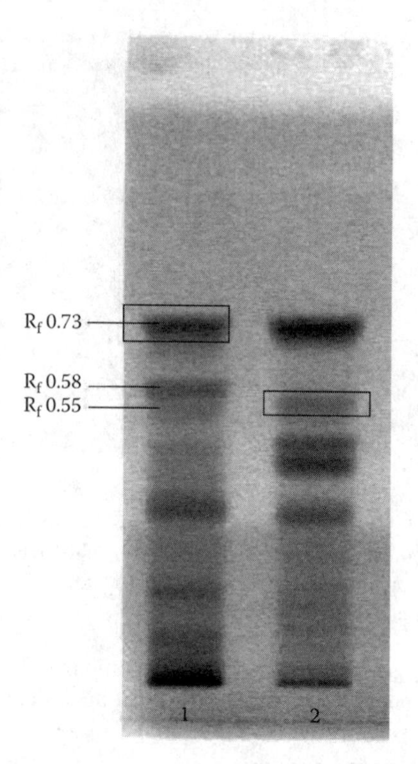

FIGURE 16.10 Development of the pyrolyzed terpenes of *B. carterii* (lane 1) and *B. serrata* (lane 2). The marked violet zone (R_f 0.73) of lane 1 and the light violet zone (R_f 0.55) of lane 2 were used for further purification steps.

the upper zones (R_f 0.73) gave a violet color, the yellowish-ochre zone below (R_f 0.58) that only appeared in *B. carterii* originated from incensol acetate, and the zone with a light violet color (R_f 0.55) occurred in the fume of the resins of both *B. carterii* and *B. serrata*.

16.4.2.1 Separation of the Nortriterpenes with Carbohydrate Structure of *B. carterii*

The upper violet zones (R_f 0.73) were removed from the TLC plate, eluted from the sorbent with ethyl acetate, and prepared for GC–MS.

The GC–MS data (Figure 16.11) of the violet zone of *B. carterii* revealed that the unchanged diterpenes (verticillatriene, cembrene A, and cembrene C) and the nortriterpenes with carbohydrate structure originated from the pyrolyzed triterpenes (Figure 16.12) of the α- and β-boswellic acids, named 24-norursa-3,12-diene (compound 7), 24-norursa-3,9(11),12-triene (compound 8), 24-noroleana-3,12-diene (compound 9), and 24-noroleana-3,9(11),12-triene (compound 10).

The corresponding TLC separation of the violet zones was achieved with pentane as mobile phase at –25°C. The compounds are visible under $\lambda = 254$ nm UV light. This system was accepted to get higher amounts of compound (compound 7) by

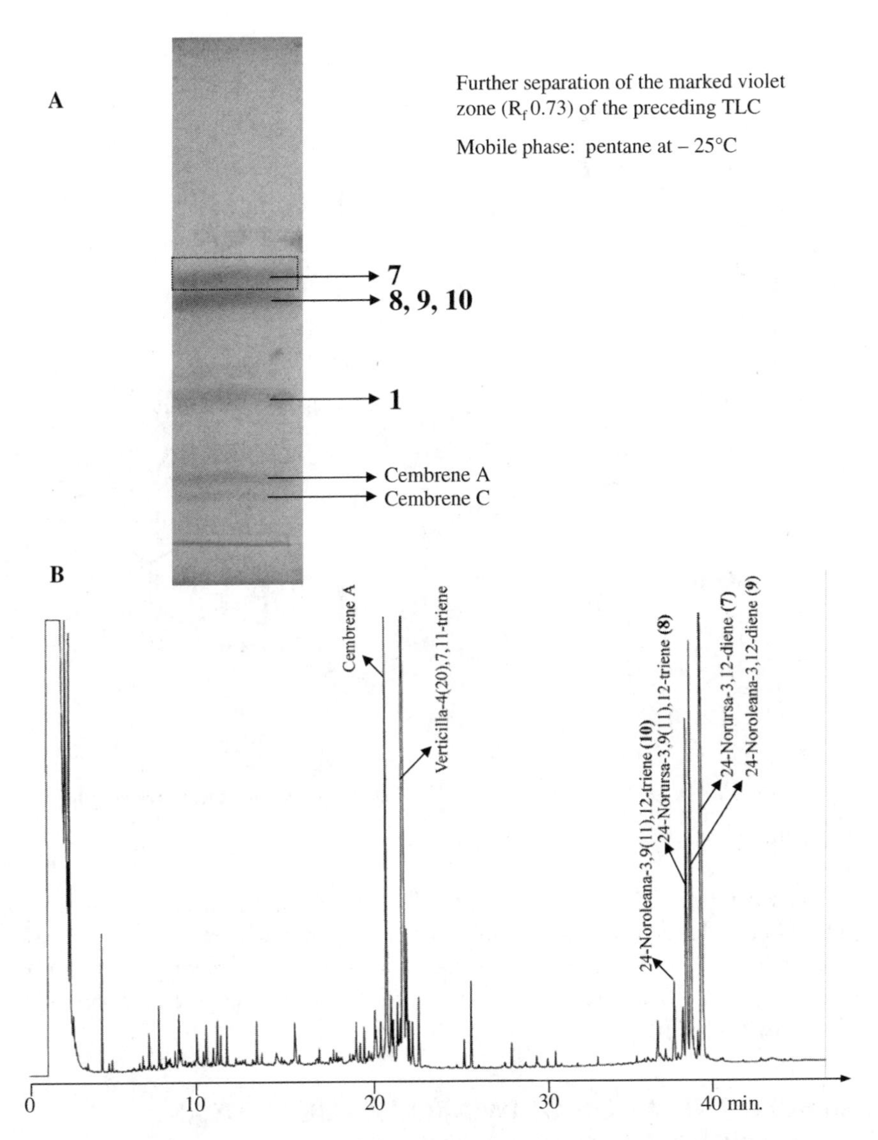

FIGURE 16.11 (A) TLC separation of the violet zone at –25°C and the marking of compound 7 for separation by PLC. (B) The GC of the violet zone (R_f 0.73) of the preceding TLC.

PLC for pharmacological and toxicological investigations. The separation of compound 8, compound 9, and compound 10 did not materialize completely.

16.4.2.2 Separation and Isolation of an Oxidized Notriterpene of *B. serrata*

The isolation of the second violet zone (R_f 0.55) succeeded more easily with *B. serrata* resins because there was no interference with incensol acetate.

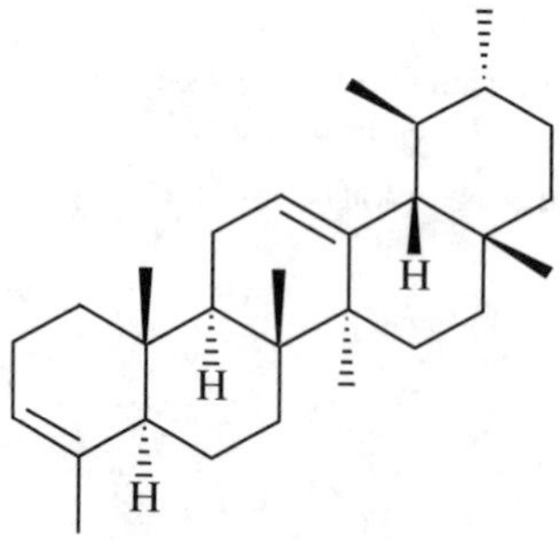

24-Norursa-3,12-diene (7)

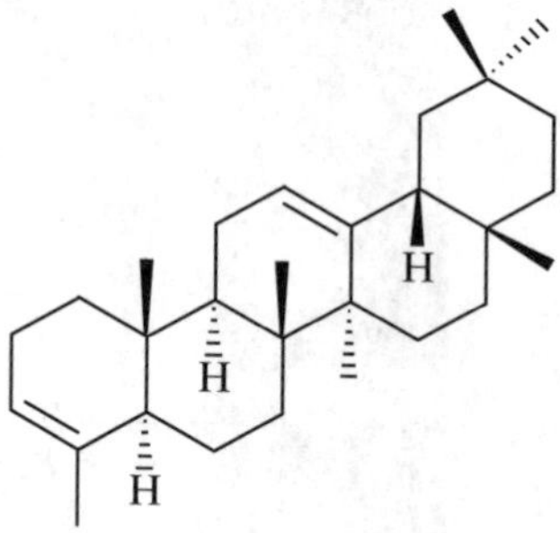

24-Noroleana-3,12-diene (9)

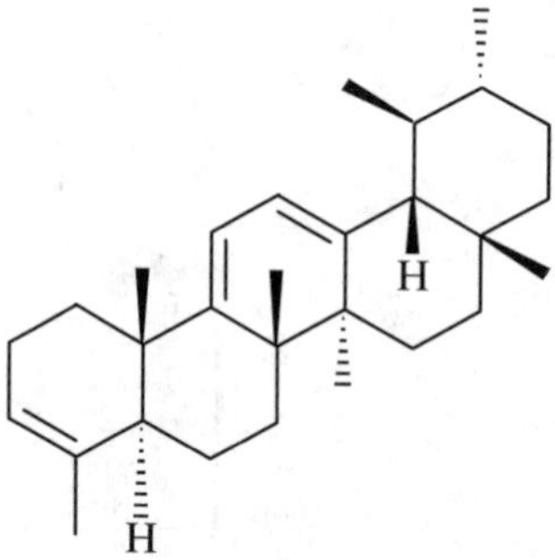

24-Norursa-3, 9(11), 12-triene (8)

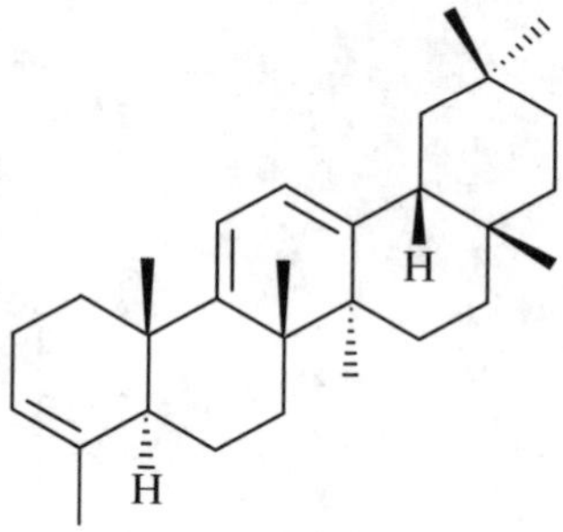

24-Noroleana-3, 9(11), 12-triene (10)

FIGURE 16.12 Structures of triterpenic pyrolysis products of olibanum resins.

PLC separations (Figure 16.13) started with toluene–ethyl acetate (95 + 5; v/v) as the mobile phase, whereas the final purification was achieved with cyclohexane–diethyl ether (80 + 20; v/v). The isolated and extracted compound matched the GC–MS of 24-norursa-3,12-dien-11-one (compound 11) and is confirmed as an oxidized pyrolyzed boswellic acid.

16.5 DETERMINATION OF TWO BOSWELLIC ACIDS AS MARKER OF ALTERATIONS DURING PROCESSING OF THE RESINS

Common pharmaceutical products of olibanum and salai guggul are tablets prepared from dried extracts of boswellic acids, which are obtained by processes involving treatment of the resins with alkali and acid. The stress involved in this treatment is expected to lead to alteration of some triterpenes as, e.g., the conversion of the unstable 3-*O*-acetyl-11-hydroxy-β-boswellic acid (compound 12) to the stable compound 3-*O*-acetyl-9,11-dehydro-β-boswellic acid (compound 13). Two-dimensional TLC is an excellent means of observing this conversion [5]. For verification of this process, the substances have to be isolated by PLC and identified by GC–MS.

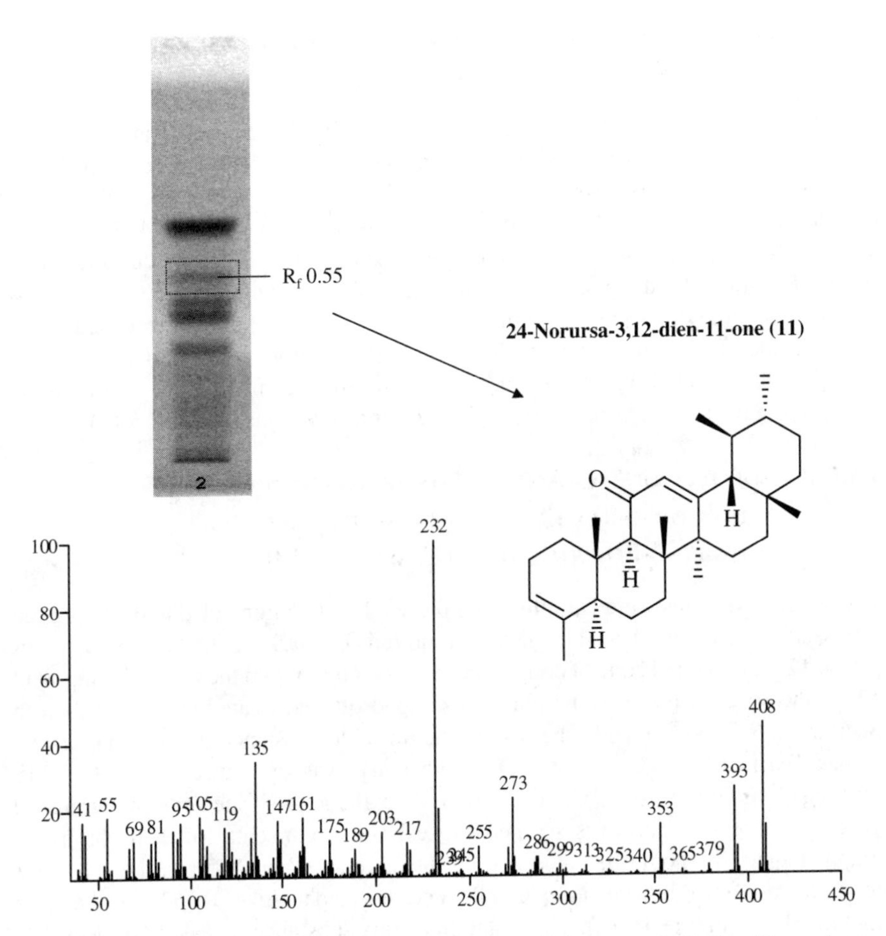

FIGURE 16.13 Mass spectrum of the isolated compound 24-norursa-3,12-dien-11-one (compound 11) (R_f 0.55), an oxidized product from the pyrolysate of *B. serrata*.

FIGURE 16.14 Structures of 3-*O*-acetyl-11-hydroxy-β-boswellic acid (compound 12) and 3-*O*-acetyl-9,11-dehydro-β-boswellic acid (compound 13).

16.5.1 Determination of the "Indicator" Boswellic Acids (Compound 12 and Compound 13)

A methanolic solution of the resin sample was prepared at a concentration of 2.5% to 5%. The sample was applied as an 8-cm band to the plate and developed to a distance of 7 cm in a twin-trough chamber, without chamber saturation, with toluene–ethyl acetate–formic acid–heptane (80 + 20 + 3 + 10; v/v) as the mobile phase. The developed plate was observed under λ = 254 nm UV light and cut into equal pieces. One piece of the plate was scanned in multiwavelength mode (λ = 254 and 285 nm), and the other piece was first heated at 80°C for 15 min and subsequently scanned, too. The comparison of both scans of the chromatogram under λ = 254 nm UV light revealed the conversion of the "invisible" compound 12 to compound 13, which gave a strong quenching of λ = 254 nm UV light (Figure 16.15).

16.5.2 Isolation of 3-*O*-Acetyl-11-Hydroxy-β-Boswellic Acid (Compound 12) and 3-*O*-Acetyl-9,11-Dehydro-β-Boswellic Acid (Compound 13)

The developed plates were cut into equal pieces. It is recommended to deal with the "stressed" pieces first, because they are required for marking the "invisible" compound 12. The marked zone of compound 12, by comparing to the zone of compound 13 on the stressed piece of the plate, is scraped off and treated as is described in Section 16.6. To set limits to the loss of the unstable substance (compound 12), no further purification was executed. The substance was used directly for GC–MS investigations (Figure 16.15). On the other hand, the appropriate zone of compound 13 on the stressed plate was scraped off, extracted, concentrated, and chromatographed again. It should be noted that by the "stressing" procedures, the substance (compound 13) had become more unpolar and revealed a higher R_f value in the same mobile phase (Figure 16.16). The isolation of this substance revealed GC–MS data which match the structure of the altered 3-*O*-acetyl-9,11-dehydro-β-boswellic acid (compound 13) (Figure 16.17).

16.6 GENERAL ISOLATION INSTRUCTIONS

Here, we give some general directions and explanations for the procedures. We recommend the use of TLC plates with a layer thickness of 0.25 mm, occasionally of 0.5 mm, to increase the separation profile. Commercially available precoated silica gel 60 F_{254} plates with a size of 20 × 20 cm (Merck #105744 or #105715) were cut into the format of the development chambers (10 × 10 cm). We used a 10-μl syringe and an automatic application system (Desaga AS 30). The applied volumes are due to the concentration of the sample solution, but in doing so the stickiness of the sample should always be considered. A lower concentration of the sample solution and a frequently practiced rinsing step would prevent blocked needles. We normally applied 5 × 10 μl/8 cm (according to a concentration of 2.5% resin per solute). Generally, the applied lane must not look glossy. The prewashed plates were developed twice for improving the resolution in the appropriate mobile phase in a

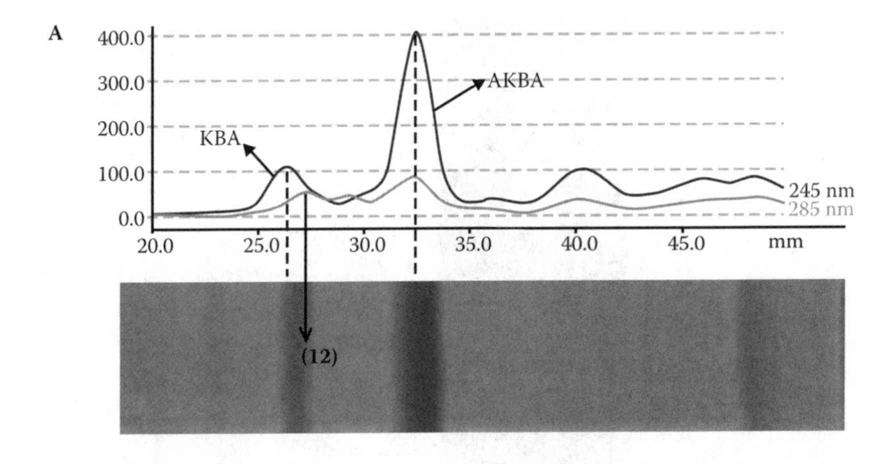

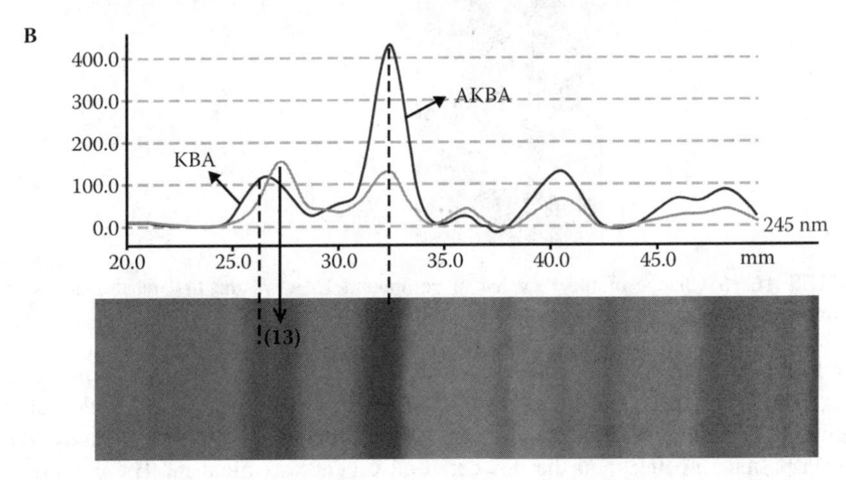

FIGURE 16.15 Determination of the unstressed compound 12 on the TLC plate. (a) Without stressing compound 12 reveals no quenching of λ = 254 nm UV light on the plate and no remarkable peak when scanned at λ = 285 nm. (b) After stressing by heat compound 12 converts to compound 13, which is confirmed by a quenching of λ = 254 nm. UV light and by an intense peak after scanning the lane at λ = 285 nm (AKBA: 3-*O*-acetyl-11-keto-boswellic acid; KBA: 11-keto-β-boswellic acid).

vertical chamber. The quality of the separation was confirmed by scanning (Desaga CD 60) the separated lanes. The results were transferred to the plates, and the marked zones were removed from the plate by scraping. If no scanner is available, the zones should be marked under UV light or, if the substances are not visible under UV light, a piece of the lane on the plate has to be cut off and derivatized, e.g., with anisaldehyde reagent. The position of the colored zones has to be transferred to the remaining plate. The use of a face mask is highly recommended for health reasons and to prevent the scientist from blowing away powdered isolates. The removed zones of 20 plates on average were collected, crushed into powder, and transferred into a centrifuge tube. The appropriate eluent was added, and the tube was sonicated

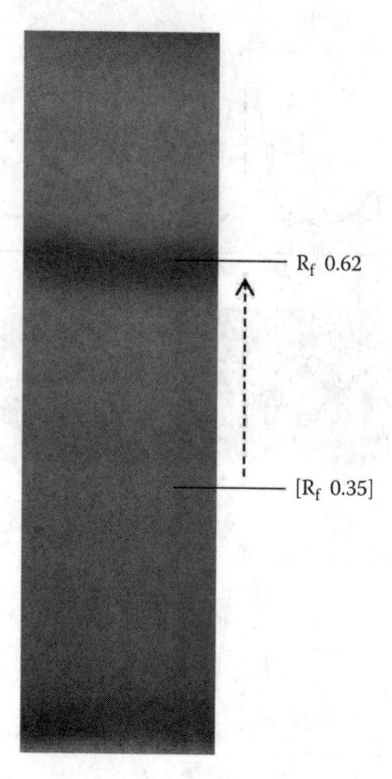

FIGURE 16.16 Change of the R_f value of compound 13 when the first purification steps have completed.

for 15 min before being placed in a centrifuge. After centrifugation, the supernatant was carefully poured and filtered into a vial. For small amounts, we used another technique for eluting the substance from the adsorbent: a cotton swab was inserted into a pasteur pipette, then the powdered adsorbent was filled in. The compound was extracted with the appropriate eluent with the aid of slight pressure (rubber ball). To search artifacts caused by the isolation and analyzing procedures, the fractions were constantly investigated by 2-D TLC.

REFERENCES

1. Helfer, W., Die Tränen der Götter, Jemen, Oman, V.A. Emirate, *Merian,* 59, 1996.
2. Hahn-Deinstrop, E., Koch, A., and Müller, M., Guidelines for the assessment of the traditional herbal medicine "Olibanum" by application of HPTLC and DESAGA ProViDoc® video documentation; *J. Planar Chromatogr.*, 11, 404, 1998.
3. Basar, S., Koch, A., and König, W.A., A verticillane-type diterpene from *Boswellia carterii* essential oil, *Flavour Fragrances J.*, 16, 315, 2001.
4. Basar, S., Phytochemical Investigations on Boswellia Species, Dissertation Hamburg, 2005.
5. Basar, S. and Koch, A., Test of the stability of Olibanum resins and extracts, *J. Planar Chromatogr.,* 17, 479, 2004.

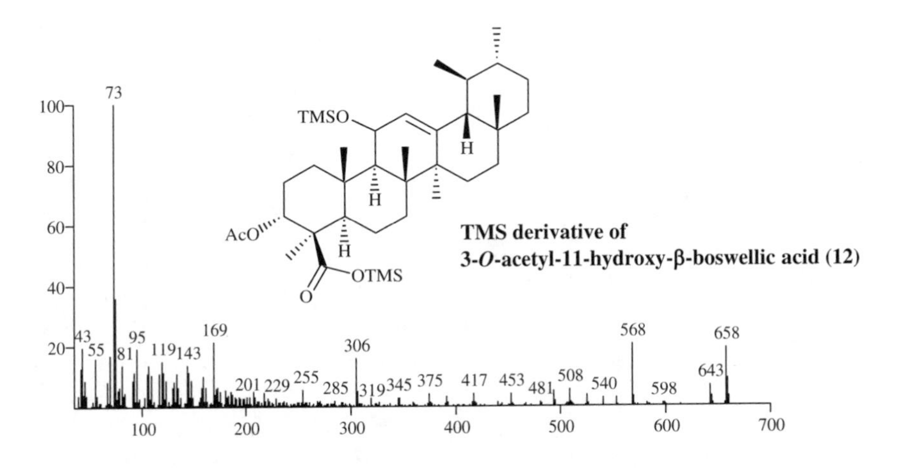

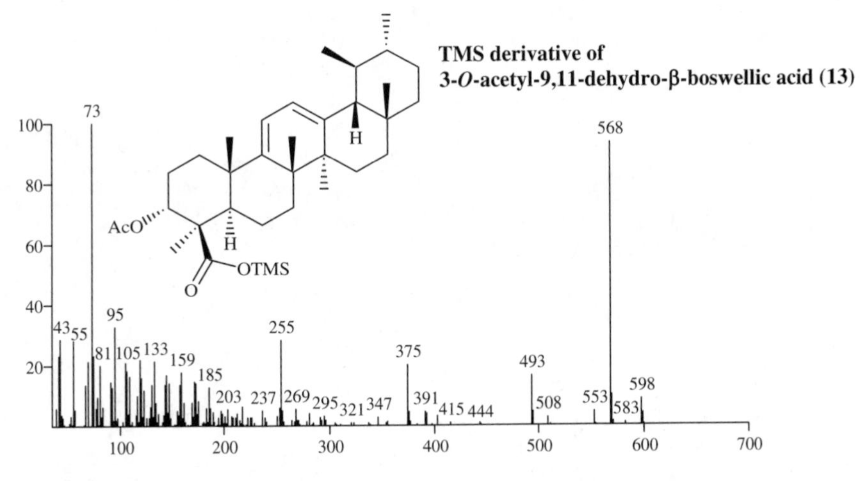

FIGURE 16.17 Mass spectra of the sililated 3-*O*-acetyl-11-OH-β-boswellic acid (compound 12) and 3-*O*-acetyl-9,11-dehydro-β-boswellic acid (compound 13).

Index

C

D

E

F

G

H

I

J

K

L

M

N

O

P

Q

R

S

T

U

V

W

X

Y

Z